ENCYCLOPEDIA OF SPECIAL FUNCTIONS: THE ASKEY–BATEMAN PROJECT

Volume 1: Univariate Orthogonal Polynomials

This is the first of three volumes that form the *Encyclopedia of Special Functions*, an extensive update of the Bateman Manuscript Project.

Volume 1 contains most of the material on orthogonal polynomials, from the classical orthogonal polynomials of Hermite, Laguerre and Jacobi to the Askey–Wilson polynomials, which are the most general basic hypergeometric orthogonal polynomials. Separate chapters cover orthogonal polynomials on the unit circle, zeros of orthogonal polynomials and matrix orthogonal polynomials, with detailed results about matrix-valued Jacobi polynomials. A chapter on moment problems provides many examples of indeterminate moment problems. A thorough bibliography rounds off what will be an essential reference.

MOURAD E. H. ISMAIL is Research Professor at the University of Central Florida.

ENCYCLOPEDIA OF SPECIAL FUNCTIONS: THE ASKEY–BATEMAN PROJECT

Volume 1: Univariate Orthogonal Polynomials

Edited by

MOURAD E. H. ISMAIL
University of Central Florida

with assistance by

WALTER VAN ASSCHE
KU Leuven, Belgium

CAMBRIDGE
UNIVERSITY PRESS

CAMBRIDGE
UNIVERSITY PRESS

University Printing House, Cambridge CB2 8BS, United Kingdom
One Liberty Plaza, 20th Floor, New York, NY 10006, USA
477 Williamstown Road, Port Melbourne, VIC 3207, Australia
314–321, 3rd Floor, Plot 3, Splendor Forum, Jasola District Centre, New Delhi – 110025, India
79 Anson Road, #06-04/06, Singapore 079906

Cambridge University Press is part of the University of Cambridge.

It furthers the University's mission by disseminating knowledge in the pursuit of education, learning, and research at the highest international levels of excellence.

www.cambridge.org

Information on this title: www.cambridge.org/9780521197427

DOI: 10.1017/9780511979156

First published 2020

Printed in the United Kingdom by TJ International Ltd. Padstow Cornwall

A catalogue record for this publication is available from the British Library.

Library of Congress Cataloging-in-Publication Data
Title: Encyclopedia of special functions: the Askey-Bateman project.
Description: Cambridge; New York, NY: Cambridge University Press, 2020- |
Includes bibliographical references and index. | Contents: Volume I.
Univariate orthogonal polynomials / edited by Mourad E. H. Mourad --
Identifiers: LCCN 2020007276 | ISBN 9780521197427 (hardback; v. 1)
Subjects: LCSH: Functions, Special--Encyclopedias.
Classification: LCC QA351 .E63 2020 | DDC 515/.503--dc23
LC record available at https://lccn.loc.gov/2020007276

ISBN – 3 Volume Set 978-1-108-88244-6 Hardback
ISBN – Volume 1 978-0-521-19742-7 Hardback
ISBN – Volume 2 978-1-107-00373-6 Hardback
ISBN – Volume 3 978-0-521-19039-8 Hardback

Contents

Contributors

Leonid Golinskii *Institute for Low Temperature Physics and Engineering, Ukrainian Academy of Sciences, 47 Lenin Avenue, Kharkov 61103, Ukraine*
Andrea Laforgia *Department of Mathematics, Università degli Studi Roma Tre, Largo San Leonardo Murialdo 1, IT–00146 Rome, Italy.*
Martin Muldoon *Department of Mathematics & Statistics, York University, Toronto, Ontario M3J 1P3, Canada.*
Christian Berg *Institute for Mathematical Sciences, University of Copenhagen, DK–2100 Copenhagen, Denmark.*
Jacob S. Christiansen *Lund University, Centre for Mathematical Sciences, Box 118, SE–22100, Lund, Sweden.*
Antonio J. Durán *Departamento de Análisis Matemático, Universidad de Sevilla, Apdo (P.O. Box) 1160, ES–41080 Sevilla, Spain.*
F. Alberto Grünbaum *Department of Mathematics, University of California, Berkeley, CA 94720, USA.*
Ines Pacharoni *CIEM-FaMAF, Universidad Nacional de Córdoba, 5000 Córdoba, Argentina*
Juan A. Tirao *CIEM-FaMAF, Universidad Nacional de Córdoba, 5000 Córdoba, Argentina*

Preface

In the early 1970s, Richard Askey proposed to update the Bateman Project (Erdélyi et al., 1953a–1955). The rationale was that, at the time, he believed that the one-variable theory was more or less complete and it was time to update the Bateman Project and incorporate the results missed by it, or discovered after its publication. At the time this was indeed how many people felt. The big surprise came in 1977 when Wilson, under Askey's guidance, discovered what are now known as the Wilson polynomials, which are today familiar across mathematics. This was followed by many q-polynomials and cumulated in the Askey–Wilson polynomials. At the same time, the q-ultraspherical polynomials were identified and a new scheme of orthogonal polynomials emerged. Some other polynomials, which do not share the properties of classical polynomials, were discovered. These include the Pollaczek polynomials (Pollaczek, 1956) and the random walk polynomials of Askey and Ismail (1984) together with their q-analogues, as well as the Lommel-type polynomials of Wimp (1985) and Ismail (1998). Then the associated polynomials of the classical orthogonal polynomials and their q-analogues attracted a lot of interest and this also led to the sieved polynomials (Al-Salam, Allaway, and Askey, 1984). At the same time, a combinatorial theory of orthogonal polynomials was developed by many notable mathematicians, spearheaded by the French school of Foata and Viennot (Cartier and Foata, 1969; Viennot, 1983). These remarks outline the developments in special systems of orthogonal polynomials. They indirectly led to developments in the theory of continued fractions and other special functions. We also started to see developments in the theory of multivariate orthogonal polynomials and their q-analogues, together with parallel developments in multivariate statistics. The emergence of quantum groups then led to developments in orthogonal polynomials and special functions (Koornwinder, 1990). Since the 1990s we have seen a strong interaction between integrable systems and special functions and orthogonal polynomials. The Riemann–Hilbert approach of Fokas, Its, and Kitaev, combined with the nonlinear steepest descent method of Deift and Zhou (Deift, 1999), revolutionized the asymptotic theory of orthogonal polynomials.

This volume is the first in a series of volumes which form an *Encyclopedia of Special Functions*. This volume contains most of the material on univariate orthogonal polynomials. The combinatorial results on orthogonal polynomials, as well as related topics in continued fractions, will be in a later volume, which will also contain the material on exceptional orthogonal polynomials, as well as a coverage of multiple orthogonal polynomials.

The series will be edited jointly with Walter Van Assche, who will also contribute some future chapters. Many people have worked on this volume. The chapter authors are indicated at the beginning of each chapter and we are grateful to all of them for their contributions. Mourad Ismail wrote the remaining chapters. Erik Koelink and Tom Koornwinder wrote the material on addition theorems in Section 2.7.

The subject of orthogonal polynomials is still an active research area and we have no doubt that this will continue for years to come. Our beloved Mizan Rahman passed away on January 5, 2015, Martin Muldoon on August 1, 2019 and Richard Askey on October 9, 2019. We are sure they are with us in spirit and we are sorry that they did not live to see the completed project.

I gratefully acknowledge the help of Denise Marks of the University of South Florida in putting the material together. David Tranah of Cambridge University Press has been very encouraging and patient over the years; thanks David. The subject and this project owe a great deal to the mathematical contributions and leadership of George Andrews and Richard Askey. I would like to thank the King Saud University for generous research support during the last five years through the Distinguished Scientist Fellowship Program. This part-time affiliation with King Saud University has been very beneficial. Last but not least, I thank the University of Central Florida for my research appointment which gave me ample time to concentrate on research and writing.

Orlando, FL
Leuven, Belgium

Mourad E. H. Ismail
Walter Van Assche

1
Preliminaries

In this chapter we define the special functions used in this volume and state the properties relevant to the treatment of orthogonal polynomials. We also state a few facts from complex analysis used in the later parts.

1.1 Analytic Facts

Two important special cases of the Lagrange expansion are

$$e^{\alpha z} = 1 + \sum_{n=1}^{\infty} \frac{\alpha(\alpha+n)^{n-1}}{n!} w^n, \quad w = ze^{-z}, \tag{1.1.1}$$

$$(1+z)^{\alpha} = 1 + \alpha \sum_{n=1}^{\infty} \binom{\alpha+\beta n-1}{n-1} \frac{w^n}{n}, \quad w = z(1+z)^{-\beta}. \tag{1.1.2}$$

The Perron–Stieltjes inversion formula (see Stone, 1932, Lemma 5.2) is

$$F(z) = \int_{\mathbb{R}} \frac{d\mu(t)}{z-t}, \quad z \notin \mathbb{R} \tag{1.1.3}$$

if and only if

$$\mu(t) - \mu(s) = \lim_{\epsilon\to 0^+} \int_s^t \frac{F(x-i\epsilon) - F(x+i\epsilon)}{2\pi i}\, dx. \tag{1.1.4}$$

Formula (1.1.4) shows that the absolutely continuous component of μ is

$$\mu'(x) = \left[F\left(x - i0^+\right) - F\left(x + i0^+\right)\right]/(2\pi i). \tag{1.1.5}$$

Here μ is normalized by $\mu(x) = \left[\mu\left(x+0^+\right) + \mu\left(x+0^-\right)\right]/2$.

Definition 1.1.1 Let f be an entire function. The maximum modulus is

$$M(r; f) := \sup\left\{|f(z)|\colon |z| \le r\right\}, \quad r > 0. \tag{1.1.6}$$

The order $\rho(f)$ of f is defined by

$$\rho(f) := \limsup_{r\to\infty} \frac{\ln\ln M(r,f)}{\ln r}. \tag{1.1.7}$$

Theorem 1.1.2 (Boas, 1954) *If $\rho(f)$ is finite and is not equal to a positive integer, then f has infinitely many zeros.*

If f has finite order, its type σ is

$$\sigma = \inf\{K: M(r) < \exp(Kr^{\rho})\}. \tag{1.1.8}$$

The Phragmén–Lindelöf indicator of an entire function of finite order and type is

$$h(\theta) = \limsup_{r\to\infty} \frac{\ln\left|f(re^{i\theta})\right|}{r^{\rho}}. \tag{1.1.9}$$

Theorem 1.1.3 *Given two differential equations in the form*

$$\frac{d^2u}{dz^2} + f(z)u(z) = 0, \qquad \frac{d^2v}{dz^2} + g(z)v(z) = 0,$$

then $y = uv$ satisfies

$$\frac{d}{dz}\left\{\frac{y''' + 2(f+g)y' + (f'+g')y}{f-g}\right\} + (f-g)y = 0 \quad \text{if } f \neq g, \tag{1.1.10}$$

$$y''' + 4fy' + 2f'y = 0 \quad \text{if } f = g. \tag{1.1.11}$$

Watson (1944, §5.4), attributes Theorem 1.1.3 to P. Appell.

Lemma 1.1.4 *Let $y = y(x)$ satisfy the differential equation*

$$\phi(x)y''(x) + y(x) = 0, \quad a < x < b, \tag{1.1.12}$$

where $\phi(x) > 0$, and $\phi'(x)$ is positive (negative) and continuous on (a, b). Then the successive relative maxima of $|y|$ increase (decrease) with x in (a, b) if ϕ increases (decreases) on (a, b).

The Wronskian of f and g is

$$W(f, g) := fg' - gf' = \det\begin{pmatrix} f & g \\ f' & g' \end{pmatrix}. \tag{1.1.13}$$

1.2 Hypergeometric Functions

Standard references in the area of special functions are Andrews, Askey, and Roy (1999), Bailey (1935), Rainville (1960), Erdélyi et al. (1953a), Slater (1966), and the real classic Whittaker and Watson (1927). The **shifted factorial** is

$$(a)_0 := 1, \quad (a)_n := a(a+1)\cdots(a+n-1), \quad n > 0, \tag{1.2.1}$$

so that

$$(a)_n = \frac{\Gamma(a+n)}{\Gamma(a)}. \tag{1.2.2}$$

Note that (1.2.2) is meaningful for any complex n, when $a+n$ is not a pole of the gamma function. The **multishifted factorial** is

$$(a_1, \ldots, a_m)_n = \prod_{j=1}^{m} \left(a_j\right)_n .$$

The difference operators are

$$\begin{aligned} \Delta f(x) &= (\Delta f)(x) := f(x+1) - f(x), \\ \nabla f(x) &= (\nabla f)(x) := f(x) - f(x-1). \end{aligned} \tag{1.2.3}$$

We will also use the symmetric difference operator

$$\left(\tilde{\Delta}_h\right) f(x) = [f(x+h/2) - f(x-h/2)]/h. \tag{1.2.4}$$

It is clear that $\tilde{\Delta}_h$ is a discrete analogue of the derivative. Another divided difference operator is the **Wilson operator**,

$$(\mathcal{W} f)(x) = \frac{\breve{f}(y+i/2) - \breve{f}(y-i/2)}{\breve{e}(y+i/2) - \breve{e}(y-i/2)}, \tag{1.2.5}$$

where

$$y = \sqrt{x}, \quad \breve{e}(y) = x, \quad \text{and} \quad \breve{f}(y) := f(x). \tag{1.2.6}$$

It is easy to see that $\breve{e}(y+i/2) - \breve{e}(y-i/2) = 2i\sqrt{x}$. It is a fact that

$$\mathcal{W}\psi_n(x;a) = n\psi_{n-1}(x; a+1/2), \tag{1.2.7}$$

where

$$\psi_n(x;a) := \left(a + i\sqrt{x}\right)_n \left(a - i\sqrt{x}\right)_n . \tag{1.2.8}$$

A hypergeometric series is

$${}_rF_s\left(\left.\begin{matrix} a_1, \ldots, a_r \\ b_1, \ldots, b_s \end{matrix}\right| z\right) = {}_rF_s\left(a_1, \ldots, a_r; b_1, \ldots, b_s; z\right) = \sum_{n=0}^{\infty} \frac{(a_1, \ldots, a_r)_n}{(b_1, \ldots, b_s)_n} \frac{z^n}{n!}. \tag{1.2.9}$$

If one of the numerator parameters is a negative integer, say $-k$, then the series (1.2.9) becomes a finite sum, $0 \le n \le k$, and the ${}_rF_s$ series is called **terminating**. As a function of z, the nonterminating series is an entire function if $r \le s$, and it is analytic in $\{|z| < 1\}$ if $r = s+1$. The Gauss hypergeometric function ${}_2F_1(a, b; c; z)$ (Erdélyi et al., 1953a, §2.1) satisfies the **hypergeometric differential equation**

$$z(1-z)\frac{d^2y}{dz^2} + [c - (a+b+1)z]\frac{dy}{dz} - aby = 0, \tag{1.2.10}$$

and has the **Euler integral representation**

$${}_2F_1(a, b; c; z) = \frac{\Gamma(c)}{\Gamma(b)\Gamma(c-b)} \int_0^1 t^{b-1}(1-t)^{c-b-1}(1-zt)^{-a}\, dt, \tag{1.2.11}$$

for $\mathrm{Re}\, b > 0$, $\mathrm{Re}(c-b) > 0$.

The **confluent hypergeometric function** (Erdélyi et al., 1953a, §6.1)

$$\Phi(a, c; z) := {}_1F_1(a; c; z) \tag{1.2.12}$$

satisfies the differential equation

$$z\frac{d^2y}{dz^2} + (c-z)\frac{dy}{dz} - ay = 0, \tag{1.2.13}$$

and $\lim_{b\to\infty} {}_2F_1(a, b; c; z/b) = {}_1F_1(a; c; z)$. The Tricomi Ψ function is a second linearly independent solution of (1.2.13) and is defined by (Erdélyi et al., 1953a, §6.5)

$$\Psi(a, c; x) := \frac{\Gamma(1-c)}{\Gamma(a-c+1)}\Phi(a, c; x) + \frac{\Gamma(c-1)}{\Gamma(a)}x^{1-c}\Phi(a-c+1, 2-c; x). \tag{1.2.14}$$

The function Ψ has the integral presentation (Erdélyi et al., 1953b, §6.5)

$$\Psi(a, c; x) = \frac{1}{\Gamma(a)}\int_0^\infty e^{-xt}t^{a-1}(1+t)^{c-a-1}\, dt, \quad \mathrm{Re}\, a > 0,\ \mathrm{Re}\, x > 0. \tag{1.2.15}$$

The Bessel function J_ν and the modified Bessel function I_ν (Watson, 1944) are

$$J_\nu(z) = \sum_{n=0}^\infty \frac{(-1)^n(z/2)^{\nu+2n}}{\Gamma(n+\nu+1)n!}, \tag{1.2.16}$$

$$I_\nu(z) = e^{-i\pi\nu/2}J_\nu\left(ze^{i\pi/2}\right) = \sum_{n=0}^\infty \frac{(z/2)^{\nu+2n}}{\Gamma(n+\nu+1)n!}. \tag{1.2.17}$$

Observe the special cases

$$J_{1/2}(z) = \sqrt{\frac{2}{\pi z}}\sin z, \quad J_{-1/2}(z) = \sqrt{\frac{2}{\pi z}}\cos z. \tag{1.2.18}$$

The Bessel functions satisfy the recurrence relation

$$\frac{2\nu}{z}J_\nu(z) = J_{\nu+1}(z) + J_{\nu-1}(z). \tag{1.2.19}$$

The Bessel functions J_ν and $J_{-\nu}$ satisfy the **Bessel differential equation**

$$x^2\frac{d^2y}{dx^2} + x\frac{dy}{dx} + \left(x^2 - \nu^2\right)y = 0. \tag{1.2.20}$$

Another solution, $Y_\nu(z)$, to both (1.2.19) and (1.2.20) is

$$\begin{aligned} Y_\nu(z) &= \frac{J_\nu(z)\cos\nu\pi - J_{-\nu}(z)}{\sin\nu\pi}, \quad \nu \neq 0, \pm 1, \pm 2, \ldots, \\ Y_n(z) &= \lim_{\nu\to n} Y_\nu(z), \quad n = 0, \pm 1, \pm 2, \ldots. \end{aligned} \tag{1.2.21}$$

The functions $J_\nu(z)$ and $Y_\nu(z)$ are linearly independent solutions of (1.2.20).

The function I_ν satisfies the differential equation

$$x^2\frac{d^2y}{dx^2}+x\frac{dy}{dx}-\left(x^2+\nu^2\right)y=0, \tag{1.2.22}$$

whose second solution is

$$\begin{aligned} K_\nu(x) &= \frac{\pi}{2}\frac{I_{-\nu}(x)-I_\nu(x)}{\sin(\pi\nu)}, \qquad \nu\neq 0,\pm1,\pm2,\ldots, \\ K_n(x) &= \lim_{\nu\to n} K_\nu(x), \qquad n=0,\pm1,\pm2,\ldots. \end{aligned} \tag{1.2.23}$$

We also have the recursion relations

$$I_{\nu-1}(x)-I_{\nu+1}(x)=\frac{2\nu}{x}I_\nu(x), \qquad K_{\nu+1}(x)-K_{\nu-1}(x)=\frac{2\nu}{x}K_\nu(x). \tag{1.2.24}$$

Two important integrals are **Sonine's first integral**,

$$J_{\alpha+\beta+1}(z)=\frac{2^{-\beta}z^{\beta+1}}{\Gamma(\beta+1)}\int_0^1 x^{2\beta+1}\left(1-x^2\right)^{\alpha/2}J_\alpha\left(z\sqrt{1-x^2}\right)dx, \tag{1.2.25}$$

for $\operatorname{Re}\alpha>-1$ and $\operatorname{Re}\beta>-1$, and **Sonine's second integral**,

$$x^\mu y^\nu\frac{J_{\mu+\nu+1}\left(\sqrt{x^2+y^2}\right)}{(x^2+y^2)^{(\mu+\nu+1)/2}}=\int_0^{\pi/2} J_\mu(x\sin\theta)J_\nu(y\cos\theta)\sin^{\mu+1}\theta\sin^{\nu+1}\theta\,d\theta \tag{1.2.26}$$

(Andrews, Askey, and Roy, 1999, Theorem 4.11.1).

Theorem 1.2.1 *When $\nu>-1$, the function $z^{-\nu}J_\nu(z)$ has only real and simple zeros. Furthermore, the positive (negative) zeros of $J_\nu(z)$ and $J_{\nu+1}(z)$ interlace for $\nu>-1$.*

We shall denote the positive zeros of $J_\nu(z)$ by $(j_{\nu,k})_k$, that is,

$$0<j_{\nu,1}<j_{\nu,2}<\cdots<j_{\nu,n}<\cdots. \tag{1.2.27}$$

The Bessel functions satisfy the differential recurrence relations (Watson, 1944)

$$zJ_\nu'(z)=\nu J_\nu(z)-zJ_{\nu+1}(z), \tag{1.2.28}$$

$$zY_\nu'(z)=\nu Y_\nu(z)-zY_{\nu+1}(z), \tag{1.2.29}$$

$$zI_\nu'(z)=zI_{\nu+1}(z)+\nu I_\nu(z), \tag{1.2.30}$$

$$zK_\nu'(z)=\nu K_\nu(z)-zK_{\nu+1}(z). \tag{1.2.31}$$

The Bessel functions are special cases of ${}_1F_1$ since (Erdélyi et al., 1953a, §6.9.1)

$$e^{-iz}\,{}_1F_1(\nu+1/2;2\nu+1;2iz)=\Gamma(\nu+1)(z/2)^{-\nu}J_\nu(z). \tag{1.2.32}$$

The Airy function $A(x)$ is a combination of

$$\begin{aligned} k(x) &:= \frac{\pi}{3}\left(\frac{x}{3}\right)^{\frac{1}{2}} J_{-1/3}\left(2(x/3)^{3/2}\right) = \frac{\pi}{3}\sum_{n=0}^{\infty}\frac{(-x/3)^{3n}}{n!\Gamma(n+2/3)}, \\ \ell(x) &:= \frac{\pi}{3}\left(\frac{x}{3}\right)^{\frac{1}{2}} J_{1/3}\left(2(x/3)^{3/2}\right) = \frac{\pi}{9}x\sum_{n=0}^{\infty}\frac{(-x/3)^{3n}}{n!\Gamma(n+4/3)}. \end{aligned} \tag{1.2.33}$$

Indeed $\{k(x), \ell(x)\}$ is a basis of solutions of the Airy equation

$$\frac{d^2y}{dx^2} + \frac{1}{3}xy = 0. \tag{1.2.34}$$

The only solution of (1.2.34) which is bounded as $x \to -\infty$ is $k(x) + \ell(x)$. Set

$$A(x) := k(x) + \ell(x). \tag{1.2.35}$$

The function $A(x)$ is called the Airy function and has the asymptotic behavior

$$A(x) = \frac{\sqrt{\pi}}{2\,3^{1/4}}|x|^{-\frac{1}{4}}\exp\left\{-2(|x|/3)^{3/2}\right\}(1 + o(1)), \quad x \to -\infty. \tag{1.2.36}$$

Nowadays it is more common to use the Airy function $\mathrm{Ai}(x)$ which is a solution of $y'' = xy$ that remains bounded as $x \to \infty$. The relation with (1.2.35) is $A(x) = 3^{-1/3}\pi\mathrm{Ai}(-3^{-1/3}x)$. The Airy function plays an important role in the theory of orthogonal polynomials with exponential weights, random matrix theory, as well as other parts of mathematical physics and applied mathematics. The function $A(x)$ is positive on $(-\infty, 0)$ and has only positive simple zeros. We shall denote the zeros of $A(x)$ by

$$0 < i_1 < i_2 < \cdots. \tag{1.2.37}$$

The Appell functions generalize the hypergeometric function to two variables. They are defined by (Appell and Kampé de Fériet, 1926; Erdélyi et al., 1953a)

$$F_1(a; b, b'; c; x, y) = \sum_{m,n=0}^{\infty}\frac{(a)_{n+m}(b)_m(b')_n}{(c)_{m+n}m!n!}x^m y^n, \tag{1.2.38}$$

$$F_2(a; b, b'; c, c'; x, y) = \sum_{m,n=0}^{\infty}\frac{(a)_{n+m}(b)_m(b')_n}{(c)_m(c')_n m!n!}x^m y^n, \tag{1.2.39}$$

$$F_3(a, a'; b, b'; c; x, y) = \sum_{m,n=0}^{\infty}\frac{(a)_m(a')_n(b)_m(b')_n}{(c)_{m+n}m!n!}x^m y^n, \tag{1.2.40}$$

$$F_4(a, b; c, c'; x, y) = \sum_{m,n=0}^{\infty}\frac{(a)_{n+m}(b)_{m+n}}{(c)_m(c')_n m!n!}x^m y^n. \tag{1.2.41}$$

The complete elliptic integrals of the first and second kinds are (Erdélyi et al., 1953b)

$$\mathbf{K} = \mathbf{K}(k) = \int_0^1 \frac{du}{\sqrt{(1-u^2)(1-k^2u^2)}}, \tag{1.2.42}$$

$$\mathbf{E} = \mathbf{E}(k) = \int_0^1 \sqrt{\frac{1-k^2u^2}{1-u^2}}\, du, \tag{1.2.43}$$

respectively. One has

$$\mathbf{K}(k) = \frac{\pi}{2}\, {}_2F_1\left(1/2, 1/2; 1; k^2\right), \tag{1.2.44}$$

$$\mathbf{E}(k) = \frac{\pi}{2}\, {}_2F_1\left(-1/2, 1/2; 1; k^2\right). \tag{1.2.45}$$

We refer to k as the modulus, while the complementary modulus k' is

$$k' = \left(1-k^2\right)^{1/2}. \tag{1.2.46}$$

1.3 Summation Theorems and Transformations

In the shifted factorial notation the binomial theorem is

$$\sum_{n=0}^{\infty} \frac{(a)_n}{n!} z^n = (1-z)^{-a}, \quad |z| < 1. \tag{1.3.1}$$

If a is a negative integer then the sum is finite and gives the familiar binomial formula

$$\sum_{k=0}^{n} \binom{n}{k} (-1)^k z^k = (1-z)^n.$$

The **Gauss sum** is

$${}_2F_1\left(\left.\begin{matrix} a, b \\ c \end{matrix}\right| 1\right) = \frac{\Gamma(c)\Gamma(c-a-b)}{\Gamma(c-a)\Gamma(c-b)}, \quad \mathrm{Re}\{c-a-b\} > 0. \tag{1.3.2}$$

The terminating version of (1.3.2) is the **Chu–Vandermonde sum**

$${}_2F_1\left(\left.\begin{matrix} -n, b \\ c \end{matrix}\right| 1\right) = \frac{(c-b)_n}{(c)_n}. \tag{1.3.3}$$

A hypergeometric series (1.2.9) is called **balanced** if $r = s + 1$ and

$$1 + \sum_{k=1}^{s+1} a_k = \sum_{k=1}^{s} b_k. \tag{1.3.4}$$

The sum of a terminating balanced ${}_3F_2$ of unit argument is the **Pfaff–Saalschütz theorem**

$${}_3F_2\left(\left.\begin{matrix} -n, a, b \\ c, d \end{matrix}\right| 1\right) = \frac{(c-a)_n(c-b)_n}{(c)_n(c-a-b)_n} \quad \text{if } c + d = 1 - n + a + b. \tag{1.3.5}$$

Stirling's formula for the gamma function is

$$\operatorname{Log}\Gamma(z) = \left(z - \frac{1}{2}\right)\operatorname{Log} z - z + \frac{1}{2}\ln(2\pi) + O\left(z^{-1}\right), \tag{1.3.6}$$

$|\arg z| \le \pi - \epsilon$, $\epsilon > 0$. An important consequence of Stirling's formula is

$$\lim_{z\to\infty} z^{b-a}\frac{\Gamma(z+a)}{\Gamma(z+b)} = 1. \tag{1.3.7}$$

The **Pfaff–Kummer transformation** is

$${}_2F_1\left(\left.\begin{matrix} a, b \\ c \end{matrix}\right| z\right) = (1-z)^{-a}{}_2F_1\left(\left.\begin{matrix} a, c-b \\ c \end{matrix}\right| \frac{z}{z-1}\right), \tag{1.3.8}$$

and is valid for $|z| < 1$, $|z| < |z-1|$. An iterate of (1.3.8) is

$${}_2F_1\left(\left.\begin{matrix} a, b \\ c \end{matrix}\right| z\right) = (1-z)^{c-a-b}{}_2F_1\left(\left.\begin{matrix} c-a, c-b \\ c \end{matrix}\right| z\right), \tag{1.3.9}$$

for $|z| < 1$. Since ${}_1F_1(a;c;z) = \lim_{b\to\infty} {}_2F_1(a,b;c;z/b)$, (1.3.9) yields

$${}_1F_1(a;c;z) = e^z {}_1F_1(c-a;c;-z). \tag{1.3.10}$$

1.4 q-Series

An analogue of the derivative is the q-difference operator

$$\left(D_q f\right)(x) = \left(D_{q,x} f\right)(x) = \frac{f(x) - f(qx)}{(1-q)x}. \tag{1.4.1}$$

It is clear that

$$D_{q,x}x^n = \frac{1-q^n}{1-q}x^{n-1}, \tag{1.4.2}$$

and for differentiable functions $\lim_{q\to 1^-}(D_q f)(x) = f'(x)$. The **product rule for D_q** is

$$\left(D_q fg\right)(x) = f(x)\left(D_q g\right)(x) + g(qx)\left(D_q f\right)(x). \tag{1.4.3}$$

For finite a and b, $0 < q < 1$ the ***q*-integral** is

$$\int_0^a f(x)\,d_qx := \sum_{n=0}^{\infty}\left[aq^n - aq^{n+1}\right] f\left(aq^n\right), \tag{1.4.4}$$

$$\int_a^b f(x)\,d_qx := \int_0^b f(x)\,d_qx - \int_0^a f(x)\,d_qx. \tag{1.4.5}$$

Moreover,

$$\int_0^\infty f(x)\,d_qx := (1-q)\sum_{n=-\infty}^{\infty} q^n f\left(q^n\right). \tag{1.4.6}$$

The analogue of a change of variable is

$$\int_a^b f(x)g(qx)\,d_qx = q^{-1}\int_a^b g(x)f(x/q)\,d_qx + q^{-1}(1-q)[ag(a)f(a/q) - bg(b)f(b/q)]. \quad (1.4.7)$$

Let

$$x_k := aq^k, \quad y_k := bq^k, \quad (1.4.8)$$

and $w(x_k) > 0$ and $w(y_k) > 0$ for $k = 0, 1, \ldots$. We will take $a \le 0 \le b$ and use the inner product

$$\langle f, g\rangle_q = \int_a^b f(t)\overline{g(t)}w(t)\,d_qt. \quad (1.4.9)$$

A **q-analogue of integration by parts** for $D_{q,x}$ is

$$\begin{aligned}\int_a^b D_{q,x}f(t)\overline{g(t)}w(t)\,d_qt &= -f(x_0)\overline{g(x_{-1})}w(x_{-1}) + f(y_0)\overline{g(y_{-1})}w(y_{-1}) \\ &\quad - q^{-1}\left\langle f, \frac{1}{w(x)}D_{q^{-1},x}(g(x)w(x))\right\rangle_q,\end{aligned} \quad (1.4.10)$$

provided that

$$\lim_{n\to\infty} w(x_n)f(x_{n+1})\overline{g(x_n)} = \lim_{n\to\infty} w(y_n)f(y_{n+1})\overline{g(y_n)} = 0. \quad (1.4.11)$$

We also let

$$\begin{aligned}\langle f, g\rangle_{q^{-1}} :=& \frac{(1-q)}{q}\sum_{n=0}^{\infty} f(r_n)\overline{g(r_n)}r_nw(r_n) \\ &+ \frac{(1-q)}{q}\sum_{n=0}^{\infty} f(s_n)\overline{g(s_n)}s_nw(s_n),\end{aligned} \quad (1.4.12)$$

with

$$r_n := \alpha q^{-n}, \quad s_n = \beta q^{-n}, \quad (1.4.13)$$

and w a function positive at r_n and s_n. The analogue of integration by parts is

$$\begin{aligned}\left\langle D_{q^{-1},x}f, g\right\rangle_{q^{-1}} &= -f(r_0)\frac{\overline{g(r_{-1})r_{-1}w(r_{-1})}}{r_{-1} - r_0} + f(s_0)\frac{\overline{g(s_{-1})s_{-1}w(s_{-1})}}{s_{-1} - s_0} \\ &\quad - q\left\langle f, \frac{x}{w(x)}D_{q,x}(g(x)w(x))\right\rangle_{q^{-1}},\end{aligned} \quad (1.4.14)$$

provided that both sides are well defined and

$$\lim_{n\to\infty}\left[-w(r_n)r_nf(r_{n+1})\overline{g(r_n)} + w(s_n)f(s_{n+1})\overline{g(s_n)}\right] = 0. \quad (1.4.15)$$

The ***q*-shifted factorials** are

$$(a;q)_0 := 1, \quad (a;q)_n := \prod_{k=1}^{n}\left(1 - aq^{k-1}\right), \quad n = 1,2,\ldots, \tag{1.4.16}$$

and if $|q| < 1$,

$$(a;q)_\infty := \prod_{k=1}^{\infty}\left(1 - aq^{k-1}\right). \tag{1.4.17}$$

The **multiple *q*-shifted factorials** are defined by

$$(a_1, a_2, \ldots, a_k; q)_n := \prod_{j=1}^{k}\left(a_j; q\right)_n. \tag{1.4.18}$$

We shall also use

$$(a;q)_\alpha = \frac{(a;q)_\infty}{(aq^\alpha;q)_\infty}, \tag{1.4.19}$$

which agrees with (1.4.16) when $\alpha = 0, 1, 2, \ldots$ but holds for general α when $aq^\alpha \neq q^{-n}$ for a nonnegative integer n. The q-binomial coefficient is

$$\begin{bmatrix} n \\ k \end{bmatrix}_q := \frac{(q;q)_n}{(q;q)_k(q;q)_{n-k}}. \tag{1.4.20}$$

Unless we say otherwise we shall always assume that

$$0 < q < 1. \tag{1.4.21}$$

A **basic hypergeometric series** (Gasper and Rahman, 2004) is

$$\begin{aligned} {}_r\phi_s\left(\left.\begin{matrix} a_1, \ldots, a_r \\ b_1, \ldots, b_s \end{matrix}\right| q, z\right) &= {}_r\phi_s\left(a_1, \ldots, a_r; b_1, \ldots, b_s; q, z\right) \\ &= \sum_{n=0}^{\infty} \frac{(a_1, \ldots, a_r; q)_n}{(q, b_1, \ldots, b_s; q)_n} z^n \left(-q^{(n-1)/2}\right)^{n(s+1-r)}. \end{aligned} \tag{1.4.22}$$

Note that $(q^{-k};q)_n = 0$ for $n = k+1, k+2, \ldots$. If one of the numerator parameters is of the form q^{-k} then the sum on the right-hand side of (1.4.22) is a finite sum and we say that the series in (1.4.22) is **terminating**. A series that does not terminate is called **nonterminating**. The radius of convergence of the series in (1.4.22) is 1, 0, or ∞ for $r = s+1$, $r > s+1$, and $r < s+1$, respectively, as can be seen from the ratio test.

A basic hypergeometric series is called **balanced** if $r = s+1$ and $q\prod_{j=1}^{s+1} a_j = \prod_{j=1}^{s} b_j$. The Sears transformation connects two balanced terminating ${}_4\phi_3$. It is (Gasper and Rahman, 2004, (III.15))

$${}_4\phi_3\left(\left.\begin{matrix} q^{-n}, a, b, c \\ d, e, f \end{matrix}\right| q, q\right) = \left(\frac{bc}{d}\right)^n \frac{(de/bc, df/bc; q)_n}{(e, f; q)_n}\, {}_4\phi_3\left(\left.\begin{matrix} q^{-n}, a, d/b, d/c \\ d, de/bc, df/bc \end{matrix}\right| q, q\right), \tag{1.4.23}$$

provided that $abc = defq^{n-1}$. A limiting case of (1.4.23) and a further special case are

$$ {}_3\phi_2\left(\begin{matrix} q^{-n}, a, b \\ c, d \end{matrix} \middle| q, q\right) = \frac{b^n (d/b; q)_n}{(d; q)_n} \, {}_3\phi_2\left(\begin{matrix} q^{-n}, b, c/a \\ c, q^{1-n} b/d \end{matrix} \middle| q, aq/d\right), \tag{1.4.24} $$

$$ {}_3\phi_2\left(\begin{matrix} q^{-n}, a, b \\ c, 0 \end{matrix} \middle| q, q\right) = \frac{(b; q)_n a^n}{(c; q)_n} \, {}_2\phi_1\left(\begin{matrix} q^{-n}, c/b \\ q^{1-n}/b \end{matrix} \middle| q, q/a\right). \tag{1.4.25} $$

An important q-difference operator is the **Askey–Wilson operator** $\mathcal{D}_q$, which we now define. Given a polynomial f we set $\breve{f}(e^{i\theta}) := f(x)$, $x = \cos\theta$, that is,

$$ \breve{f}(z) = f((z + 1/z)/2), \quad z = e^{\pm i\theta}. \tag{1.4.26} $$

In other words, we think of $f(\cos\theta)$ as a function of $e^{i\theta}$ or $e^{-i\theta}$. The **Askey–Wilson divided difference operator** is

$$ \left(\mathcal{D}_q f\right)(x) = \frac{\breve{f}\left(q^{1/2} z\right) - \breve{f}\left(q^{-1/2} z\right)}{(q^{1/2} - q^{-1/2})(z - 1/z)/2}, \quad x = (z + 1/z)/2. \tag{1.4.27} $$

It is important to note that although we use $x = \cos\theta$, θ is not necessarily real. In fact, z and z^{-1} are defined as

$$ z = x + \sqrt{x^2 - 1}, \quad z^{-1} = x - \sqrt{x^2 - 1}, \quad |z| \geq 1. \tag{1.4.28} $$

The branch of the square root is taken such that $\sqrt{x+1} > 0$, $x > -1$. This makes $z = e^{\pm i\theta}$ if $\operatorname{Im} x \lessgtr 0$. It is not difficult to see that for all polynomials f,

$$ \lim_{q \to 1} \left(\mathcal{D}_q f\right)(x) = f'(x). \tag{1.4.29} $$

In the calculus of the Askey–Wilson operator the basis $\{\phi_n(x; a) \colon 0 \leq n < \infty\}$,

$$ \phi_n(x; a) := \left(ae^{i\theta}, ae^{-i\theta}; q\right)_n = \prod_{k=0}^{n-1} \left[1 - 2axq^k + a^2 q^{2k}\right], \tag{1.4.30} $$

plays the role played by the monomials $\{x^n \colon 0 \leq n < \infty\}$ in the differential and integral calculus. Indeed,

$$ \mathcal{D}_q \left(ae^{i\theta}, ae^{-i\theta}; q\right)_n = -\frac{2a(1 - q^n)}{1 - q} \left(aq^{1/2} e^{i\theta}, aq^{1/2} e^{-i\theta}; q\right)_{n-1}. \tag{1.4.31} $$

The **averaging operator** $\mathcal{A}_q$ is

$$ \left(\mathcal{A}_q f\right)(x) = \frac{1}{2}\left[\breve{f}\left(zq^{1/2}\right) + \breve{f}\left(zq^{-1/2}\right)\right]. \tag{1.4.32} $$

The product rule for $\mathcal{D}_q$ is

$$ \mathcal{D}_q(fg) = \left(\mathcal{A}_q f\right)\left(\mathcal{D}_q g\right) + \left(\mathcal{A}_q g\right)\left(\mathcal{D}_q f\right). \tag{1.4.33} $$

An induction argument implies (Cooper, 2002)

$$ \mathcal{D}_q^n f(x) = \frac{(2z)^n q^{n(3-n)/4}}{(q-1)^n} \sum_{k=0}^{n} \begin{bmatrix} n \\ k \end{bmatrix}_q \frac{q^{k(n-k)} z^{-2k} q^{2k-n} \breve{f}\left(zq^{k-n/2}\right)}{(q^{n-2k+1} z^{-2}; q)_k \, (z^2 q^{2k+1-n}; q)_{n-k}}. \tag{1.4.34} $$

Ismail (1995) used the Askey–Wilson operator to derive many summation theorems of q-series.

We shall now use the inner product

$$\langle f, g\rangle := \int_{-1}^{1} f(x)\overline{g(x)}\frac{dx}{\sqrt{1-x^2}}. \tag{1.4.35}$$

The operator $\mathcal{D}_q$ is well defined on $H_{1/2}$, where

$$H_\nu := \{f : f((z+1/z)/2) \text{ is analytic for } q^\nu \le |z| \le q^{-\nu}\}. \tag{1.4.36}$$

The operator $\mathcal{D}_q$ is well defined on polynomials and we shall see that on the Askey–Wilson polynomials it plays the role played by $D = \frac{d}{dx}$ on the classical polynomials of Jacobi, Hermite, and Laguerre. The analogue of integration by parts is the following theorem (Brown, Evans, and Ismail, 1996; Ismail, 2005b).

Theorem 1.4.1 *The Askey–Wilson operator $\mathcal{D}_q$ satisfies*

$$\begin{aligned}\langle \mathcal{D}_q f, g\rangle = {} & \frac{\pi\sqrt{q}}{1-q}\left[f\left(\frac{1}{2}\left(q^{1/2}+q^{-1/2}\right)\right)\overline{g(1)} - f\left(-\frac{1}{2}\left(q^{1/2}+q^{-1/2}\right)\right)\overline{g(-1)}\right] \\ & - \left\langle f, \sqrt{1-x^2}\,\mathcal{D}_q\left(g(x)\left(1-x^2\right)^{-1/2}\right)\right\rangle,\end{aligned} \tag{1.4.37}$$

for $f, g \in H_{1/2}$.

The ***q*-gamma function** is

$$\Gamma_q(z) := \frac{(q;q)_\infty}{(1-q)^{z-1}\,(q^z;q)_\infty}. \tag{1.4.38}$$

It satisfies the functional equation

$$\Gamma_q(z+1) = \frac{1-q^z}{1-q}\Gamma_q(z). \tag{1.4.39}$$

There are three ***q*-analogues of the exponential function**. The first two are due to Euler and have infinite product representations. They are

$$e_q(z) := \sum_{n=0}^{\infty} \frac{z^n}{(q;q)_n} = \frac{1}{(z;q)_\infty}, \quad |z| < 1 \tag{1.4.40}$$

and

$$E_q(z) := \sum_{n=0}^{\infty} \frac{z^n}{(q;q)_n} q^{n(n-1)/2} = (-z;q)_\infty. \tag{1.4.41}$$

It readily follows that $e_q((1-q)x) \to e^x$, and $E_q((1-q)x) \to e^x$ as $q \to 1^-$. A third q-exponential function

$$\mathcal{E}_q(\cos\theta, \cos\phi; \alpha) := \frac{\left(\alpha^2; q^2\right)_\infty}{\left(q\alpha^2; q^2\right)_\infty} \sum_{n=0}^{\infty} \frac{\left(\alpha e^{-i\phi}\right)^n}{(q;q)_n} q^{n^2/4} \times \left(-e^{i(\phi+\theta)} q^{(1-n)/2}, -e^{i(\phi-\theta)} q^{(1-n)/2}; q\right)_n \tag{1.4.42}$$

was introduced by Ismail and Zhang (1994). In view of (1.4.31), we have

$$\mathcal{D}_q \mathcal{E}_q(x, y; \alpha) = \frac{2\alpha q^{1/4}}{1-q} \mathcal{E}_q(x, y; \alpha). \tag{1.4.43}$$

Ismail and Zhang also introduced the function

$$\mathcal{E}_q(\cos\theta; \alpha) = \mathcal{E}_q(x, 0; \alpha) = \frac{\left(\alpha^2; q^2\right)_\infty}{\left(q\alpha^2; q^2\right)_\infty} \times \sum_{n=0}^{\infty} \left(-ie^{i\theta} q^{(1-n)/2}, -ie^{-i\theta} q^{(1-n)/2}; q\right)_n \frac{(-i\alpha)^n}{(q;q)_n} q^{n^2/4}. \tag{1.4.44}$$

It is a fact that $\lim_{q\to 1} \mathcal{E}_q(x; (1-q)t/2) = \exp(tx)$, and $\mathcal{E}_q(0; \alpha) = 1$. Ismail and Stanton (2003a) proved that $\mathcal{E}_q$ has the q-hypergeometric representation

$$\mathcal{E}_q(\cos\theta; t) = \frac{\left(-t; q^{1/2}\right)_\infty}{\left(qt^2; q^2\right)_\infty} {}_2\phi_1\left(\left.\begin{matrix} q^{1/4}e^{i\theta}, q^{1/4}e^{-i\theta} \\ -q^{1/2} \end{matrix}\right| q^{1/2}, -t\right). \tag{1.4.45}$$

One can prove that for real θ and t, we have

$$\begin{aligned} \operatorname{Re} \mathcal{E}_q(\cos\theta; it) &= {}_2\phi_1\left(\left.\begin{matrix} -e^{2i\theta}, -e^{-2i\theta} \\ q \end{matrix}\right| q^2, qt^2\right), \\ \operatorname{Im} \mathcal{E}_q(\cos\theta; it) &= \frac{2tq^{1/4}\cos\theta}{1-q} {}_2\phi_1\left(\left.\begin{matrix} -qe^{2i\theta}, -qe^{-2i\theta} \\ q^3 \end{matrix}\right| q^2, qt^2\right). \end{aligned} \tag{1.4.46}$$

The functions on the right-hand sides of (1.4.46) are q-analogues of the cosine and sine functions, respectively (Atakishiyev and Suslov, 1992).

F. H. Jackson introduced the ***q*-Bessel functions**

$$J_\nu^{(2)}(z; q) = \frac{\left(q^{\nu+1}; q\right)_\infty}{(q;q)_\infty} \sum_{n=0}^{\infty} \frac{(-1)^n (z/2)^{\nu+2n}}{\left(q, q^{\nu+1}; q\right)_n} q^{n(\nu+n)}, \tag{1.4.47}$$

$$J_\nu^{(3)}(z; q) = \frac{\left(q^{\nu+1}; q\right)_\infty}{(q;q)_\infty} \sum_{n=0}^{\infty} \frac{(-1)^n (z/2)^{\nu+2n}}{\left(q, q^{\nu+1}; q\right)_n} q^{\binom{n+1}{2}}. \tag{1.4.48}$$

Jackson (1903; 1903–1904; 1904–1905) studied the cases of integer ν, which are called the q-Bessel coefficients. The q-Bessel functions satisfy the recurrence relation

$$q^\nu J_{\nu+1}^{(2)}(z; q) = \frac{2\left(1-q^\nu\right)}{z} J_\nu^{(2)}(z; q) - J_{\nu-1}^{(2)}(z; q). \tag{1.4.49}$$

It follows from (1.4.47) that

$$\lim_{q\to 1^-} J_\nu^{(2)}(x(1-q);q) = J_\nu(x). \tag{1.4.50}$$

Moreover, $J_\nu^{(2)}$ satisfies the q-difference equation

$$\left(1+qx^2/4\right) J_\nu^{(2)}(qx;q) + J_\nu^{(2)}(x;q) = \left(q^{\nu/2}+q^{-\nu/2}\right) J_\nu^{(2)}\left(\sqrt{q}\,x;q\right). \tag{1.4.51}$$

The functions $I_\nu^{(2)}(z;q)$ and $K_\nu^{(2)}(z;q)$ were introduced in Ismail (1981) in a way similar to the definition of $I_\nu(z)$ and $K_\nu(z)$ of (1.2.17) and (1.2.23):

$$I_\nu^{(2)}(z;q) = e^{-i\pi\nu/2} J_\nu^{(2)}\left(ze^{i\pi/2};q\right), \tag{1.4.52}$$

$$K_\nu^{(2)}(z;q) = \frac{\pi}{2}\frac{I_{-\nu}^{(2)}(z;q) - I_\nu^{(2)}(z;q)}{\sin(\pi\nu)}, \tag{1.4.53}$$

with $K_n^{(2)}(z;q) = \lim_{\nu\to n} K_\nu^{(2)}(z;q)$, $n = 0, \pm1, \pm2, \ldots$. Observe that $K_\nu^{(2)}(z;q)$ is an even function of ν. The functions $K_\nu^{(2)}$ and $I_\nu^{(2)}$ satisfy the three-term relations

$$\begin{aligned} 2\frac{1-q^\nu}{z} K_\nu^{(2)}(z;q) &= q^\nu K_{\nu+1}^{(2)}(z;q) - K_{\nu-1}^{(2)}(z;q), \\ 2\frac{1-q^\nu}{z} I_\nu^{(2)}(z;q) &= I_{\nu-1}^{(2)}(z;q) - q^\nu I_{\nu+1}^{(2)}(z;q). \end{aligned} \tag{1.4.54}$$

Recall that $(a;q)_n$ for $n<0$ has been defined in (1.4.19). A **bilateral basic hypergeometric function** is

$${}_m\psi_m\left(\left.\begin{matrix} a_1,\ldots,a_m \\ b_1,\ldots,b_m \end{matrix}\right| q,z\right) = \sum_{-\infty}^{\infty} \frac{(a_1,\ldots,a_m;q)_n}{(b_1,\ldots,b_m;q)_n} z^n. \tag{1.4.55}$$

It is easy to see that the series in (1.4.55) converges if $\left|\frac{b_1b_2\cdots b_m}{a_1a_2\cdots a_m}\right| < |z| < 1$. The **Ramanujan ${}_1\psi_1$ sum** is

$${}_1\psi_1(a;b;q,z) = \frac{(b/a, q, q/az, az;q)_\infty}{(b, b/az, q/a, z;q)_\infty}, \quad |b/a| < |z| < 1. \tag{1.4.56}$$

It contains the **Jacobi triple product identity**

$$\sum_{-\infty}^{\infty} q^{n^2} z^n = \left(q^2, -qz, -q/z; q^2\right)_\infty. \tag{1.4.57}$$

1.5 Theta Functions

We follow the notation in Whittaker and Watson (1927, Chapter 21). The four **theta functions** have the infinite product representations (Whittaker and Watson, 1927, §21.3)

$$\vartheta_1(z,q) = 2q^{1/4}\sin z\left(q^2, q^2e^{2iz}, q^2e^{-2iz}; q^2\right)_\infty, \tag{1.5.1}$$

$$\vartheta_3(z,q) = \left(q^2, -qe^{2iz}, -qe^{-2iz}; q^2\right)_\infty, \tag{1.5.2}$$

$$\vartheta_2(z,q) = 2q^{1/4}\cos z\left(q^2, -q^2e^{2iz}, -q^2e^{-2iz}; q^2\right)_\infty, \tag{1.5.3}$$

$$\vartheta_4(z,q) = \left(q^2, qe^{2iz}, qe^{-2iz}; q^2\right)_\infty. \tag{1.5.4}$$

We usually drop q when there is no ambiguity. The Jacobi triple product identity yields

$$\vartheta_1(z,q) = q^{1/4}\sum_{-\infty}^{\infty}(-1)^n q^{n^2+n}\sin(2n+1)z, \tag{1.5.5}$$

$$\vartheta_3(z,q) = 2\sum_{-\infty}^{\infty} q^{n^2}\cos(2nz), \tag{1.5.6}$$

$$\vartheta_2(z,q) = q^{1/4}\sum_{-\infty}^{\infty} q^{n^2+n}\cos(2n+1)z, \tag{1.5.7}$$

$$\vartheta_4(z,q) = 2\sum_{-\infty}^{\infty}(-1)^n q^{n^2}\cos(2nz). \tag{1.5.8}$$

1.6 Orthogonality

The orthogonality of the classical orthogonal polynomials is often proved using the **bootstrap** method where we may obtain new orthogonal functions from old ones. This method is formalized in Ismail (2005b). Assume that $(P_n(x))_n$ satisfies (2.1.3) and

$$\sum_{n=0}^{\infty}\frac{P_n(x)t^n}{c_n} = G(x,t), \tag{1.6.1}$$

with $(c_n)_n$ a suitable numerical sequence of nonzero elements. The orthogonality relation (2.1.3) is equivalent to

$$\int_{-\infty}^{\infty} G(x,t_1)\,G(x,t_2)\,d\mu(x) = \sum_{n=0}^{\infty}\zeta_n\frac{(t_1t_2)^n}{c_n^2}, \tag{1.6.2}$$

provided that we can justify the interchange of integration and summation. The idea is to use

$$G(x,t_1)\,G(x,t_2)\,d\mu(x)$$

as a new measure, the total mass of which is given by (1.6.2), and then look for a system of functions (preferably polynomials) orthogonal or biorthogonal with respect to this new measure. If such a system is found one can then repeat the process until the functions involved become too complicated.

2

General Orthogonal Polynomials

2.1 Basic Facts

Suppose we are given a positive Borel measure μ on $\mathbb{R}$ with infinite support whose moments

$$m_n := \int_{\mathbb{R}} x^n \, d\mu(x)$$

exist for $n = 0, 1, \ldots$. We normalize μ by $m_0 = 1$. The **distribution function** F_μ is right continuous and defined by

$$F_\mu(x) = \mu((-\infty, x]) = \int_{-\infty}^{x} d\mu(t). \tag{2.1.1}$$

A polynomial sequence $(\varphi_n(x))_n$ is a sequence of polynomials such that φ_n has exact degree n. Such a sequence is monic if $\varphi_n(x) - x^n$ has degree at most $n-1$.

The moments of μ generate a positive-definite quadratic form $\sum_{j,k=0}^{n} m_{k+j} \overline{x_j} x_k$, hence the **Hankel determinants** D_n,

$$D_n := \begin{vmatrix} m_0 & m_1 & \cdots & m_n \\ m_1 & m_2 & \cdots & m_{n+1} \\ \vdots & \vdots & & \vdots \\ m_n & m_{n+1} & \cdots & m_{2n} \end{vmatrix}, \tag{2.1.2}$$

are positive for $n = 0, 1, \ldots$. The corresponding Hankel matrices $H_n = (m_{k+j})_{0 \le k, j \le n}$ are therefore positive definite. The monic orthogonal polynomials $(P_n(x))_n$ satisfy the orthogonality relation

$$\int_{\mathbb{R}} P_m(x) P_n(x) \, d\mu(x) = \zeta_n \, \delta_{m,n}, \quad \zeta_n := \frac{D_n}{D_{n-1}}. \tag{2.1.3}$$

They have the determinant representations

$$P_n(x) = \frac{1}{D_{n-1}} \begin{vmatrix} m_0 & m_1 & \cdots & m_n \\ m_1 & m_2 & \cdots & m_{n+1} \\ \vdots & \vdots & & \vdots \\ m_{n-1} & m_n & \cdots & m_{2n-1} \\ 1 & x & \cdots & x^n \end{vmatrix}, \tag{2.1.4}$$

$$P_n(x) = \frac{1}{D_{n-1}} \begin{vmatrix} m_0x - m_1 & m_1x - m_2 & \cdots & m_{n-1}x - m_n \\ m_1x - m_2 & m_2x - m_3 & \cdots & m_nx - m_{n+1} \\ \vdots & \vdots & & \vdots \\ m_{n-1}x - m_n & m_nx - m_{n+1} & \cdots & m_{2n-2}x - m_{2n-1} \end{vmatrix}. \tag{2.1.5}$$

It is clear that the polynomial P_n depends on μ only through its moments.

The **Heine integral representation** is

$$P_n(x) = \frac{1}{n!D_{n-1}} \int_{\mathbb{R}^n} \prod_{i=1}^{n} (x - x_i) \prod_{1 \le j < k \le n} \left(x_k - x_j\right)^2 d\mu(x_1) \cdots d\mu(x_n). \tag{2.1.6}$$

By equating coefficients of x^n on both sides of (2.1.6) we conclude that

$$\int_{\mathbb{R}^n} \prod_{1 \le j < k \le n} (x_k - x_j)^2 \, d\mu(x_1) \cdots d\mu(x_n) = n!D_{n-1}. \tag{2.1.7}$$

Both (2.1.6) and (2.1.7) have been used extensively in the theory of random matrices (Mehta, 2004; Deift, 1999). Note that (2.1.7) continues to hold when μ is a signed measure. Also (2.1.3)–(2.1.6) hold for signed measures if $D_n \neq 0$ for all $n \geq 0$.

The **orthonormal polynomials** will be denoted by $(p_n(x))_n$, that is,

$$p_n(x) = \frac{P_n(x)}{\sqrt{\zeta_n}} = \frac{1}{\sqrt{D_n D_{n-1}}} \begin{vmatrix} m_0 & m_1 & \cdots & m_n \\ m_1 & m_2 & \cdots & m_{n+1} \\ \vdots & \vdots & & \vdots \\ m_{n-1} & m_n & \cdots & m_{2n-1} \\ 1 & x & \cdots & x^n \end{vmatrix}. \tag{2.1.8}$$

We shall use the notation

$$p_n(x) = \gamma_n x^n + \text{ lower-order terms} \tag{2.1.9}$$

with

$$\gamma_n = \sqrt{D_{n-1}/D_n} \quad \text{and} \quad D_n = 1/\left[\gamma_1^2 \cdots \gamma_n^2\right]. \tag{2.1.10}$$

Observe that $\gamma_0 = 1$ by our normalization $m_0 = 1$.

For a general polynomial basis $\{\phi_n(x)\}$ with real coefficients and with $\phi_0(x) = 1$ and a given probability measure with finite moments set

$$\phi_{j,k} = \int_{\mathbb{R}} \phi_j(x)\phi_k(x)\, d\mu(x). \tag{2.1.11}$$

The matrices $\{\phi_{j,k}: 0 \le j, k \le n\}$, $n = 0, 1, \ldots$ are positive definite. With

$$\tilde{D}_n = \begin{vmatrix} \phi_{0,0} & \phi_{0,1} & \cdots & \phi_{0,n} \\ \phi_{1,0} & \phi_{1,1} & \cdots & \phi_{1,n} \\ \vdots & \vdots & & \vdots \\ \phi_{n,0} & \phi_{n,1} & \cdots & \phi_{n,n} \end{vmatrix}, \quad n \ge 0, \tag{2.1.12}$$

the polynomials orthonormal with respect to μ are

$$p_n(x) = \frac{1}{\sqrt{\tilde{D}_n \tilde{D}_{n-1}}} \begin{vmatrix} \phi_{0,0} & \phi_{0,1} & \cdots & \phi_{0,n} \\ \phi_{1,0} & \phi_{1,1} & \cdots & \phi_{1,n} \\ \vdots & \vdots & & \vdots \\ \phi_{n-1,0} & \phi_{n-1,1} & \cdots & \phi_{n-1,n} \\ \phi_0(x) & \phi_1(x) & \cdots & \phi_n(x) \end{vmatrix}. \tag{2.1.13}$$

Theorem 2.1.1 *Let $f \in L^2(\mu, \mathbb{R})$ and N be a positive integer. If $\{p_j(x): 0 \le j \le N\}$ are orthonormal with respect to μ, then*

$$\inf\left\{ \int_{\mathbb{R}} \left| f(x) - \sum_{j=0}^{N} f_j p_j(x) \right|^2 d\mu(x) : f_j \in \mathbb{R},\ 0 \le j \le N \right\}$$

is attained if and only if $f_j = \int_{\mathbb{R}} f(x) p_j(x)\, d\mu(x)$. The infimum is

$$\int_{\mathbb{R}} |f(x)|^2\, d\mu(x) - \sum_{j=0}^{N} |f_j|^2 .$$

Theorem 2.1.2 *Let N be a positive integer and let $q_N(x)$ be a monic polynomial with real coefficients. Then the infimum of $\int_{\mathbb{R}} |q_N(x)|^2\, d\mu(x)$ over such q_N is attained if and only if $q_N(x) = p_N(x)/\gamma_N$. The infimum is $\zeta_N = 1/\gamma_N^2$.*

Since a Hankel matrix H_N formed by moments of a positive measure is positive definite, it is natural to study the limiting behavior of its smallest eigenvalue, $\lambda_1(N)$. Szegő (1936) proved that $\lim_{N\to\infty} \lambda_1(N) = 0$ in the cases of Legendre, Hermite, and Laguerre polynomials. Berg, Chen, and Ismail (2002) proved that $\lim_{N\to\infty} \lambda_1(N) > 0$ if and only if the moments m_n do not determine a unique measure μ; see Chapter 11. When μ is not unique, Berg, Chen, and Ismail also gave a positive lower bound for the limit.

Monic orthogonal polynomials satisfy a **three-term recurrence relation**

$$xP_n(x) = P_{n+1}(x) + \alpha_n P_n(x) + \beta_n P_{n-1}(x), \quad n > 0, \tag{2.1.14}$$

with

$$P_0(x) = 1, \quad P_1(x) = x - \alpha_0, \tag{2.1.15}$$

where α_n is real, for $n \geq 0$ and $\beta_n > 0$ for $n > 0$. Moreover,

$$\zeta_n = \beta_1 \cdots \beta_n. \tag{2.1.16}$$

The corresponding orthonormal polynomials satisfy

$$xp_n(x) = a_{n+1}p_{n+1}(x) + b_n p_n(x) + a_n p_{n-1}(x), \quad n \geq 0, \tag{2.1.17}$$

with initial conditions

$$p_0(x) = 1, \quad p_{-1}(x) = 0, \tag{2.1.18}$$

and with recurrence coefficients $a_n = \sqrt{\beta_n}$ and $b_n = \alpha_n$.

The **Christoffel–Darboux identities** are

$$\sum_{k=0}^{N-1} p_k(x)p_k(y) = \sum_{k=0}^{N-1} \frac{P_k(x)P_k(y)}{\zeta_k} = \frac{P_N(x)P_{N-1}(y) - P_N(y)P_{N-1}(x)}{\zeta_{N-1}(x-y)}, \tag{2.1.19}$$

$$\sum_{k=0}^{N-1} p_k^2(x) = \sum_{k=0}^{N-1} \frac{P_k^2(x)}{\zeta_k} = \frac{P'_N(x)P_{N-1}(x) - P_N(x)P'_{N-1}(x)}{\zeta_{N-1}}. \tag{2.1.20}$$

The Christoffel–Darboux formulas follow from (2.1.14)–(2.1.15), hence they will hold for any solution of (2.1.14), with possibly an additional term $c/(x-y)$ depending on the initial conditions. A consequence of (2.1.19)–(2.1.20) is the following result.

Theorem 2.1.3 *Assume that α_{n-1} is real and $\beta_n > 0$ for all $n = 1, 2, \ldots$. Then the zeros of the polynomials generated by* (2.1.14)–(2.1.15) *are real and simple. Furthermore, the zeros of P_n and P_{n-1} interlace.*

The monic polynomials $(P_n)_n$ are characteristic polynomials of real symmetric matrices. Let $J_n = (a_{j,k} : 0 \leq j, k < n)$ be the tridiagonal matrix

$$\begin{aligned} a_{j+1,j} &= \sqrt{\beta_{j+1}} = a_{j,j+1}, \\ a_{j,j} &= \alpha_j, \quad a_{j,k} = 0 \quad \text{for } |j-k| > 1; \end{aligned} \tag{2.1.21}$$

then $P_n(\lambda)$ is the determinant of $\lambda I - J_n$ for $n > 0$:

$$P_n(x) = \begin{vmatrix} x-\alpha_0 & -\sqrt{\beta_1} & 0 & 0 & \cdots & 0 \\ -\sqrt{\beta_1} & x-\alpha_1 & -\sqrt{\beta_2} & 0 & \cdots & 0 \\ 0 & -\sqrt{\beta_2} & x-\alpha_2 & -\sqrt{\beta_3} & \cdots & 0 \\ \vdots & & \ddots & \ddots & \ddots & \vdots \\ 0 & \cdots & 0 & -\sqrt{\beta_{n-2}} & x-\alpha_{n-2} & -\sqrt{\beta_{n-1}} \\ 0 & \cdots & 0 & 0 & -\sqrt{\beta_{n-1}} & x-\alpha_{n-1} \end{vmatrix}. \tag{2.1.22}$$

The matrix J_n is the **Jacobi matrix** of order n. The interlacing property of the zeros of P_n and P_{n-1} is an instance of the Cauchy interlacing theorem (Parlett, 1980; Horn and Johnson, 1992) for the matrices J_n and J_{n-1}.

Theorem 2.1.4 *Let $\{P_n\}$ be a sequence of orthogonal polynomials satisfying* (2.1.3); *then any closed interval containing the support of μ will contain all the zeros of P_n for $n = 0, 1, \ldots$.*

Power sums of zeros of orthogonal polynomials are

$$s_k = \sum_{j=1}^{n} \left(x_{n,j}\right)^k, \tag{2.1.23}$$

where

$$x_{n,1} > x_{n,2} > \cdots > x_{n,n} \tag{2.1.24}$$

are the zeros of P_n. Clearly,

$$\frac{P_n'(x)}{P_n(x)} = \sum_{j=1}^{n} \frac{1}{x - x_{n,j}} = \sum_{k=0}^{\infty} \frac{1}{x^{k+1}} \sum_{j=1}^{n} \left(x_{n,j}\right)^k, \quad |x| > \max\{|x_{n,j}| : 1 \le j \le n\}.$$

Thus for $|z| > \max\{|x_{n,j}| : 1 \le j \le n\}$, we have

$$P_n'(z)/P_n(z) = \sum_{k=0}^{\infty} s_k z^{-k-1}. \tag{2.1.25}$$

Power sums for various special orthogonal polynomials can be evaluated using (2.1.25). If no $x_{n,j} = 0$, we can define s_k for $k < 0$ and conclude that

$$P_n'(z)/P_n(z) = -\sum_{k=0}^{\infty} z^k s_{-k-1}, \tag{2.1.26}$$

for $|z| < \min\{|x_{n,j}| : 1 \le j \le n\}$. Formula (2.1.26) also holds when P_n is replaced by a function with the factor product representation

$$f(z) = \prod_{k=1}^{\infty} (1 - z/x_k).$$

An example is $f(z) = \Gamma(\nu + 1)(2/z)^\nu J_\nu(z)$. Examples of (2.1.25) and (2.1.26) and their applications are in Ahmed et al. (1979), Ahmed, Laforgia, and Muldoon (1982), and Ahmed and Muldoon (1983). The power sums of zeros of Bessel polynomials have a remarkable property, as we shall see in Chapter 3.

The **Poisson kernel** $P_r(x, y)$ of a system of orthogonal polynomials is

$$P_r(x, y) = \sum_{n=0}^{\infty} P_n(x) P_n(y) \frac{r^n}{\zeta_n}. \tag{2.1.27}$$

One would expect $\lim_{r \to 1^-} P_r(x, y)$ to be a Dirac function $\delta(x - y)$. Indeed, under certain conditions,

$$\lim_{r \to 1^-} \int_{\mathbb{R}} P_r(x, y) f(y)\, d\mu(y) = f(x), \tag{2.1.28}$$

for $f \in L^2(\mu)$. A crucial step in establishing (2.1.28) for a specific system of orthogonal polynomials is the nonnegativity of the Poisson kernel on the support of μ.

Definition 2.1.5 The **kernel polynomials** $K_n(x, y)$ of a measure μ are

$$K_n(x, y) = \sum_{k=0}^{n} \overline{p_k(x)} p_k(y) = \sum_{k=0}^{n} \overline{P_k(x)} P_k(y)/\zeta_k, \quad n = 0, 1, \ldots. \tag{2.1.29}$$

Theorem 2.1.6 *Let π be a polynomial of degree at most n and normalized by*

$$\int_{\mathbb{R}} |\pi(x)|^2 \, d\mu(x) = 1. \tag{2.1.30}$$

Then the maximum of $|\pi(x_0)|^2$ taken over all such π is attained when

$$\pi(x) = cK_n(x_0, x) / \sqrt{K_n(x_0, x_0)}, \quad |c| = 1.$$

The maximum is $K_n(x_0, x_0)$.

Theorem 2.1.7 *Let π be a polynomial of degree at most n. Then*

$$\int_{\mathbb{R}} K_n(x, y)\pi(x) \, d\mu(x) = \pi(y). \tag{2.1.31}$$

Conversely, if $J_n(x, y)$ is a polynomial in x of degree at most n and

$$\int_{\mathbb{R}} J_n(x, y)\pi(x) \, d\mu(x) = \pi(y)$$

holds for every such polynomial π of degree at most n, then $J_n(x, y) = K_n(x, y)$.

An immediate corollary to Theorem 2.1.7 is the representation

$$K_n(x, y) = -\frac{1}{D_n} \begin{vmatrix} 0 & 1 & x & \cdots & x^n \\ 1 & m_0 & m_1 & \cdots & m_n \\ y & m_1 & m_2 & \cdots & m_{n+1} \\ \vdots & \vdots & \vdots & & \vdots \\ y^n & m_{n-1} & m_n & \cdots & m_{2n-1} \end{vmatrix}. \tag{2.1.32}$$

The determinant representations (2.1.8) and (2.1.32) imply the following theorem.

Theorem 2.1.8 *Let $H_n = (m_{i+j})_{0 \le i, j \le n}$ be the Hankel matrix with determinant D_n given in (2.1.2) and let*

$$K_n(x, y) = \sum_{j,k=0}^{n} a_{j,k}(n) x^j y^k. \tag{2.1.33}$$

Then $a_{j,k}(n) = a_{k,j}(n)$ and the matrix $A_n = (a_{j,k}(n))_{0 \le j,k \le n}$ is the inverse of H_n.

Theorem 2.1.9 *When $d\mu(x) = w(x)\,dx$ we have*

$$\prod_{1\le j<k\le n} \left(x_k - x_j\right)^2 = D_{n-1} \det\left(K_{n-1}(x_j, x_k)\right)_{1\le j,k\le n}. \tag{2.1.34}$$

For a proof see Deift (1999).

Remark 2.1.10 Sometimes it is convenient to use neither the monic nor the orthonormal polynomials. If $(\phi_n(x))_n$ satisfy

$$\int_{\mathbb{R}} \phi_m(x)\phi_n(x)\,d\mu(x) = h_n\,\delta_{m,n}, \tag{2.1.35}$$

then

$$x\phi_n(x) = A_n\phi_{n+1}(x) + B_n\phi_n(x) + C_n\phi_{n-1}(x) \tag{2.1.36}$$

and

$$h_n = \frac{C_1 \cdots C_n}{A_0 \cdots A_{n-1}} h_0. \tag{2.1.37}$$

2.2 Numerators and Quadratures

Consider (2.1.14) as a difference equation in the variable n,

$$xy_n = y_{n+1} + \alpha_n y_n + \beta_n y_{n-1}. \tag{2.2.1}$$

A solution is $y_n = P_n(x)$. Define a second solution $(P_n^*(x))_n$ by the initial conditions

$$P_0^*(x) = 0, \quad P_1^*(x) = 1. \tag{2.2.2}$$

Clearly $P_n^*(x)$ is a monic polynomial of degree $n-1$ for all $n > 0$. The Casorati determinant (Milne-Thomson, 1933) is the discrete analogue of the Wronskian,

$$\Delta_n(x) := P_n(x)P_{n-1}^*(x) - P_n^*(x)P_{n-1}(x) = \det\begin{pmatrix} P_n(x) & P_{n-1}(x) \\ P_n^*(x) & P_{n-1}^*(x) \end{pmatrix} = -\beta_1 \cdots \beta_{n-1}. \tag{2.2.3}$$

Two solutions of a second-order linear difference equation are linearly independent if and only if their Casorati determinant is not zero (Jordan, 1965; Milne-Thomson, 1933). Thus P_n and P_n^* are linearly independent solutions of (2.2.1). We shall refer to $(P_n^*)_n$ as the **numerator polynomials**.

Theorem 2.2.1 *For $n > 0$, the zeros of P_n^* are all real and simple and interlace with the zeros of P_n.*

Definition 2.2.2 Let $(P_n)_n$ be a family of monic orthogonal polynomials generated by (2.1.14) and (2.1.15). The **associated polynomials** $(P_n(x;c))_n$ of order c are polynomials satisfying (2.1.14) with n replaced by $n+c$ and given initially by

$$P_0(x;c) := 1, \quad P_1(x;c) = x - \alpha_c. \tag{2.2.4}$$

The above procedure is always applicable when $c = 1, 2, \ldots$. Moreover,

$$P_n^*(z) = P_{n-1}(z; 1) = \int_{\mathbb{R}} \frac{P_n(z) - P_n(y)}{z - y} \, d\mu(y), \quad n \geq 0. \tag{2.2.5}$$

Let $(P_n)_n$ be a sequence of monic orthogonal polynomials satisfying (2.1.3) with zeros as in (2.1.24). Then the **quadrature formula**

$$\int_{\mathbb{R}} p(x) \, d\mu(x) = \sum_{k=1}^{N} \lambda_k p(x_{N,k}) \tag{2.2.6}$$

holds for all polynomials p of degree at most $2N - 1$ and the **Christoffel numbers** $\lambda_1, \ldots, \lambda_N$, which depend on N, are given by

$$\lambda_k = \lambda_{N,k} = \int_{\mathbb{R}} \frac{P_N(x) \, d\mu(x)}{P_N'(x_{N,k})(x - x_{N,k})} = \int_{\mathbb{R}} \left[\frac{P_N(x)}{P_N'(x_{N,k})(x - x_{N,k})} \right]^2 d\mu(x). \tag{2.2.7}$$

Moreover, if (2.2.6) holds for all p of degree at most $2N - 1$ then the $(\lambda_k)_{1<k<N}$ are unique and are given by (2.2.7). The Christoffel numbers have the properties

$$\begin{gathered} \sum_{k=1}^{N} \lambda_k = \mu(\mathbb{R}) = m_0, \qquad \lambda_k = -\zeta_N / \left[P_{N+1}(x_{N,k}) P_N'(x_{N,k}) \right], \\ \frac{1}{\lambda_k} = \sum_{j=0}^{N} P_j^2(x_{N,k}) / \zeta_j = K_N(x_{N,k}, x_{N,k}). \end{gathered} \tag{2.2.8}$$

Let $[a, b]$ be the convex hull of the support of μ. For $N \geq 2$, we let

$$u_k = x_{N,N-k}, \tag{2.2.9}$$

so that $u_1 < u_2 < \cdots < u_N$ are the zeros of P_N in increasing order. Let F_μ be the distribution function of μ, that is, $F_\mu(x) = \int_{-\infty}^{x} d\mu(y)$. In view of the positivity of the $(\lambda_k)_{1\leq k\leq N}$ and (2.2.8), there exist numbers $y_1 < y_2 < \cdots < y_{N-1}$, with $a < y_1, y_{N-1} < b$, such that

$$\lambda_k = F_\mu(y_k) - F_\mu(y_{k-1}), \quad 1 \leq k \leq N, \quad y_0 := a, \quad y_N := b. \tag{2.2.10}$$

Theorem 2.2.3 (Separation theorem) *The points $(y_k)_{1\leq k\leq N-1}$ interlace with the zeros $(u_k)_{1\leq k\leq N}$; that is,*

$$u_k < y_k < u_{k+1}, \quad 1 \leq k \leq N - 1.$$

Equivalently,

$$F_\mu(u_k) < F_\mu(y_k) = \sum_{j=1}^{k} \lambda_j < F_\mu\left(u_{k+1}^-\right), \quad 1 \leq k < N.$$

The following is an immediate consequence of the separation theorem.

Corollary 2.2.4 *Let I be an open interval formed by two consecutive zeros of P_N. Then $\mu(I) > 0$.*

Corollary 2.2.5 (Density of zeros) *Assume that μ has compact support and I is an interval with $\mu(I) > 0$. Then for n sufficiently large P_n has at least one zero in I.*

Szegő ([1939] 1975, §3.41) gives several proofs of the separation theorem.

2.3 The Spectral Theorem

The following theorem is called the **spectral theorem for orthogonal polynomials**.

Theorem 2.3.1 (Spectral theorem for monic orthogonal polynomials) *Given a sequence of monic polynomials $(P_n)_n$ generated by* (2.1.14)–(2.1.15) *with $\alpha_{n-1} \in \mathbb{R}$ and $\beta_n > 0$ for all $n > 0$, there exists a measure μ such that*

$$\int_{\mathbb{R}} P_m(x)P_n(x)\,d\mu(x) = \zeta_n\,\delta_{m,n}, \tag{2.3.1}$$

where ζ_n is given by (2.1.16).

There is a counterpart to Theorem 2.3.1 for orthonormal polynomials.

Theorem 2.3.2 (Spectral theorem for orthonormal polynomials) *Given two sequences $(a_n)_{n\geq 1}$, $(b_n)_{n\geq 0}$ with $a_n > 0$ and $b_n \in \mathbb{R}$, let $(p_n)_{n\geq 0}$ be the system of polynomials determined by* (2.1.17) *and* (2.1.18). *Then there exists a probability measure μ on $\mathbb{R}$ with infinite support and moments of any order such that*

$$\int_{\mathbb{R}} p_m(x)p_n(x)\,d\mu(x) = \delta_{m,n}. \tag{2.3.2}$$

The name "spectral theorem for orthogonal polynomials" was suggested by Ismail in earlier lecture notes and was used in Ismail (2005b). In the older literature (Chihara, 1978; Szegő, [1939] 1975) it is called Favard's theorem, because Favard (1935) proved it. Shohat (1936) claimed to have had an unpublished proof several years before Favard published his paper. The theorem is stated and proved in Wintner (1929) on spectral theory of Jacobi matrices, and also appeared in Stone (1932, Thm. 10.27) without attributing it to any particular author. For proofs see Shohat and Tamarkin (1950), Ismail (2005b), and Stone (1932).

Shohat (1936) proved that if $\alpha_n \in \mathbb{R}$ and $\beta_{n+1} \neq 0$ for all $n \geq 0$, then there is a real signed measure μ with total mass 1 such that (2.3.1) holds, with $\zeta_0 = 1$, $\zeta_n = \beta_1 \cdots \beta_n$; see also Shohat (1938).

The measure μ in Theorems 2.3.1 and 2.3.2 may not be unique. For example the **Stieltjes weight function**

$$w(x;\alpha) = [1 + \alpha \sin(2\pi c \ln x)] \exp\left(-c \ln^2 x\right), \quad x \in (0,\infty),$$

for $\alpha \in [-1,1]$ and $c > 0$ has moments

$$m_n = \int_0^\infty x^n w(x;\alpha)\,dx = \sqrt{\frac{\pi}{c}}\exp\left(\frac{(n+1)^2}{4c}\right), \tag{2.3.3}$$

which are independent of α. Therefore the weight functions $w(x;\alpha)$, for all $\alpha \in [-1,1]$, have the same moments (Ismail, 2005b).

Definition 2.3.3 The **true interval of orthogonality** of a sequence of polynomials $(P_n)_n$ generated by (2.1.14)–(2.1.15) is the interval $[\xi, \eta]$, where

$$\xi := \lim_{n\to\infty} x_{n,n}, \quad \eta := \lim_{n\to\infty} x_{n,1}. \tag{2.3.4}$$

The limits in (2.3.4) exist since $(x_{n,1})_n$ is an increasing sequence and $(x_{n,n})_n$ is a decreasing sequence. The cases $\xi = -\infty$ and $\eta = +\infty$ are allowed. Moreover, it is clear from Theorem 2.2.3 that $[\xi, \eta]$ is a subset of the convex hull of $\text{supp}(\mu)$.

Theorem 2.3.4 *If $(\alpha_n)_n$ and $(\beta_n)_n$ are bounded sequences, then μ is unique and supported on a compact set. The convex hull of the support of μ is $[\xi, \eta]$.*

Some authors prefer to work with a positive linear functional $\mathcal{L}$ defined by $\mathcal{L}(x^n) = m_n$. The question of constructing the orthogonality measure then becomes a question of finding a representation of $\mathcal{L}$ as an integral with respect to a positive measure. This approach is used in Chihara (1978). Some authors even prefer to work with a linear functional which is not necessarily positive, but the determinants D_n of (2.1.2) are assumed to be nonzero. Such functionals are called regular. An extensive theory of polynomials orthogonal with respect to regular linear functionals has been developed by Brezinski (1980), Draux (1983), and Maroni (1987), and their students and collaborators.

Theorem 2.3.5 (Nevai) *If in addition to the assumptions of the spectral theorem we assume that*

$$\sum_{n=1}^{\infty} \left(\left| \sqrt{\beta_n} - \frac{1}{2} \right| + |\alpha_n| \right) < \infty, \tag{2.3.5}$$

then μ has an absolutely continuous component μ_{ac} supported on $[-1, 1]$. Furthermore, if μ has a discrete part then it will lie outside $(-1, 1)$ with accumulation points (if any) at ± 1. In addition, the limiting relation

$$\lim_{n\to\infty} \left(\sqrt{1-x^2}\, \frac{P_n(x)}{\sqrt{\zeta_n}} - \sqrt{\frac{2\sqrt{1-x^2}}{\pi\mu'(x)}}\, \sin((n+1)\theta - \varphi(\theta)) \right) = 0 \tag{2.3.6}$$

holds, with $x = \cos\theta \in (-1, 1)$. The function $\varphi(\theta)$ in (2.3.6) does not depend on n.

Boas (1939) proved that any sequence of real numbers $(c_n)_n$ has a moment representation $c_n = \int_0^\infty x^n\, d\mu(x)$ for a nontrivial finite signed measure μ. A nontrivial signed measure whose moments are all zero is called a **polynomial killer**.

The following examples are due to Stieltjes (1894):

$$0 = \int_0^\infty x^n \sin(2\pi c \ln x) \exp\left(-c \ln x^2\right) dx, \quad c > 0, \tag{2.3.7}$$

$$0 = \int_0^\infty x^n \sin x^{1/4} \exp\left(-x^{1/4}\right) dx. \tag{2.3.8}$$

There is a close connection between constructing orthogonality measures and the spectral problem on a certain Hilbert space. Consider the operator T_0 which is multiplication by x on the vector space of polynomials. Using the basis $\{p_n(x)\}$, the three-term recurrence relation (2.1.17) together with the initial conditions (2.1.18) gives a realization of T_0 as a tridiagonal matrix, called the Jacobi matrix:

$$J = \begin{pmatrix} b_0 & a_1 & 0 & 0 & 0 & \cdots \\ a_1 & b_1 & a_2 & 0 & 0 & \cdots \\ 0 & a_2 & b_2 & a_3 & 0 & \cdots \\ \vdots & & \ddots & \ddots & \ddots & \end{pmatrix}. \tag{2.3.9}$$

This matrix defines an operator on the dense subset of ℓ^2 consisting of sequences which are eventually zero. It is clear that T_0 is symmetric, so it has a symmetric closure T. When T is self-adjoint there exists a unique orthogonality measure. In the non-self-adjoint case, T has deficiency indices $(1,1)$ and by von Neumann's theory there exists a one-parameter family of self-adjoint extensions as operators in ℓ^2. However, also, self-adjoint extensions S of T defined as operators on a Hilbert space H containing ℓ^2 as a closed subspace are important for the construction of orthogonality measures. For these extensions of the Hilbert space, see Akhiezer and Glazman (1950, App. 1). If S is a self-adjoint extension of T, whether on the Hilbert space ℓ^2 or on a bigger Hilbert space H, the spectral theorem shows that S is an integral with respect to a projection-valued measure E on the spectrum $\sigma(S) \subseteq \mathbb{R}$, that is,

$$S = \int_{\sigma(S)} \lambda \, dE(\lambda).$$

In other words,

$$(y, p(S)x) = \int_{\sigma(S)} p(\lambda)\,(y, dE(\lambda)x) \tag{2.3.10}$$

holds for polynomials p, and for all $x, y \in H$. By choosing the standard basis $e_0, e_1, \ldots$ for ℓ^2, that is, $e_n = (u_0, u_1, \ldots)$ with $u_k = \delta_{kn}$, we see that $(e_0, dE(\lambda)e_0)$ is a positive measure. This is a measure of orthonormality for $(p_n)_n$. The details of this theory are in Akhiezer (1965), Simon (2005b), and Stone (1932).

The operator T is a **discrete Schrödinger operator** (Cycon et al., 1987) and the diagonal entries $(b_n)_n$ represent a discrete potential. An extensive theory of doubly infinite Jacobi matrices with $a_n = 1, n \geq 1$ and a random potential $(b_n)_n$ is given in Cycon et al. (1987, Chapter 9). The theory of general doubly infinite tridiagonal matrices is treated in Berezanskii (1968).

2.4 Continued Fractions

A **continued J-fraction** is

$$\frac{A_0}{A_0z+B_0-}\,\frac{C_1}{A_1z+B_1-}\,\frac{C_2}{A_2z+B_2-}\cdots=\cfrac{A_0}{A_0z+B_0-\cfrac{C_1}{A_1z+B_1-\cfrac{C_2}{A_2z+B_2-\ddots}}}. \tag{2.4.1}$$

The nth convergent of the above continued fraction is the rational function

$$\frac{A_0}{A_0z+B_0-}\,\frac{C_1}{A_1z+B_1-}\cdots\frac{C_{n-1}}{A_{n-1}z+B_{n-1}}=\frac{N_n(z)}{D_n(z)}. \tag{2.4.2}$$

We call this the nth convergent, and the first convergent is $A_0/(A_0z+B_0)$.

Definition 2.4.1 The J-fraction (2.4.1) is of positive type if $A_nA_{n-1}C_n>0$, $n=1,2,\ldots$.

Theorem 2.4.2 *Assume that $A_nC_{n+1}\neq 0$, $n=0,1,\ldots$. Then the polynomials $N_n(z)$ and $D_n(z)$ are solutions of the recurrence relation*

$$y_{n+1}(z)=[A_nz+B_n]\,y_n(z)-C_ny_{n-1}(z),\quad n>0, \tag{2.4.3}$$

with the initial values

$$D_0(z):=1,\ \ D_1(z):=A_0z+B_0,\quad N_0(z):=0,\ \ N_1(z):=A_0. \tag{2.4.4}$$

When $A_n=1$, $B_n=-\alpha_n$, and $C_n=\beta_n$ then D_n and N_n become P_n and P_n^* of Section 2.2, respectively.

Theorem 2.4.3 (Markov) *Assume that the true interval of orthogonality $[\xi,\eta]$ is bounded. Then*

$$\lim_{n\to\infty}\frac{P_n^*(z)}{P_n(z)}=\int_\xi^\eta\frac{d\mu(t)}{z-t},\quad z\notin[\xi,\eta], \tag{2.4.5}$$

and the limit is uniform on compact subsets of $\mathbb{C}\setminus[\xi,\eta]$.

Proofs are in Szegő ([1939] 1975), Wall (1948), and Ismail (2005b). **Markov's theorem** can be used to compute orthogonality measures for orthogonal polynomials from their recurrence coefficients.

Definition 2.4.4 A solution $(u_n(z))_n$ of (2.4.3) is called a **minimal solution** at ∞ if $\lim_{n\to\infty}u_n(z)/v_n(z)=0$ for any other linear independent solution $(v_n(z))_n$. A minimal solution at $-\infty$ is similarly defined.

It is clear that if a minimal solution exists then it is unique, up to multiplication by a function of z.

Theorem 2.4.5 (Pincherle) *The continued fraction* (2.4.2) *converges at* $z = z_0$ *if and only if the recurrence relation* (2.4.3) *has a minimal solution* $(y_n^{(\min)}(z_0))_n$. *Furthermore, if a minimal solution* $(y_n^{(\min)}(z_0))_n$ *exists, then the continued fraction converges to* $y_0^{(\min)}(z_0)/y_{-1}^{(\min)}(z_0)$.

A proof can be found in Jones and Thron (1980, pp. 164–166) or Lorentzen and Waadeland (1992, pp. 202–203). In many interesting cases the minimal solutions were found by David Masson and his collaborators. Walter Gautschi (1967) seems to have been the first to promote the use of Pincherle's theorem in the theory of orthogonal polynomials.

Theorem 2.4.6 *Let* $(u_n(z))_n$ *be a solution to* (2.1.14) *and assume that* $u_n(\zeta) \neq 0$ *for all* n *and a fixed* ζ. *Then* $(u_n(\zeta))_n$ *is a minimal solution to* (2.4.3) *at* ∞ *if and only if*

$$\sum_{n=1}^{\infty} \frac{\prod_{m=1}^{n} \beta_m}{u_n(\zeta)u_{n+1}(\zeta)} = \infty.$$

A proof is in Lorentzen and Waadeland (1992, §4.2.2).

The following theorem gives an interesting connection between addition theorems and continued fractions.

Theorem 2.4.7 (Wall, 1948, Theorem 53.1) *Let* $(\beta_n)_{n>0}$ *be a sequence of real numbers and* $\beta_n \neq 0$ *for all* $n > 0$. *The coefficients in the continued J-fraction*

$$\frac{1}{z-\alpha_0-}\,\frac{\beta_1}{z-\alpha_1-}\cdots\frac{\beta_n}{z-\alpha_n-}\cdots = \cfrac{1}{z-\alpha_0-\cfrac{\beta_1}{z-\alpha_1-\cdots-\cfrac{\beta_n}{z-\alpha_n-\ddots}}} \tag{2.4.6}$$

and its power series expansion

$$P(1/z) = \sum_{n=0}^{\infty} \frac{m_n}{z^{n+1}} \tag{2.4.7}$$

are connected by the relationships

$$m_{p+q} = k_{0,p}k_{0,q} + \sum_{j=1}^{\infty} \beta_1\beta_2\cdots\beta_j k_{j,p}k_{j,q}, \tag{2.4.8}$$

with

$$k_{0,0} = 1, \quad k_{r,s} = 0 \quad \text{if } r > s,$$

and where the $k_{r,s}$ are recursively generated by the matrix equation

$$\begin{pmatrix} k_{0,0} & 0 & 0 & 0 & \cdots \\ k_{0,1} & k_{1,1} & 0 & 0 & \cdots \\ k_{0,2} & k_{1,2} & k_{2,2} & 0 & \cdots \\ \vdots & \vdots & \vdots & \vdots & \end{pmatrix} \begin{pmatrix} \alpha_0 & 1 & 0 & 0 & 0 & \cdots \\ \beta_1 & \alpha_1 & 1 & 0 & 0 & \cdots \\ 0 & \beta_2 & \alpha_2 & 1 & 0 & \cdots \\ \vdots & \vdots & \vdots & \vdots & \vdots & \end{pmatrix}$$
$$= \begin{pmatrix} k_{0,1} & k_{1,1} & 0 & 0 & 0 & \cdots \\ k_{0,2} & k_{1,2} & k_{2,2} & 0 & 0 & \cdots \\ k_{0,3} & k_{1,3} & k_{2,3} & k_{3,3} & 0 & \cdots \\ \vdots & \vdots & \vdots & \vdots & \vdots & \end{pmatrix}. \tag{2.4.9}$$

Moreover, there is a formal decomposition into sums of squares

$$\sum_{p,q=0}^{\infty} m_{p+q} x_p x_q = \left(\sum_{j=0}^{\infty} k_{0,j} x_j\right)^2 + \sum_{s=1}^{\infty} \beta_1 \beta_2 \cdots \beta_s \left(\sum_{j=s}^{\infty} k_{s,j} x_j\right)^2. \tag{2.4.10}$$

Conversely, if there is a decomposition (2.4.10), *where the β_n are nonzero, then $P(1/z)$ is the power series expansion of the continued J-fraction* (2.4.6), *where*

$$\alpha_0 = k_{0,1}, \quad \alpha_p = k_{p,p+1} - k_{p-1,p}, \quad p = 1, 2, \ldots. \tag{2.4.11}$$

Finally, the problem of expanding the power series (2.4.7) *into the continued J-fraction* (2.4.6) *is equivalent to the problem of obtaining a formal addition theorem of the form*

$$Q(x+y) = Q(x)Q(y) + \sum_{s=1}^{\infty} \beta_1 \beta_2 \cdots \beta_s Q_s(x) Q_s(y), \tag{2.4.12}$$

where the β are different from zero, and

$$Q(z) = \sum_{n=0}^{\infty} \frac{m_n}{n!} z^n \ (m_0 = 1); \quad Q_r(z) = \sum_{n=0}^{\infty} \frac{k_{r,n}}{n!} z^n. \tag{2.4.13}$$

The coefficients $(\beta_n)_n$ in the continued J-fraction (2.4.6) *are the β_n of* (2.4.12), *and the α_n of the continued J-fraction* (2.4.6) *are given by* (2.4.11) *in terms of the $k_{r,s}$ of* (2.4.13).

Observe that (2.4.9) and the initial conditions show that $k_{j,j} = 1$ for $j = 0, 1, \ldots$. Moreover, the addition theorem in (2.4.12) is equivalent to (2.4.10).

Wall (1948) attributes Theorem 2.4.7 to Stieltjes and the formulation of (2.4.10) as an addition theorem to Rogers.

Let K be the matrix with entries $k_{i,j}$ where $k_{i,j} = 0$ if $i > j$. It must be noted that the matrix K is also the matrix of the connection coefficients

$$x^n = \sum_{k=0}^{n} k_{j,n} P_j(x), \tag{2.4.14}$$

where $(P_j)_j$ are the corresponding monic orthogonal polynomials. Many interesting examples of (2.4.12) are in Ismail and Zeng (2010).

2.5 Modifications of Measures and Recursions

Given a measure μ and the corresponding orthogonal polynomials $(P_n)_n$, an interesting question is: what can we say about the polynomials orthogonal with respect to $\Phi(x)\,d\mu(x)$ where Φ is a function positive on the support of μ?

Theorem 2.5.1 (Christoffel) *Let $(P_n)_n$ be monic orthogonal polynomials with respect to μ and let*

$$\Phi(x) = \prod_{k=1}^{m} (x - x_k) \tag{2.5.1}$$

be nonnegative on the support of μ. If the x_k are simple zeros then the monic polynomials q_n defined by

$$C_{n,m}\Phi(x)q_n(x) = \begin{vmatrix} P_n(x_1) & P_{n+1}(x_1) & \cdots & P_{n+m}(x_1) \\ P_n(x_2) & P_{n+1}(x_2) & \cdots & P_{n+m}(x_2) \\ \vdots & \vdots & & \vdots \\ P_n(x_m) & P_{n+1}(x_m) & \cdots & P_{n+m}(x_m) \\ P_n(x) & P_{n+1}(x) & \cdots & P_{n+m}(x) \end{vmatrix}, \tag{2.5.2}$$

with

$$C_{n,m} = \begin{vmatrix} P_n(x_1) & P_{n+1}(x_1) & \cdots & P_{n+m-1}(x_1) \\ P_n(x_2) & P_{n+1}(x_2) & \cdots & P_{n+m-1}(x_2) \\ \vdots & \vdots & & \vdots \\ P_n(x_m) & P_{n+1}(x_m) & \cdots & P_{n+m-1}(x_m) \end{vmatrix}, \tag{2.5.3}$$

are orthogonal with respect to $\Phi(x)\,d\mu(x)$, and q_n has degree n. If the zero x_k has multiplicity $r > 1$, then we replace the corresponding rows of (2.5.2) *and* (2.5.3) *by derivatives of order $0, 1, \dots, r-1$ at x_k.*

Observe that the special case $m = 1$ of (2.5.2) shows that the kernel polynomials $K_n(x, c)$ (see Definition 2.1.5) are orthogonal with respect to $(x - c)\,d\mu(x)$. Theorem 2.5.1 deals with a **polynomial modification of orthogonality measures** while Theorem 2.5.2 below deals with a **rational modification of orthogonality measures**.

Given a measure μ define

$$d\nu(x) = \frac{\prod_{i=1}^{m}(x - x_i)}{\prod_{j=1}^{k}\left(x - y_j\right)}\,d\mu(x), \tag{2.5.4}$$

where the products $\prod_{i=1}^{m}(x - x_i)$, and $\prod_{j=1}^{k}(x - y_j)$ are positive for x in the support of μ. We now construct the polynomials orthogonal with respect to ν.

Theorem 2.5.2 (Uvarov) *Let $(P_n)_n$ be the monic orthogonal polynomials for μ. Let ν be as in* (2.5.4) *and assume that $(P_n(x; m, k))_n$ are orthogonal polynomials for the measure ν. Set*

$$\tilde{Q}_n(x) := \int_{\mathbb{R}} \frac{P_n(y)}{x-y}\, d\mu(y). \tag{2.5.5}$$

Then for $n \geq k$ we have

$$\left[\prod_{i=1}^{m}(x-x_i)\right] P_n(x;m,k) = \begin{vmatrix} P_{n-k}(x_1) & P_{n-k+1}(x_1) & \cdots & P_{n+m}(x_1) \\ \vdots & \vdots & & \vdots \\ P_{n-k}(x_m) & P_{n-k+1}(x_m) & \cdots & P_{n+m}(x_m) \\ \tilde{Q}_{n-k}(y_1) & \tilde{Q}_{n-k+1}(y_1) & \cdots & \tilde{Q}_{n+m}(y_1) \\ \vdots & \vdots & & \vdots \\ \tilde{Q}_{n-k}(y_k) & \tilde{Q}_{n-k+1}(y_k) & \cdots & \tilde{Q}_{n+m}(y_k) \\ P_{n-k}(x) & P_{n+1}(x) & \cdots & P_{n+m}(x) \end{vmatrix}. \tag{2.5.6}$$

If $n < k$ then

$$\left[\prod_{i=1}^{m}(x-x_i)\right] P_n(x;m,k) = \begin{vmatrix} 0 & \cdots & 0 & P_0(x_1) & \cdots & P_{n+m}(x_1) \\ \vdots & & \vdots & \vdots & & \vdots \\ 0 & \cdots & 0 & P_0(x_m) & \cdots & P_{n+m}(x_m) \\ b_{1,1} & \cdots & b_{1,k-n} & \hat{Q}_0(y_1) & \cdots & \tilde{Q}_{n+m}(y_1) \\ \vdots & & \vdots & \vdots & & \vdots \\ b_{k,1} & \cdots & b_{k,k-n} & \tilde{Q}_0(y_k) & \cdots & \tilde{Q}_{n+m}(y_k) \\ 0 & \cdots & 0 & P_0(x) & \cdots & P_{n+m}(x) \end{vmatrix}, \tag{2.5.7}$$

where

$$b_{ij} = y_i^{j-1}, \quad 1 \leq i \leq k,\ 1 < j \leq k-n.$$

If an x_j (or y_l) is repeated r times, then the corresponding r rows will contain $P_s(x_j), \ldots, P_s^{(r-1)}(x_j)$ ($\tilde{Q}_s(x_j), \ldots, \tilde{Q}_s^{(r-1)}(x_j)$), respectively.

Uvarov outlined this result in a brief announcement (Uvarov, 1959) and later gave the details in Uvarov (1969). The modification of a measure by multiplication by a polynomial or a rational function can also be explained through the Darboux transformation of integrable systems. For detailed references see Bueno and Marcellán (2004).

The **Toda lattice equations** describe the oscillations of an infinite system of points joined by spring masses, where the interaction is exponential in the distance between two spring masses (Toda, 1989). The semi-infinite Toda lattice equations in one time variable are

$$\begin{aligned} \frac{d\alpha_n(t)}{dt} &= \beta_n(t) - \beta_{n+1}(t), \quad n \geq 0, \\ \frac{d\beta_n(t)}{dt} &= \beta_n(t)\left[\alpha_{n-1}(t) - \alpha_n(t)\right], \quad n > 0. \end{aligned} \tag{2.5.8}$$

Orthogonal polynomials can be used to provide an explicit solution to (2.5.8).

Theorem 2.5.3 *Let μ be a probability measure with finite moments, and let $(\alpha_n)_{n\geq 0}$ and $(\beta_n)_{n>0}$ be the recursion coefficients of the corresponding monic orthogonal polynomials.*

Let $(P_n(x,t))_n$ be the monic polynomials orthogonal with respect to $\exp(-xt)\,d\mu(x)$ under the additional assumption that the moments $\int_{\mathbb{R}} x^n \exp(-xt)\,d\mu(x)$ exist for all n, $n \geq 0$. Then $(\alpha_n(t))_{n\geq 0}$ and $(\beta_n(t))_{n>0}$, the recursion coefficients for $(P_n(x,t))_n$, solve the system (2.5.8) with $\alpha_n(0) = \alpha_n$ and $\beta_n(0) = \beta_n$.

The multitime Toda lattice equations can be written in the form

$$\partial_{t_j} Q = \left[Q, \left(Q^j\right)_+\right], \quad j = 1, \ldots, M, \tag{2.5.9}$$

where Q is the tridiagonal matrix with entries (q_{ij}), $q_{ii} = \alpha_i$, $q_{i,i+1} = 1$, $q_{i+1,i} = \beta_{i+1}$, $i = 0, 1, 2, \ldots$, and $q_{ij} = 0$ if $|i - j| > 1$. For a matrix A, $(A)_+$ means that all the entries below the main diagonal are replaced by zeros.

We start with a tridiagonal matrix Q formed by the initial values of α_n and β_n and find a measure of the orthogonal polynomials. We form a new probability measure according to

$$d\mu\,(x;\mathbf{t}) = \frac{1}{\zeta_0(\mathbf{t})} \exp\left(-\sum_{s=1}^{M} t_s x^s\right) d\mu(x), \tag{2.5.10}$$

where $\mathbf{t}$ stands for $(t_1, \ldots, t_M)$ and $\zeta_0(\mathbf{t}) = \int_{\mathbb{R}} \exp\left(-\sum_{s=1}^{M} t_s x^s\right) d\mu(x)$. Let the corresponding monic orthogonal polynomials be $(P_n(x;\mathbf{t}))_n$, and $(\alpha_n(\mathbf{t}))_{n\geq 0}$ and $(\beta_n(\mathbf{t}))_{n>0}$ be their recursion coefficients. Then the matrix $Q(\mathbf{t})$ formed by the new recursion coefficients solves (2.5.9).

The **partition function** is

$$Z_n(\mathbf{t}) := \frac{1}{\zeta_0^n(\mathbf{t})} \int_{\mathbb{R}^n} \exp\left(-\sum_{j=1}^{n}\sum_{s=1}^{M} t_s x_j^s\right) \prod_{1\leq i<j\leq n} \left(x_i - x_j\right)^2 d\mu\,(x_1) \cdots d\mu\,(x_n) \tag{2.5.11}$$

and the **tau function** is

$$\tau_n(\mathbf{t}) = Z_n(\mathbf{t})/n!. \tag{2.5.12}$$

Formulas (2.1.7) and (2.1.10) establish

$$\tau_{n+1}(\mathbf{t}) = D_n = \prod_{j=1}^{n} \beta_j^{n-j+1}. \tag{2.5.13}$$

The next theorem provides the **modification of orthogonality measures by a discrete part.**

Theorem 2.5.4 (Uvarov, 1969) *Let*

$$\nu = \mu + \sum_{j=1}^{r} c_j \delta_{x_j}, \tag{2.5.14}$$

where δ_u is a unit atomic measure concentrated at $x = u$. Let $(P_n)_n$ and $(R_n)_n$ be the monic polynomials orthogonal with respect to μ and ν, respectively. Then

$$R_n(x) = P_n(x) - \sum_{j=1}^{r} c_j R_n(x_j) a_j(x), \tag{2.5.15}$$

where

$$a_j(x) = \sum_{s=0}^{n-1} P_s(x)P_s(x_j)/\zeta_s = K_{n-1}(x, x_j).$$

Consider the Jacobi polynomials where $d\mu$ is $(1-x)^\alpha(1+x)^\beta\,dx$; see Chapter 3. Let

$$A\,d\nu(x) = (1-x)^\alpha(1+x)^\beta\,dx + M_1\delta_1 + M_2\delta_{-1},$$

where A is the normalization constant

$$A = 2^{\alpha+\beta+1}\frac{\Gamma(\alpha+1)\Gamma(\beta+1)}{\Gamma(\alpha+\beta+2)} + M_1 + M_2,$$

and δ_s is a Dirac measure at $x = s$. The monic Jacobi polynomials are (3.1.1) and (3.1.4),

$$\begin{aligned} P_n(x) &= \frac{(\alpha+1)_n 2^n}{(\alpha+\beta+n+1)_n}\, {}_2F_1\left(\left.\begin{matrix} -n, \alpha+\beta+n+1 \\ \alpha+1 \end{matrix}\right| \frac{1-x}{2}\right) \\ &= \frac{(-2)^n(\beta+1)_n}{(\alpha+\beta+n+1)_n}\, {}_2F_1\left(\left.\begin{matrix} -n, \alpha+\beta+n+1 \\ \beta+1 \end{matrix}\right| \frac{1+x}{2}\right). \end{aligned}$$

One can then find an explicit formula for $R_n(x)$: if $M_1, M_2 > 0$ then

$$R_n(x) = C_n\, {}_4F_3\left(\left.\begin{matrix} -n, n+\alpha+\beta+1, -a_n+1, b_n+1 \\ \alpha+2, -a_n, b_n \end{matrix}\right| \frac{1-x}{2}\right) \tag{2.5.16}$$

where $C_n = C_n(\alpha, \beta, M_1, M_2)$ is a constant to make R_n a monic polynomial, $a_n > n, b_n > 0$ satisfy

$$\begin{aligned} a_n b_n &= \frac{(\alpha+1)(\alpha+\beta+1)(\beta+1)_n n! A_n}{(\alpha+1)_n(\alpha+\beta+1)_n M_2 B_n}, \\ a_n - b_n &= \beta\frac{M_1 A_n}{M_2 B_n} + \frac{(\alpha+\beta+1)A_n}{M_2} - \alpha - 1, \end{aligned}$$

and

$$\begin{aligned} A_n &= \frac{(\alpha+1)_n n!}{(\beta+1)_n(\alpha+\beta+1)_n} + \frac{n(n+\alpha+\beta+1)M_2}{(\beta+1)(\alpha+\beta+1)}, \\ B_n &= \frac{(\beta+1)_n n!}{(\alpha+1)_n(\alpha+\beta+1)_n} + \frac{n(n+\alpha+\beta+1)M_1}{(\alpha+1)(\alpha+\beta+1)}. \end{aligned}$$

The above example was studied in Koornwinder (1984b) without the use of Uvarov's theorem.

The following theorem deals with modifications of recursion coefficients. It was proved in Wendroff (1961), but it seems to have been known to Geronimus in the 1940s (see Geronimus, 1977). It is stated without proof as a footnote in Geronimus (1946) where the corresponding result for orthogonal polynomials on the circle is proved.

Theorem 2.5.5 *Given sequences $(\alpha_n)_{n\ge N}$, $(\beta_n)_{n\ge N}$, with $\alpha_n \in \mathbb{R}$, $\beta_n > 0$, and two finite sequences $x_1 > x_2 > \cdots > x_N$, $y_1 > y_2 > \cdots > y_{N-1}$ such that $x_{k-1} > y_{k-1} > x_k$, $1 < k \le N$, then there is a sequence of monic orthogonal polynomials $(P_n)_{n\ge 0}$ such that*

$$P_N(x) = \prod_{j=1}^{N}\left(x - x_j\right), \quad P_{N-1}(x) = \prod_{j=1}^{N-1}\left(x - y_j\right), \tag{2.5.17}$$

and

$$xP_n(x) = P_{n+1}(x) + \tilde{\alpha}_n P_n(x) + \tilde{\beta}_n P_{n-1}(x), \quad n > 0, \tag{2.5.18}$$

with $\tilde{\beta}_n > 0$, $n > 0$, and $\tilde{\alpha}_n = \alpha_n$, $\tilde{\beta}_n = \beta_n$, for $n \ge N$.

Remark 2.5.6 It is clear that Theorem 2.5.5 can be stated in terms of eigenvalues of tridiagonal matrices instead of zeros of P_N and P_{N-1}; see (2.1.21). The process of changing finitely many entries in a Jacobi matrix corresponds to finite rank perturbations in operator theory.

2.6 Linearization and Connection Relations

Given a system of orthogonal polynomials $(P_n(x;\mathbf{a}))_n$ depending on parameters $a_1, \ldots, a_s$, the **connection coefficients** are the coefficients $c_{n,k}(\mathbf{a},\mathbf{b})$ in the expansion

$$P_n(x;\mathbf{b}) = \sum_{k=0}^{n} c_{n,k}(\mathbf{a},\mathbf{b}) P_k(x;\mathbf{a}). \tag{2.6.1}$$

We used the vector notation

$$\mathbf{a} = (a_1, \ldots, a_s), \quad \mathbf{b} = (b_1, \ldots, b_s). \tag{2.6.2}$$

The **linearization coefficients** $c_{m,n,k}$ are the coefficients in

$$P_m(x;\mathbf{a})P_n(x;\mathbf{a}) = \sum_{k=0}^{m+n} c_{m,n,k}(\mathbf{a}) P_k(x;\mathbf{a}). \tag{2.6.3}$$

When the coefficients cannot be found explicitly, one usually tries to find sign patterns, or unimodality properties satisfied by the coefficients. Clearly,

$$c_{n,k}(\mathbf{a},\mathbf{b})\zeta_k(\mathbf{a}) = \int_{\mathbb{R}} P_n(x;\mathbf{b})P_k(x;\mathbf{a})\, d\mu(x;\mathbf{a}), \tag{2.6.4}$$

$$c_{m,n,k}(\mathbf{a})\zeta_k(\mathbf{a}) = \int_{\mathbb{R}} P_m(x;\mathbf{a})P_n(x;\mathbf{a})P_k(x;\mathbf{a})\, d\mu(x;\mathbf{a}), \tag{2.6.5}$$

where

$$\int_{\mathbb{R}} P_m(x;\mathbf{a})P_n(x;\mathbf{a})\, d\mu(x;\mathbf{a}) = \zeta_n(\mathbf{a})\, \delta_{m,n}. \tag{2.6.6}$$

This leads to the problem of studying sign behavior of the integrals

$$I(n_1, \dots, n_k) := \int_{\mathbb{R}} P_{n_1}(x; \mathbf{a}) \cdots P_{n_k}(x; \mathbf{a})\, d\mu(x, \mathbf{a}). \tag{2.6.7}$$

When $d\mu(x; \mathbf{a}) = w(x; \mathbf{a})\, dx$ then (2.6.1) implies the orthogonal expansion

$$w(x; \mathbf{b}) P_k(x; \mathbf{b}) \sim \sum_{n=k}^{\infty} c_{n,k}(\mathbf{b}, \mathbf{a}) \frac{\zeta_k(\mathbf{b})}{\zeta_n(\mathbf{a})} P_n(x; \mathbf{a}) w(x; \mathbf{a}). \tag{2.6.8}$$

Polynomials with nonnegative linearization coefficients usually have very special properties and lead to convolution structures.

Theorem 2.6.1 (Ismail, 2005b) *Let $(p_n)_n$ be orthonormal with respect to μ and assume μ is supported on a subset of $(-\infty, \xi]$. Also assume that*

$$(p_n(x))^N = \sum_k c(k, N, n) p_k(x), \quad c(k, N, n) \geq 0. \tag{2.6.9}$$

Then

$$|p_n(x)| \leq p_n(\xi), \quad \mu\text{-almost everywhere.} \tag{2.6.10}$$

A general expansion formula is (Fields and Wimp, 1961; Verma, 1972)

$$\begin{aligned} \sum_{m=0}^{\infty} a_m b_m \frac{(zw)^m}{m!} &= \sum_{n=0}^{\infty} \frac{(-z)^n}{n!(\gamma+n)_n} \left(\sum_{r=0}^{\infty} \frac{b_{n+r} z^r}{r!(\gamma+2n+1)_r} \right) \\ &\quad \times \left(\sum_{s=0}^{n} \frac{(-n)_s (n+\gamma)_s}{s!} a_s w^s \right). \end{aligned} \tag{2.6.11}$$

A companion formula is

$$\sum_{m=0}^{\infty} a_m b_m (zw)^m = \sum_{n=0}^{\infty} \frac{(-z)^n}{n!} \left[\sum_{j=0}^{\infty} \frac{b_{n+j}}{j!} z^j \right] \left[\sum_{k=0}^{n} (-n)_k a_k w^k \right]. \tag{2.6.12}$$

Fields and Ismail (1975) showed how to derive (2.6.11) and other identities from generating functions of Boas and Buck type. This essentially uses Lagrange inversion formulas. Some q-analogues are in Gessel and Stanton (1983, 1986).

Theorem 2.6.2 (Wilson, 1970) *Let $(p_n)_n$ and $(s_n)_n$ be polynomial sequences with positive leading terms and assume that $(p_n)_n$ is orthonormal with respect to μ. If*

$$\int_{\mathbb{R}} s_m(x) s_n(x)\, d\mu(x) \leq 0, \quad n \neq m,$$

then

$$p_n(x) = \sum_{k=0}^{n} a_{n,k} s_k(x), \tag{2.6.13}$$

with $a_{n,k} \geq 0$ for $0 \leq k \leq n$ and $n \geq 0$.

Theorem 2.6.3 (Askey, 1971) *Let $(P_n)_n$ and $(Q_n)_n$ be monic orthogonal polynomials satisfying*

$$\begin{aligned} xQ_n(x) &= Q_{n+1}(x) + A_nQ_n(x) + B_nQ_{n-1}(x), \\ xP_n(x) &= P_{n+1}(x) + \alpha_nP_n(x) + \beta_nP_{n-1}(x). \end{aligned} \tag{2.6.14}$$

If

$$A_k \le \alpha_n, \quad B_{k+1} \le \beta_{n+1}, \quad 0 \le k \le n,\ n \ge 0, \tag{2.6.15}$$

then

$$Q_n(x) = \sum_{k=0}^{n} c_{n,k}P_k(x), \tag{2.6.16}$$

with $c_{n,k} \ge 0$ for $0 \le k \le n$ and $n \ge 0$.

Theorem 2.6.4 (Szwarc, 1992) *Let $(r_n)_n$ and $(s_n)_n$ satisfy $r_0(x) = s_0(x) = 1$, and*

$$\begin{aligned} xr_n(x) &= A_nr_{n+1}(x) + B_nr_n(x) + C_nr_{n-1}(x), \\ xs_n(x) &= A'_ns_{n+1}(x) + B'_ns_n(x) + C'_ns_{n-1}(x), \end{aligned} \tag{2.6.17}$$

for $n \ge 0$ with $C_0r_{-1}(x) = C'_0s_{-1}(x) = 0$. Assume that

(i) $C'_m \ge C_n$ *for* $m \le n$,
(ii) $B'_m \ge B_m$ *for* $m \le n$,
(iii) $C'_m + A'_m \ge C_n + A_n$ *for* $m \le n$,
(iv) $A'_m \ge C_n$ *for* $m < n$.

Then the connection coefficients $c(n,k)$ in

$$r_n(x) = \sum_{k=0}^{n} c(n,k)s_k(x) \tag{2.6.18}$$

are nonnegative.

Corollary 2.6.5 (Szwarc, 1992) *Assume that $(r_n)_n$ are generated by (2.6.17), for $n \ge 0$, and $r_0(x) = 1$, $r_{-1}(x) = 0$. If $C_n \le 1/2$, $A_n + C_n \le 1$, $B_n \le 0$, then $r_n(x)$ can be represented as a linear combination of Chebyshev polynomials with nonnegative coefficients.*

Corollary 2.6.6 (Szwarc, 1992) *Let E denote the closure of the area enclosed by any ellipse whose foci are ± 1. Under the assumptions of Corollary 2.6.5, the maximum of $|r_n(z)|$ for $z \in E$ is attained at the right endpoint of the major axis.*

Theorem 2.6.7 (Szwarc, 1992) *Let $(r_n)_n$ and $(s_n)_n$ be as in Theorem 2.6.4 with $B_n = B'_n = 0$, $n \ge 0$. Assume that*

(i) $C'_{2m} \ge C_{2n}$ *and* $C'_{2m+1} \ge C_{2n+1}$, *for* $0 < m \le n$,
(ii) $A'_{2m} + C'_{2m} \ge A_{2n} + C_{2n}$, *and* $A'_{2m+1} + C'_{2m+1} \ge A_{2n+1} + C_{2n+1}$, *for* $m \le n$,
(iii) $A'_{2m} > A_{2n}$ *and* $A'_{2n+1} \ge A_{2m+1}$ *for* $m < n$.

Then the connection coefficients in (2.6.18) *are nonnegative. The same conclusion holds if* (i)–(iii) *are replaced by*

(i′) $C_1 \geq C_1' \geq C_2 \geq C_2' \geq \cdots$, $B_0 \geq B_0' \geq B_1 \geq B_1' \geq \cdots$,
(ii′) $A_0 + C_0 \geq A_0' + C_0' \geq A_1 + C_1 \geq A_1' + C_1' \geq \cdots$,
(iii′) $A_m' \geq C_n$ *for* $m < n$.

2.7 Addition Theorems[1]

An algebraic addition theorem for a function f is an identity of the form

$$P(f(x), f(y), f(x+y)) = 0 \tag{2.7.1}$$

for some polynomial P in three variables. According to a theorem by Weierstrass (Forsyth, 1918) an analytic function satisfying an algebraic addition theorem is a rational function in z, a rational function in $e^{\lambda z}$ for some λ, or an elliptic function. This notion is too restricted for use in special functions.

In general we say that a family, say ϕ_λ, of special functions satisfies an addition formula if for some fixed elementary continuous function Λ of three variables x, y, t there exists an expansion in terms of a (usually different) family of special functions ψ_μ such that the expansion coefficients factor as products in x and y,

$$\phi_\lambda(\Lambda(x, y, t)) = \sum_\mu C(\lambda, \mu)\phi_\lambda^\mu(x)\phi_\lambda^\mu(y)\psi_\mu(t), \quad C(\lambda, \mu) \in \mathbb{C}, \tag{2.7.2}$$

or alternatively,

$$\phi_\lambda(z) = \sum_\mu{}' C(\lambda, \mu)\phi_\lambda^\mu(x)\phi_\lambda^\mu(y)\psi_\mu\left(\Lambda(x, y, z)\right), \quad C(\lambda, \mu) \in \mathbb{C}. \tag{2.7.3}$$

Here we assume that ϕ_λ^μ are special functions associated to the original family ϕ_λ and we usually assume that $\mu = 0$ occurs in the sum and that $\psi_0(t) = 1$, $\phi_\lambda^0 = \phi_\lambda$. In the case that we also assume that ψ_μ is an orthogonal family, for example $\int \psi_\mu(t)\psi_\nu(t)\, d\alpha(t) = \delta_{\mu,\nu}$, the addition formula (2.7.2) leads to the corresponding **product formula**

$$\int_{\mathbb{R}} \phi_\lambda(\Lambda(x, y, t))\, d\alpha(t) = C(\lambda, 0)\phi_\lambda(x)\phi_\lambda(y), \tag{2.7.4}$$

or alternatively, by changing variables,

$$\int_{\mathbb{R}} \phi_\lambda(z)\, d\nu_{x,y}(z) = C(\lambda, 0)\phi_\lambda(x)\phi_\lambda(y), \tag{2.7.5}$$

for some measure $\nu_{x,y}$ depending on x and y.

The basic example of the addition formula (2.7.2) is Laplace's 1782 addition formula for the Legendre polynomials (the case $\nu = 1/2$ of Theorem 3.5.1), where the special functions

[1] This section was written by Erik Koelink and Tom Koornwinder.

ϕ_λ are Legendre polynomials, ψ_μ are Chebyshev polynomials, ϕ^μ_λ are Gegenbauer (or ultraspherical) polynomials times an elementary factor, and $\Lambda(x,y,t) = xy + t\sqrt{1-x^2}\sqrt{1-y^2}$. For the original papers one can consult Laplace (1782) and Legendre (1785, p. 420; 1789, p. 432). Heine (1961, pp. 2, 313) discusses the priority question of the addition formula. The general addition theorem, Theorem 3.5.1, is due to Gegenbauer (1874, 1893)

The form (2.7.2) not only fits orthogonal polynomials, but can also be used to state addition formulas for generalized orthogonal systems such as Bessel functions (see Watson, 1944) and Jacobi functions (Koornwinder, 1984a).

For fixed x and y the system $\{t \mapsto \phi_\lambda(\Lambda(x,y,t))\}_\lambda$ is orthogonal and (2.7.2) can also be viewed as a connection coefficient formula between this orthogonal system and the orthogonal system ψ_μ. Also, (2.7.4) can be viewed as the dual of a linearization formula. Moreover, addition formulas can give rise to positive linearization coefficients.

Theorem 2.7.1 (Koornwinder, 1978) *Assume that the families of special functions ϕ_λ, ϕ^μ_λ, and ψ_μ are continuous and real valued, and that the orthogonality measure for ψ_μ is compactly supported. Assume, moreover, that ϕ_λ satisfies the orthogonality relations*

$$\int \phi_\lambda(t)\phi_\mu(t)\, d\beta(t) = \delta_{\lambda,\mu}$$

for a compactly supported measure β. If ϕ_λ satisfies an addition formula of the form (2.7.2) *such that the sum is finite and $C(\lambda,\mu) \geq 0$, $C(\lambda,0) > 0$ then the linearization coefficients in $\phi_\lambda(x)\phi_\mu(x) = \sum_\nu a_\nu(\lambda,\mu)\phi_\nu(x)$ are nonnegative: $a_\nu(\lambda,\mu) \geq 0$.*

The product formula gives rise to a **generalized translation operator**

$$\left(T_y f\right)(x) = \int f(\Lambda(x,y,t))\, d\alpha(t), \tag{2.7.6}$$

so that the product formula is $(T_y\phi_\lambda)(x) = C(\lambda,0)\phi_\lambda(x)\phi_\lambda(y)$. Associated to the generalized translation operator we define the **convolution structure**

$$(f \star g)(y) = \int f(x)\left(T_y g\right)(x)\, d\beta(x) \tag{2.7.7}$$

for $f, g \in L^1(\beta)$, so that $\phi_\lambda \star \phi_\mu = \delta_{\lambda,\mu} C(\lambda,0)\phi_\lambda$; see Connett and Schwartz (2000). Another **dual convolution structure** can be associated to orthogonal polynomials with nonnegative linearization coefficients; see Szwarc (2005).

A more general type of addition theorem is sometimes needed, especially for special functions of basic hypergeometric type. As a generalization of (2.7.3) we consider

$$\phi_\lambda(t)\psi_\nu(t;x,y) = \sum_\mu C(\lambda,\mu,\nu)\phi^\mu_\lambda(x)\phi^\mu_\lambda(y)\psi_\mu(t;x,y), \tag{2.7.8}$$

where $\psi_\mu(\cdot;x,y)$ is a class of special functions depending on x, y.

Addition and product formulas for special functions can often be obtained in a natural way in the case that these functions have an interpretation on a group (Lie group, finite group, quantum group); see Askey (1975b), Helgason (1978), Koelink (1997), Stanton (1984), Vilenkin (1968), and Vilenkin and Klimyk (1991–1993).

2.8 Differential Equations

Assume that μ is absolutely continuous and let

$$d\mu(x) = w(x)\,dx, \quad x \in (a,b), \quad w(x) = \exp(-v(x)), \tag{2.8.1}$$

where a can be $-\infty$ and/or b can be $+\infty$. We require v to be twice differentiable and assume that the integrals

$$\int_a^b y^n \frac{v'(x) - v'(y)}{x-y} w(y)\,dy, \quad n = 0, 1, \ldots \tag{2.8.2}$$

exist. We shall also use the orthonormal form $(p_n)_n$ of (2.1.8). In terms of the orthonormal polynomials, (2.1.14) and (2.1.15) become

$$p_0(x) = 1, \quad p_1(x) = (x - b_0)/a_1, \tag{2.8.3}$$

$$x p_n(x) = a_{n+1} p_{n+1}(x) + b_n p_n(x) + a_n p_{n-1}(x), \quad n > 0. \tag{2.8.4}$$

Next define functions A_n and B_n via

$$\frac{A_n(x)}{a_n} = \left.\frac{w(y)p_n^2(y)}{y-x}\right|_a^b + \int_a^b \frac{v'(x) - v'(y)}{x-y} p_n^2(y) w(y)\,dy, \tag{2.8.5}$$

$$\frac{B_n(x)}{a_n} = \left.\frac{w(y)p_n(y)p_{n-1}(y)}{y-x}\right|_a^b + \int_a^b \frac{v'(x) - v'(y)}{x-y} p_n(y)p_{n-1}(y) w(y)\,dy. \tag{2.8.6}$$

It is assumed that the boundary terms in (2.8.5) and (2.8.6) exist.

Theorem 2.8.1 *Let v be a twice continuously differential function on $[a,b]$. Then the orthonormal polynomials $(p_n)_n$ satisfy the **lowering** and **raising** relations*

$$\left(\frac{d}{dx} p_n(x) + B_n(x) p_n(x)\right) = A_n(x) p_{n-1}(x), \tag{2.8.7}$$

$$\left(-\frac{d}{dx} + B_n(x) + v'(x)\right) p_{n-1}(x) = A_{n-1}(x) \frac{a_n}{a_{n-1}} p_n(x). \tag{2.8.8}$$

Theorem 2.8.1 was proved for symmetric polynomials in Mhaskar (1990); see also Bonan and Clark (1990) and Bauldry (1990). It was rediscovered in the present form in Chen and Ismail (1997). Note that $A_n(x)$ and $B_n(x)$ are polynomials (rational functions) in x if v is a polynomial (rational function). Moreover when v' is a rational function, then there exists a fixed polynomial π and constants $a_{n,j}$ such that

$$\pi(x) p_n'(x) = \sum_{j=0}^{M} a_{n,j} p_{n+m-j-1}(x), \tag{2.8.9}$$

where m is the degree of π and M is a fixed positive integer independent of n. Moreover, π does not depend on n.

In monic polynomial notation, (2.8.5) and (2.8.6) become

$$\frac{\tilde{A}_n(x)}{\beta_n} = \left.\frac{w(y)P_n^2(y)}{(y-x)\zeta_n}\right|_a^b + \int_a^b \frac{v'(x)-v'(y)}{(x-y)\zeta_n} P_n^2(y)w(y)\,dy, \tag{2.8.10}$$

$$\tilde{B}_n(x) = \frac{w(y)P_n(y)P_{n-1}(y)}{(y-x)\zeta_{n-1}} + \int_a^b \frac{v'(x)-v'(y)}{(x-y)\zeta_{n-1}} P_n(y)P_{n-1}(y)w(y)\,dy, \tag{2.8.11}$$

and ζ_n is given by (2.1.16). Moreover,

$$P_n'(x) = \tilde{A}_n(x)P_{n-1}(x) - \tilde{B}_n(x)P_n(x). \tag{2.8.12}$$

The functions A_n and B_n satisfy the recurrence relation

$$B_n(x) + B_{n+1}(x) = \frac{x-b_n}{a_n}A_n(x) - v'(x) \tag{2.8.13}$$

(Mhaskar, 1990; Chen and Ismail, 1997) and

$$B_{n+1}(x) - B_n(x) = \frac{a_{n+1}A_{n+1}(x)}{x-b_n} - \frac{a_n^2 A_{n-1}(x)}{a_{n-1}\,(x-b_n)} - \frac{1}{x-b_n} \tag{2.8.14}$$

(Ismail and Wimp, 1998). Additional recursive properties of A_n and B_n will be listed below. For proofs see Ismail (2005b). We have

$$2B_{n+1}(x) = \frac{x-b_n}{a_n}A_n(x) + \frac{a_{n+1}A_{n+1}(x)}{x-b_n} - \frac{a_n^2 A_{n-1}(x)}{a_{n-1}\,(x-b_n)} - v'(x) - \frac{1}{x-b_n}, \tag{2.8.15}$$

$$2B_n(x) = \frac{x-b_n}{a_n}A_n(x) - \frac{a_{n+1}A_{n+1}(x)}{x-b_n} + \frac{a_n^2 A_{n-1}(x)}{a_{n-1}\,(x-b_n)} - v'(x) + \frac{1}{x-b_n}, \tag{2.8.16}$$

$$\begin{aligned}\frac{a_{n+2}A_{n+2}(x)}{x-b_{n+1}} &= \left[\frac{x-b_{n+1}}{a_{n+1}} - \frac{a_{n+1}}{x-b_n}\right]A_{n+1}(x) + \left[\frac{a_{n+1}^2}{a_n\,(x-b_{n+1})} - \frac{(x-b_n)}{a_n}\right]A_n(x)\\ &\quad + \frac{a_n^2}{a_{n-1}\,(x-b_n)}A_{n-1}(x) + \frac{1}{x-b_n} + \frac{1}{x-b_{n+1}}, \qquad n > 1.\end{aligned} \tag{2.8.17}$$

One may extend the validity of (2.8.17) to the cases $n = 0$ and $n = 1$ through $a_0 := 1$, $p_{-1} := 0$ so that $B_0 = A_{-1} = 0$.

The B_n also satisfy the inhomogeneous four-term recurrence relation

$$\begin{aligned}B_{n+2}(x) &= \left[\frac{(x-b_n)\,(x-b_{n+1})}{a_{n+1}^2} - 1\right]B_{n+1}(x)\\ &\quad + \left[\frac{a_n^2\,(x-b_{n+1})}{a_{n+1}^2\,(x-b_{n-1})} - \frac{(x-b_n)\,(x-b_{n+1})}{a_{n+1}^2}\right]B_n(x)\\ &\quad + \frac{a_n^2\,(x-b_{n+1})}{a_{n+1}^2\,(x-b_{n-1})}B_{n-1}(x) + \frac{(x-b_{n+1})}{a_{n+1}^2}\\ &\quad + \left[\frac{a_n^2\,(x-b_{n+1})}{a_{n+1}^2\,(x-b_{n-1})} - 1\right]v'(x), \qquad n > 1.\end{aligned} \tag{2.8.18}$$

Rewrite (2.8.7)–(2.8.8) as

$$L_{1,n}p_n(x) = A_n(x)p_{n-1}(x), \quad L_{2,n}p_{n-1}(x) = A_{n-1}(x)\frac{a_n}{a_{n-1}}p_n(x), \tag{2.8.19}$$

where $L_{1,n}$ and $L_{2,n}$ are the differential operators

$$L_{1,n} = \frac{d}{dx} + B_n(x), \quad L_{2,n} = -\frac{d}{dx} + B_n(x) + v'(x). \tag{2.8.20}$$

In this form $L_{2,n}$, and $L_{1,n}$ are **raising (creation)** and **lowering (annihilation)** operators for p_n. Define the weighted inner product

$$(f,g)_w := \int_a^b f(x)\overline{g(x)}w(x)\,dx, \tag{2.8.21}$$

and consider the function space where $(f,f)_w$ is finite and $f(x)\sqrt{w(x)}$ vanishes at $x = a, b$. With respect to the inner product (2.8.21), $L_{2,n}$ is the adjoint of $L_{1,n}$. A discussion of raising and lowering operators for specific one-variable polynomials as they relate to two-variable theory is in Koornwinder (2006).

Theorem 2.8.2 *Assume that v is a polynomial of degree $2m$ and w is supported on $\mathbb{R}$. Then the Lie algebra generated by $L_{1,n}$ and $L_{2,n}$ has dimension $2m+1$ for all $n > 0$.*

Ismail (2005b) conjectures that if the Lie algebra generated by $L_{1,n}$ and $L_{2,n}$ is finite-dimensional then v is a polynomial. It follows that the $(p_n)_n$ satisfy the factored equation

$$L_{2,n}\left(\frac{1}{A_n(x)}\left(L_{1,n}p_n(x)\right)\right) = \frac{a_n}{a_{n-1}}A_{n-1}(x)p_n(x). \tag{2.8.22}$$

Equivalently, (2.8.22) is

$$p_n''(x) + R_n(x)p_n'(x) + S_n(x)p_n(x) = 0, \tag{2.8.23}$$

where

$$R_n(x) = -\left[v'(x) + \frac{A_n'(x)}{A_n(x)}\right], \tag{2.8.24}$$

$$S_n(x) = B_n'(x) - B_n(x)\frac{A_n'(x)}{A_n(x)} - B_n(x)\left[v'(x) + B_n(x)\right] + \frac{a_n}{a_{n-1}}A_n(x)A_{n-1}(x). \tag{2.8.25}$$

The so-called Schrödinger form of (2.8.22)–(2.8.23) is

$$\Psi_n''(x) + V(x,n)\Psi_n(x) = 0, \tag{2.8.26}$$

$$\Psi_n(x) = \frac{\exp[-v(x)/2]}{\sqrt{A_n(x)}}p_n(x), \tag{2.8.27}$$

where

$$V(x,n) = A_n(x)\left(\frac{B_n(x)}{A_n(x)}\right)' - B_n(x)\left[v'(x) + B_n(x)\right] + A_n(x)A_{n-1}(x)\frac{a_n}{a_{n-1}} + \frac{v''(x)}{2} + \frac{1}{2}\left(\frac{A_n'(x)}{A_n(x)}\right)' - \frac{1}{4}\left[v'(x) + \frac{A_n'(x)}{A_n(x)}\right]^2. \tag{2.8.28}$$

Observe that $A_n(x) > 0$ for real x if v is convex.

Shohat gave a procedure to derive (2.8.7) when $w'(x)/w(x)$ is a rational function (Shohat, 1939). Another derivation of the same result is in Atkinson and Everitt (1981), who also established (2.8.7) when $\int_{\mathbb{R}}(x-t)^{-1}\,d\mu(t)$ satisfies a linear first-order differential equation, without assuming that μ is absolutely continuous.

It is important to note that (2.8.8) leads to

$$a_n p_n(x) = \left[\mathbb{L}_n\mathbb{L}_{n-1}\cdots\mathbb{L}_1\right]1, \tag{2.8.29}$$

$$(\mathbb{L}_n f)(x) = \frac{1}{A_{n-1}(x)}\left[-\frac{d}{dx} + B_n(x) + v'(x)\right]f(x). \tag{2.8.30}$$

Formula (2.8.30) is a generalization of the Rodrigues formula.

Chen and Ismail observed that the compatibility conditions for A_n and B_n encode all the properties of the orthogonal polynomials. They applied this approach to Laguerre and Jacobi polynomials (Chen and Ismail, 2005).

An example of an exponential weight is $w(x) = ce^{-(x^4-2tx^2)}$, $x \in \mathbb{R}$, and c such that $\int_{\mathbb{R}} w(x)\,dx = 1$. In this case

$$A_n(x) = 4a_n\left[x^2 - t + a_n^2 + a_{n+1}^2\right], \quad B_n(x) = 4a_n^2 x. \tag{2.8.31}$$

Therefore

$$R_n(x) = -4x^3 - \frac{2x}{x^2 + a_n^2 + a_{n+1}^2}, \tag{2.8.32}$$

$$S_n(x) = 4a_n^2 - \frac{8a_n^2x^2}{x^2 + a_n^2 + a_{n+1}^2} + 16x^2a_n^2\left(a_{n-1}^2 + a_n^2 + a_{n+1}^2\right) + 16a_n^2\left(a_n^2 + a_{n+1}^2\right)\left(a_n^2 + a_{n-1}^2\right). \tag{2.8.33}$$

By equating the coefficients of x^{n-1} on both sides of (2.8.7) we find the so-called string equation (Ismail, 2005b)

$$n = 4a_n^2\left(a_{n+1}^2 + a_n^2 + a_{n-1}^2 - t\right). \tag{2.8.34}$$

On the other hand, (2.8.14) gives the more general result

$$4a_{n+1}^2\left(a_{n+1}^2 + a_{n+2}^2 - t\right) - 4a_n^2\left(a_n^2 + a_{n+1}^2 - t\right) = -1. \tag{2.8.35}$$

Additional nonlinear relations can be obtained by equating coefficients of other powers of x in (2.8.7). Freud proved (2.8.34) when $t = 0$ and derived a similar nonlinear relation for the weight $C \exp(-x^6)$ in Freud (1976). Lew and Quarles (1983) and Nevai (1983) studied asymptotics of solutions to the nonlinear recursion

$$c_n^2 = x_n \left(x_{n+1} + x_n + x_{n-1}\right), \quad n > 0, \tag{2.8.36}$$

with $x_0 \geq 0$, where $(c_n)_n$ is a given sequence of real numbers.

The following theorem describes, in a special case, how the string equation characterizes the orthogonal polynomials.

Theorem 2.8.3 (Bonan and Nevai, 1984) *Let $(p_n)_n$ be orthonormal with respect to a probability measure μ. Then the following are equivalent:*

(i) *There exist nonnegative integers j, k and two sequences $(e_n)_{n\geq 1}$ and $(c_n)_{n\geq 1}$ such that $j < k$ and*

$$p_n'(x) = e_n p_{n-j}(x) + c_n p_{n-k}(x), \quad n = 1, 2, \ldots.$$

(ii) *There exists a nonnegative constant c such that*

$$p_n'(x) = \frac{n}{a_n} p_{n-1}(x) + c a_n a_{n-1} a_{n-2} p_{n-3}(x), \quad n = 1, 2, \ldots,$$

where $(a_n)_n$ are as in (2.8.4).

(iii) *There exist real numbers c, b, and K such that $c \geq 0$ (if $c = 0$ then $K > 0$) and the recursion coefficients in (2.8.4) satisfy*

$$n = c a_n^2 \left[a_{n+1}^2 + a_n^2 + a_{n-1}^2\right] + K a_n^2, \quad n = 1, 2, \ldots,$$

and $b_n = b$ for $n = 0, 1, \ldots.$

(iv) *The measure μ is absolutely continuous with $\mu'(x) = e^{-v(x)}$ and*

$$v(x) = \frac{c}{4}(x-b)^4 - \frac{K}{2}(x-b)^2 + d, \quad b, c, d, K \in \mathbb{R}.$$

Moreover, $c \geq 0$ and if $c = 0$ then $K > 0$.

Theorem 2.8.4 (Bonan, Lubinsky, and Nevai, 1987) *Let $(p_n)_n$ be orthonormal polynomials with respect to μ. Then*

$$p_n'(x) = \sum_{k=n-m+1}^{n-1} c_{k,n} p_k(x) \tag{2.8.37}$$

holds for constants $c_{k,n}$ if and only if μ is absolutely continuous, $\mu'(x) = e^{-v(x)}$, and v is a polynomial of exact degree m.

Theorem 2.8.5 (Ismail, 2000b) *Let $\mu = \mu_{ac} + \mu_s$, where μ_{ac} is absolutely continuous on $[a, b]$, $\mu_{ac}'(x) = e^{-v(x)}$, and μ_s is a discrete measure with finite support contained in $\mathbb{R} \setminus [a, b]$.*

Assume that $(p_n)_n$ are orthonormal polynomials with respect to μ and let $[A, B]$ be the true interval of orthogonality of $(p_n)_n$. Define functions

$$\begin{aligned}\frac{\mathcal{A}_n(x)}{a_n} = \frac{w(y)p_n^2(y)}{y-x}\bigg|_{y=a}^{b} - \int_A^B \frac{d}{dy}\left[\frac{p_n^2(y)}{x-y}\right] d\mu_s(y) \\ + v'(x)\int_A^B \frac{p_n^2(y)}{x-y}\, d\mu_s(y) + \int_a^b \frac{v'(x)-v'(y)}{x-y} p_n^2(y)w(y)\, dy,\end{aligned} \tag{2.8.38}$$

$$\begin{aligned}\frac{\mathcal{B}_n(x)}{a_n} = \frac{w(y)p_n(y)p_{n-1}(y)}{y-x}\bigg|_{y=a}^{b} - \int_A^B \frac{d}{dy}\left[\frac{p_n(y)p_{n-1}(y)}{x-y}\right] d\mu_s(y) \\ - v'(x)\int_A^B \frac{p_n(y)p_{n-1}(y)}{x-y}\, d\mu_s(y) \\ + \int_a^b \frac{v'(x)-v'(y)}{x-y} p_n(y)p_{n-1}(y)w(y)\, dy,\end{aligned} \tag{2.8.39}$$

and assume that all the above quantities are defined for all $n \geq 1$. Then $(p_n)_n$ satisfies (2.8.7), (2.8.8), *and* (2.8.22) *with A_n, B_n replaced by $\mathcal{A}_n$ and $\mathcal{B}_n$.*

Discrete and q-analogues of Theorem 2.8.1 will be stated in Sections 6.3 and 6.6, respectively.

2.9 Discriminants and Electrostatics

The discriminant D of a polynomial g,

$$g(x) = \gamma \prod_{j=1}^{n} \left(x - x_j\right), \tag{2.9.1}$$

is defined by (Dickson, 1939)

$$D(g) := \gamma^{2n-2} \prod_{1\leq j<k\leq n} \left(x_j - x_k\right)^2. \tag{2.9.2}$$

Observe that $D(g) = 0$ if and only if g has at least two equal zeros. An equivalent representation for the discriminant is

$$D(g) = (-1)^{n(n-1)/2}\gamma^{n-2} \prod_{j=1}^{n} g'\left(x_j\right). \tag{2.9.3}$$

The resultant of two polynomials f and g is

$$\operatorname{Res}\{f, g\} = \gamma^n \prod_{j=1}^{n} f\left(x_j\right), \tag{2.9.4}$$

where g is given in (2.9.1). Ismail introduced the concept of a discriminant relative to a linear operator T. If T reduces the degree of a polynomial by 1, he defined the **discriminant relative to T** by

$$D(g;T) = (-1)^{n(n-1)/2}\gamma^{n-2}\prod_{j=1}^{n}(Tg)\left(x_j\right), \tag{2.9.5}$$

where g is as in (2.9.1). If $T = \frac{d}{dx}$ (differential operator) then formula (2.9.5) becomes (2.9.3). When $T = \Delta$ (forward difference operator) the generalized discriminant becomes the **discrete discriminant**

$$D(g;\Delta) = \gamma^{2n-2}\prod_{1\le j<k\le n}\left(x_j - x_k - 1\right)\left(x_j - x_k + 1\right). \tag{2.9.6}$$

Note that if in (2.9.5) we use the difference operator δ_h,

$$(\delta_h f)(x) = [f(x+h) - f(x)]/h, \tag{2.9.7}$$

then (2.9.6) becomes

$$D(g,\delta_h) = h^{-n}\gamma^{2n-2}\prod_{1\le j<k\le n}\left(x_j - x_k - h\right)\left(x_j - x_k + h\right) \tag{2.9.8}$$

and it is clear that $h^n D(g;\Delta_h)$ reduces to the usual discriminant when $h \to 0$. When $T = D_q$ (q-difference operator) we have the **q-discriminant** (Ismail, 2003)

$$D(f;q) = D\left(f;D_q\right) = \gamma^{2n-2}q^{\binom{n}{2}}\prod_{1\le i<j\le n}\left(q^{-\frac{1}{2}}x_i - q^{\frac{1}{2}}x_j\right)\left(q^{\frac{1}{2}}x_i - q^{-\frac{1}{2}}x_j\right) \tag{2.9.9}$$

$$= \gamma^{2n-2}q^{\binom{n}{2}}\prod_{1\le i<j\le 4}\left(x_i^2 + x_j^2 - x_i x_j\left(q+q^{-1}\right)\right). \tag{2.9.10}$$

Ismail and Jing (2001) showed how the expression in (2.9.5) arises as a correlation function or expectation value in certain models involving vertex operators when $T = \Delta$ or $T = D_q$.

Theorem 2.9.1 (Ismail, 2003) *Assume that $(p_n)_n$ satisfies* (2.8.7) *and* (2.8.3)–(2.8.4). *Then the discriminant of p_n is given by*

$$D(p_n) = \left\{\prod_{j=1}^{n}\frac{A_n(x_{n,j})}{a_n}\right\}\left[\prod_{k=1}^{n}a_k^{2k-2n+2}\right]. \tag{2.9.11}$$

The proof uses a theorem of Schur (Ismail, 2005b, Lemma 3.4.1). Note that the term in square brackets in (2.9.11) is the Hankel determinant since $\beta_k = a_k^2$. Therefore

$$D(p_n) = D_n\prod_{j=1}^{n}\frac{A_n(x_{n,j})}{a_n}. \tag{2.9.12}$$

Theorem 2.9.2 *Under the assumptions of Theorem 2.8.5, the discriminant of the monic polynomial P_n is given by*

$$D(P_n) = \left\{\prod_{j=1}^{n} \frac{A_n(x_{n,j})}{a_n}\right\}\left[\prod_{k=1}^{n} a_k^{2k}\right]. \tag{2.9.13}$$

Stieltjes (1885a,b), Hille (1993), and later Schur (1931) gave explicit evaluations of the discriminant of a Jacobi polynomial $P_n^{(\alpha,\beta)}$. Their result is

$$D\left(P_n^{(\alpha,\beta)}\right) = 2^{-n(n-1)} \prod_{j=1}^{n} j^{j-2n+2}(j+\alpha)^{j-1}(j+\beta)^{j-1}(n+j+\alpha+\beta)^{n-j}. \tag{2.9.14}$$

This contains evaluations of the discriminants of the Hermite and Laguerre polynomials as limiting cases:

$$D(H_n) = 2^{3n(n-1)/2} \prod_{k=1}^{n} k^k, \tag{2.9.15}$$

$$D\left(L_n^{(\alpha)}\right) = \prod_{k=1}^{n} k^{k+2-2n}(k+\alpha)^{k-1}. \tag{2.9.16}$$

Consider a **logarithmic potential** on the line, where the potential energy at x of a point charge e located at c is $-2e \ln|x-c|$. Consider a system of n movable unit charged particles confined to $[a,b]$, in the presence of an external potential V. Let $x_1, \ldots, x_n$ be the positions of the particles, and

$$\mathbf{x} = (x_1, x_2, \ldots, x_n), \quad x_1 > x_2 > \cdots > x_n. \tag{2.9.17}$$

The total energy of the system is

$$E(\mathbf{x}) = \sum_{k=1}^{n} V(x_k) - 2 \sum_{1 \le j < k \le n} \ln\left|x_j - x_k\right|. \tag{2.9.18}$$

Let

$$T(\mathbf{x}) = \exp(-E(\mathbf{x})). \tag{2.9.19}$$

A weight function w generates an external potential v via (2.8.1). In the presence of the n charged particles, the external field is modified to become V,

$$V(x) = v(x) + \ln\left(A_n(x)/a_n\right). \tag{2.9.20}$$

Theorem 2.9.3 (Electrostatic equilibrium theorem) *Assume $w(x) = e^{-v(x)} > 0$, $x \in (a,b)$ and let v and V of (2.9.20) be twice continuously differentiable functions whose second derivative is nonnegative on (a,b). Then the equilibrium position of n movable unit charges in $[a,b]$ in the presence of the external potential V is unique and attained at the zeros of p_n, provided* *that the particle interaction obeys a logarithmic potential and that* $T(\mathbf{x}) \to 0$ *as* $\mathbf{x}$ *tends to any boundary point of* $[a,b]^n$, *where*

$$T(\mathbf{x}) = \left[\prod_{j=1}^{n} \frac{\exp\left(-v\left(x_j\right)\right)}{A_n(x_j)/a_n}\right] \prod_{1 \le \ell < k \le n} (x_\ell - x_k)^2. \tag{2.9.21}$$

Theorem 2.9.4 *Let* $T_{\max}$ *be the maximum value of* $T(\mathbf{x})$ *and* E_n *the equilibrium energy of the n particle system. Then*

$$T_{\max} = \exp\left(-\sum_{j=1}^{n} v\left(x_{n,j}\right)\right) \prod_{k=1}^{n} a_k^{2k}, \tag{2.9.22}$$

$$E_n = \sum_{j=1}^{n} v\left(x_{n,j}\right) - 2\sum_{j=1}^{n} j \ln a_j. \tag{2.9.23}$$

Stieltjes proved Theorem 2.9.3 when $e^{-v} = (1-x)^\alpha(1+x)^\beta$, $x = [-1,1]$, so that the modification term $\ln(A_n(x)/a_n)$ is $-\ln(1-x^2)$ plus a constant. In this model, V is due to fixed charges $(\alpha+1)/2$ and $(\beta+1)/2$ located at $x = 1, -1$, respectively. The equilibrium is attained at the zeros of $P_n^{(\alpha,\beta)}$. Theorems 2.9.3 and 2.9.4 are from Ismail (2000a). Discrete and q-analogues of Theorems 2.9.1 and 2.9.2 will be stated in Sections 6.3 and 6.6, respectively.

2.10 Functions of the Second Kind

The **functions of the second kind** associated with polynomials $(p_n)_n$ orthonormal with respect to μ satisfying (2.8.1) are

$$q_n(z) = \frac{1}{w(z)} \int_{\mathbb{R}} \frac{p_n(y)}{z-y} w(y)\, dy, \quad z \in \mathbb{C} \setminus \operatorname{supp}(w),\ n \ge 0. \tag{2.10.1}$$

The sequence $(q_n)_n$ satisfies

$$z q_n(z) = a_{n+1} q_{n+1}(z) + b_n q_n(z) + a_n q_{n-1}(z), \quad n \ge 0, \tag{2.10.2}$$

provided that $a_0 q_{-1}(z) = 1/w(z)$, $z \notin \operatorname{supp}(w)$.

Theorem 2.10.1 *Let* $(p_n)_n$ *be orthonormal with respect to* $w(x) = e^{-v(x)}$ *on* $[a,b]$, *and assume* $w(a^+) = w(b^-) = 0$. *Then for* $n \ge 0$ *both* p_n *and* q_n *have the same raising and lowering operators, that is,*

$$q_n'(z) = A_n(z) q_{n-1}(z) - B_n(z) q_n(z), \tag{2.10.3}$$

$$\left(-\frac{d}{dz} + B_n(z) + v'(z)\right) q_{n-1}(z) = A_{n-1}(z) \frac{a_n}{a_{n-1}} q_n(z). \tag{2.10.4}$$

Furthermore, p_n *and* q_n *are linearly independent solutions of the differential equation* (2.8.23) *on* $[c,d]$ *if* $A_n(x) \ne 0$ *for every* $x \in [c,d]$.

The Wronskian of p_n and q_n is related to the **Casorati determinant** via

$$p_n'(x)q_n(x) - q_n'(x)p_n(x) = A_n(x)\left[q_n(x)p_{n-1}(x) - p_n(x)q_{n-1}(x)\right]. \tag{2.10.5}$$

Formula (2.2.5) implies

$$P_n^*(z) = P_n(z)w(z)Q_0(z) - w(z)Q_n(z), \tag{2.10.6}$$

with

$$Q_n(z) = \frac{1}{w(z)}\int_{\mathbb{R}} \frac{P_n(y)}{z-y} w(y)\,dy. \tag{2.10.7}$$

If w is supported on a compact set $\subset [a,b]$, then $Q_n(z)$ is a minimal solution of (2.1.14) for $z \in \mathbb{C} \setminus [a,b]$.

2.11 Dual Systems

A discrete system of orthogonal polynomials induces another orthogonal system where the roles of the variable and the degree are interchanged.

Theorem 2.11.1 (Ismail, 2005b) *Let the coefficients $(\beta_n)_n$ in (2.1.14) be bounded and assume that the moment problem has a unique solution. If the orthogonality measure μ has isolated point masses at α and $\beta \le \alpha$ then we have a dual orthogonality relation*

$$\sum_{n=0}^{\infty} P_n(\alpha)P_n(\beta)/\zeta_n = \frac{1}{\mu\{\alpha\}}\,\delta_{\alpha,\beta}. \tag{2.11.1}$$

Theorem 2.11.2 (Berg, 2004) *Let $\mu = \sum_{\lambda\in\Lambda} a_\lambda \delta_\lambda$ be a discrete probability measure and assume that $(p_n)_n$ is a polynomial system orthonormal with respect to μ and complete in $L^2(\mu;\mathbb{R})$. Then the dual system $(p_n(\lambda)\sqrt{a_\lambda})_{\lambda\in\Lambda}$ is a dual orthonormal basis for ℓ^2, that is,*

$$\sum_{n=0}^{\infty} p_n(\lambda_1)\,p_n(\lambda_2) = \delta_{\lambda_1,\lambda_2}/a_{\lambda_1} \quad \text{for } \lambda_1, \lambda_2 \in \Lambda.$$

Let $\{\phi_n(x)\colon 0 \le n \le N\}$ be a finite system of orthogonal polynomials satisfying

$$\sum_{x=0}^{N} \phi_m(x)\phi_n(x)w(x) = \delta_{m,n}/h_n. \tag{2.11.2}$$

From linear algebra we see that

$$\sum_{n=0}^{N} \phi_n(x)\phi_n(y)h_n = \delta_{x,y}/w(x) \tag{2.11.3}$$

holds for $x, y = 0, 1, \ldots, N$.

De Boor and Saff (1986) introduced the concept of deB–S duality in. Given a sequence of polynomials satisfying (2.1.14)–(2.1.15), define a deB–S dual system $(Q_n)_{0\le n\le N}$ by

$$Q_0(x) = 1, \quad Q_1(x) = x - \alpha_{N-1}, \tag{2.11.4}$$

$$Q_{n+1}(x) = (x - \alpha_{N-n-1})\, Q_n(x) - \beta_{N-n} Q_{n-1}(x), \tag{2.11.5}$$

for $0 \le n < N$. From (2.1.14)–(2.1.15) and (2.11.4)–(2.11.5) it follows that

$$Q_n(x) = P_n(x; N - n).$$

The mapping $(P_n)_{0\le n\le N} \to (Q_n)_{0\le n\le N}$ is an involution. Induction shows that

$$Q_{N-n-1}(x)P_{n+1}(x) - \beta_n Q_{N-n-2}(x)P_n(x) = P_N(x), \tag{2.11.6}$$

$0 \le n < N$. One can show that $(P_n)_{0\le n<N}$ are orthogonal on $\{x_{N,j}\colon 1 \le j \le N\}$ with respect to a discrete measure with masses $\rho(x_{N,j})$ at $x_{N,j}$, $1 \le j \le N$ where

$$\frac{P_N^*(x)}{P_N(x)} = \sum_{k=1}^{N} \frac{\rho(x_{N,j})}{x - x_{N,j}}.$$

Define the dual weights $\rho_Q(x_{N,j})$ by

$$\frac{P_{N-1}(x)}{P_N(x)} = \frac{Q_N^*(x)}{Q_N(x)} = \sum_{j=1}^{N} \frac{\rho_Q(x_{N,j})}{x - x_{N,j}},$$

so that the numbers $\rho_Q(x_{N,j}) = P_{N-1}(x_{N,j})/P_N'(x_{N,j})$ have the properties

$$\sum_{j=1}^{N} \rho_Q(x_{N,j}) = 1, \quad \sum_{j=1}^{N} \rho_Q(x_{N,j})Q_m(x_{N,j})Q_n(x_{N,j}) = \zeta_n(Q)\,\delta_{m,n}, \tag{2.11.7}$$

and

$$\zeta_n(Q) = \beta_{N-1}\beta_{N-2}\cdots\beta_{N-n}, \quad \text{or} \quad \zeta_n(Q) = \zeta_{N-1}/\zeta_{N-n-1}. \tag{2.11.8}$$

Theorem 2.11.3 (Vinet and Zhedanov, 2004) *The Jacobi, Hermite, and Laguerre polynomials are the only orthogonal polynomials where $\rho_Q(x_{N,j}) = \pi(x_{N,j})/c_N$ for all N, where π is a polynomial of degree at most 2 and c_N is a constant.*

A sequence of orthogonal polynomials is called **semiclassical** if it is orthogonal with respect to $e^{-v}\,dx$, and v' is a rational function.

Theorem 2.11.4 (Vinet and Zhedanov, 2004) *Assume that $(Q_n)_{0\le n\le N}$ is the deB–S dual to $(P_n)_{0\le n\le N}$ and let $\{x_{n,i}\colon 1 \le i \le N\}$ be the zeros of P_N. Then $(P_N)_N$ are semiclassical if and only if, for every $N > 0$, the polynomials $(Q_i)_{0\le i<N}$ are orthogonal on $\{x_{N,i}\colon 1 \le i \le N\}$ with respect to the weights*

$$\frac{q(x_{N,i})}{\tau(x_{N,i}, N)},$$

where q is a polynomial of fixed degree and its coefficients do not depend on N, and $\tau(\cdot, N)$ is a polynomial of fixed degree whose coefficients may depend on N.

2.12 Moment Representations and Determinants

The Laplace method for solving difference equations is to represent solutions of difference equations as moments (Jordan, 1965; Milne-Thomson, 1933). Since we are interested only in orthogonal polynomials we let

$$xA_nS_n(x) = B_nS_{n+1}(x) + C_nS_n(x) + D_nS_{n-1}(x), \tag{2.12.1}$$

where A_n, B_n, C_n, and D_n are polynomials in n. We look for suitable a and b to make $S_n(x) = \int_a^b u^n w(u,x)\,du$ satisfy (2.12.1). In the analysis we replace $(n+1)_k u^n$ by $\frac{d^k}{du^k}u^{n+k}$ and integrate by parts to find a differential equation for w. Ismail and Stanton extended this to q-integral representations when A_n, B_n, C_n, and D_n are polynomials in q^n (Ismail and Stanton, 2002).

Let

$$S_n(x) = \int_a^b u^n\,d\mu(u,x), \tag{2.12.2}$$

where $d\mu(u,x) = w(u,x)\,dx$ or $d\mu(u,x) = w(u,x)d_qx$. If we start with a power series identity $f(t) = \sum_{n=0}^{\infty} a_n t^n$, then replace t by ut and integrate, this replaces u^n by $S_n(x)$ and leads to a generating function for $(S_n(x))_n$. This enables us to derive bilinear generating functions from linear generating functions. This is the spirit of the umbral calculus (Roman and Rota, 1978) and the symbolic method of Kaplansky (1944).

Another application is to evaluate determinants of orthogonal polynomials because the moment representation identifies such determinants as Hankel determinants of some other measure. We define the Turán determinants

$$T(S_m(x), S_{m+1}(x), \ldots, S_{m+n}(x)) := \begin{vmatrix} S_m(x) & S_{m+1}(x) & \cdots & S_{m+n}(x) \\ S_{m+1}(x) & S_{m+2}(x) & \cdots & S_{m+n+1}(x) \\ \vdots & \vdots & & \vdots \\ S_{m+n}(x) & S_{m+n+1}(x) & \cdots & S_{m+2n}(x) \end{vmatrix}. \tag{2.12.3}$$

When we have a moment representation (2.12.2) then the Hankel determinant

$$D_n(a) = T(S_0(a), S_1(a), \ldots, S_n(a)) \tag{2.12.4}$$

is given by (2.1.10).

3

Jacobi and Related Polynomials

3.1 Recursions and Representations

The Jacobi polynomials are defined by

$$P_n^{(\alpha,\beta)}(x) = \frac{(\alpha+1)_n}{n!}\, {}_2F_1\left(-n, \alpha+\beta+n+1; \alpha+1; (1-x)/2\right), \tag{3.1.1}$$

and satisfy the **orthogonality relations**

$$\int_{-1}^{1} P_m^{(\alpha,\beta)}(x) P_n^{(\alpha,\beta)}(x) (1-x)^\alpha (1+x)^\beta \, dx = h_n^{(\alpha,\beta)}\, \delta_{m,n}, \tag{3.1.2}$$

$$h_n^{(\alpha,\beta)} = \frac{2^{\alpha+\beta+1}\Gamma(\alpha+n+1)\Gamma(\beta+n+1)}{n!\Gamma(\alpha+\beta+n+1)(\alpha+\beta+2n+1)}. \tag{3.1.3}$$

The idea of the proof is to use the **attachment procedure**, we start by evaluating the integral $\int_{-1}^{1} w(x;\alpha,\beta)\,dx$, which is a beta integral. Then put $P_n^{(\alpha,\beta)}(x) = \sum_{k=0}^{n} c_{n,k}(1-x)^k$ and integrate:

$$\int_{-1}^{1} (1+x)^m \sum_{k=0}^{n} c_{n,k}(1-x)^k w(x;\alpha,\beta)\,dx = \sum_{k=0}^{n} c_{n,k} \int_{-1}^{1} w(x;\alpha+k,\beta+m)\,dx.$$

The choice $c_{n,k} = 2^{-k}(-1)^k(-n)_k(n+\alpha+\beta+1)_k/(\alpha+1)_k k!$ then gives the desired orthogonality, taking into account the Chu–Vandermonde sum (1.3.3).

The invariance of w under $(x,\alpha,\beta) \to (-x,\beta,\alpha)$ implies

$$P_n^{(\alpha,\beta)}(x) = (-1)^n P_n^{(\beta,\alpha)}(-x) = (-1)^n \frac{(\beta+1)_n}{n!}\, {}_2F_1\left(-n, \alpha+\beta+n+1; \beta+1; (1+x)/2\right). \tag{3.1.4}$$

Furthermore, it is clear from (3.1.1) and (3.1.4) that

$$P_n^{(\alpha,\beta)}(x) = \frac{(\alpha+\beta+n+1)_n}{n!2^n} x^n + \text{lower-order terms}, \tag{3.1.5}$$

$$P_n^{(\alpha,\beta)}(1) = \frac{(\alpha+1)_n}{n!}, \quad P_n^{(\alpha,\beta)}(-1) = (-1)^n \frac{(\beta+1)_n}{n!}. \tag{3.1.6}$$

The Jacobi polynomials have the **convolution-type representation**

$$\frac{P_n^{(\alpha,\beta)}(x)}{(\alpha+1)_n(\beta+1)_n} = \sum_{k=0}^{n} \frac{((x-1)/2)^k}{(\alpha+1)_k k!} \frac{((x+1)/2)^{n-k}}{(\beta+1)_{n-k}(n-k)!}, \tag{3.1.7}$$

$$(-1)^n P_n^{(\alpha,\beta)}(x) = \sum_{k=0}^{n} \frac{(-n-\beta)_k}{k!} \left(\frac{x-1}{2}\right)^k \frac{(-n-\alpha)_{n-k}}{(n-k)!} \left(\frac{x+1}{2}\right)^{n-k}, \tag{3.1.8}$$

which follow from (3.1.1) and (3.1.4).

Formulas (3.1.4) and (3.1.1) have inverses

$$\frac{((1+x)/2)^n}{(\beta+1)_n n!} = \sum_{k=0}^{n} \frac{(\alpha+\beta+2k+1)\,\Gamma(\alpha+\beta+k+1)}{(\beta+1)_k\,\Gamma(\alpha+\beta+n+2+k)\,(n-k)!}, P_k^{(\alpha,\beta)}(x) \tag{3.1.9}$$

and

$$\frac{((1-x)/2)^m}{(\alpha+1)_m m!} = \sum_{k=0}^{m} \frac{(\alpha+\beta+2k+1)\,\Gamma(\alpha+\beta+k+1)}{(\alpha+1)_k\,\Gamma(\alpha+\beta+m+2+k)\,(m-k)!} (-1)^k P_k^{(\alpha,\beta)}(x), \tag{3.1.10}$$

respectively. A common generalization of (3.1.9) and (3.1.10) is

$$\begin{aligned}\left(\frac{1-x}{2}\right)^m \left(\frac{1+x}{2}\right)^n &= \frac{(\alpha+1)_m(\beta+1)_n}{\Gamma(\alpha+\beta+m+n+2)} \\ &\times \sum_{k=0}^{m+n} (\alpha+\beta+2k+1)\frac{\Gamma(\alpha+\beta+k+1)}{(\beta+1)_k}\, {}_3F_2\left(\left.\begin{matrix}-k, k+\alpha+\beta+1, \alpha+m+1\\ \alpha+1, \alpha+\beta+m+n+2\end{matrix}\right| 1\right) P_k^{(\alpha,\beta)}(x).\end{aligned} \tag{3.1.11}$$

Formulas (3.1.1) and (3.1.10) lead to the **inverse relations**

$$u_n = \sum_{k=0}^{n} \frac{n!(\beta+1)_n(\alpha+\beta+1)_k(2k+\alpha+\beta+1)}{(n-k)!(\beta+1)_k(\alpha+\beta+1)_{n+k+1}} v_k \tag{3.1.12}$$

if and only if

$$v_n = (-1)^n \frac{(\beta+1)_n}{n!} \sum_{k=0}^{n} \frac{(-n)_k(\alpha+\beta+n+1)_k}{k!(\beta+1)_k} u_k. \tag{3.1.13}$$

The **lowering** (annihilation) and **raising** (creation) operator formulas are

$$\frac{d}{dx} P_n^{(\alpha,\beta)}(x) = \frac{1}{2}(n+\alpha+\beta+1)\, P_{n-1}^{(\alpha+1,\beta+1)}(x), \tag{3.1.14}$$

$$\frac{1}{(1-x)^\alpha(1+x)^\beta}\frac{d}{dx}\left[(1-x)^{\alpha+1}(1+x)^{\beta+1} P_{n-1}^{(\alpha+1,\beta+1)}(x)\right] = -2n P_n^{(\alpha,\beta)}(x). \tag{3.1.15}$$

Combining (3.1.14) and (3.1.15) we obtain the **differential equation**

$$\begin{aligned}&\frac{1}{(1-x)^\alpha(1+x)^\beta}\frac{d}{dx}\left[(1-x)^{\alpha+1}(1+x)^{\beta+1}\frac{d}{dx} P_n^{(\alpha,\beta)}(x)\right] \\ &\quad = -n(n+\alpha+\beta+1)P_n^{(\alpha,\beta)}(x).\end{aligned} \tag{3.1.16}$$

Thus $y = P_n^{(\alpha,\beta)}$ is a solution to the **Jacobi differential equation**

$$\left(1 - x^2\right) y''(x) + [\beta - \alpha - x(\alpha + \beta + 2)]y'(x) = -n(n + \alpha + \beta + 1)y(x). \tag{3.1.17}$$

Observe that (3.1.16) can be written in the factored form $T^*Ty = \lambda y$, known as the Infeld–Hull factorization. Equation (3.1.17) can be transformed to

$$\begin{gathered}\frac{d^2u}{dx^2} + \left\{\frac{1-\alpha^2}{4(1-x)^2} + \frac{1-\beta^2}{4(1+x)^2} + \frac{n(n+\alpha+\beta+1) + (\alpha+1)(\beta+1)}{1-x^2}\right\} u = 0, \\ u = u(x) = (1-x)^{(\alpha+1)/2}(1+x)^{(1+\beta)/2} P_n^{(\alpha,\beta)}(x),\end{gathered} \tag{3.1.18}$$

and

$$\begin{gathered}\frac{d^2u}{d\theta^2} + \left\{\frac{1/4-\alpha^2}{4\sin^2\theta/2} + \frac{1/4-\beta^2}{4\cos^2\theta/2} + \left(n + \frac{\alpha+\beta+1}{2}\right)^2\right\} u = 0, \\ u = u(\theta) = ((\sin(\theta/2))^{\alpha+1/2}(\cos(\theta/2))^{\beta+1/2} P_n^{(\alpha,\beta)}(\cos\theta).\end{gathered} \tag{3.1.19}$$

Note that (3.1.17) is the most general second-order differential equation of the form

$$\pi_2(x)y''(x) + \pi_1(x)y'(x) + \lambda y = 0,$$

with a polynomial solution of degree n, where $\pi_j(x)$ denotes a generic polynomial in x of degree at most j. To see this observe that λ is uniquely determined by the requirement that y is a polynomial of degree n for every $n = 0, 1, 2, \ldots$. Thus we need to determine five coefficients in π_1 and π_2. But by dividing the differential equation by a constant we can make one of the nonzero coefficients equal to 1, so we have only four parameters left. On the other hand, the change of variable $x \to ax + b$ will absorb two of the parameters, so we have only two free parameters at our disposal. The differential equation (3.1.17) does indeed have two free parameters.

By iterating (3.1.15) we find

$$2^k(-1)^k n! P_n^{(\alpha,\beta)}(x) = \frac{(n-k)!}{(1-x)^\alpha(1+x)^\beta} \frac{d^k}{dx^k}\left[(1-x)^{\alpha+k}(1+x)^{\beta+k} P_{n-k}^{(\alpha+k,\beta+k)}(x)\right]. \tag{3.1.20}$$

In particular, the case $k = n$ is the **Rodrigues formula**,

$$2^n(-1)^n n! P_n^{(\alpha,\beta)}(x) = \frac{1}{(1-x)^\alpha(1+x)^\beta} \frac{d^n}{dx^n}\left[(1-x)^{\alpha+n}(1+x)^{\beta+n}\right]. \tag{3.1.21}$$

The **three-term recurrence relation** for Jacobi polynomials is

$$\begin{aligned}&2(n+1)(n+\alpha+\beta+1)(\alpha+\beta+2n)P_{n+1}^{(\alpha,\beta)}(x) \\ &= (\alpha+\beta+2n+1)\left[\left(\alpha^2-\beta^2\right) + x(\alpha+\beta+2n+2)(\alpha+\beta+2n)\right] P_n^{(\alpha,\beta)}(x) \\ &- 2(\alpha+n)(\beta+n)(\alpha+\beta+2n+2)P_{n-1}^{(\alpha,\beta)}(x),\end{aligned} \tag{3.1.22}$$

for $n \geq 0$, with $P_{-1}^{(\alpha,\beta)}(x) = 0$, $P_0^{(\alpha,\beta)}(x) = 1$.

In the case of Jacobi polynomials, (2.8.7) becomes

$$\begin{aligned}&\left(1-x^2\right)(2n+\alpha+\beta)\frac{d}{dx}P_n^{(\alpha,\beta)}(x)\\&=\frac{1}{2}(n+\alpha+\beta+1)(2n+\alpha+\beta)\left(1-x^2\right)P_{n-1}^{(\alpha+1,\beta+1)}(x)\\&=2(n+\alpha)(n+\beta)P_{n-1}^{(\alpha,\beta)}(x)-n[\beta-\alpha+x(2n+\alpha+\beta)]P_n^{(\alpha,\beta)}(x).\end{aligned} \tag{3.1.23}$$

The **discriminant of a Jacobi polynomial** is given by (2.9.14).

The Christoffel formula (2.5.2) and (3.1.6) imply

$$P_n^{(\alpha+1,\beta)}(x)=\frac{2\left[(n+\alpha+1)P_n^{(\alpha,\beta)}(x)-(n+1)P_{n+1}^{(\alpha,\beta)}(x)\right]}{(2n+\alpha+\beta+2)(1-x)}, \tag{3.1.24}$$

$$P_n^{(\alpha,\beta+1)}(x)=\frac{2\left[(n+\beta+1)P_n^{(\alpha,\beta)}(x)+(n+1)P_{n+1}^{(\alpha,\beta)}(x)\right]}{(2n+\alpha+\beta+2)(1+x)}, \tag{3.1.25}$$

and

$$\begin{aligned}\frac{\left(1-x^2\right)}{(\alpha+\beta+n+1)}\frac{d}{dx}P_n^{(\alpha,\beta)}(x)&=\frac{2(\alpha+n)(\beta+n)}{(2n+\alpha+\beta)(2n+\alpha+\beta+1)}P_{n-1}^{(\alpha,\beta)}(x)\\&+\frac{2n(\alpha-\beta)}{(2n+\alpha+\beta)(2n+\alpha+\beta+2)}P_n^{(\alpha,\beta)}(x)+\frac{2n(n+1)}{(2n+\alpha+\beta+1)(2n+\alpha+\beta+2)}P_{n+1}^{(\alpha,\beta)}(x).\end{aligned} \tag{3.1.26}$$

Note that (3.1.25) also follows from (3.1.24) and (3.1.4).

Lemma 3.1.1 *Let $\alpha>-1,\beta>-1$, and set*

$$x_0=\frac{\beta-\alpha}{\alpha+\beta+1},\qquad M_n=\max\left\{\left|P_n^{(\alpha,\beta)}(x)\right|:\ -1\le x\le 1\right\}.$$

Then

$$M_n=\begin{cases}(s)_n/n! & \text{if } s\ge -1/2,\\ P_n^{(\alpha,\beta)}(x') & \text{if } s<-1/2,\end{cases} \tag{3.1.27}$$

where $s=\min\{\alpha,\beta\}$ and x' is one of the two maximum points closest to x_0.

We next record the **connection relation** for Jacobi polynomials.

Theorem 3.1.2 *The connection relation for Jacobi polynomials is*

$$P_n^{(\gamma,\delta)}(x)=\sum_{k=0}^{n}c_{n,k}(\gamma,\delta;\alpha,\beta)P_k^{(\alpha,\beta)}(x), \tag{3.1.28}$$

with

$$\begin{aligned}c_{n,k}(\gamma,\delta;\alpha,\beta)&=\frac{(\gamma+k+1)_{n-k}(n+\gamma+\delta+1)_k}{(n-k)!\Gamma(\alpha+\beta+2k+1)}\Gamma(\alpha+\beta+k+1)\\&\times{}_3F_2\left(\left.\begin{matrix}-n+k,n+k+\gamma+\delta+1,\alpha+k+1\\ \gamma+k+1,\alpha+\beta+2k+2\end{matrix}\right|1\right).\end{aligned} \tag{3.1.29}$$

Theorem 3.1.2 follows from (2.6.11); see Ismail (2005b, §9).

Corollary 3.1.3 *We have the* ***connection relation***

$$P_n^{(\alpha,\delta)}(x) = \sum_{k=0}^{n} d_{n,k} P_k^{(\alpha,\beta)}(x) \tag{3.1.30}$$

with

$$d_{n,k} = \frac{(\alpha+k+1)_{n-k}(n+\alpha+\delta+1)_k\Gamma(\alpha+\beta+k+1)(\beta-\delta+2k-n)_{n-k}}{(n-k)!\Gamma(\alpha+\beta+n+k+1)}. \tag{3.1.31}$$

Unlike the connection coefficients, the linearization coefficients $c_{k,m,n}^{(\alpha,\beta)}$ in

$$P_m^{(\alpha,\beta)}(x)P_n^{(\alpha,\beta)}(x) = \sum_{k=|m-n|}^{m+n} c_{k,m,n}^{(\alpha,\beta)} P_k^{(\alpha,\beta)}(x) \tag{3.1.32}$$

do not have a closed form which exhibits the parameter domain where the coefficients $c_{k,m,n}^{(\alpha,\beta)}$ are nonnegative.

Theorem 3.1.4 (Gasper, 1970) *The linearization coefficients $c_{k,m,n}^{(\alpha,\beta)}$ are nonnegative for $k, m, n = 0, 1, \ldots$ if and only if $\alpha \geq \beta$ and*

$$a(a+5)(a+3)^2 \geq \left(a^2 - 7a - 24\right) b^2,$$

where $a = \alpha + \beta + 1$ and $b = \alpha - \beta$.

In particular, $c_{k,m,n}^{(\alpha,\beta)} \geq 0$ when $\alpha \geq \beta > -1$ and $\alpha + \beta + 1 \geq 0$. This result was then used by Gasper to construct a convolution structure, prove the positivity of a generalized translation operator, and to derive a Wiener–Lévy-type theorem (Gasper, 1971, 1972).

We now record the **expansion of a plane wave** e^{ixy} in a series of Jacobi polynomials.

Theorem 3.1.5 *For $\alpha > -1, \beta > -1$, we have*

$$e^{xy} = \sum_{n=0}^{\infty} \frac{\Gamma(\alpha+\beta+n+1)}{\Gamma(\alpha+\beta+2n+1)} (2y)^n e^{-y} {}_1F_1\left(\left.\begin{matrix}\beta+n+1\\ \alpha+\beta+2n+2\end{matrix}\right| 2y\right) P_n^{(\alpha,\beta)}(x). \tag{3.1.33}$$

Equivalently,

$$\begin{aligned}
&\int_{-1}^{1} e^{xy}(1-x)^\alpha(1+x)^\beta P_n^{(\alpha,\beta)}(x)\,dx \\
&\quad = \frac{y^n}{n!} e^{-y} \frac{2^{\alpha+\beta+n+1}\Gamma(\alpha+n+1)\Gamma(\beta+n+1)}{\Gamma(\alpha+\beta+2n+1)} {}_1F_1\left(\left.\begin{matrix}\beta+n+1\\ \alpha+\beta+2n+2\end{matrix}\right| 2y\right).
\end{aligned} \tag{3.1.34}$$

Exceptional Case: When $\alpha > -1, \beta > -1$, and $\alpha + \beta \neq 0$, the Jacobi polynomials are well defined through (3.1.22) with

$$P_0^{(\alpha,\beta)}(x) = 1, \quad P_1^{(\alpha,\beta)}(x) = [x(\alpha+\beta+2) + \alpha - \beta]/2. \tag{3.1.35}$$

When $\alpha + \beta = 0$, there is indeterminacy in defining $P_1^{(\alpha,\beta)}(x)$. If we use (3.1.35), then $P_1^{(\alpha,-\alpha)}(x) = x + \alpha$. On the other hand, if we set $\alpha + \beta = 0$ then apply (3.1.22) and the initial conditions $P_{-1}(x) = 0$, $P_0(x) = 1$, we will see in addition to the option $P_1(x) = x + \alpha$ we may also choose $\mathcal{P}_1(x) = x$. The first choice leads to $(P_n^{(\alpha,-\alpha)}(x))_n$, while the second option leads to the polynomials $(\mathcal{P}_n^{(\alpha)}(x))_n$,

$$\mathcal{P}_n^{(\alpha)}(x) = \frac{1}{2}\left[P_n^{(\alpha,-\alpha)}(x) + P_n^{(-\alpha,\alpha)}(x)\right]. \tag{3.1.36}$$

This observation is due to Ismail and Masson (1991) who also established the orthogonality relation

$$\int_{-1}^{1} \mathcal{P}_m^{(\alpha)}(x)\mathcal{P}_n^{(\alpha)}(x)w(x;\alpha)\,dx = \frac{(1+\alpha)_n(1-\alpha)_n}{(2n+1)(n!)^2}\,\delta_{m,n}, \tag{3.1.37}$$

$$w(x;\alpha) = \frac{2\sin(\pi\alpha)}{\pi\alpha}\,\frac{\left(1-x^2\right)^{\alpha}}{(1-x)^{2\alpha} + 2\cos(\pi\alpha)\left(1-x^2\right)^{\alpha} + (1+x)^{2\alpha}}, \tag{3.1.38}$$

for $-1 < \alpha < 1$. They satisfy (3.1.22) with the initial conditions

$$\mathcal{P}_0^{(\alpha)}(x) = 1, \quad \mathcal{P}_1^{(\alpha)}(x) = x. \tag{3.1.39}$$

Define the differential operator

$$D(\alpha,\beta;n) := \left(1-x^2\right)\frac{d^2}{dx^2} + [\beta - \alpha - (\alpha+\beta+2)x]\frac{d}{dx} + n(n+\alpha+\beta+1). \tag{3.1.40}$$

Then $\mathcal{P}_n^{(\alpha)}$ satisfies the fourth-order differential equation

$$D(1-\alpha, 1+\alpha; n-1)D(\alpha,-\alpha;n)\mathcal{P}_n^{(\alpha)}(x) = 0. \tag{3.1.41}$$

One can show that

$$\mathcal{P}_n^{(\alpha)}(x) = \lim_{c\to 0^+} P_n^{(\alpha,-\alpha)}(x;c), \tag{3.1.42}$$

where $(P_n^{(\alpha,\beta)}(x;c))_n$ are the polynomials in Section 4.4. Using the representation (4.4.7) one can also confirm (3.1.36).

3.2 Generating Functions

The Jacobi polynomials have the generating functions

$$\sum_{n=0}^{\infty} \frac{(\alpha+\beta+1)_n}{(\alpha+1)_n} t^n P_n^{(\alpha,\beta)}(x) = (1-t)^{-\alpha-\beta-1}\,{}_2F_1\left(\left.\begin{matrix}(\alpha+\beta+1)/2,\ 1+(\alpha+\beta)/2\\ \alpha+1\end{matrix}\,\right|\,\frac{2t(x-1)}{(1-t)^2}\right), \tag{3.2.1}$$

$$\sum_{n=0}^{\infty} \frac{(\alpha+\beta+1)_n}{(\alpha+1)_n} \frac{(\alpha+\beta+1+2n)}{(\alpha+\beta+1)} t^n P_n^{(\alpha,\beta)}(x) = \frac{(1+t)}{(1-t)^{\alpha+\beta+2}} {}_2F_1\left(\begin{matrix}(\alpha+\beta+2)/2, (\alpha+\beta+3)/2 \\ \alpha+1\end{matrix}\middle| \frac{2t(x-1)}{(1-t)^2}\right), \tag{3.2.2}$$

$$\sum_{n=0}^{\infty} \frac{P_n^{(\alpha,\beta)}(x)t^n}{(\beta+1)_n(\alpha+1)_n} = {}_0F_1(—;\alpha+1, t(x-1)/2)\, {}_0F_1(—;\beta+1; t(x+1)/2). \tag{3.2.3}$$

Rainville (1960) calls (3.2.3) Bateman's generating function. The generating function (3.2.1) is a special case of the following generating function for hypergeometric polynomials:

$$\sum_{n=0}^{\infty} \frac{(\lambda)_n}{n!} \phi_n(x)t^n = (1-t)^{-\lambda} {}_{p+2}F_s\left(\begin{matrix}\lambda/2, (\lambda+1)/2, a_1, \dots, a_p \\ b_1, \dots, b_s\end{matrix}\middle| \frac{-4tx}{(1-t)^2}\right), \tag{3.2.4}$$

when $s \ge p+1$, $\left|tx/(1-t)^2\right| < 1/4$, $|t| < 1$, where

$$\phi_n(x) = {}_{p+2}F_s\left(\begin{matrix}-n, n+\lambda, a_1, \dots, a_p \\ b_1, \dots, b_s\end{matrix}\middle| x\right). \tag{3.2.5}$$

Formula (3.12.1) leads to the **generating function**

$$\sum_{n=0}^{\infty} P_n^{(\alpha,\beta)}(x)t^n = \frac{2^{\alpha+\beta}R^{-1}}{(1-t+R)^{\alpha}(1+t+R)^{\beta}}, \qquad R = R(x,t) := \sqrt{1-2tx+t^2}, \quad \text{and} \quad R(x,0) = 1. \tag{3.2.6}$$

Additional generating functions are

$$\sum_{n=0}^{\infty} \frac{\alpha+\beta+1}{\alpha+\beta+1+n} t^n P_n^{(\alpha,\beta)}(x) = \left(\frac{2}{1+t+R}\right)^{\alpha+\beta+1} {}_2F_1\left(\begin{matrix}\alpha+1, \alpha+\beta+1 \\ \alpha+\beta+2\end{matrix}\middle| \frac{2t}{1+t+R}\right), \tag{3.2.7}$$

$$\sum_{n=0}^{\infty} \frac{(\gamma)_n(\alpha+\beta+1-\gamma)_n}{(\alpha+1)_n(\beta+1)_n} P_n^{(\alpha,\beta)}(x)t^n = {}_2F_1\left(\begin{matrix}\gamma, \alpha+\beta+1-\gamma \\ \alpha+1\end{matrix}\middle| \frac{1-R-t}{2}\right) {}_2F_1\left(\begin{matrix}\gamma, \alpha+\beta+1-\gamma \\ \beta+1\end{matrix}\middle| \frac{1-R+t}{2}\right). \tag{3.2.8}$$

Formula (3.1.7) implies another generating function, namely

$$\sum_{n=0}^{\infty} \frac{(\gamma)_n(\delta)_n t^n}{(\alpha+1)_n(\beta+1)_n} P_n^{(\alpha,\beta)}(x) = F_4\left(\gamma, \delta; \alpha+1, \beta+1; \frac{t}{2}(x-1), \frac{t}{2}(x+1)\right), \tag{3.2.9}$$

where F_4 is defined in (1.2.41). Another important generating function is (Srivastava and Singhal, 1973)

$$\sum_{n=0}^{\infty} P_n^{(\alpha+\lambda n, \beta+\mu n)}(x)t^n = \frac{(1-\zeta)^{\alpha+1}(1+\zeta)^{\beta+1}}{(1-x)^{\alpha}(1+x)^{\beta}} \times \frac{(1-x)^{\lambda}(1+x)^{\mu}}{(1-x)^{\lambda}(1+x)^{\mu} + \frac{1}{2}t(1-\zeta)^{\lambda}(1+\zeta)^{\nu}[\mu-\lambda-z(\lambda+\mu+2)]} \tag{3.2.10}$$

for $\operatorname{Re} x \in (-1, 1)$, where

$$\zeta = x - t\frac{(1-\zeta)^{\lambda+1}(1+\zeta)^{\mu+1}}{2(1-x)^{\alpha}(1+x)^{\beta}}. \tag{3.2.11}$$

This is proved using Lagrange inversion and has many applications.

We also have (Cohen, 1977)

$$\begin{aligned}\sum_{n=0}^{\infty} \frac{(n+k)!(\alpha+\beta+k+1)_n}{n!(\alpha+\beta+2k+2)_m} P_{n+k}^{(\alpha,\beta)}(x) \\ = 2\frac{k!\Gamma(2k+\alpha+\beta+2)t^{-\alpha-\beta-k-1}}{\Gamma(k+\alpha+1)\Gamma(k+\beta+1)} P_k^{(\alpha,\beta)}\left(\frac{1-R}{t}\right) Q_k^{(\alpha,\beta)}\left(\frac{1+R}{t}\right),\end{aligned} \tag{3.2.12}$$

where R is as in (3.2.6), $|t| < 1$, and $Q_k^{(\alpha,\beta)}$ is the Jacobi function of the second kind.

Theorem 3.2.1 (Bateman, 1905) *We have the functional relation*

$$\left(\frac{x+y}{2}\right)^n P_n^{(\alpha,\beta)}\left(\frac{1+xy}{x+y}\right) = \sum_{k=0}^{n} c_{n,k} P_k^{(\alpha,\beta)}(x) P_k^{(\alpha,\beta)}(y), \tag{3.2.13}$$

with

$$c_{n,k} = \frac{(\alpha+1)_n(\beta+1)_n(\alpha+\beta+1)_k(\alpha+\beta+1+2k)k!}{(\alpha+1)_k(\beta+1)_k(\alpha+\beta+1)_{n+k+1}(n-k)!}. \tag{3.2.14}$$

Moreover, (3.2.13) *has inverse*

$$\begin{aligned}\frac{(-1)^n n!n!}{(\alpha+1)_n(\beta+1)_n} P_n^{(\alpha,\beta)}(x) P_n^{(\alpha,\beta)}(y) \\ = \sum_{k=0}^{n} \frac{(-n)_k(\alpha+\beta+n+1)_k}{(\alpha+1)_k(\beta+1)_k}\left(\frac{x+y}{2}\right)^k P_k^{(\alpha,\beta)}\left(\frac{1+xy}{x+y}\right).\end{aligned} \tag{3.2.15}$$

Theorem 3.2.2 *The Jacobi polynomials have the bilinear generating functions*

$$\begin{aligned}\sum_{n=0}^{\infty} \frac{n!(\alpha+\beta+1)_n}{(\alpha+1)_n(\beta+1)_n} t^n P_n^{(\alpha,\beta)}(x) P_n^{(\alpha,\beta)}(y) \\ = (1+t)^{-\alpha-\beta-1} F_4\left(\left.\begin{matrix}(\alpha+\beta+1)/2, (\alpha+\beta+2)/2 \\ \alpha+1, \beta+1\end{matrix}\right| A, B\right)\end{aligned} \tag{3.2.16}$$

and

$$\begin{aligned}\sum_{n=0}^{\infty} \frac{n!(\alpha+\beta+1)_n}{(\alpha+1)_n(\beta+1)_n}(2n+\alpha+\beta+1) t^n P_n^{(\alpha,\beta)}(x) P_n^{(\alpha,\beta)}(y) \\ = \frac{(\alpha+\beta+1)(1-t)}{(1+t)^{\alpha+\beta+2}} F_4\left(\left.\begin{matrix}(\alpha+\beta+2)/2, (\alpha+\beta+3)/2 \\ \alpha+1, \beta+1\end{matrix}\right| A, B\right),\end{aligned} \tag{3.2.17}$$

where

$$A = \frac{t(1-x)(1-y)}{(1+t)^2}, \quad B = \frac{t(1+x)(1+y)}{(1+t)^2}. \tag{3.2.18}$$

The generating function (3.2.17) is a constant multiple of the Poisson kernel for Jacobi polynomials.

3.3 Jacobi Functions of the Second Kind

The following function of the second kind for the Jacobi polynomials differs from the general definition in Section 2.10 by a phase factor

$$Q_n^{(\alpha,\beta)}(x) = \frac{1}{2}(x-1)^{-\alpha}(x+1)^{-\beta}\int_{-1}^{1}(1-t)^{\alpha}(1+t)^{\beta}\frac{P_n^{(\alpha,\beta)}(t)}{x-t}\,dt. \tag{3.3.1}$$

Indeed, $Q_n^{(\alpha,\beta)}(x)$ satisfies (3.1.16) and (3.1.22). The Rodrigues formula (3.1.20) and integration by parts transform (3.3.1) into the equivalent form

$$Q_n^{(\alpha,\beta)}(x) = \frac{(n-k)!k!}{2^{k+1}n!}(x-1)^{-\alpha}(x+1)^{-\beta}\int_{-1}^{1}\frac{(1-t)^{\alpha+k}(1+t)^{\beta+k}}{(x-t)^{k+1}}P_{n-k}^{(\alpha+k,\beta+k)}(t)\,dt. \tag{3.3.2}$$

In particular,

$$Q_n^{(\alpha,\beta)}(x) = \frac{(x-1)^{-\alpha}(x+1)^{-\beta}}{2^{n+1}}\int_{-1}^{1}\frac{(1-t)^{\alpha+n}(1+t)^{\beta+n}}{(x-t)^{n+1}}\,dt. \tag{3.3.3}$$

Formulas (3.3.1)–(3.3.3) hold when $\mathrm{Re}\,\alpha > -1$, $\mathrm{Re}\,\beta > -1$, $n + |\alpha+\beta+1| \neq 0$, and $x \in \mathbb{C}\setminus[-1,1]$. In the exceptional case $n = 0$ and $\alpha+\beta+1 = 0$, $Q_0^{(\alpha,\beta)}(x)$ is a constant. This makes $P_0^{(\alpha,\beta)}(x)$ and $Q_0^{(\alpha,\beta)}(x)$ linearly dependent solutions of (3.1.17). A nonconstant solution of (3.1.17) is

$$Q^{(\alpha)}(x) = \ln(1+x) + \frac{\sin\pi\alpha}{\pi}(x-1)^{-\alpha}(x+1)^{-\beta}\int_{-1}^{1}\frac{(1-t)^{\alpha}(1+t)^{\beta}}{x-t}\ln(1+t)\,dt. \tag{3.3.4}$$

The function $Q_n^{(\alpha,\beta)}(x)$ is called the Jacobi function of the second kind. In the exceptional case $n = 0$, $\alpha+\beta+1 = 0$, the Jacobi function of the second kind is $Q^{(\alpha)}(x)$. Note that

$$Q^{(\alpha)}(x) = 2\frac{\sin\pi\alpha}{\pi}\frac{\partial}{\partial\beta}Q_0^{(\alpha,\beta)}(x)\bigg|_{\beta=-\alpha-1}. \tag{3.3.5}$$

Moreover, the function of the second kind has the hypergeometric function representations

$$\begin{aligned} Q_n^{(\alpha,\beta)}(x) = {} & \frac{\Gamma(n+\alpha+1)\Gamma(n+\beta+1)}{\Gamma(2n+\alpha+\beta+2)2^{-n-\alpha-\beta}}(x-1)^{-n-\alpha-1}(x+1)^{-\beta} \\ & \times {}_2F_1\left(\begin{matrix} n+\alpha+1,\, n+1 \\ 2n+\alpha+\beta+2 \end{matrix}\,\middle|\,\frac{2}{1-x}\right), \end{aligned} \tag{3.3.6}$$

$$Q^{(\alpha)}(x) = \ln(x+1) + c + \left(1-\frac{2}{1-x}\right)^{\alpha+1}\sum_{k=1}^{\infty}\frac{(\alpha+1)_k}{k!}\left(\sum_{j=1}^{k}\frac{1}{j}\right)\left(\frac{2}{1-x}\right)^{k}, \tag{3.3.7}$$

where

$$c = -\gamma - \frac{\Gamma'(-\alpha)}{\Gamma(-\alpha)} - \ln 2 \tag{3.3.8}$$

and γ is the Euler constant. One can also prove the following integral representation.

Theorem 3.3.1 *Assume $\alpha, \beta \in (-1, \infty)$. The following integral representations hold:*

$$Q_0^{(\alpha,\beta)}(x) = -2^{\alpha+\beta}\frac{\Gamma(\alpha+1)\Gamma(\beta+1)}{\Gamma(\alpha+\beta+1)} \int_\infty^x (t-1)^{-\alpha-1}(t+1)^{-\beta-1}\,dt \quad \text{if } \alpha+\beta+1>0, \tag{3.3.9}$$

$$Q_0^{(\alpha,\beta)}(x) = -2^{\alpha+\beta}\frac{\Gamma(\alpha+1)\Gamma(\beta+1)}{\Gamma(\alpha+\beta+1)} \left\{ \int_\infty^x \left[(t-1)^{-\alpha-1}(t+1)^{-\beta-1} - t^{-\alpha-\beta-2}\right] dt - \frac{x^{-\alpha-\beta-1}}{\alpha+\beta+1} \right\} \quad \text{if } \alpha+\beta+1<0, \tag{3.3.10}$$

$$Q^{(\alpha)}(x) = \int_\infty^x \left[(t-1)^{-\alpha-1}(t+1)^{-\beta-1} - \frac{1}{1+t}\right] dt + \log(x+1) \quad \text{if } \alpha+\beta+1=0. \tag{3.3.11}$$

In the formula below we assume $\alpha > -1, \beta > -1$:

$$Q_n^{(\alpha,\beta)}(x) = P_n^{(\alpha,\beta)}(x) Q_0^{(\alpha,\beta)}(x) - \frac{1}{2}(x-1)^{-\alpha}(x+1)^{-\beta} \int_{-1}^{1} (1-t)^\alpha (1+t)^\beta \frac{P_n^{(\alpha,\beta)}(x) - P_n^{(\alpha,\beta)}(t)}{x-t}\,dt. \tag{3.3.12}$$

Note that the integral in (3.3.12) is a multiple of the numerator polynomial. This equation also shows that the behavior of $Q_n^{(\alpha,\beta)}(x)$ as $x \to \pm 1$ can be determined from the behavior of $Q_0^{(\alpha,\beta)}(x)$ as $x \to \pm 1$.

The behavior of $Q_0^{(\alpha,\beta)}(x)$ near $x = +1$ is not difficult to determine. The following discussion is taken from Szegő ([1939] 1975, §4.62). Expanding the factor $(t+1)^{-\beta-1}$ in the integrand of (3.3.9) into a power series in $t-1$, we obtain for $\alpha+\beta+1>0$, for α not an integer,

$$Q_0^{(\alpha,\beta)} = \text{const} + (x-1)^{-\alpha} M\left(\frac{1-x}{2}\right);$$

here $M(u)$ is a power series in u, convergent for $|u| < 1$, and $M(0) \neq 0$. Similarly we can treat the case $\alpha+\beta+1<0$ using (3.3.10). In the exceptional case $\alpha+\beta+1=0$, this is not true for $Q_0^{(\alpha,\beta)}(x)$; however, it is true for $Q^{(\alpha)}(x)$, via (3.3.10). Now, let α be an integer; then we use (3.3.10) again. In view of

$$\begin{aligned}(t-1)^{-\alpha-1}(t+1)^{-\beta-1} &= 2^{-\beta-1}(t-1)^{-\alpha-1}\left(1-\frac{1-t}{2}\right)^{-\beta-1} \\ &= 2^{-\beta-1}(t-1)^{-\alpha-1}\left\{1+\cdots+\binom{\beta+\alpha}{\alpha}\left(\frac{1-t}{2}\right)^{\alpha}+\cdots\right\},\end{aligned}$$

we obtain a logarithmic term. Therefore,

$$Q_0^{(\alpha,\beta)}(x) = \begin{cases} \text{const} + (x-1)^{-\alpha} M_1\left(\dfrac{1-x}{2}\right) \\ \qquad \text{if } \alpha > -1,\ \beta > -1,\ \alpha \neq 0, 1, 2, \ldots,\ \alpha + \beta + 1 \neq 0; \\ \dfrac{(-1)^\alpha}{2} \log \dfrac{1}{x-1} + (x-1)^{-\alpha} M_2\left(\dfrac{1-x}{2}\right) \\ \qquad \text{if } \alpha = 0, 1, 2, \ldots,\ \beta > -1. \end{cases} \tag{3.3.13}$$

Here $M_1(u)$ and $M_2(u)$ are power series convergent for $|u| < 1$ with $M_1(0) \neq 0$, $M_2(0) \neq 0$. A representation similar to the first one holds for the function $Q^{(\alpha)}(x)$.

For example, we have

$$Q_0^{(0,0)}(x) = Q_0(x) = \int_\infty^x \frac{dt}{1-t^2} = \frac{1}{2} \log \frac{x+1}{x-1}. \tag{3.3.14}$$

To see that $M_2(0) \neq 0$ for $\alpha = 0$, we take $x > 1$ and then integrate by parts. Thus

$$\int_\infty^x (t-1)^{-1}(t+1)^{-\beta-1} dt = (x+1)^{-\beta-1} \log(x-1) + (\beta+1) \int_\infty^x (t+1)^{-\beta-2} \log(t-1)\, dt,$$

so that

$$\begin{aligned} M_2(0) &= \lim \left\{ Q_0^{(\alpha,\beta)}(x) + \frac{1}{2} \log(x-1) \right\} \\ &= -(\beta+1)2^\beta \int_\infty^1 (t+1)^{-\beta-2} \log(t-1)\, dt \neq 0. \end{aligned}$$

The following theorems are from Szegő ([1939] 1975).

Theorem 3.3.2 *Let x be real, $x > 1$, and take $(x-1)^\alpha$, $(x+1)^\beta$ real and positive. We then have, for $x \to 1+0$,*

$$Q_n^{(\alpha,\beta)}(x) \sim \begin{cases} (x-1)^{-\alpha}, & \alpha > 0, \\ \log(x-1), & \alpha = 0, \\ 1, & \alpha < 0. \end{cases} \tag{3.3.15}$$

More precisely,

$$Q_n^{(\alpha,\beta)}(x) \simeq \begin{cases} 2^{\alpha-1} \dfrac{\Gamma(\alpha)\Gamma(n+\beta+1)}{\Gamma(n+\alpha+\beta+1)} (x-1)^{-\alpha}, & \alpha > 0, \\ \dfrac{1}{2} \log \dfrac{1}{x-1}, & \alpha = 0. \end{cases} \tag{3.3.16}$$

The behavior near $x = -1$ of $Q_n^{(\alpha,\beta)}(x)$ is similar.

Theorem 3.3.3 *Let α be an integer, $\alpha \geq 0$. We consider $Q_n^{(\alpha,\beta)}(x)$ (real and positive for $x > 1$) in the complex plane cut along the line $[-\infty, +1]$. Then*

$$Q_n^{(\alpha,\beta)}(x + i0) - Q_n^{(\alpha,\beta)}(x - i0) = (-1)^{\alpha-1}\pi i P_n^{(\alpha,\beta)}(x), \quad -1 < x < 1. \tag{3.3.17}$$

In general, if α and β are both integers, the function $Q_n^{(\alpha,\beta)}(x)$ is regular and single-valued in the whole plane cut along $[-1, +1]$.

The functions of the second kind satisfy the same recurrence formula as $P_n^{(\alpha,\beta)}(x)$, that is,

$$\begin{aligned} &2n(n + \alpha + \beta)(2n + \alpha + \beta - 2)Q_n^{(\alpha,\beta)}(x) \\ &\quad = (2n + \alpha + \beta - 1)\left\{(2n + \alpha + \beta)(2n + \alpha + \beta - 2)x + \alpha^2 - \beta^2\right\} Q_{n-1}^{(\alpha,\beta)}(x) \\ &\qquad - 2(n + \alpha - 1)(n + \beta - 1)(2n + \alpha + \beta)Q_{n-2}^{(\alpha,\beta)}(x), \quad n = 2, 3, 4, \dots. \end{aligned} \tag{3.3.18}$$

Moreover,

$$\begin{aligned} Q_1^{(\alpha,\beta)}(x) = \frac{1}{2}\left[(\alpha + \beta + 2)x + \alpha - \beta\right] Q_0^{(\alpha,\beta)}(x) \\ - 2^{\alpha+\beta-1}(\alpha + \beta + 2)\frac{\Gamma(\alpha + 1)\Gamma(\beta + 1)}{\Gamma(\alpha + \beta + 2)}(x - 1)^{-\alpha}(x + 1)^{-\beta}. \end{aligned} \tag{3.3.19}$$

The function $Q_n^{(\alpha,\beta)}(x)$ also satisfies the differential equation (3.1.17). Similarly, we have the Christoffel–Darboux formula

$$\begin{aligned} &\frac{k_n}{k_{n+1}} \frac{p_{n+1}(x)q_n(y) - p_n(x)q_{n+1}(y)}{x - y} \\ &\qquad = p_n(x)q_n(y) + \frac{k_{n-1}}{k_n} \frac{p_n(x)q_{n-1}(y) - p_{n-1}(x)q_n(y)}{x - y}. \end{aligned} \tag{3.3.20}$$

Here k_n denotes the coefficient of x^n in the orthonormal polynomial

$$p_n(x) = \frac{P_n^{(\alpha,\beta)}(x)}{\sqrt{h_n^{(\alpha,\beta)}}}$$

and q_n is given by (2.10.1). This formula also holds for $n = 0$ if we modify it as follows:

$$\frac{k_0}{k_1} \frac{p_1(x)q_0(y) - p_0(x)q_1(y)}{x - y} = p_0(x)q_0(y) + \text{const}\, \frac{q_{-1}(y)}{x - y}. \tag{3.3.21}$$

Adding, we obtain the following result:

$$\begin{aligned} &\sum_{\nu=0}^{n} \frac{2\nu + \alpha + \beta + 1}{2^{\alpha+\beta+1}} \frac{\Gamma(\nu + 1)\Gamma(\nu + \alpha + \beta + 1)}{\Gamma(\nu + \alpha + 1)\Gamma(\nu + \beta + 1)} P_\nu^{(\alpha,\beta)}(x)Q_\nu^{(\alpha,\beta)}(y) \\ &\quad = \frac{1}{2} \frac{(y - 1)^{-\alpha}(y + 1)^{-\beta}}{y - x} + \frac{2^{-\alpha-\beta}}{2n + \alpha + \beta + 2} \\ &\qquad \times \frac{\Gamma(n + 2)\Gamma(n + \alpha + \beta + 2)}{\Gamma(n + \alpha + 1)\Gamma(n + \beta + 1)} \frac{P_{n+1}^{(\alpha,\beta)}(x)Q_n^{(\alpha,\beta)}(y) - P_n^{(\alpha,\beta)}(x)Q_{n+1}^{(\alpha,\beta)}(y)}{x - y}. \end{aligned} \tag{3.3.22}$$

3.4 Routh–Jacobi Polynomials

In 1884, E. J. Routh (1884) observed that $\{P_n^{(a+ib,a-ib)}(ix)\}$ form a finite family of orthogonal polynomials when $a < -1$. Let

$$R_n(x;a,b) = (-i)^n P_n^{(-a+ib,-a-ib)}(ix), \tag{3.4.1}$$

then the R_n satisfy

$$\begin{aligned}(n+1)(a+n)(n+1-2a)&R_{n+1}(x;a,b)\\ &= [ab(2a-2n-1)+x(n-a)_2]\,R_n(x;a,b)\\ &\quad-(n+1-a)\left[(n-a)^2+b^2\right]R_{n-1}(x;a,b), \quad n>0,\end{aligned} \tag{3.4.2}$$

$$\int_{\mathbb{R}} R_m(x;a,b)R_n(x;a,b)\frac{\exp(2b\arctan x)}{(1+x^2)^a}\,dx = h_n(a,b)\,\delta_{m,n}, \tag{3.4.3}$$

$$h_n(a,b) := C\frac{(1-2a)(1-a+ib)_n(1-a-ib)_n}{n!(1-2a)_n(2n+1-2a)}, \tag{3.4.4}$$

for $m,n = 0,1,\ldots,N$, with $N =$ the largest integer $< a - 1/2$, $b \in \mathbb{R}$, and

$$C := \int_{\mathbb{R}} \frac{\exp(2b\arctan x)}{(1+x^2)^a}\,dx. \tag{3.4.5}$$

The differential equation (3.1.16) becomes

$$\frac{\exp(-2b\arctan x)}{(1+x^2)^{-a}}\frac{d}{dx}\left[\left(1+x^2\right)^{1-a}\exp(2b\arctan x)y_n'\right] = n(n-2a+1)y_n. \tag{3.4.6}$$

In other words, $y = R_n(x;a,b)$ solves

$$\left(1+x^2\right)y'' + 2[b+x(1-a)]y' - n(n+1-2a)y = 0. \tag{3.4.7}$$

The Routh–Jacobi polynomials with $b = 0$ were studied in Askey (1989a), where the orthogonality relation was proved by direct evaluation of integrals. It is clear that when $b = 0$, the weight function reduces to the probability density function of the student t-distribution.

3.5 Ultraspherical (Gegenbauer) Polynomials

The **ultraspherical polynomials** are

$$\begin{aligned}C_n^\nu(x) &= \frac{(2\nu)_n}{(\nu+1/2)_n}P_n^{(\nu-1/2,\nu-1/2)}(x)\\ &= \frac{(2\nu)_n}{n!}\,{}_2F_1\left(-n,n+2\nu;\nu+1/2;(1-x)/2\right)\\ &= \frac{(2\nu)_n}{n!}(-1)^n\,{}_2F_1\left(-n,n+2\nu;\nu+1/2;(1+x)/2\right),\end{aligned} \tag{3.5.1}$$

hence

$$C_n^\nu(x) = \frac{(\nu)_n}{n!}2^n x^n + \text{lower-order terms}. \tag{3.5.2}$$

Other representations are

$$\begin{aligned} C_{2n}^{\nu}(x) &= \frac{(\nu)_n}{(1/2)_n} P_n^{(\nu-1/2,-1/2)}\left(2x^2-1\right) \\ &= \frac{(2\nu)_n}{(2n)!} {}_2F_1\left(-n, n+\nu; \nu+1/2; 1-x^2\right) \\ &= (-1)^n \frac{(\nu)_n}{n!} {}_2F_1\left(-n, n+\nu; 1/2; x^2\right) \end{aligned} \tag{3.5.3}$$

and

$$\begin{aligned} C_{2n+1}^{\nu}(x) &= \frac{(\nu)_{n+1}}{(1/2)_{n+1}} x P_n^{(\nu-1/2,1/2)}\left(2x^2-1\right) \\ &= \frac{(2\nu)_{2n+1}}{(2n+1)!} x\, {}_2F_1\left(-n, n+\nu+1; \nu+1/2; 1-x^2\right) \\ &= 2(-1)^n \frac{(\nu)_{n+1}}{n!} {}_2F_1\left(-n, n+\nu+1; 1/2; x^2\right). \end{aligned} \tag{3.5.4}$$

The ultraspherical polynomials are also known as **Gegenbauer polynomials**. The **Legendre polynomials** $\{P_n(x)\}$ correspond to the choice $\nu = 1/2$. The ultraspherical polynomials $\{C_n^{\nu}(x)\}$ are the spherical harmonics on $\mathbb{R}^m$, $\nu = -1 + m/2$. In the case of ultraspherical polynomials, the **generating function** (3.2.1) is

$$\sum_{n=0}^{\infty} C_n^{\nu}(x) t^n = \left(1 - 2xt + t^2\right)^{-\nu}. \tag{3.5.5}$$

The three-term recurrence relation is

$$2x(n+\nu) C_n^{\nu}(x) = (n+1) C_{n+1}^{\nu}(x) + (n+2\nu-1) C_{n-1}^{\nu}(x). \tag{3.5.6}$$

The monic form is

$$x y_n(x) = y_{n+1}(x) + \frac{n(n+2\nu-1)}{4(n+\nu)(n+\nu-1)} y_{n-1}(x). \tag{3.5.7}$$

The orthogonality relation (3.1.2)–(3.1.3) becomes, for $\operatorname{Re}\nu > -1/2$ and $\nu \neq 0$,

$$\int_{-1}^{1} \left(1-x^2\right)^{\nu-1/2} C_m^{\nu}(x) C_n^{\nu}(x)\, dx = \frac{(2\nu)_n \sqrt{\pi}\, \Gamma(\nu+1/2)}{n!(n+\nu)\Gamma(\nu)} \delta_{m,n}. \tag{3.5.8}$$

The ultraspherical polynomials satisfy the **differential recursion relations**

$$\frac{d}{dx} C_n^{\nu}(x) = 2\nu C_{n-1}^{\nu+1}(x), \tag{3.5.9}$$

$$2\nu\left(1-x^2\right) \frac{d}{dx} C_n^{\nu}(x) = (n+2\nu) x C_{n-1}^{\nu}(x) - (n+1) C_{n+1}^{\nu}(x), \tag{3.5.10}$$

$$4\left(1-x^2\right) \frac{d}{dx} C_n^{\nu}(x) = \frac{(n+2\nu)(n+2\nu-1)}{\nu(n+\nu)} C_{n-1}^{\nu}(x) - \frac{n(n+1)}{\nu(n+\nu)} C_{n+1}^{\nu}(x), \tag{3.5.11}$$

$$\left(1-x^2\right)^{\nu-1/2} C_n^{\nu}(x) = \frac{(n-k)!(-2)^n (\nu)_k}{n!(2\nu+n)_k} \frac{d^k}{dx^k}\left[\left(1-x^2\right)^{\nu+k-1/2} C_{n-k}^{\nu+k}(x)\right]. \tag{3.5.12}$$

Moreover, we have the **recursion relations**

$$(n+1)C_{n+1}^{\nu}(x) = 2\nu x C_{n}^{\nu+1}(x) - 2\nu C_{n-1}^{\nu+1}(x), \tag{3.5.13}$$

$$(n+\nu)C_{n}^{\nu}(x) = \nu\left[C_{n}^{\nu+1} - C_{n-2}^{\nu+1}(x)\right]. \tag{3.5.14}$$

The **Rodrigues formula** is the case $k = n$ in (3.5.12), namely

$$\left(1-x^2\right)^{\nu-1/2} C_n^{\nu}(x) = \frac{(-1)^n(2\nu)_n}{2^n n!(\nu+1/2)_n}\frac{d^n}{dx^n}\left(1-x^2\right)^{\nu+n-1/2}. \tag{3.5.15}$$

The **ultraspherical differential equation** is

$$\left(1-x^2\right)y'' - x(2\nu+1)y' + n(n+2\nu)y = 0. \tag{3.5.16}$$

We also have

$$\begin{gathered}\frac{d^2u}{dx^2} - \left\{\frac{\nu^2-\nu-3/4}{4(1-x)^2} + \frac{\nu^2-\nu-3/4}{4(1+x)^2} + \frac{n(n+2\nu)+(\nu+1/2)^2}{1-x^2}\right\}u = 0,\\ u = u(x) = \left(1-x^2\right)^{(2\nu+1)/4} C_n^{\nu}(x)\end{gathered} \tag{3.5.17}$$

and

$$\begin{gathered}\frac{d^2u}{d\theta^2} - \left\{\frac{\nu^2-\nu}{4\sin^2\theta/2} + \frac{\nu^2-\nu}{4\cos^2\theta/2} + (n+2\nu)^2\right\}u = 0,\\ u = u(\theta) = [\sin(\theta/2)\cos(\theta/2)]^{\nu}\, C_n^{\nu}(\cos\theta).\end{gathered} \tag{3.5.18}$$

The ultraspherical polynomials have the **explicit representations**

$$C_n^{\nu}(\cos\theta) = \sum_{j=0}^{n}\frac{(\nu)_j(\nu)_{n-j}}{j!(n-j)!}e^{i(n-2j)\theta}, \tag{3.5.19}$$

$$C_n^{\nu}(x) = \sum_{k=0}^{\lfloor n/2\rfloor}\frac{(2\nu)_n x^{n-2k}\left(x^2-1\right)^k}{2^{2k}k!(\nu+1/2)_k(n-2k)!}, \tag{3.5.20}$$

$$C_n^{\nu}(x) = \sum_{k=0}^{\lfloor n/2\rfloor}\frac{(-1)^k(\nu)_{n-k}(2x)^{n-2k}}{k!(n-2k)!}. \tag{3.5.21}$$

Formulas (3.5.19) and (3.5.21) follow from the generating function (3.5.5) by writing its right-hand side as

$$\left(1-te^{i\theta}\right)^{-\nu}\left(1-te^{-i\theta}\right)^{-\nu} \quad \text{and} \quad (1+t^2)^{-\nu}\left(1-\frac{2xt}{1+t^2}\right)^{-\nu}$$

respectively. Formula (3.5.20) leads to the integral representation (**Laplace first integral**)

$$C_n^{\nu}(x) = \frac{(2\nu)_n\Gamma(\nu+1/2)}{n!\Gamma(1/2)\Gamma(\nu)}\int_0^{\pi}\left[x+\sqrt{x^2-1}\,\cos\varphi\right]^n \sin^{2\nu-1}\varphi\, d\varphi. \tag{3.5.22}$$

Fejer's generalization of the **Dirichlet–Mehler** integral representation is

$$C_n^\nu(\cos\theta) = \frac{1}{\pi}\int_0^\pi F(\phi)|2\cos\phi - 2\cos\theta|^{-\nu}d\phi, \qquad F(\phi) = \begin{cases}\cos(n+\nu)\phi & \text{if } 0 \le \phi < \theta,\\ \cos[(n+\nu)\phi - \nu\pi] & \text{if } \theta \le \phi < \pi.\end{cases} \tag{3.5.23}$$

The **connection relation** between two ultraspherical polynomials is

$$C_n^\gamma(x) = \sum_{k=0}^{\lfloor n/2\rfloor} \frac{(\gamma-\beta)_k(\gamma)_{n-k}}{k!(\beta+1)_{n-k}}\left(\frac{\beta+n-2k}{\beta}\right) C_{n-2k}^\beta(x), \tag{3.5.24}$$

while the **linearization of products** is

$$\begin{aligned} C_m^\nu(x)C_n^\nu(x) &= \sum_{k=0}^{m\wedge n} \frac{(m+n+\nu-2k)}{m+n+\nu-k}\frac{(\nu)_k}{k!}\\ &\times \frac{(\nu)_{m-k}(\nu)_{n-k}(2\nu)_{m+n-k}(m+n-2k)!}{(m-k)!(n-k)!(\nu)_{m+n-k}(2\nu)_{m+n-2k}} C_{m+n-2k}^\nu(x).\end{aligned} \tag{3.5.25}$$

The special case $\alpha = \beta$ of (3.1.33) is the **ultraspherical plane wave expansion**

$$e^{ixy} = \Gamma(\nu)(x/2)^{-\nu}\sum_{n=0}^\infty i^n(\nu+n)J_{\nu+n}(y)C_n^\nu(x), \tag{3.5.26}$$

which is equivalent to

$$J_{\nu+n}(z) = \frac{(-i)^n n!(z/2)^\nu}{\Gamma(\nu+1/2)\Gamma(1/2)(2\nu)_n}\int_{-1}^1 e^{izy}\left(1-y^2\right)^{\nu-1/2} C_n^\nu(y)\,dy, \tag{3.5.27}$$

for $\operatorname{Re}\nu > -1/2$. Formula (3.5.27) is called "Gegenbauer's generalization of Poisson's integral" in Watson (1944). Note that (3.5.27) can be proved directly from (3.5.1) and (1.2.32). The cases n even and n odd of (3.5.27) are

$$J_{\nu+2n}(z) = \frac{(-1)^n(2n)!(z/2)^\nu}{\Gamma(\nu+1/2)\Gamma(1/2)(2\nu)_{2n}}\int_0^\pi \cos(z\cos\phi)(\sin\varphi)^{2\nu}C_{2n}^\nu(\cos\varphi)\,d\varphi \tag{3.5.28}$$

and

$$J_{\nu+2n+1}(z) = \frac{(2n+1)!(z/2)^\nu}{\Gamma(\nu+1/2)\Gamma(1/2)(2\nu)_{2n+1}}\int_0^\pi \sin(z\cos\varphi)(\sin\varphi)^{2\nu}C_{2n+1}^\nu(\cos\varphi)\,d\varphi. \tag{3.5.29}$$

Additional **generating functions** are

$$\begin{aligned}&\sum_{n=0}^\infty \frac{C_n^\nu(x)}{(2\nu)_n(\nu+1/2)_n}t^n\\ &\quad = {}_0F_1(-;\lambda+1/2;(x-1)t/4)\,{}_0F_1(-;\lambda+1/2;(x+1)t/2),\end{aligned} \tag{3.5.30}$$

$$\sum_{n=0}^{\infty} \frac{(\gamma)_n}{(2\nu)_n} C_n^{\nu}(x) t^n = (1-xt)^{-\gamma} {}_2F_1\left(\left. \begin{matrix} \gamma/2, (\gamma+1)/2 \\ \nu+1/2 \end{matrix} \right| \frac{(x^2-1)t^2}{(1-xt)^2} \right), \tag{3.5.31}$$

$$\sum_{n=0}^{\infty} \frac{C_n^{\nu}(x)}{(2\nu)_n} t^n = e^{xt} {}_0F_1\left(-; \nu+1/2; \left(x^2-1\right) t^2/4\right), \tag{3.5.32}$$

$$\sum_{n=0}^{\infty} \frac{(\nu+1/2)_n}{(2\nu)_n} C_n^{\nu}(x) t^n = \frac{1}{R} \left(\frac{1+R-xt}{2} \right)^{-\nu+1/2} \tag{3.5.33}$$

and

$$\begin{aligned} &\sum_{n=0}^{\infty} \frac{(\gamma)_n (2\nu-\gamma)_n}{(2\nu)_n (\nu+1/2)_n} C_n^{\nu}(x) t^n \\ &= {}_2F_1\left(\left. \begin{matrix} \gamma, 2\nu-\gamma \\ \nu+1/2 \end{matrix} \right| \frac{1-R-t}{2} \right) {}_2F_1\left(\left. \begin{matrix} \gamma, 2\nu-\gamma \\ \nu+1/2 \end{matrix} \right| \frac{1-R+t}{2} \right), \end{aligned} \tag{3.5.34}$$

where $R = R(t)$ is defined in (3.2.6).

Theorem 3.5.1 *The ultraspherical polynomials satisfy the **(Gegenbauer) addition theorem***

$$\begin{aligned} &C_n^{\nu}(\cos\theta\cos\varphi + \sin\theta\sin\varphi\cos\psi) \\ &= \sum_{k=0}^{n} a_{k,n}^{\nu} (\sin\theta)^k C_{n-k}^{\nu+k}(\cos\theta) (\sin\varphi)^k C_{n-k}^{\nu+k}(\cos\varphi) C_k^{\nu-1/2}(\cos\psi), \end{aligned} \tag{3.5.35}$$

with

$$a_{k,n}^{\nu} = \frac{\Gamma(\nu-1/2)(\nu)_k}{\Gamma(2\nu+n+k)} \frac{(n-k)!\Gamma(2\nu+2k)}{\Gamma(\nu+k-1/2)}. \tag{3.5.36}$$

The next theorem is due to Beckenbach, Seidel, and Szász (1951); see also Karlin and Szegő (1960/1961, (14.3)) and Ismail (2005c, Theorem 5). The proof in Ismail (2005c) uses the idea of Section 2.12 and the Laplace first integral (3.5.22). Recall the definition of a Turánian in (2.12.3).

Theorem 3.5.2 *Let $\mathcal{C}_n^{\nu}(x) = C_n^{\nu}(x)/C_n^{\nu}(1)$, that is,*

$$\mathcal{C}_n^{\nu}(x) := \frac{n!}{(2\nu)_n} C_n^{\nu}(x). \tag{3.5.37}$$

Then

$$T(\mathcal{C}_0^{\nu}(x), \mathcal{C}_1^{\nu}(x), \ldots, \mathcal{C}_n^{\nu}(x)) = \left(\frac{x^2-1}{4} \right)^{n(n+1)/2} \prod_{j=1}^{n} \frac{j!(2\nu-1)_j}{(\nu-1/2)_j (\nu+1/2)_j}. \tag{3.5.38}$$

3.6 Chebyshev Polynomials

The Chebyshev polynomials of the first and second kinds are

$$T_n(x) = \cos(n\theta), \quad U_n(x) = \frac{\sin(n+1)\theta}{\sin\theta}, \quad x := \cos\theta, \tag{3.6.1}$$

respectively. Their orthogonality relations are

$$\int_{-1}^{1} T_m(x)T_n(x)\frac{dx}{\sqrt{1-x^2}} = \begin{cases} \frac{\pi}{2}\,\delta_{m,n}, & m \neq 0, \\ \pi\,\delta_{0,n} \end{cases} \tag{3.6.2}$$

and

$$\int_{-1}^{1} U_m(x)U_n(x)\sqrt{1-x^2}\,dx = \frac{\pi}{2}\,\delta_{m,n}. \tag{3.6.3}$$

Moreover,

$$\begin{aligned} T_n(x) &= \frac{n!}{(1/2)_n} P_n^{(-1/2,-1/2)}(x) = {}_2F_1(-n, n; 1/2; (1-x)/2), \\ U_n(x) &= \frac{(n+1)!}{(3/2)_n} P_n^{(1/2,1/2)}(x) = (n+1)\,{}_2F_1(-n, n+2; 3/2; (1-x)/2). \end{aligned} \tag{3.6.4}$$

Therefore,

$$\begin{aligned} T_n(x) &= 2^{n-1}x^n + \text{ lower-order terms}, \\ U_n(x) &= 2^n x^n + \text{ lower-order terms}. \end{aligned} \tag{3.6.5}$$

It is easy to see that

$$\mathcal{D}_q T_n(x) = \frac{q^{n/2} - q^{-n/2}}{q^{1/2} - q^{-1/2}} U_{n-1}(x), \tag{3.6.6}$$

and its $q \to 1$ limit is

$$\frac{d}{dx} T_n(x) = n U_{n-1}(x). \tag{3.6.7}$$

In terms of ultraspherical polynomials, the Chebyshev polynomials are

$$U_n(x) = C_n^1(x), \quad T_n(x) = \lim_{\nu \to 0} \frac{n+2\nu}{2\nu} C_n^\nu(x), \quad n \geq 0. \tag{3.6.8}$$

The polynomials $\{U_n(x)\}$ and $\{T_n(x)\}$ have the **generating functions**

$$\sum_{n=0}^{\infty} U_n(x)t^n = \frac{1}{1-2xt+t^2}, \quad \sum_{n=0}^{\infty} T_n(x)t^n = \frac{1-xt}{1-2xt+t^2}, \tag{3.6.9}$$

$$\sum_{n=0}^{\infty} \frac{T_n(x)}{n!} t^n = e^{xt} \cosh\left(t\sqrt{x^2-1}\right), \tag{3.6.10}$$

$$\sum_{n=0}^{\infty} \frac{U_n(x)}{(n+1)!} t^n = e^{xt} \frac{\sinh\left(t\sqrt{x^2-1}\right)}{\sqrt{x^2-1}}, \tag{3.6.11}$$

$$\sum_{n=0}^{\infty} \frac{T_n(x)}{(2n)!} t^n = \cosh\left(\sqrt{t(x+1)/2}\right) \cosh\left(\sqrt{t(x-1)/2}\right), \tag{3.6.12}$$

$$\sum_{n=0}^{\infty} \frac{U_n(x)}{(2n+2)!} t^n = \frac{\sinh\left(\sqrt{t(x+1)/2}\right) \sinh\left(\sqrt{t(x-1)/2}\right)}{\sqrt{x^2-1}\, t/2}. \tag{3.6.13}$$

It is clear that

$$\begin{aligned} T_n(z) &= \frac{1}{2}\left[\left(z+\sqrt{z^2-1}\right)^n + \left(z-\sqrt{z^2-1}\right)^n\right], \\ U_n(z) &= \frac{\left(z+\sqrt{z^2-1}\right)^{n+1} - \left(z-\sqrt{z^2-1}\right)^{n+1}}{2\sqrt{z^2-1}}. \end{aligned} \tag{3.6.14}$$

Formulas (3.5.20) and (3.6.8) yield

$$T_n(x) = \sum_{k=0}^{\lfloor n/2 \rfloor} \frac{(-n)_{2k}}{(2k)!} x^{n-2k} \left(x^2-1\right)^k, \tag{3.6.15}$$

$$U_n(x) = (n+1) \sum_{k=0}^{\lfloor n/2 \rfloor} \frac{(-n)_{2k}}{(2k+1)!} x^{n-2k} \left(x^2-1\right)^k. \tag{3.6.16}$$

The representations (3.6.15)–(3.6.16) also follow from (3.6.14). Both $U_n(x)$ and $T_n(x)$ satisfy the **three-term recurrence relation**

$$2xy_n(x) = y_{n+1}(x) + y_{n-1}(x), \quad n > 0, \tag{3.6.17}$$

with

$$T_0(x) = 1, \;\; T_1(x) = x, \qquad U_0(x) = 1, \;\; U_1(x) = 2x. \tag{3.6.18}$$

Formula (3.6.6) leads to a definition of a right inverse to $\mathcal{D}_q$. Let $\mathcal{D}_q f = g$ so that $f(x) \sim \sum_{n=1}^{\infty} f_n T_n(x)$, $g(x) \sim \sum_{n=0}^{\infty} g_n U_{n-1}(x)$, and

$$f_n = g_n \frac{\left(q^{1/2} - q^{-1/2}\right)}{\left(q^{n/2} - q^{-n/2}\right)}, \quad n > 0. \tag{3.6.19}$$

Using this we are led to the following definition

Definition 3.6.1 The inverse operator $\mathcal{D}_q^{-1}$ is defined as the integral operator

$$\left(\mathcal{D}_q^{-1} g\right)(\cos\theta) = \frac{1-q}{4\pi\sqrt{q}} \int_{-\pi}^{\pi} \frac{\vartheta_4'\left((\theta+\phi)/2, \sqrt{q}\right)}{\vartheta_4\left((\theta+\phi)/2, \sqrt{q}\right)} g(\cos\phi) \sin\phi \, d\phi \tag{3.6.20}$$

on the space $L^2[\left(1-x^2\right)^{1/2}, [-1,1]]$.

The kernel of the integral operator (3.6.20) is uniformly bounded when $(x, y) = (\cos\theta, \cos\phi) \in [-1, 1] \times [-1, 1]$. Thus the operator $\mathcal{D}_q^{-1}$ is well defined and bounded on $L^2[(1-x^2)^{1/2}, [-1, 1]]$. Furthermore, $\mathcal{D}_q^{-1}$ is a one-to-one mapping from $L^2[(1-x^2)^{1/2}, [-1, 1]]$ into $L^2[(1-x^2)^{-1/2}, [-1, 1]]$. Moreover,

$$\mathcal{D}_q\mathcal{D}_q^{-1} = \text{the identity operator} \tag{3.6.21}$$

on $L^2[(1-x^2)^{1/2}, [-1, 1]]$. The details of this construction are in Brown and Ismail (1995). See also Ismail (2005b, §16.1). One can also show that if $\breve{f}(z)$ is analytic in $q^{1/2} \le |z| \le q^{-1/2}$ then

$$\left(\mathcal{D}_q^{-1}\mathcal{D}_q - \mathcal{D}_q\mathcal{D}_q^{-1}\right) f(x) = \int_0^\pi f(\cos\theta)\, d\theta. \tag{3.6.22}$$

A variant of the Chebyshev polynomials is to study a solution of (3.6.17) with the initial conditions

$$u_0(x) = 1, \quad u_1(x) = ax + b. \tag{3.6.23}$$

One can show that $\{u_n(x)\}$ are orthogonal with respect to a positive measure μ, μ' supported on $[-1, 1]$, and μ_s has at most two masses and they are outside $[-1, 1]$; see Geronimus (1977). Moreover, $u_n(x)$ is a sum of at most three Chebyshev polynomials. One can generalize this further by defining v_m as a solution of (3.6.17) for $n > m$ and

$$v_m(x) := \phi(x), \quad v_{m+1}(x) := \psi(x). \tag{3.6.24}$$

Here ϕ, ψ have degrees m, $m + 1$, respectively, and have real simple and interlacing zeros. The Wendroff theorem, Theorem 2.5.5, shows that $v_n(x)$ can be completed to an orthogonal polynomial system. Indeed, one can show that

$$v_{n+m}(x) = \phi(x)\, T_n(x) + [\psi(x) - x\phi(x)]\, U_{n-1}(x). \tag{3.6.25}$$

Recall the definition of a resultant in Section 2.9. Dilcher and Stolarsky (2005) proved that

$$\begin{aligned} &\operatorname{Res}\{U_n(x) + kU_{n-1}(x), U_{n-1}(x) + hU_{n-2}(x)\} \\ &\quad = (-1)^{\binom{n}{2}} 2^{n(n-1)} h^n \left[U_n\left(\frac{1+kh}{2h}\right) - kU_{n-1}\left(\frac{1+kh}{2h}\right)\right]. \end{aligned} \tag{3.6.26}$$

Gishe and Ismail (2008) gave a simple proof of (3.6.26) and gave the following extensions of it.

Theorem 3.6.2 (Gishe and Ismail, 2008) *Let*

$$T_n(x; a, k) := T_n(x) + (ax + k)T_{n-1}(x), \tag{3.6.27}$$

$$U_n(x; a, k) := U_n(x) + (ax + k)U_{n-1}(x), \tag{3.6.28}$$

and

$$f(x) := 1 + (bx + h)(2x + ax + k). \tag{3.6.29}$$

Then the following evaluations of resultants hold:

$$\begin{aligned} &\mathrm{Res}\{T_n(x) + kT_{n-1}(x), T_{n-1}(x) + hT_{n-2}(x)\} \\ &\quad = \frac{2^{n^2-3n+3}}{(-1)^{n(n-1)/2}} h^n \left[T_n((1+hk)/(2h)) - kT_{n-1}((1+hk)/(2h))\right], \end{aligned} \tag{3.6.30}$$

$$\mathrm{Res}\{U_n(x;a,k), U_{n-1}(x;b,h)\} = \frac{(-1)^{\binom{n}{2}}}{(2+a)^{n+1}} 2^{(n-1)(n-2)} \,\mathrm{Res}\,(f(x), U_n(x;a,k)), \tag{3.6.31}$$

and

$$\mathrm{Res}\,\{T_n(x;a,k), T_{n-1}(x;b,h)\} = (-1)^{n(n-1)/2} \frac{2^{n^2-4n+3}}{(a+2)^2} \,\mathrm{Res}\,(f, T_n(x;a,k)). \tag{3.6.32}$$

Theorem 3.6.3 *Let E denote the closure of the area enclosed by an ellipse whose foci are at ± 1. Then $\max\{|T_n(x)| : x \in E\}$ is attained at the right of the major axis. Moreover, the same property holds for the ultraspherical polynomials $C_n^\nu(x)$ for $\nu \geq 0$.*

Ortiz and Rivlin (1983) superimposed the graphs of Chebyshev polynomials of different degrees and observed many interesting patterns. One would expect the same picture if we plotted the graphs of polynomials orthonormal on $[-1, 1]$ with respect to a weight w with

$$\int_{-1}^{1} \log w(x) \frac{dx}{\sqrt{1-x^2}} > \ \infty.$$

3.7 Legendre Polynomials

Since the Legendre polynomials are special cases of the Jacobi and ultraspherical polynomials we will mostly state their properties with minimal commentary. The hypergeometric representations are

$$\begin{aligned} P_n(x) &= P_n^{(0,0)}(x) = C_n^{1/2}(x) \\ &= {}_2F_1(-n, n+1; 1; (1-x)/2) = {}_2F_1(-n, n+1; 1; (1+x)/2). \end{aligned} \tag{3.7.1}$$

Hence the leading term is

$$P_n(x) = \frac{(1/2)_n}{n!} 2^n x^n + \text{lower-order terms.} \tag{3.7.2}$$

A **generating function** is

$$\sum_{n=0}^{\infty} P_n(x) t^n = \left(1 - 2xt + t^2\right)^{-1/2} \tag{3.7.3}$$

the **three-term recurrence relation** is

$$x(2n+1)P_n(x) = (n+1)P_{n+1}(x) + nP_{n-1}(x), \tag{3.7.4}$$

and its monic form is

$$xy_n(x) = y_{n+1}(x) + \frac{n^2}{(2n-1)(2n+1)} y_{n-1}(x). \tag{3.7.5}$$

The orthogonality relation is

$$\int_{-1}^{1} P_m(x) P_n(x)\, dx = \frac{2}{2n+1}\, \delta_{m,n}. \tag{3.7.6}$$

Some **differential recursion relations** are

$$\frac{d}{dx} P_n(x) = C_{n-1}^{3/2}(x), \tag{3.7.7}$$

$$\left(1 - x^2\right) \frac{d}{dx} P_n(x) = (n+1)xP_{n-1}(x) - (n+1)P_{n+1}(x), \tag{3.7.8}$$

$$\left(1 - x^2\right) \frac{d}{dx} P_n(x) = \frac{n(n+1)}{(2n+1)} P_{n-1}(x) - \frac{n(n+1)}{2n+1} P_{n+1}(x), \tag{3.7.9}$$

$$P_n^{\nu}(x) = \frac{(n-k)!(-2)^n (1/2)_k}{(n+k)!} \frac{d^k}{dx^k} \left[\left(1 - x^2\right)^k C_{n-k}^{k+1/2}(x)\right]. \tag{3.7.10}$$

The **Rodrigues formula** is

$$P_n(x) = \frac{(-1)^n}{2^n n!} \frac{d^n}{dx^n} \left(1 - x^2\right)^n. \tag{3.7.11}$$

The **Legendre differential equation** is

$$\left(1 - x^2\right) y'' - 2xy' + n(n+1)y = 0. \tag{3.7.12}$$

Transformed equations are

$$\begin{gathered} \frac{d^2u}{dx^2} + \left\{ \frac{1}{4(1-x)^2} + \frac{1}{4(1+x)^2} + \frac{n(n+1)+1}{1-x^2} \right\} u = 0, \\ u = u(x) = \left(1 - x^2\right)^{1/2} P_n(x) \end{gathered} \tag{3.7.13}$$

and

$$\begin{gathered} \frac{d^2u}{d\theta^2} + \left\{ \frac{1}{16 \sin^2 \theta/2} + \frac{1}{16 \cos^2 \theta/2} + (n+1)^2 \right\} u = 0, \\ u = u(\theta) = [\sin(\theta/2) \cos(\theta/2)]^{1/2} P_n(\cos\theta). \end{gathered} \tag{3.7.14}$$

There are the **explicit representations**

$$P_n(\cos\theta) = \sum_{j=0}^{n} \frac{(1/2)_j(1/2)_{n-j}}{j!(n-j)!} e^{i(n-2j)\theta}, \tag{3.7.15}$$

$$P_n(x) = \sum_{k=0}^{\lfloor n/2 \rfloor} \frac{n!x^{n-2k}\left(x^2-1\right)^k}{2^{2k}(k!)^2(n-2k)!}, \tag{3.7.16}$$

$$P_n(x) = \sum_{k=0}^{\lfloor n/2 \rfloor} \frac{(-1)^k(n-k)!(2x)^{n-2k}}{k!(n-2k)!}. \tag{3.7.17}$$

The **linearization of product** formula is

$$\begin{aligned} P_m(x)P_n(x) = \sum_{k=0}^{m\wedge n} & \frac{(m+n-2k+1/2)}{m+n-k+1/2}\frac{(1/2)_k}{k!} \\ & \times \frac{(1/2)_{m-k}(1/2)_{n-k}(m+n-k)!}{(m-k)!(n-k)!(1/2)_{m+n-k}} P_{m+n-2k}(x). \end{aligned} \tag{3.7.18}$$

The **Laplace first integral** is

$$P_n(x) = \frac{1}{\pi}\int_0^{\pi} \left[x + \sqrt{x^2-1}\cos\varphi\right]^n d\varphi. \tag{3.7.19}$$

The **Dirichlet–Mehler** integral representation is

$$P_n(\cos\theta) = \frac{2}{\pi}\int_0^{\theta} \frac{\cos(n+1/2)\phi}{\sqrt{\cos\phi - 2\cos\theta}}\, d\phi \tag{3.7.20}$$

$$= \frac{2}{\pi}\int_{\theta}^{\pi} \frac{\sin(n+1/2)\phi}{\sqrt{\cos\theta - 2\cos\phi}}\, d\phi. \tag{3.7.21}$$

The **addition theorem for Legendre polynomials** is the special case $\nu = 1/2$ of (3.5.35).

The **function of the second kind** associated with Legendre polynomials has the integral representation (Laplace integral)

$$Q_n(z) = \int_0^{\infty} \left\{z + \left(z^2-1\right)^{1/2}\cosh\theta\right\}^{-n-1} d\theta, \tag{3.7.22}$$

$n = 0, 1, \ldots,$ where $z + (z^2-1)^{1/2}\cosh\theta$ has its principal value when $\theta \neq 0$.

3.8 Laguerre and Hermite Polynomials

In the older literature (for example, Bateman, 1932), Laguerre polynomials were called Sonine polynomials. Askey (1975a) pointed out that the Hermite, Laguerre (Sonine), Jacobi, and Hahn polynomials are not named after the first person to define or use them.

The Laguerre polynomials $\{L_n^{(\alpha)}(x)\}$ and the Hermite polynomials $\{H_n(x)\}$ are the following limiting cases of Jacobi polynomials:

$$L_n^{(\alpha)}(x) = \lim_{\beta\to\infty} P_n^{(\alpha,\beta)}(1 - 2x/\beta), \tag{3.8.1}$$

$$H_n(x) = 2^n n! \lim_{\alpha\to\infty} \alpha^{-n/2} P_n^{(\alpha,\alpha)}(x/\sqrt{\alpha}\,). \tag{3.8.2}$$

The Laguerre polynomials have the properties

$$L_n^{(\alpha)}(x) = \frac{(\alpha+1)_n}{n!}\,{}_1F_1(-n;\alpha+1;x) = \frac{(\alpha+1)_n}{n!}\sum_{k=0}^{n}\binom{n}{k}\frac{(-x)^k}{(\alpha+1)_k}, \tag{3.8.3}$$

$$\int_0^\infty x^\alpha e^{-x} L_m^{(\alpha)}(x) L_n^{(\alpha)}(x)\,dx = \frac{\Gamma(a+n+1)}{n!}\,\delta_{m,n}, \quad \mathrm{Re}(\alpha) > -1, \tag{3.8.4}$$

$$L_n^{(\alpha)}(x) = \frac{(-1)^n}{n!}x^n + \text{lower-order terms}. \tag{3.8.5}$$

The Laguerre polynomials arise from the Ψ function as

$$\begin{aligned} L_n^{(\alpha)}(x) &= \frac{(-1)^n}{n!}\Psi(-n, n+1; x), \\ \Psi(a, a+n+1; x) &= (a)_n x^{-a-n} L_n^{(-a-n)}(x). \end{aligned} \tag{3.8.6}$$

We shall use the notation

$$L_n(x) = L_n^{(0)}(x). \tag{3.8.7}$$

The Hermite polynomials satisfy

$$H_n(x) := \sum_{k=0}^{\lfloor n/2\rfloor} \frac{n!(-1)^k(2x)^{n-2k}}{k!(n-2k)!} = 2^n x^n + \text{lower-order terms}, \tag{3.8.8}$$

$$\int_{\mathbb{R}} H_m(x)H_n(x)e^{-x^2}\,dx = 2^n n!\sqrt{\pi}\,\delta_{m,n}. \tag{3.8.9}$$

Moreover,

$$H_{2n}(x) = (-1)^n 2^{2n} n! L_n^{(-1/2)}\left(x^2\right), \tag{3.8.10}$$

$$H_{2n+1}(x) = (-1)^n 2^{2n+1} n! x L_n^{(1/2)}\left(x^2\right). \tag{3.8.11}$$

Furthermore,

$$L_n^{(\alpha)}(0) = \frac{(\alpha+1)_n}{n!}, \quad H_{2n+1}(0) = 0, \quad H_{2n}(0) = (-1)^n 4^n (1/2)_n. \tag{3.8.12}$$

Formulas (3.8.3) and (3.8.8) have the inverses

$$x^n = \sum_{k=0}^{n} \frac{(-1)^k n!(\alpha+1)_n}{(n-k)!(\alpha+1)_k} L_k^{(\alpha)}(x), \tag{3.8.13}$$

$$x^n = \frac{n!}{2^n} \sum_{k=0}^{\lfloor n/2 \rfloor} \frac{1}{k!} \frac{H_{n-2k}(x)}{(n-2k)!}, \tag{3.8.14}$$

respectively. The Hermite and Laguerre polynomials are related via

$$H_n(x) = 2^n n! \lim_{\beta\to\infty} \beta^{-n} L_n^{(\beta^2/2)}\left(-\beta x + \beta^2/2\right). \tag{3.8.15}$$

Moreover,

$$(1-x)^n/n! = \lim_{\alpha\to\infty} \alpha^{-n} L_n^{(\alpha)}(\alpha x). \tag{3.8.16}$$

A combinatorial proof of (3.8.15) is in Labelle and Yeh (1989).

Formula (3.8.3) implies

$$\frac{d}{dx} L_n^{(\alpha)}(x) = -L_{n-1}^{(\alpha+1)}(x), \tag{3.8.17}$$

and the adjoint relation is

$$\frac{d}{dx}\left[x^{\alpha+1}e^{-x}L_n^{(\alpha+1)}(x)\right] = (n+1)x^\alpha e^{-x} L_{n+1}^{(\alpha)}(x). \tag{3.8.18}$$

Combining (3.8.17) and (3.8.18) we establish the factored differential equation

$$\frac{d}{dx}\left[x^{\alpha+1}e^{-x}\frac{d}{dx}L_n^{(\alpha)}(x)\right] + nx^\alpha e^{-x} L_n^{(\alpha)}(x) = 0. \tag{3.8.19}$$

In other words, $y = L_n^{(\alpha)}(x)$ is a solution to the **Laguerre differential equation**

$$xy'' + (1+\alpha-x)y' + ny = 0. \tag{3.8.20}$$

Three transformed equations are

$$\begin{aligned} &xz'' + (x+1)z' + \left(n+1+\frac{\alpha}{2}-\frac{\alpha^2}{4x}\right)z = 0, \quad z = e^{-x}x^{\alpha/2}L_n^{(\alpha)}(x), \\ &u'' + \left(\frac{n+(\alpha+1)/2}{x} + \frac{1-\alpha^2}{4x^2} - \frac{1}{4}\right)u = 0, \quad u = e^{-x/2}x^{(\alpha+1)/2}L_n^{(\alpha)}(x), \\ &v'' + \left(4n+2\alpha+2-x^2-\frac{\alpha^2-1/4}{x^2}\right)v = 0, \quad v = e^{-x^2/2}x^{(\alpha+1/2)}L_n^{(\alpha)}(x^2). \end{aligned} \tag{3.8.21}$$

Another recurrence relation is

$$(x+1)L_{n+1}^{(\alpha-1)}(x) = (\alpha-x)L_n^{(\alpha)}(x) - xL_{n-1}^{(\alpha+1)}(x). \tag{3.8.22}$$

Equation (3.8.18) leads to

$$L_n^{(\alpha)}(x) = \frac{(n-k)!}{n!} x^{-\alpha} e^x \frac{d^k}{dx^k}\left[x^{\alpha+k}e^{-x}L_{n-k}^{(\alpha+k)}(x)\right]. \tag{3.8.23}$$

In particular, we have the **Rodrigues formula**

$$L_n^{(\alpha)}(x) = \frac{1}{n!} x^{-\alpha} e^x \frac{d^n}{dx^n} \left[x^{\alpha+n} e^{-x}\right]. \tag{3.8.24}$$

Similarly, from (3.8.8) one derives

$$\frac{d}{dx} H_n(x) = 2n H_{n-1}(x), \tag{3.8.25}$$

and (3.8.9) gives the adjoint relation

$$H_{n+1} = -e^{x^2} \frac{d}{dx} \left[e^{-x^2} H_n(x)\right]. \tag{3.8.26}$$

The Hermite differential equation is

$$e^{x^2} \frac{d}{dx} \left[e^{-x^2} \frac{d}{dx} H_n(x)\right] + 2n H_n(x) = 0. \tag{3.8.27}$$

Indeed,

$$y'' - 2xy' + 2ny = 0, \quad y = H_n(x), \tag{3.8.28}$$

$$z'' + \left(2n + 1 - x^2\right) z = 0, \quad z = e^{-x^2/2} H_n(x). \tag{3.8.29}$$

Furthermore, (3.8.26) leads to

$$H_n(x) = (-1)^k e^{x^2} \frac{d^k}{dx^k} \left[e^{-x^2} H_{n-k}(x)\right] \tag{3.8.30}$$

and the case $k = n$ is the **Rodrigues formula**

$$H_n(x) = (-1)^n e^{x^2} \frac{d^n}{dx^n} e^{-x^2}. \tag{3.8.31}$$

The **three-term recurrence relations** of the Laguerre and Hermite polynomials are

$$\begin{aligned} x L_n^{(\alpha)}(x) = {} & -(n+1) L_{n+1}^{(\alpha)}(x) \\ & + (2n + \alpha + 1) L_n^{(\alpha)}(x) - (n + \alpha) L_{n-1}^{(\alpha)}(x) \end{aligned} \tag{3.8.32}$$

and

$$2x H_n(x) = H_{n+1}(x) + 2n H_{n-1}(x), \tag{3.8.33}$$

respectively. Their monic recursion relations are

$$x y_n(x) = y_{n+1}(x) + \frac{n}{2} y_{n-1}(x), \quad y_n(x) = 2^{-n} H_n(x), \tag{3.8.34}$$

$$x y_n(x) = y_{n+1}(x) + (2n + \alpha + 1) y_n(x) + n(n + \alpha) y_{n-1}(x), \quad y_n(x) = \frac{(-1)^n}{n!} L_n^{(\alpha)}(x). \tag{3.8.35}$$

The **discriminants** of Hermite and Laguerre polynomials are given by (2.9.15) and (2.9.16), respectively.

The **linearization of products** of Hermite polynomials is

$$H_m(x)H_n(x) = \sum_{k=0}^{m\wedge n} \frac{m!n!2^k}{k!(m-k)!(n-k)!} H_{m+n-2k}(x). \tag{3.8.36}$$

The linearization of products of Laguerre polynomials is related to the derangement problem (Even and Gillis, 1976).

The Hermite and Laguerre polynomials have the **generating functions**

$$\sum_{n=0}^{\infty} \frac{H_n(x)}{n!} t^n = \exp\left(2xt - t^2\right), \tag{3.8.37}$$

$$\sum_{n=0}^{\infty} \frac{H_{n+k}(x)}{n!} t^n = \exp\left(2xt - t^2\right) H_k(x-t), \tag{3.8.38}$$

$$\sum_{n=0}^{\infty} \frac{L_n^{(\alpha)}(x)}{(\alpha+1)_n} t^n = e^t \, {}_0F_1(-;\alpha+1;-xt), \tag{3.8.39}$$

$$\sum_{n=0}^{\infty} \frac{(c)_n}{(\alpha+1)_n} L_n^{(\alpha)}(x) t^n = (1-t)^{-c} \, {}_1F_1(c;1+\alpha;-xt/(1-t)), \tag{3.8.40}$$

$$\sum_{n=0}^{\infty} \frac{(n+k)!}{n!k!} L_{n+k}^{(\alpha)}(x) t^n = (1-t)^{-\alpha-1-k} \exp\left(-\frac{xt}{1-t}\right) L_k^{(\alpha)}\left(\frac{x}{1-t}\right). \tag{3.8.41}$$

The additional **generating functions**

$$\sum_{n=0}^{\infty} \frac{H_{2n}(x)t^n}{2^{2n}n!} = (1+t)^{-1/2} \exp\left(x^2 t/(1+t)\right), \tag{3.8.42}$$

$$\sum_{n=0}^{\infty} \frac{H_{2n+1}(x)t^n}{x2^{2n+1}n!} = (1+t)^{-3/2} \exp\left(x^2 t/(1+t)\right), \tag{3.8.43}$$

$$\sum_{n=0}^{\infty} \frac{(-1)^n}{(2n)!} t^n H_{2n}(x) = e^t \cos\left(2x\sqrt{t}\right), \tag{3.8.44}$$

$$\sum_{n=0}^{\infty} \frac{(-1)^n}{(2n+1)!} t^n H_{2n+1}(x) = e^t \frac{\sin\left(2x\sqrt{t}\right)}{\sqrt{t}}, \tag{3.8.45}$$

follow from (3.8.10), (3.8.11), and (3.8.37). Moreover, one can show that

$$\sum_{n=0}^{\infty} \frac{H_n(x)}{2^n \lfloor n/2 \rfloor} t^n = \frac{1+xt+t^2}{(1+t^2)^{3/2}} \exp\left(\frac{x^2t^2}{1+t^2}\right), \tag{3.8.46}$$

$$\sum_{n=0}^{\infty} \frac{(\gamma)_n}{(2n)!} H_{2n}(x) t^n = \left(1+t^2\right)^{-\gamma} {}_1F_1\left(\left.\begin{matrix}\gamma \\ 1/2\end{matrix}\right| \frac{x^2t^2}{1+t^2}\right), \tag{3.8.47}$$

$$\sum_{n=0}^{\infty} \frac{(\gamma+1/2)_n}{(2n+1)!} H_{2n+1}(x) t^n = \frac{xt}{\sqrt{1+t^2}} \, {}_1F_1\left(\left.\begin{matrix}\gamma+1/2 \\ 3/2\end{matrix}\right| \frac{x^2t^2}{1+t^2}\right), \tag{3.8.48}$$

for arbitrary γ.

The following scaled expansions, or **multiplication formulas**, hold:

$$L_n^{(\alpha)}(cx) = (\alpha+1)_n \sum_{k=0}^{n} \frac{c^k(1-c)^{n-k}}{(n-k)!(\alpha+1)_k} L_k^{(\alpha)}(x), \tag{3.8.49}$$

$$H_n(cx) = \sum_{k=0}^{\lfloor n/2 \rfloor} \frac{n!(-1)^k}{k!(n-2k)!} \left(1-c^2\right)^k c^{n-2k} H_{n-2k}(x). \tag{3.8.50}$$

Observe that (3.8.49) implies that for $c \geq 0$, the coefficients of $L_k^{\alpha}(x)$ in the expansion of $L_n^{(\alpha)}(x)$ are positive if and only if $c < 1$. Formula (3.8.50) has a similar interpretation. We also have the **connection relations**

$$\begin{aligned} C_n^{\nu}(x) &= \sum_{k=0}^{\lfloor n/2 \rfloor} {}_2F_0(-k, \nu+n-k; —; 1) \frac{(-1)^k (\nu)_{n-k}}{k!(n-2k)!} H_{n-2k}(x), \\ \frac{H_n(x)}{n!} &= \sum_{k=0}^{\lfloor n/2 \rfloor} \frac{(-1)^k (\nu+n-2k)}{k!(\nu)_{n+1-2k}} {}_1F_1(-k; \nu+n+1-2k; 1) C_{n-2k}^{\nu}(x). \end{aligned} \tag{3.8.51}$$

The effect of a translation on Hermite and Laguerre polynomials is

$$L_n^{(\alpha)}(x+y) = \sum_{k,m=0}^{n} L_k^{(\alpha)}(x) L_m^{(\alpha)}(y) \frac{(-\alpha-1)_{n-k-m}}{(n-k-m)!}, \tag{3.8.52}$$

$$L_n^{(\alpha+\beta+1)}(x+y) = \sum_{k=0}^{n} L_k^{(\alpha)}(x) L_{n-k}^{(\beta)}(y), \tag{3.8.53}$$

$$H_n(x+y) = \sum_{k,s=0}^{n} \frac{H_s(x)}{s!} \frac{H_{n-2k-s}(y)}{(n-2k-s)!} \frac{1}{k!}, \tag{3.8.54}$$

where the sum is over $k, s \geq 0$, with $2k + s \leq n$. Additional results are

$$H_n(x\cos\alpha + y\sin\alpha) = n! \sum_{k=0}^{n} \frac{H_k(x) H_{n-k}(y)}{k!(n-k)!} \cos^k\alpha \sin^{n-k}\alpha, \tag{3.8.55}$$

$$L_n(x^2+y^2) = (-1)^n 2^{-2n} \sum_{k=0}^{n} \frac{H_{2k}(x) H_{2n-2k}(y)}{k!(n-k)!}. \tag{3.8.56}$$

A dual to (3.8.53) is

$$\frac{L_{m+n}^{(\alpha)}(x)}{L_{m+n}^{(\alpha)}(0)} = \frac{\Gamma(\alpha+1)}{\Gamma(\beta+1)\Gamma(\alpha-\beta)} \int_0^1 t^{\alpha}(1-t)^{\alpha-\beta-1} \frac{L_m^{(\alpha)}(xt)}{L_m^{(\alpha)}(0)} \frac{L_n^{(\alpha-\beta-1)}(x(1-t))}{L_n^{(\alpha-\beta-1)}(0)}\, dt \tag{3.8.57}$$

for $\operatorname{Re}\alpha > -1$, $\operatorname{Re}\alpha > \operatorname{Re}\beta$, and is due to Feldheim (Andrews, Askey, and Roy, 1999, §6.2). Formulas (3.8.53) and (3.8.57) are analogues of Sonine's second integral (1.2.26) since

$$\lim_{n\to\infty} n^{-\alpha} L_n^{(\alpha)}\left(\frac{x^2}{4n}\right) = (2/x)^{\alpha} J_{\alpha}(x). \tag{3.8.58}$$

A generalization of (3.8.53) was proved in Van der Jeugt (1997) and was further generalized in Koelink and Van der Jeugt (1998).

The Hermite and Laguerre polynomials have the moment integral representations

$$\frac{H_n(ix)}{(2i)^n} = \frac{1}{\sqrt{\pi}} \int_{-\infty}^{\infty} e^{-(y-x)^2} y^n \, dy, \tag{3.8.59}$$

$$n! L_n^{\alpha}(x) = x^{-\alpha/2} \int_0^{\infty} e^{x-y} y^{n+\alpha/2} J_\alpha \left(2\sqrt{xy}\right) dy, \tag{3.8.60}$$

valid for $n = 0, 1, \ldots,$ and $\alpha > -1$. **Integral relations** connecting Hermite and Laguerre polynomials are

$$\int_0^t L_n(x(t-x)) \, dx = \frac{(-1)^n H_{2n+1}(t/2)}{2^{2n}(3/2)_n}, \tag{3.8.61}$$

$$\int_0^t \frac{H_{2n}\left(\sqrt{x(t-x)}\right)}{\sqrt{x(t-x)}} \, dx = (-1)^n \pi 2^{2n} (1/2)_n L_n\left(t^2/4\right), \tag{3.8.62}$$

where $L_n(x)$ is as in (3.8.7). Moreover,

$$L_n^{(\alpha)}(x) = \frac{(-1)^n \Gamma(n+\alpha+1)}{\Gamma(\alpha+1/2)\sqrt{\pi}\,(2n)!} \int_1^1 \left(1-t^2\right)^{\alpha-1/2} H_{2n}\left(tx^{1/2}\right) dt, \tag{3.8.63}$$

for $\alpha > -1/2$. Furthermore, we have the inverse relations

$$e^{-x/2} L_n(x) = \frac{2^{1-n}}{n!\sqrt{\pi}} \int_0^{\infty} e^{-t^2} H_n^2(t) \cos\left(\sqrt{xt}\right) dt, \tag{3.8.64}$$

$$e^{-x^2} H_n^2(x) = \frac{2^n n!}{\sqrt{\pi}} \int_0^{\infty} e^{-s^2/4} L_n(s^2/2) \cos(sx) \, ds. \tag{3.8.65}$$

The Hermite polynomials have the **operational representation**

$$\exp\left((-1/4)\partial_x^2\right)(2x)^n = H_n(x). \tag{3.8.66}$$

The Hermite functions $\{e^{-x^2/2} H_n(x)\}$ are the eigenfunctions of the Fourier transform,

$$e^{-x^2/2} H_n(x) = \frac{i^{-n}}{\sqrt{2\pi}} \int_{\mathbb{R}} e^{ixy} e^{-y^2/2} H_n(y) \, dy, \tag{3.8.67}$$

$n = 0, 1, \ldots$. In other words,

$$e^{-x^2} H_n(x) = \frac{(-1)^{\lfloor n/2 \rfloor}}{\sqrt{\pi}} 2^{n+1} \int_0^\infty e^{-t^2} t^n \cos(2xt)\, dt, \quad n \text{ even}, \tag{3.8.68}$$

$$e^{-x^2} H_n(x) = \frac{(-1)^{\lfloor n/2 \rfloor}}{\sqrt{\pi}} 2^{n+1} \int_0^\infty e^{-t^2} t^n \sin(2xt)\, dt, \quad n \text{ odd}. \tag{3.8.69}$$

The Laguerre functions are also eigenfunctions of the Hankel transform. Indeed,

$$e^{-x/2} x^{\alpha/2} L_n^{(\alpha)}(x) = \frac{(-1)^n}{2} \int_0^\infty J_\alpha(\sqrt{xy})\, e^{-y/2} y^{\alpha/2} L_n^{(\alpha)}(y)\, dy, \tag{3.8.70}$$

for $\alpha > -1$, $n = 0, 1, \ldots$. Another eigenfunction relation is

$$x^\alpha e^{-x} \left[L_n^{(\alpha)}(x)\right]^2 = \int_0^\infty J_{2\alpha}(2\sqrt{xy})\, y^\alpha e^{-y} \left[L_n^{(\alpha)}(y)\right]^2 dy, \tag{3.8.71}$$

$\alpha > -1/2$, and its special case

$$e^{-x^2} \frac{H_{2n+1}^2(\sqrt{x})}{\sqrt{x}} = \int_0^\infty J_1(2\sqrt{xy})\, e^{-y} \frac{H_{2n+1}^2(\sqrt{y})}{\sqrt{y}}\, dy. \tag{3.8.72}$$

Other integral relations are

$$P_n(x) = \frac{2}{n!\sqrt{\pi}} \int_0^\infty \exp\left(-t^2\right) t^n H_n(xt)\, dt, \tag{3.8.73}$$

$$\int_x^\nu e^{-x/2} J_\nu(ax) L_n^{(2\nu)}(x)\, dx = \frac{2^\nu \Gamma(\nu + 1/2)}{\sqrt{\pi a}} (\sin\theta)^{\nu+1/2} C_n^{\nu+1/2}(\cos\theta), \tag{3.8.74}$$

for $a > 0$, $\nu > -1/2$, $\cos\theta = (a^2 - 1/4)/(a^2 + 1/4)$. Moreover, Erdélyi (1938) proved that

$$\begin{aligned} &\int_0^\infty J_{\alpha+\beta}(2\sqrt{xy})\, e^{-y} y^{(\alpha+\beta)/2} L_m^{(\alpha)}(y) L_n^{(\beta)}(y)\, dy \\ &\quad = (-1)^{m+n} e^{-x} x^{(\alpha+\beta)/2} L_m^{(\beta+n-m)}(x) L_n^{(\alpha+m-n)}(x), \end{aligned} \tag{3.8.75}$$

for $\text{Re}(\alpha + \beta) > -1$.

The **arithmetic properties of the zeros** of Laguerre polynomials have been studied since the early part of the twentieth century. Schur proved that $\{L_m^{(0)}(x)\}$ are irreducible over the rationals for $m > 1$, and later proved the same result for $\{L_m^{(1)}(x)\}$ (Schur, 1929, 1931). Recently, Filaseta and Lam (2002) proved that $L_m^{(\alpha)}(x)$ is irreducible over the rationals for all but finitely many m, when α is rational but is not a negative integer.

We now state **positivity results**. Let

$$A^{(\alpha)}(n_1,\ldots,n_k;\mu) := \int_0^\infty \frac{x^\alpha e^{-\mu x}}{\Gamma(\alpha+1)} L_{n_1}^{(\alpha)}(x)\cdots L_{n_k}^{(\alpha)}(x)\,dx, \tag{3.8.76}$$

with $\alpha > -1$. It is known that

$$(-1)^{n_1+\cdots+n_k} A^{(\alpha)}(n_1,\ldots,n_k;1) \geq 0; \quad \alpha = 0,1,\ldots, \tag{3.8.77}$$

$\alpha = 0$ is in Even and Gillis (1976), and general α is in Askey and Ismail (1976). Moreover, Askey and Gasper (1977) established the inequality

$$A^{(\alpha)}(n_1,n_2,n_3;2) \geq 0 \quad \text{if and only if} \quad \alpha \geq \left(\sqrt{17}-5\right)/2. \tag{3.8.78}$$

Another proof is in Gillis, Reznick, and Zeilberger (1983). A combinatorial proof for nonnegative integer α is in Ismail and Tamhankar (1979). The special case $k = 3$ and $\alpha = 3$ goes back to Szegő (1926). A key to the analysis is the generating function

$$\begin{aligned}\sum_{j=1}^{k}\sum_{n_j=0}^{\infty} A^{(\alpha)}(n_1,\ldots,n_k;\mu)\,t_1^{n_1}\cdots t_k^{n_k} &= \prod_{j=1}^{k}(1-t_j)^{-\alpha-1}\left[\mu+\sum_{j=1}^{k} t_j/(1-t_j)\right]^{-\alpha-1}\\ &= \left[\mu+\sum_{j=1}^{k}(-1)^j(\mu-j)e_j\right]^{-\alpha-1},\end{aligned} \tag{3.8.79}$$

where e_j is the jth elementary symmetric function of $t_1,\ldots,t_k$. Combinatorial techniques were introduced in Askey and Ismail (1976) and Ismail and Tamhankar (1979).

For $k > 2$ let

$$C^{(\alpha)}(n_1,\ldots,n_k,b_1,\ldots,b_k) = \int_0^\infty \frac{x^\alpha e^{-x}}{\Gamma(\alpha+1)} \prod_{j=1}^{k} L_{n_j}^{(\alpha)}(b_j x)\,dx. \tag{3.8.80}$$

Koornwinder (1978) studied the case $k = 3$, $b_1 = 1$, $b_2 + b_3 = 1$, $b_1 \geq 0$, $b_2 \geq 0$. Askey, Ismail, and Koornwinder (1978) treated the general case.

Theorem 3.8.1 *We have the generating function*

$$\begin{aligned}G^{(\alpha)}(b_1,\ldots,b_k;t_1\cdots t_k) &:= \sum_{n_1,\ldots,n_k=0}^{\infty} C^{(\alpha)}(n_1,\ldots,n_k;b_1,\ldots,b_k)\,t_1^{n_1}\cdots t_k^{n_k}\\ &= \left[\prod_{j=1}^{k}\left(1-t_j\right)+\sum_{l=1}^{k} b_l \prod_{j=1,j\neq l}^{k}\left(1-t_j\right)\right]^{-\alpha-1}.\end{aligned} \tag{3.8.81}$$

The following combinatorial result is due to Askey, Ismail, and Koornwinder.

Theorem 3.8.2 (Askey, Ismail, and Koornwinder, 1978) *Let*

$$a_{ii} = 1 - b_i, \quad a_{ij} = -\sqrt{b_i b_j}, \quad i \neq j. \tag{3.8.82}$$

Then the coefficient of $t_1^{n_1} t_2^{n_2} \cdots t_k^{n_k}$ *in the expansion*

$$\prod_{i=1}^{k} \left(\sum_{j=1}^{k} a_{ij} t_j \right)^{n_i} \tag{3.8.83}$$

is $C^{(0)}(n_1, n_2, \ldots, n_k; b_1, b_2, \ldots, b_k)$.

Theorem 3.8.3 (Koornwinder, 1978) *The inequality*

$$C^{(\alpha)}(\ell, m, n; \lambda, 1-\lambda, 1) \geq 0 \tag{3.8.84}$$

holds for $\alpha \geq 0$, $\lambda \in [0,1]$, *with strict inequality if* $\ell = 0$ *and* $\lambda \in (0,1)$.

A generating function for integrals of products of Hermite polynomials is

$$\sum_{n_1,\ldots,n_k=0}^{\infty} \frac{t_1^{n_1} \cdots t_k^{n_k}}{n_1! \cdots n_k!} \int_{\mathbb{R}} \frac{e^{-x^2}}{\sqrt{\pi}} H_{n_1}(x) \cdots H_{n_k}(x)\, dx = \exp\left(2 \sum_{1 \leq i < j \leq k} t_i t_j \right). \tag{3.8.85}$$

Therefore the integrals

$$2^{-(n_1+\cdots+n_k)/2} \int_{\mathbb{R}} \frac{e^{-x^2}}{\sqrt{\pi}} H_{n_1}(x) \cdots H_{n_k}(x)\, dx \tag{3.8.86}$$

are nonnegative integers. One special case is

$$\int_{\mathbb{R}} e^{-x^2} H_m^2(x) H_{2n}(x)\, dx = \frac{2^{m+n}(2n)!(m!)^2 \sqrt{\pi}}{(m-n)!(n!)^2}, \tag{3.8.87}$$

for $0 \leq n \leq m$, $m = 0, 1, \ldots$. The integral is zero when $m < n$. A combinatorial interpretation is in Azor, Gillis, and Victor (1982) and a q-analogue is in Ismail, Stanton, and Viennot (1987).

3.9 The Complex Hermite Polynomials

The complex Hermite polynomials are the usual Hermite polynomials $\{H_n(x+iy)\}$, $x, y \in \mathbb{R}$ when viewed as functions of the two variables x and y. Many of the algebraic properties of Hermite polynomials $\{H_n(x)\}$ stated in the previous sections hold when x is taken as a complex variable.

Theorem 3.9.1 *Assume that*

$$0 < a < b, \quad \frac{1}{a} = 1 + \frac{1}{b}. \tag{3.9.1}$$

Then the complex Hermite polynomials satisfy the orthogonality relation (Karp, 2001; van Eijndhoven and Meyers, 1990)

$$\int_{\mathbb{R}^2} H_m(x+iy) H_n(x-iy) e^{-ax^2-by^2}\, dx\, dy = \frac{\pi}{\sqrt{ab}} 2^n n! \left(\frac{a+b}{ab} \right)^n \delta_{m,n}, \tag{3.9.2}$$

as well as the orthogonality relations (Gorska, 2016)

$$\begin{aligned}&\int_{\mathbb{R}^2} H_m(x+iy)H_n(x+iy)e^{-ax^2-by^2}\,dx\,dy\\&\quad=\int_{\mathbb{R}^2} H_m(x-iy)H_n(x-iy)e^{-ax^2-by^2}\,dx\,dy=\frac{\pi}{\sqrt{ab}}2^n n!\,\delta_{m,n}.\end{aligned} \tag{3.9.3}$$

The proof follows from the generating function (3.8.37).

The weight function $e^{-ax^2-by^2}$ on $\mathbb{R}$ has the moments

$$\begin{aligned}&\int_{\mathbb{R}^2}(x\pm iy)^{2n+1}e^{-ax^2-by^2}\,dx\,dy=0,\\&\int_{\mathbb{R}^2}(x\pm iy)^{2n}e^{-ax^2-by^2}\,dx\,dy=\frac{\pi}{\sqrt{ab}}(1/2)_n,\end{aligned} \tag{3.9.4}$$

provided that condition (3.9.1) is satisfied.

The combinatorics of the complex Hermite polynomials were studied in Ismail and Simeonov (2015). The combinatorics led to the more general orthogonality relation

$$\int_{\mathbb{R}^2} H_m(\alpha x+\beta y)H_n(\gamma x+\delta y)e^{-ax^2-by^2}\,dx\,dy=\frac{\pi}{\sqrt{ab}}2^n n!\left(\frac{\alpha\gamma}{a}+\frac{\beta\delta}{b}\right)^n\delta_{m,n}, \tag{3.9.5}$$

where

$$a>0,\quad b>0,\quad \frac{\alpha^2}{a}+\frac{\beta^2}{b}=\frac{\gamma^2}{a}+\frac{\delta^2}{b}=1. \tag{3.9.6}$$

It is clear that the orthogonality relations (3.9.2) and (3.9.3) are special cases of (3.9.5).

Theorem 3.9.2 *Suppose that we have two colored multisets. The first, of color I, has size* $\mathbf{m}=(m_1,m_2,\ldots,m_k)\in\mathbb{N}_0^k$ *and the second, of color II, has size* $\mathbf{n}=(n_1,n_2,\ldots,n_k)\in\mathbb{N}_0^k$. *We match elements from different sets, and to each pair of matched elements we assign weight* 1 *if the elements have the same color and weight* $1/a+1/b$ *if the elements have different colors.*

Then the total weight of all perfect matchings of this type is the number $B(\mathbf{m},\mathbf{n})$ *defined by*

$$B(\mathbf{m},\mathbf{n})=\frac{\sqrt{ab}}{\pi}2^{-\frac12(|\mathbf{m}|+|\mathbf{n}|)}\int_{\mathbb{R}^2}\prod_{j=1}^{k}\left[H_{m_j}(x+iy)H_{n_j}(x-iy)\right]e^{-ax^2-by^2}\,dx\,dy. \tag{3.9.7}$$

The special case $k=1$ of Theorem 3.9.2 is the orthogonality relation (3.9.2).

Theorem 3.9.3 (Ismail and Simeonov, 2015) *The complex Hermite polynomials are eigenfunctions of a 2-dimensional Fourier transform in the sense that*

$$\frac{\sqrt{ab}}{\pi\sqrt{2}}\int_{\mathbb{R}^2} e^{iz(\xi-i\eta)+(z^2+(\xi-i\eta)^2)/2} H_n(\xi+i\eta)e^{-a\xi^2-b\eta^2}\,d\xi\,d\eta = \left(\frac{a+b}{ab}i\right)^n H_n(z), \tag{3.9.8}$$

$$\frac{\sqrt{ab}}{\pi\sqrt{2}}\int_{\mathbb{R}^2} e^{iz(\xi-i\eta)+(z^2+(\xi-i\eta)^2)/2} H_n(\xi-i\eta)e^{-a\xi^2-b\eta^2}\,d\xi\,d\eta = i^n H_n(z), \tag{3.9.9}$$

for $n \geq 0$, and a and b satisfy conditions (3.9.1).

3.10 Hermite Functions

The Hermite function $H_\lambda(t)$ can be defined (see for example Hayman and Ortiz, 1975/76) by

$$H_\lambda(t) = -\frac{\sin \pi\lambda \Gamma(1+\lambda)}{2\pi}\sum_{n=0}^{\infty}\frac{\Gamma((n-\lambda)/2)}{\Gamma(n+1)}(-2t)^n \tag{3.10.1}$$

or, in terms of the confluent hypergeometric functions, by (Durand, 1975, 1978)

$$\begin{aligned} H_\lambda(t) = \frac{2^\lambda}{\sqrt{\pi}}\Bigg[&\cos\frac{\lambda\pi}{2}\Gamma\left(\frac{\lambda}{2}+\frac{1}{2}\right){}_1F_1\left(-\frac{\lambda}{2},\frac{1}{2};t^2\right) \\ &+ 2t\sin\frac{\lambda\pi}{2}\Gamma\left(\frac{\lambda}{2}+1\right){}_1F_1\left(-\frac{\lambda}{2}+\frac{1}{2},\frac{3}{2};t^2\right)\Bigg]. \end{aligned} \tag{3.10.2}$$

See also Elbert and Muldoon (1999). In the case that λ is a nonnegative integer, formula (3.10.1) is to be understood in a limiting sense so that $H_\lambda(t)$ reduces to the Hermite polynomial.

We note also that

$$H_\lambda(t) = 2^\lambda \Psi\left(-\frac{\lambda}{2},\frac{1}{2};t^2\right),$$

with the notation of Section 1.2, or Erdélyi et al. (1953a, p. 257) for confluent hypergeometric functions.

For each fixed λ, $H_\lambda(t)$ is the solution of the Hermite differential equation

$$y'' - 2ty' + 2\lambda y = 0 \tag{3.10.3}$$

which grows relatively slowly as $t \to +\infty$. We consider also a solution of (3.10.3) which is linearly independent of $H_\lambda(z)$:

$$\begin{aligned} G_\lambda(t) = \frac{2^\lambda}{\sqrt{\pi}}\Bigg[&-\sin\frac{\lambda\pi}{2}\Gamma\left(\frac{\lambda+1}{2}\right){}_1F_1\left(\frac{-\lambda}{2},\frac{1}{2};t^2\right) \\ &+ 2t\cos\frac{\lambda\pi}{2}\Gamma\left(\frac{\lambda+2}{2}\right){}_1F_1\left(\frac{-\lambda+1}{2},\frac{3}{2};t^2\right)\Bigg]. \end{aligned} \tag{3.10.4}$$

The functions $e^{-t^2/2}H_\lambda(t)$ and $e^{-t^2/2}G_\lambda(t)$ are linearly independent solutions of the modified Hermite equation

$$y'' + \left(2\lambda+1-t^2\right)y = 0. \tag{3.10.5}$$

From Elbert and Muldoon (1999, §5), the Wronskian of $e^{-t^2/2}H_\lambda(t)$ and $e^{-t^2/2}G_\lambda(t)$ is given by

$$W = \pi^{-1/2}2^{\lambda+1}\Gamma(\lambda+1). \tag{3.10.6}$$

For $\lambda > -1$, we have a formula due to Durand (1975):

$$\frac{\pi\Gamma(\lambda+1)}{2^{\lambda+1}}e^{-x^2}\left[H_\lambda^2(x)+G_\lambda^2(x)\right] = \int_0^\infty e^{-(2\lambda+1)\tau+x^2\tanh\tau}\frac{d\tau}{\sqrt{\sinh\tau\cosh\tau}}. \tag{3.10.7}$$

Note that (3.10.7) is a **Nicholson-type formula**, for the Hermite functions, analogous to that for sums of squares of Bessel functions (Watson, 1944, p. 444, (1)). The right-hand side of (3.10.7) is an absolutely monotonic function of x^2, and $H_\lambda^2(x)+G_\lambda^2(x)$ a completely monotonic function of x^2 on $(0,\infty)$, for $\lambda > -1$; see Widder (1941, Ch. 4) for information on absolutely and completely monotonic functions. In addition, we see that, for each fixed $x > 0$, the right-hand side of (3.10.7) is a completely monotonic function of λ on $(-1,\infty)$.

3.11 Multilinear Generating Functions

The Poisson kernel for Hermite polynomials is (3.11.4). It is a special case of the Kibble–Slepian formula (Kibble, 1945; Slepian, 1972), which will be stated as Theorem 3.11.1. Recent proofs are in Louck (1981). An interesting combinatorial proof was given by Foata (1981). A slight modification of Louck's proof (Louck, 1981) is in Ismail (2005b).

For an $n\times n$ matrix $S=(s_{ij})$ the Euclidean norm is

$$\|S\| = \left(\sum_{i,j=1}^n |s_{ij}|^2\right)^{1/2}.$$

Theorem 3.11.1 (Kibble–Slepian) *Let $S=(s_{ij})$ be an $n\times n$ real symmetric matrix, and assume that $\|S\|<1$, I being an identity matrix. Then*

$$\begin{aligned}&[\det(I+S)]^{-1/2}\exp\left(\mathbf{x}^T S(I+S)^{-1}\mathbf{x}\right)\\ &= \sum_K\left[\prod_{1\le i\le j\le n}\left(s_{ij}/2\right)^{k_{ij}}/k_{ij}!\right]2^{-\operatorname{tr}K}H_{k_1}(x_1)\cdots H_{k_n}(x_n),\end{aligned} \tag{3.11.1}$$

where $K=(k_{ij})$, $1\le i,j\le n$, $k_{ij}=k_{ji}$, and

$$\operatorname{tr}K := \sum_{i=1}^n k_{ii}, \quad k_i := k_{ii}+\sum_{j=1}^n k_{ij}, \quad i=1,\ldots,n. \tag{3.11.2}$$

In (3.11.1) $\sum_K$ denotes the $(n(n+1)/2)$-fold sum over $k_{ij}=0,1,\ldots$ for all positive integers i,j such that $1\le i\le j\le n$.

In the case $n = 2$, and $s_{11} = a$, $s_{12} = s_{21} = t$, $s_{22} = b$, (3.11.1) becomes

$$\sum_{j,k,l=0}^{\infty} \frac{a^j t^k b^l}{j!k!l!} 2^{-2j-k-2l} H_{2j+k}(x) H_{k+2l}(y) = \frac{\exp\left(x^2\left(a+ab-t^2\right)+y^2\left(b+ab-t^2\right)+2txy\right)}{\sqrt{1+a+b+ab-t^2}}. \tag{3.11.3}$$

When $a = b = 0$ then (3.11.3) becomes the **Mehler formula**

$$\sum_{n=0}^{\infty} \frac{H_n(x)H_n(y)}{2^n n!} t^n = \left(1-t^2\right)^{-1/2} \exp\left[\left(2xyt - x^2t^2 - y^2t^2\right)/\left(1-t^2\right)\right]. \tag{3.11.4}$$

The Mehler formula is essentially the **Poisson kernel for Hermite polynomials**. The bilinear generating function

$$\sum_{n=0}^{\infty} \frac{n! r^n}{(\alpha+1)_n} L_n^{(\alpha)}(x) L_n^{(\alpha)}(y) = (1-r)^{-\alpha-1} \exp\left(-r(x+y)/(1-r)\right) {}_0F_1\left(-;\alpha+1; xyr/(1-r)^2\right) \tag{3.11.5}$$

is called the **Hille–Hardy formula** and is a constant multiple of the **Poisson kernel for Laguerre polynomials**. A general multilinear generating function for Laguerre polynomials and confluent hypergeometric functions was given in Foata and Strehl (1981). It generalizes an old result of A. Erdélyi. Other related and more general generating functions are in Koelink and Van der Jeugt (1998).

Note that the right-hand side of the Mehler formula (3.11.4) is positive for all $x, y \in \mathbb{R}$ and $t \in (-1, 1)$, while the right-hand side of the Hille–Hardy formula (3.11.5) is positive for $x, y \in (0, \infty)$ and $t \in (0, 1)$. The left-hand side of the Kibble–Slepian formula is also positive for $x_1, \dots, x_n \in \mathbb{R}$ and all S in a neighborhood of $S = 0$ defined by $\det(I + S) > 0$.

Theorem 3.11.2 (Tyan and Thomas, 1975; Tyan, Derin, and Thomas, 1976) *Let $\{p_n\}$ be orthonormal with respect to μ and assume that $f(x,y) \ge 0$, $\mu \times \mu$ almost everywhere, where $f(x,y) := \sum_{n=0}^{\infty} c_n p_n(x) p_n(y)$:*

(i) *If the support of μ is unbounded to the right and left, then there exists a positive measure ν such that $c_n = \int_{-1}^{1} t^n \, d\nu(t)$.*
(ii) *If the support of μ is unbounded and contained in $[0, \infty)$, then there exists a positive measure ν such that $c_n = \int_0^1 t^n \, d\nu(t)$.*

Case (i) of Theorem 3.11.2 for Hermite polynomials is in Sarmanov and Bratoeva (1967), while case (ii) in for Laguerre polynomials is in Sarmanov (1968).

It is clear that the sequences $\{c_n\}$ which make $f(x,y) \ge 0$, for $x, y \in \mathbb{R}$, form a convex subset of ℓ^2 which we shall denote by C_1. The extreme points of this set are moment sequences when μ is a singleton, that is, $c_n = t^n$ for some $t \in (-1, 1)$. In other words, Mehler's formula corresponds to the cases when $\{c_n\}$ is an extreme point of C_1. Similarly, in the Hille–Hardy formula, $\{c_n\}$ is an extreme point of the set of $\{c_n\}$, $\{c_n\} \in \ell^2$, and $f(x,y) \ge 0$ for all $x \ge 0$,

$y \geq 0$. The bilinear formulas for Jacobi or ultraspherical polynomials have a more complicated structure.

The special case $y = -1$ of (3.2.16) and (3.2.17) are (3.2.1) and (3.2.2), respectively.

Remark 3.11.3 It is important to note that (3.2.17) is essentially the Poisson kernel for Jacobi polynomials and is positive when $t \in [0, 1)$, and $x, y \in [-1, 1]$ when $\alpha > -1, \beta > -1$. The kernel in (3.2.16) is also positive for $t \in [0, 1)$, and $x, y \in [-1, 1]$, but in addition to $\alpha > -1$, $\beta > -1$ we also require $\alpha + \beta + 1 \geq 0$. One can generate other positive kernels by integrating (3.2.16) or (3.2.17) with respect to positive measures supported on subsets of $[0, 1]$, provided that both sides are integrable and interchanging summation and integration is justified. Taking nonnegative combinations of these kernels also produces positive kernels.

A substitute in the case of Jacobi polynomials is the following.

Theorem 3.11.4 *Let $\alpha \geq \beta$ and either $\beta \geq -1/2$ or $\alpha \geq -\beta, \beta > -1$, and assume $\sum_{n=0}^{\infty} |a_n| < \infty$. Then*

$$f(x, y) = \sum_{n=0}^{\infty} a_n \frac{P_n^{(\alpha,\beta)}(x)}{P_n^{(\alpha,\beta)}(1)} \frac{P_n^{(\alpha,\beta)}(y)}{P_n^{(\alpha,\beta)}(1)} \geq 0, \quad 1 \leq x,\ y \leq 1, \tag{3.11.6}$$

if and only if

$$f(x, 1) \geq 0, \quad x \in [-1, 1]. \tag{3.11.7}$$

When $\alpha \geq \beta \geq -1/2$, this follows from Theorem 3.12.3. Gasper (1972) proved the remaining cases when $-1 < \beta < -1/2$. The remaining cases, namely $\alpha = -\beta = 1/2$ and $\alpha = \beta = -1/2$, are easy. When $\alpha = \beta$, Weinberger proved Theorem 3.11.4 from a maximum principle for hyperbolic equations. The conditions on α, β in Theorem 3.11.4 are best possible (Gasper, 1972). For applications to discrete Banach algebras (convolution structures), see Gasper (1971). Theorem 3.11.4 gives the positivity of the generalized translation operator associated with Jacobi series.

In the case of ultraspherical polynomials, the following slight refinement is in Bochner (1954).

Theorem 3.11.5 *The inequality*

$$f_r(x, y) := \sum_{n=0}^{\infty} r^n a_n \frac{n + \nu}{\nu} C_n^{\nu}(x) C_n^{\nu}(y) \geq 0 \tag{3.11.8}$$

holds for all $x, y \in [-1, 1]$, $0 \leq r < 1$, and $\nu > 0$ if and only if

$$a_n = \int_{-1}^{1} \frac{C_n^{\nu}(x)}{C_n^{\nu}(1)} \, d\alpha(x),$$

for some positive measure α.

Theorem 3.11.6 *Let $\{p_n(x)\}$ be orthonormal on a compact set E with respect to a probability measure μ. Then*

$$\lim_{r\to 1^-} \int_E P_r(x,y) f(y)\, d\mu(y) = f(x) \tag{3.11.9}$$

for all $f \in C[E]$. Moreover, for a given f, the convergence is uniform on E.

A proof is in Ismail (2005b).

3.12 Integral Representations

We next give **integral representations** of Jacobi polynomials. The integrals involving the nth power of a function are similar in structure to the Laplace integral (3.5.22) but are double integrals. Proofs are in Ismail (2005b).

The following **contour integral representation** follows from (3.1.7):

$$P_n^{(\alpha,\beta)}(x) = \frac{1}{2\pi i} \int_C [1 + (x+1)z/2]^{n+\alpha} [1 + (x-1)z/2]^{n+\beta} \frac{dz}{z^{n+1}}, \tag{3.12.1}$$

where C is a closed contour such that the points $-2(x \pm 1)^{-1}$ are exterior to C and $x \neq \pm 1$.

Theorem 3.12.1 *The integral representation (Braaksma and Meulenbeld, 1971)*

$$\begin{aligned} P_n^{(\alpha,\beta)}(x) = {} & \frac{2^n \Gamma(\alpha+n+1)\Gamma(\beta+n+1)}{\pi \Gamma(\alpha+1/2)\Gamma(\beta+1/2)(2n)!} \\ & \times \int_0^\pi \int_0^\pi \left[i\sqrt{1-x}\, \cos\phi + \sqrt{1+x}\, \cos\psi \right]^{2n} \\ & \times (\sin\phi)^{2\alpha} (\sin\psi)^{2\beta}\, d\phi\, d\psi \end{aligned} \tag{3.12.2}$$

holds for $\operatorname{Re}\alpha > -1/2$, $\operatorname{Re}\beta > -1/2$. *Another Laplace-type integral is (Askey, 1975b)*

$$\frac{P_n^{(\alpha,\beta)}(x)}{P_n^{(\alpha,\beta)}(1)} = \int_0^\pi \int_0^1 \left[\frac{1+x-(1-x)r^2}{2} + i\sqrt{1-x^2}\, r\cos\varphi \right]^n d\mu_{\alpha,\beta}(r,\varphi), \tag{3.12.3}$$

where

$$\begin{aligned} d\mu_{\alpha,\beta}(r,\varphi) &:= c_{\alpha,\beta} \left(1-r^2\right)^{\alpha-\beta-1} r^{2\beta+1} (\sin\varphi)^{2\beta}\, dr\, d\varphi, \\ c_{\alpha,\beta} &:= 2\Gamma(\alpha+1) / \left[\sqrt{\pi}\, \Gamma(\alpha-\beta)\Gamma(\beta+1/2) \right], \end{aligned} \tag{3.12.4}$$

which holds for $\alpha > \beta > -1/2$.

Theorem 3.12.2 *We have the integral representation*

$$\frac{P_n^{(\alpha,\beta)}(x)}{P_n^{(\alpha,\beta)}(1)} = \frac{2\Gamma(\alpha+1)n!}{\Gamma(1+(\alpha+\beta)/2)\Gamma((\alpha-\beta)/2)(\alpha+\beta+1)_n} \times \int_0^1 u^{\alpha+\beta+1}\left(1-u^2\right)^{-1+(\alpha-\beta)/2} C_n^{(\alpha+\beta+1)/2}\left(1+u^2(x-1)\right) du, \tag{3.12.5}$$

valid for $\operatorname{Re}(\alpha) > \operatorname{Re}(\beta)$, $\operatorname{Re}(\alpha+\beta) > -2$, *where* $\{C_n^\nu(x)\}$ *are the ultraspherical polynomials* (3.5.1).

The integral representations in (3.12.5) are important because every integral representation for C_n^ν will lead to a double integral representation for Jacobi polynomials. Indeed, the Laplace first integral, (3.5.22), implies

$$\frac{P_n^{(\alpha,\beta)}(x)}{P_n^{(\alpha,\beta)}(1)} = \frac{2\Gamma(\alpha+1)}{\Gamma(1/2)\Gamma((\alpha-\beta)/2)} \int_0^1\int_0^\pi r^{\alpha+\beta+1}\left(1-r^2\right)^{-1+(\alpha-\beta)/2}(\sin\phi)^{\alpha+\beta} \times \left[1-r^2(1-x)+ir\cos\phi\sqrt{(1-x)(2-r^2(1-x))}\right]^n d\phi\, du, \tag{3.12.6}$$

for $\operatorname{Re}\alpha > \operatorname{Re}\beta$, and $\operatorname{Re}(\alpha+\beta) > -2$.

The Laplace first integral, (3.5.22), implies

$$\frac{P_n^{(\alpha,\beta)}(x)}{P_n^{(\alpha,\beta)}(1)} = \frac{2\Gamma(\alpha+1)}{\Gamma(1/2)\Gamma((\alpha-\beta)/2)} \int_0^1\int_0^\pi r^{\alpha+\beta+1}\left(1-r^2\right)^{-1+(\alpha-\beta)/2}(\sin\phi)^{\alpha+\beta} \times \left[1-r^2(1-x)+ir\cos\phi\sqrt{(1-x)(2-r^2(1-x))}\right]^n d\phi\, du, \tag{3.12.7}$$

for $\operatorname{Re}\alpha > \operatorname{Re}\beta$, and $\operatorname{Re}(\alpha+\beta) > -2$.

Another Laplace-type integral is

$$\frac{P_n^{(\alpha,\beta)}(x)}{P_n^{(\alpha,\beta)}(1)} = \int_0^\pi\int_0^1 \left[\frac{1+x-(1-x)r^2}{2} + i\sqrt{1-x^2}\, r\cos\varphi\right]^n d\mu_{\alpha,\beta}(r,\varphi), \tag{3.12.8}$$

where

$$\begin{aligned} d\mu_{\alpha,\beta}(r,\varphi) &:= c_{\alpha,\beta}\left(1-r^2\right)^{\alpha-\beta-1} r^{2\beta+1}(\sin\varphi)^{2\beta}\, dr\, d\varphi, \\ c_{\alpha,\beta} &:= 2\Gamma(\alpha+1)/\left[\sqrt{\pi}\,\Gamma(\alpha-\beta)\Gamma(\beta+1/2)\right], \end{aligned} \tag{3.12.9}$$

which holds for $\alpha > \beta > -1/2$.

Theorem 3.12.3 *The Jacobi polynomials have the product formula*

$$\frac{P_n^{(\alpha,\beta)}(x)P_n^{(\alpha,\beta)}(y)}{P_n^{(\alpha,\beta)}(1)} = \int_0^\pi\int_0^1 P_n^{(\alpha,\beta)}\left[\frac{1}{2}(1+x)(1+y)+\frac{1}{2}(1-x)(1-y)r^2 + \sqrt{(1-x^2)(1-y^2)}\,r\cos\varphi - 1\right]d\mu_{\alpha,\beta}(r,\varphi), \tag{3.12.10}$$

where $\mu_{\alpha,\beta}$ *is defined in* (3.12.9).

In the case of ultraspherical polynomials, the representation (3.12.10) reduces to a single integral because the Laplace first integral for the ultraspherical polynomial is a single integral; see (3.5.22). Indeed,

$$\frac{C_n^\nu(x)C_n^\nu(y)}{C_n^\nu(1)} = \frac{\Gamma(\nu+1/2)}{\sqrt{\pi}\,\Gamma(\nu)}\int_0^\pi C_n^\nu\left(xy+\sqrt{(1-x^2)(1-y^2)}\cos\varphi\right)(\sin\varphi)^{2\nu-1}\,d\varphi. \tag{3.12.11}$$

3.13 Asymptotics

Darboux's method applied to (3.2.6) establishes the following theorem.

Theorem 3.13.1 *Let* $\alpha,\beta\in\mathbb{R}$ *and set*

$$N = n+(\alpha+\beta+1)/2, \quad \gamma = -(\alpha+1/2)\pi/2.$$

Then for $0<\theta<\pi$,

$$P_n^{(\alpha,\beta)}(\cos\theta) = \frac{k(\theta)}{\sqrt{n}}\cos(N\theta+\gamma)+O\left(n^{-3/2}\right), \tag{3.13.1}$$

where

$$k(\theta) = \frac{1}{\sqrt{\pi}}[\sin(\theta/2)]^{-\alpha-1/2}[\cos(\theta/2)]^{-\beta-1/2}. \tag{3.13.2}$$

Furthermore, the error bound holds uniformly for $\theta\in[\epsilon,\pi-\epsilon]$ *and fixed* $\epsilon>0$.

The next theorem is a Mehler–Heine-type formula for Jacobi polynomials.

Theorem 3.13.2 *Let* $\alpha,\beta\in\mathbb{R}$. *Then*

$$\lim_{n\to\infty} n^{-\alpha}P_n^{(\alpha,\beta)}(\cos(z/n)) = \lim_{n\to\infty} n^{-\alpha}P_n^{(\alpha,\beta)}\left(1-z^2/2n^2\right) = (z/2)^{-\alpha}J_\alpha(z). \tag{3.13.3}$$

The limit in (3.13.3) *is uniform in* z *on compact subsets of* $\mathbb{C}$.

An important consequence of Theorem 3.13.2 is the following.

Theorem 3.13.3 *For real* α,β *we let*

$$x_{n,1}(\alpha,\beta) > x_{n,2}(\alpha,\beta) > \cdots > x_{n,n}(\alpha,\beta) \tag{3.13.4}$$

be the zeros of $P_n^{(\alpha,\beta)}(x)$ *in* $[-1,1]$. *With* $x_{n,k}(\alpha,\beta) = \cos(\theta_{n,k}(\alpha,\beta))$, $0 < \theta_{n,k}(\alpha,\beta) < \pi$, *we have*

$$\lim_{n\to\infty} n\,\theta_{n,k}(\alpha,\beta) = j_{\alpha,k}, \tag{3.13.5}$$

where $j_{\alpha,k}$ *is the kth positive zero of* $J_\alpha(z)$.

Theorem 3.13.4 *For* $\alpha \in \mathbb{R}$, *the limiting relation*

$$\lim_{n\to\infty} n^{-\alpha} L_n^{(\alpha)}(z/n) = z^{-\alpha/2} J_\nu\left(2\sqrt{2}\right) \tag{3.13.6}$$

holds uniformly for z *in compact subsets of* $\mathbb{C}$.

Theorem 3.13.5 *For* $\alpha, \beta \in \mathbb{R}$, *we have*

$$\begin{aligned} P_n^{(\alpha,\beta)}(x) &= (x-1)^{\alpha/2}(x+1)^{-\beta/2}\left\{(x+1)^{1/2} + (x-1)^{1/2}\right\}^{\alpha+\beta} \\ &\quad \times \frac{\left(x^2-1\right)^{-1/4}}{\sqrt{2\pi n}} \left\{x + \left(x^2-1\right)^{1/2}\right\}^{n+1/2} \{1 + o(1)\}, \end{aligned} \tag{3.13.7}$$

for $x \in \mathbb{C} \setminus [-1,1]$. *The above limit relation holds uniformly in* x *on compact subsets of* $\mathbb{C}$.

Proofs and references are in Szegő ([1939] 1975, §§8.1, 8.2).

Theorem 3.13.6 (Hilb-type asymptotics) *Let* $\alpha > -1, \beta \in \mathbb{R}$. *Then*

$$\left(\sin\frac{\theta}{2}\right)^\alpha \left(\cos\frac{\theta}{2}\right)^\beta P_n^{(\alpha,\beta)}(\cos\theta) = \frac{\Gamma(n+\alpha+1)}{n!\,N^\alpha}\left(\frac{\theta}{\sin\theta}\right)^{1/2} J_\alpha(N\theta) + \theta^u O(n^\nu), \tag{3.13.8}$$

as $n \to \infty$, *where* N *is as in Theorem* 3.13.1, *and*

$$\begin{aligned} &u = 1/2, \quad \nu = -1/2, \quad \text{if } c/n \le \theta \le \pi - \epsilon, \\ &u = \alpha + 2, \quad \nu = \alpha, \quad \text{if } 0 < \theta \le cn^{-1}, \end{aligned} \tag{3.13.9}$$

where c *and* ϵ *are fixed numbers*.

Theorem 3.13.7 *Let* $x = (z + 1/z)/2$, $|z| > 1$, *and* $\nu > 0$, *or* $\nu < 0$, $\nu \neq 0, -1, -2, \ldots$. *Then*

$$C_n^\nu(x) = (-1)^n \frac{z^n}{n!} \sum_{k=0}^{p-1} \frac{(\nu)_k (1-\nu)_k}{(1-\nu-n)_k} z^{-2k} (1 - 1/z^2)^{-\nu-k} + O(n^{\nu-p-1}|z|^n). \tag{3.13.10}$$

Moreover,

$$\begin{aligned} C_n^\nu(\cos\theta) &= 2(-1)^n \frac{z^n}{n!} \sum_{k=0}^{p-1} \frac{(\nu)_k (1-\nu)_k}{(1-\nu-n)_k} \\ &\quad \times \frac{\cos[(n+\nu-k)\theta - (\nu+k)\pi/2]}{(2\sin\theta)^{\nu+k}} + O\left(n^{\nu-p-1}\right), \quad 0 < \theta < \pi. \end{aligned} \tag{3.13.11}$$

Theorem 3.13.8 *For $1 > \nu > 0$, we have the asymptotic formula*

$$C_n^\nu(\cos\theta) = \frac{2\Gamma(2\nu+n)}{\Gamma(\nu)\Gamma(n+\nu+1)} \sum_{k=0}^{p-1} \frac{(\nu)_k(1-\nu)_k}{k!(n+\nu+1)_k} \times \frac{\cos[n+\nu+k)\theta - (\nu+k)\pi/2]}{(2\sin\theta)^{\nu+k}} + R_p(\theta), \tag{3.13.12}$$

where

$$\left|R_p(\theta)\right| < \frac{2\Gamma(2\nu+n)}{\Gamma(\nu)\Gamma(n+\nu+1)} \frac{(\nu)_p(1-\nu)_p}{p!(n+\nu+1)_p} \frac{M}{(2\sin\theta)^{\nu+p}}, \tag{3.13.13}$$

and M, where

$$M = \begin{cases} 1/|\cos\theta| & \text{if } \sin^2\theta < 1/2, \\ 2\sin\theta & \text{otherwise}, \end{cases} \tag{3.13.14}$$

is a number between 1 and 2.

Theorem 3.13.8 generalizes a result of Stieltjes for the Legendre polynomials. The main term in the Legendre case was proved by Laplace.

Theorem 3.13.9 (Fejer) *For $\alpha \in \mathbb{R}$, $x > 0$, we have*

$$L_n^{(\alpha)}(x) = \frac{e^{x/2}}{\sqrt{\pi}} x^{-\alpha/2-1/4} n^{\alpha/2-1/4} \cos\left\{2(nx)^{1/2} - \alpha\pi/2 - \alpha/4\right\} + O\left(n^{\alpha/2-3/4}\right), \tag{3.13.15}$$

as $n \to \infty$. The O bound is uniform for x in any compact subset of $(0, \infty)$.

Theorem 3.13.10 (Perron) *For $\alpha \in \mathbb{R}$,*

$$L_n^{(\alpha)}(x) = \frac{e^{x/2}}{2\pi} (-x)^{-\alpha/2-1/4} n^{\alpha/2-1/4} \exp\left\{2(-nx)^{1/2}\right\}, \tag{3.13.16}$$

for $x \in \mathbb{C} \setminus [0, \infty)$. In (3.13.16), the branches of $(-x)^{-\alpha/2-1/4}$ and $(-x)^{1/2}$ are real and positive for $x < 0$.

Theorem 3.13.11 (Hilb-type asymptotics) *When $\alpha > -1$ and $x > 0$,*

$$e^{-x/2} x^{\alpha/2} L_n^{(\alpha)}(x) = N^{-\alpha/2} \frac{\Gamma(\alpha+n+1)}{n!} J_\alpha\left(2(Nx)^{1/2}\right) + O\left(n^{\alpha/2-3/4}\right), \tag{3.13.17}$$

where $N = n + (\alpha+1)/2$.

Theorem 3.13.12 *Let c and C be positive constants. Then for $\alpha > -1$ and $c/n \le x \le C$, we have*

$$L_n^{(\alpha)}(x) = \frac{e^{x/2}}{\sqrt{\pi}} x^{-\alpha/2-1/4} n^{\alpha/2-1/4} \left\{\cos\left[2(nx)^{1/2} - \alpha\pi/2 - \pi/4\right] + (nx)^{-1/2} O(1)\right\}. \tag{3.13.18}$$

Theorem 3.13.13 *For x real, we have*

$$\frac{\Gamma(n/2+1)}{\Gamma(n+1)}e^{-x^2/2}H_n(x) = \cos\left(N^{\frac{1}{2}}x - n\pi/2\right) + \frac{x^3}{6}N^{-\frac{1}{2}}\sin\left(N^{\frac{1}{2}}x - n\pi/2\right) + O\left(n^{-1}\right), \tag{3.13.19}$$

where $N = 2n+1$. *The bound for the error term holds uniformly in x on every compact interval.*

Theorem 3.13.14 *The asymptotic formula in* (3.13.19) *holds in the complex x-plane if we replace the remainder term by* $\exp\{N^{\frac{1}{2}}|\operatorname{Im}(x)|\}O(n^{-p})$. *This is true uniformly for* $|x| \leqq R$ *where R is an arbitrary fixed positive number.*

Next we record the Plancherel–Rotach asymptotic formulas.

Theorem 3.13.15 (Plancherel–Rotach type) *Let* $\alpha \in \mathbb{R}$ *and* ϵ *and* ω *be fixed positive numbers. We have*

(a) *for* $x = (4n+2\alpha+2)\cos^2\phi$, $\epsilon \leqq \phi \leqq \pi/2 - \epsilon n^{-\frac{1}{2}}$,

$$e^{-x/2}L_n^{(\alpha)}(x) = (-1)^n(\pi\sin\phi)^{-\frac{1}{2}}x^{-\alpha/2-\frac{1}{4}}n^{\alpha/2-\frac{1}{4}} \times\left\{\sin[(n+(\alpha+1)/2)(\sin 2\phi - 2\phi)+3\pi/4] + (nx)^{-\frac{1}{2}}O(1)\right\}; \tag{3.13.20}$$

(b) *for* $x = (4n+2\alpha+2)\cosh^2\phi$, $\epsilon \leq \phi \leq \omega$,

$$e^{-x/2}L_n^{(\alpha)}(x) = \frac{1}{2}(-1)^n(\pi\sinh\phi)^{-\frac{1}{2}}x^{-\alpha/2-\frac{1}{4}}n^{\alpha/2-\frac{1}{4}} \times\exp\{(n+(\alpha+1)/2)(2\phi-\sinh 2\phi)\}\left\{1+O\left(n^{-1}\right)\right\}; \tag{3.13.21}$$

(c) *for* $x = 4n+2\alpha+2-2(2n/3)^{\frac{1}{3}}t$, *t complex and bounded,*

$$e^{-x/2}L_n^{(\alpha)}(x) = (-1)^n\pi^{-1}2^{-\alpha-\frac{1}{3}}3^{\frac{1}{3}}n^{-\frac{1}{3}}\left\{A(t)+O\left(n^{-\frac{2}{3}}\right)\right\} \tag{3.13.22}$$

where $A(t)$ *is Airy's function defined in* (1.2.33), (1.2.35).

Moreover, in the above formulas the O-terms hold uniformly.

Theorem 3.13.16 *Let* ϵ *and* ω *be fixed positive numbers. We have*

(a) *for* $x = (2n+1)^{\frac{1}{2}}\cos\phi$, $\epsilon \leqq \phi \leqq \pi - \epsilon$,

$$e^{-x^2/2}H_n(x) = 2^{n/2+\frac{1}{4}}(n!)^{\frac{1}{2}}(\pi n)^{-\frac{1}{4}}(\sin\phi)^{-\frac{1}{2}} \times\left\{\sin\left[\left(\frac{n}{2}+\frac{1}{4}\right)(\sin(2\phi)-2\phi)+\frac{3\pi}{4}\right]+O\left(n^{-1}\right)\right\}; \tag{3.13.23}$$

(b) *for* $x = (2n+1)^{\frac{1}{2}}\cosh\phi$, $\epsilon \leqq \phi \leqq \omega$,

$$e^{-x^2/2}H_n(x) = 2^{n/2-\frac{3}{4}}(n!)^{\frac{1}{2}}(\pi n)^{-\frac{1}{4}}(\sinh\phi)^{-\frac{1}{2}} \times\exp\left[\left(\frac{n}{2}+\frac{1}{4}\right)(2\phi-\sinh 2\phi)\right]\left\{1+O\left(n^{-1}\right)\right\}; \tag{3.13.24}$$

(c) *for* $x = (2n+1)^{\frac{1}{2}} - 2^{-\frac{1}{2}}3^{-\frac{1}{3}}n^{-\frac{1}{6}}t$, *t complex and bounded,*

$$e^{-x^2/2}H_n(x) = 3^{\frac{1}{3}}\pi^{\frac{3}{4}}2^{n/2+\frac{1}{4}}(n!)^{\frac{1}{2}}n^{1/12}\left\{A(t) + O\left(n^{-\frac{2}{3}}\right)\right\}. \tag{3.13.25}$$

In all these formulas, the O-terms hold uniformly.

For complete asymptotic expansions, proofs, and references to the literature, the reader may consult Szegő ([1939] 1975, §8.22).

Baratella and Gatteschi proved the following uniform asymptotics of $P_n^{(\alpha,\beta)}(\cos\theta)$ using the Liouville–Stekloff method (Szegő, [1939] 1975, §8.6).

Theorem 3.13.17 (Baratella and Gatteschi, 1988) *Let*

$$N = n + (\alpha+\beta+1)/2, \quad A = 1 - 4\alpha^2, \quad B = 1 - 4\beta^2,$$

$$a(\theta) = \frac{2}{\theta} - \cot\left(\frac{\theta}{2}\right), \quad b(\theta) = \tan\left(\frac{\theta}{2}\right), \quad f(\theta) = N\theta + \frac{1}{16N}\left[Aa(\theta) + Bb(\theta)\right],$$

$$u_n^{(\alpha,\beta)}(\theta) = \left(\sin\left(\frac{\theta}{2}\right)\right)^{\alpha+1/2}\left(\cos\left(\frac{\theta}{2}\right)\right)^{\beta+1/2} P_n^{(\alpha,\beta)}(\cos\theta),$$

$$F(\theta) = F_1(\theta) + F_2(\theta),$$

$$F_1(\theta) = \frac{1}{2}\frac{Aa'''(\theta) + Bb'''(\theta)}{16N^2 + Aa'(\theta) + Bb'(\theta)} - \frac{3}{4}\left[\frac{Aa''(\theta) + Bb''(\theta)}{16N^2 + Aa'(\theta) + Bb'(\theta)}\right]^2,$$

$$\begin{aligned} F_2(\theta) = {} & \frac{A}{2\theta^2}\frac{\theta Aa'(\theta) + \theta Bb'(\theta) - Aa(\theta) - Bb(\theta)}{16N^2\theta + Aa(\theta) + Bb(\theta)} \\ & \times\left[1 + \frac{1}{2}\frac{\theta Aa'(\theta) + \theta Bb'(\theta) - Aa(\theta) - Bb(\theta)}{16N^2\theta + Aa(\theta) + Bb(\theta)}\right] + \frac{[Aa'(\theta) + Bb'(\theta)]^2}{256N^2}. \end{aligned}$$

With

$$\Delta(t,\theta) := J_\alpha(f(\theta))Y_\alpha(f(t)) - J_\alpha(f(t))Y_\alpha(f(\theta)),$$

$$I^{(\alpha,\beta)} := \frac{\pi}{2}\int_0^\theta \left[\frac{f(t)}{f'(t)}\right]^{1/2} \Delta(t,\theta)F(t)u_n^{(\alpha,\beta)}(t)\,dt,$$

we have

$$\left[\frac{f'(\theta)}{f(\theta)}\right]^{1/2} u_n^{(\alpha,\beta)}(\theta) = C_1 J_\alpha(f(\theta)) - I^{(\alpha,\beta)},$$

where

$$C_1 = \frac{\Gamma(n+\alpha+1)}{\sqrt{2}\,N^\alpha n!}\left[1 + \frac{1}{16N^2}\left(\frac{A}{6} + \frac{B}{2}\right)\right]^{-\alpha}.$$

Furthermore, for $\alpha, \beta \in (-1/2, 1/2)$, $I^{(\alpha,\beta)}$ has the estimate

$$\left|I^{(\alpha,\beta)}\right| \le \begin{cases} \dfrac{\theta^\alpha}{N^4}\dbinom{n+\alpha}{n}(0.00812A + 0.0828B), & 0 < \theta < \theta^*, \\ \dfrac{\theta^{1/2}}{N^{\alpha+7/2}}\dbinom{n+\alpha}{n}(0.00526A + 0.535B), & \theta^* \le \theta \le \pi/2, \end{cases}$$

where θ^* is the root of the equation $f(\theta) = \pi/2$.

3.14 Relative Extrema of Classical Polynomials

Theorem 3.14.1 *Let $\mu_{n,1}, \dots, \mu_{n,\lfloor n/2\rfloor}$ be the relative extrema of $C_n^\nu(x)$ in $(0, 1)$ arranged in decreasing order of x. Then, for $n > 1$, we have*

$$1 > \mu_{n,1}^{(\nu)} > \mu_{n,2}^{(\nu)} > \cdots > \mu_{\lfloor n/2\rfloor}^{(\nu)}, \quad n \ge 2, \tag{3.14.1}$$

when $\nu > 0$. When $\nu < 0$, then

$$\mu_{n,1}^{(\nu)} < \mu_{n,2}^{(\nu)} < \cdots < \mu_{\lfloor n/2\rfloor}^{(\nu)}. \tag{3.14.2}$$

The corresponding result for Hermite polynomials is a limiting case of Theorem 3.14.1. In the case $\nu = 0$, all the inequality signs in (3.14.1) and (3.14.2) become equals signs as can be seen from (3.6.8).

Theorem 3.14.2 *Assume that $n > 1$. The successive maxima of $(\sin\theta)^\nu\left|C_n^\nu(\cos\theta)\right|$ for $\theta \in (0, \pi/2)$ form an increasing sequence if $\nu \in (0, 1)$, and a decreasing sequence if $\nu > 1$.*

Theorem 3.14.3 *The relative maxima $\mu_{n,k}^{(\nu)}$ decrease with n for $\nu > -1/2$, that is,*

$$\mu_{n,k}^{(\nu)} > \mu_{n+1,k}^{(\nu)}, \quad n = k+1, k+2, \dots. \tag{3.14.3}$$

Theorem 3.14.3 was first proved in Szegő (1950b) for $\nu = 1/2$ and in Szász (1950) for general ν. Theorem 3.14.3 has been generalized to orthogonal Laguerre functions $\{e^{-x/2}L_n(x)\}$ in Todd (1950) and to orthogonal Hermite functions in Szász (1951).

Let $\mu_{n,k}^{(\alpha,\beta)}$ be the relative extrema of $\left|P_n^{(\alpha,\beta)}(x)\right|$. Askey conjectured that

$$\mu_{n+1,k}^{(\alpha,\beta)} < \mu_{n,k}^{(\alpha,\beta)}, \quad k = 1, \dots, n-1, \quad \text{for } \alpha > \beta > -1/2, \tag{3.14.4}$$

in his comments on Szegő (1950b); see p. 221 of volume 3 of Szegő's collected papers. Askey also conjectured that when $\alpha = 0, \beta = -1$, the inequalities in (3.14.4) are reversed. Askey also noted that

$$P_n^{(0,-1)}(x) = \frac{1}{2}\left[P_n(x) + P_{n-1}(x)\right],$$

$\{P_n(x)\}$ being the Legendre polynomials. Both conjectures are also stated in Askey (1990). Wong and Zhang confirmed Askey's second conjecture by proving the desired result asymptotically for $n \ge 25$, then established the cases $n \le 24$ by direct comparison of numerical values. This was done in Wong and Zhang (1994b). Askey's first conjecture has been verified for n sufficiently large by the same authors in Wong and Zhang (1994a).

3.15 The Bessel Polynomials

Formulas (1.2.17)–(1.2.18) and (1.2.24) imply

$$K_{1/2}(z) = K_{-1/2}(z) = \sqrt{\frac{\pi}{2z}}\, e^{-z}, \tag{3.15.1}$$

and by induction, we see that $K_{n+1/2}(z)e^z\sqrt{z}$ is a polynomial in $1/z$. Set

$$y_n(1/z) = e^z z^{1/2} K_{n+1/2}(z)/\sqrt{\pi}. \tag{3.15.2}$$

Substitute for $K_{n+1/2}(z)$ from (3.15.2) in (1.2.22) to see that

$$z^2 y_n''(z) + (2z+1)y_n'(z) - n(n+1)y_n(z) = 0. \tag{3.15.3}$$

The only polynomial solution to (3.15.3) is a constant multiple of

$$y_n(z) = {}_2F_0(-n, n+1; —; -z/2). \tag{3.15.4}$$

The reverse polynomial

$$\theta_n(z) = z^n y_n(1/z) \tag{3.15.5}$$

also plays an important role. More general polynomials are

$$y_n(z; a, b) = {}_2F_0(-n, n+a-1; —; -z/b), \quad \theta_n(z; a, b) = z^n y_n(1/z; a, b). \tag{3.15.6}$$

The corresponding differential equations are

$$\begin{aligned} &z^2 y'' + (az+b)y' - n(n+a-1)y = 0, \quad y = y_n(z; a, b), \\ &z\theta'' - (2n-2+a+bz)\theta' + bn\theta = 0, \quad \theta = \theta_n(z; a, b). \end{aligned} \tag{3.15.7}$$

The polynomials $\{y_n(z)\}$ or $\{y_n(z; a, b)\}$ are the Bessel polynomials, while $\{\theta_n(z)\}$ and $\{\theta_n(z; a, b)\}$ are the reverse Bessel polynomials. Clearly, $y_n(z) = y_n(z; 2, 2)$, $\theta_n(z) = \theta_n(z; 2, 2)$. The polynomials $\{y_n(x; a, b)\}$ also arise from the confluent hypergeometric Ψ function, as can be seen from (3.8.6) and (3.15.10).

The notation and terminology was introduced in Krall and Frink (1949). However, the same polynomials appeared over 15 years earlier in different notation in Burchnall and Chaundy (1931).

Define w_B by

$$w_B(z; a, b) = \sum_{n=0}^{\infty} \frac{(-b/z)^n}{(a-1)_n} = {}_1F_1(1; a-1; -b/z). \tag{3.15.8}$$

In the case $a = b = 2$, w_B becomes $\exp(-2/z)$.

Theorem 3.15.1 *The Bessel polynomials satisfy the orthogonality relation*

$$\frac{1}{2\pi i} \oint_C y_m(z; a, b) y_n(z; a, b) w_B(z; a, b)\, dz = \frac{(-1)^{n+1} b}{a+2n-1} \frac{n!}{(a-1)_n} \delta_{m,n}, \tag{3.15.9}$$

where C is a closed contour containing $z = 0$ in its interior and w_B is as in (3.15.8).

The Bessel polynomials are related to Laguerre and Jacobi polynomials through

$$\begin{aligned} y_n(x;a,b) &= n!(-x/b)^n L_n^{(-2n-a+1)}(-b/x) \\ &= \frac{n!}{(a+n)_n}\left(-\frac{b}{x}\right)^{n+a-1} \Psi(a+n,a+2n+1;-b/x), \end{aligned} \tag{3.15.10}$$

$$y_n(z;a,b) = \lim_{\gamma\to\infty} \frac{n!}{(\gamma+1)_n} P_n^{(\gamma,a-\gamma)}(1+2\gamma z/b). \tag{3.15.11}$$

A limiting case of formula (3.1.23) gives the **lowering relation**

$$\begin{aligned} &(a+2n-2)z^2 y_n'(z;a,b) \\ &\quad = n[(-a+2n-2)z-b]y_n(z;a,b) + bny_{n-1}(z;a,b), \end{aligned} \tag{3.15.12}$$

and for θ_n, (3.15.12) takes the form

$$(a+2n-2)\theta_n'(z;a,b) = bn\theta_n(z;a,b) - bnz\theta_{n-1}(z;a,b). \tag{3.15.13}$$

Furthermore, (3.1.22) establishes the **three-term recurrence relation**

$$\begin{aligned} &(a+n-1)(a+2n-2)y_{n+1}(z;a,b) - n(a+2n)y_{n-1}(z;a,b) \\ &\quad = (a+2n-1)[a-2-(a+2n)(a+2n-2)z]y_n(z;a,b). \end{aligned} \tag{3.15.14}$$

The recursion (3.15.14) indicates that $\{y_n(z;a,b)\}$ are not orthogonal with respect to a positive measure. Theorems of Boas (1939) and Shohat (1939) show that they are orthogonal with respect to a signed measure supported in $[0,\infty)$. Antonio Duran was the first to construct such a signed measure in Durán (1989, 1993). A detailed exposition of the constructions of signed orthogonality measures for $\{y_n(z;a,b)\}$ and references to the literature are in Kwon (2002).

Theorem 3.15.2 *The discriminant of the Bessel polynomial is given by*

$$D\left(y_n(x;a,b)\right) = (n!)^{2n-2}\left(-b^2\right)^{-n(n-1)/2} \prod_{j-1}^{n} j^{j-2n+2}(n+j+a-2).$$

The Rodrigues formula is

$$y_n(x;a,b) = b^{-n}x^{2-a}e^{b/x}\frac{d^n}{dx^n}\left(x^{2n+a-2}e^{-b/x}\right). \tag{3.15.15}$$

Theorem 3.15.3 (Zeros of Bessel polynomials)

(a) *All zeros of $y_n(z;a,b)$ are simple.*
(b) *No two consecutive polynomials $y_n(z;a,b)$, $y_{n+1}(z;a,b)$ have a common zero.*
(c) *All zeros of y_{2n} are complex, while y_{2n+1} has only one real zero, $n=0,1,2,\ldots$.*

Observe that (c) also follows from a similar result for K_ν, $\nu>0$ (Watson, 1944).

Theorem 3.15.4 *Let $\{z_{n,j}\colon j=1,\ldots,n\}$ be the zeros of $y_n(x)$. Then*

$$\sum_{j=1}^{n} z_{n,j} = -1, \quad \sum_{j=1}^{n} z_{n,j}^{2m-1} = 0, \quad m=2,3,\ldots,n.$$

The vanishing power sums in Theorem 3.15.4 appeared as the first terms in an asymptotic expansion; see Ismail and Kelker (1976). Theorem 3.15.4 was first proved in Burchnall (1951) and independently discovered in Ismail and Kelker (1976), where an induction proof was given. Moreover,

$$\sum_{j=1}^{n} z_{n,j}^{2n+1} = \frac{(-1/4)^n}{(3/2)_n^2}, \quad \sum_{j=1}^{n} z_{n,j}^{2n+3} = \frac{(-1/4)^n}{(2n-1)(3/2)_n^2}$$

were also proved in Ismail and Kelker (1976).

Theorem 3.15.5 (Burchnall, 1951) *The system of equations*

$$\sum_{j=1}^{n} x_j = -1, \quad \sum_{j=1}^{n} x_j^{2m-1} = 0, \quad m = 2, 3, \dots, n \tag{3.15.16}$$

has a unique solution given by the zeros of $y_n(x)$.

Theorem 3.15.6 *The Bessel polynomials have the* ***generating function***

$$\sum_{n=0}^{\infty} y_n(z; a, b)\frac{t^n}{n!} = (1 - 4zt/b)^{-\frac{1}{2}} \left(\frac{2}{1 + \sqrt{1 - 4zt/b}}\right)^{a-2} \exp\left(\frac{2t}{1 + \sqrt{1 - 4zt/b}}\right). \tag{3.15.17}$$

The special case $a = b = 2$ of (3.15.17) gives an **exponential generating function** for $\{y_n(x)\}$. Another generating function is

$$\sum_{n=0}^{\infty} y_n(z)\frac{t^{n+1}}{(n+1)!} = \exp\left(\frac{2t}{1 + (1 - 2zt)^{\frac{1}{2}}}\right) - 1. \tag{3.15.18}$$

The parameter b in $y_n(z; a, b)$ scales the variable z, so there is no loss of generality in assuming $b = 2$.

Definition 3.15.7 For $a \in \mathbb{R}$, $a + n > 1$, let

$$C(n, a) := \left\{z = re^{i\theta} \in \mathbb{C}\colon 0 < r < \tfrac{1-\cos\theta}{n+a-1}\right\} \bigcup \left\{\tfrac{-2}{n+a-1}\right\}. \tag{3.15.19}$$

Theorem 3.15.8 (Saff and Varga, 1977) *All the zeros of* $y_n(z; a, b)$ *lie in the cardioidal region* $C(n, a)$.

Theorem 3.15.9 (Underhill, 1972; Saff and Varga, 1977) *For any integers* a *and* $n \geq 1$, *with* $n + a \geq 2$, *the zeros of* $y_n(x; a, 2)$ *satisfy*

$$|z| < \frac{2}{\mu(2n + a - 2)}, \tag{3.15.20}$$

where $\mu \approx 0.278465$ *is the unique positive root of* $\mu e^{\mu+1} = 1$.

Let $\mathcal{L}$ be the set of all zeros of the scaled polynomials

$$\left\{y_n\left(\tfrac{2z}{n+a-1}; a, 2\right) : n = N, a \in \mathbb{R},\ n + a > 1\right\}. \tag{3.15.21}$$

Under $z \to 2z/(n+a-1)$ the cardioidal region (3.15.19) is mapped onto

$$C := \left\{z = re^{i\theta} \in \mathbb{C} \colon 0 < r < (1-\cos\theta)/2\right\} \bigcup \{-1\}. \tag{3.15.22}$$

Theorem 3.15.10 *Each boundary point of C of* (3.15.22) *is an accumulation point of the set $\mathcal{L}$ of all zeros of the normalized polynomials in* (3.15.21).

Theorem 3.15.11 *For every $a \in \mathbb{R}$, there exists an integer $N = N(a)$ such that all the zeros of $y_n(z; a, 2)$ lie in $\{z\colon \operatorname{Re} z < 0\}$ for $n > N$. For $a < -2$, one can take $N = \lceil 2^{3-a} \rceil$.*

De Bruin, Saff, and Varga (1981a,b) proved Theorems 3.15.10 and 3.15.11. They also proved that the zeros of $y_n(z; a, 2)$ lie in the annulus

$$A(n,a) := \left\{z \in \mathbb{C}\colon \tfrac{2}{2n+a-2/3} < |z| \le \tfrac{2}{n+a-1}\right\}, \tag{3.15.23}$$

which is sharper than the disk of Theorem 3.15.9.

Theorem 3.15.12 (De Bruin, Saff, and Varga, 1981a,b) *Let $a \in \mathbb{R}$ and let ρ be the unique (negative) root of*

$$-\rho \exp\left(\sqrt{1+\rho^2}\right) = 1 + \sqrt{1+\rho^2} \quad (\rho \approx -0.662743419), \tag{3.15.24}$$

and let

$$K(\rho, a) := \frac{\rho\sqrt{1+\rho^2} + (2-a)\ln\left(\rho + \sqrt{1+\rho^2}\right)}{\sqrt{1+\rho^2}}.$$

Then for n odd, $\alpha_n(a)$, the unique negative zero of $y_n(z; a, 2)$, satisfies the asymptotic relationship

$$\frac{2}{\alpha_n(a)} = (2n+a-2)\rho + K(\rho, a) + O\left(\frac{1}{2n+a-2}\right), \quad \text{as } n \to \infty. \tag{3.15.25}$$

Grosswald (1978) contains broad applications of the Bessel polynomials, from proving the irrationality of π and e^r, for r rational, to probabilistic problems and electrical networks. A combinatorial model for the Bessel polynomials is in Dulucq and Favreau (1991).

4

Recursively Defined Polynomials

4.1 Birth and Death Process Polynomials

A birth and death process is a stationary Markov process whose states are labeled by nonnegative integers and whose transition probabilities

$$p_{m,n}(t) = \Pr\{X(t) = n \mid X(0) = m\} \tag{4.1.1}$$

satisfy certain conditions as $t \to 0^+$:

$$\begin{cases} p_{n,n+1}(t) = \lambda_n t = o(t), \\ p_{n,n-1}(t) = \mu_n t + o(t), \\ p_{n,n}(t) = 1 - (\lambda_n + \mu_n)t + o(t), \\ p_{n,m}(t) = o(t), & |m-n| > 1. \end{cases}$$

It is assumed that

$$\lambda_n > 0, \ n \geq 0 \quad \text{and} \quad \mu_0 \geq 0, \ \mu_n > 0, n > 0. \tag{4.1.2}$$

The λ_n are the birth rates and the μ_n are the death rates.

Every birth and death process leads to polynomials recursively defined by

$$\begin{gathered} F_{-1}(x) = 0, \quad F_0(x) = 1, \\ -xF_n(x) = \lambda_{n-1}F_{n-1}(x) + \mu_{n+1}F_{n+1}(x) - (\lambda_n + \mu_n)F_n(x), \quad n > 0. \end{gathered} \tag{4.1.3}$$

It can be proved that (Karlin and McGregor, 1957a,b)

$$p_{m,n}(t) = \frac{1}{\zeta_m} \int_0^\infty e^{-xt} F_m(x) F_n(x) \, d\mu(x), \tag{4.1.4}$$

where $\{F_n(x)\}$ satisfies the orthogonality relation

$$\int_{\mathbb{R}} F_m(x) F_n(x) \, d\mu(x) = \zeta_n \, \delta_{m,n}, \quad \zeta_0 := 1, \quad \zeta_n = \prod_{j=1}^{n} \frac{\lambda_{j-1}}{\mu_j}. \tag{4.1.5}$$

A related sequence of polynomials is $\{Q_n(x)\}$ with $Q_n(x) = F_n(x)/\zeta_n$. They satisfy

$$Q_0(x) = 1, \quad Q_1(x) = (\lambda_0 + \mu_0 - x)/\lambda_0, \tag{4.1.6}$$

$$-xQ_n(x) = \lambda_n Q_{n+1}(x) + \mu_n Q_{n-1}(x) - (\lambda_n + \mu_n) Q_n(x), \quad n > 0, \tag{4.1.7}$$

$$\int_0^\infty Q_m(x)Q_n(x)\, d\mu(x) = \delta_{m,n}/\zeta_n. \tag{4.1.8}$$

The corresponding **random walk polynomials** are defined by

$$\begin{gathered} R_{-1}(x) = 0, \quad R_0(x) = 1, \\ xR_n(x) = m_n R_{n+1}(x) + \ell_n R_{n-1}(x), \\ m_n = \lambda_n/(\lambda_n + \mu_n), \quad \ell_n = \mu_n/(\lambda_n + \mu_n); \end{gathered} \tag{4.1.9}$$

see Karlin and McGregor (1958, 1959). They are orthogonal on $[-1, 1]$ with respect to an even measure,

$$\int_{-1}^1 r_n(x) r_m(x)\, d\mu(x) = \delta_{m,n}/h_n, \tag{4.1.10}$$

$$h_0 = 1, \quad h_n = \frac{\lambda_0\lambda_1 \cdots \lambda_{n-1}(\lambda_n + \mu_n)}{\mu_1\mu_2 \cdots \mu_n(\lambda_0 + \mu_0)}, \quad n > 0. \tag{4.1.11}$$

Many classical polynomials are random walk polynomials or birth and death process polynomials, or limits of them, in some normalization.

Remark 4.1.1 When $\mu_0 > 0$, there are two natural families of birth and death polynomials. The first is the family $\{Q_n(x)\}$ defined by (4.1.6)–(4.1.7). Another family is $\{\tilde{Q}_n(x)\}$ defined by

$$\tilde{Q}_0(x) = 1, \quad \tilde{Q}_1 = (\lambda_0 - x)/\lambda_0, \tag{4.1.12}$$

$$-x\tilde{Q}_n(x) = \lambda_n \tilde{Q}_{n+1}(x) + \mu_n \tilde{Q}_{n-1}(x) - (\lambda_n + \mu_n) \tilde{Q}_n(x), \quad n > 0. \tag{4.1.13}$$

In effect, we redefine μ_0 to be zero. Observe that $\tilde{Q}_n(0) = 1$ for $n \geq 0$.

The relations (3.8.10)–(3.8.11) between Hermite polynomials and $\{L_n^{(\pm 1/2)}(x)\}$ carry over to general birth and death process polynomials. Given a symmetric family of polynomials $\{\mathcal{F}_n(x)\}$,

$$\mathcal{F}_0(x) = 1, \quad \mathcal{F}_1(x) = x, \tag{4.1.14}$$

$$\mathcal{F}_{n+1}(x) = x\mathcal{F}_n(x) - \beta_n \mathcal{F}_{n-1}(x), \tag{4.1.15}$$

we define two families of orthogonal polynomials by

$$\rho_n(x) = \mathcal{F}_{2n}\left(\sqrt{x}\right), \quad \sigma_n(x) = x^{-1/2}\mathcal{F}_{2n+1}\left(\sqrt{x}\right). \tag{4.1.16}$$

They are generated by

$$x\rho_n(x) = \rho_{n+1}(x) + (\beta_{2n} + \beta_{2n+1})\rho_n(x) + \beta_{2n}\beta_{2n-1}\rho_{n-1}(x), \tag{4.1.17}$$

$$x\sigma_n(x) = \sigma_{n+1}(x) + (\beta_{2n+1} + \beta_{2n+2})\sigma_n(x) + \beta_{2n}\beta_{2n+1}\sigma_{n-1}(x), \tag{4.1.18}$$

and the initial conditions

$$\rho_0(x) = 1, \quad \rho_1(x) = x - \beta_1, \tag{4.1.19}$$

$$\sigma_0(x) = 1, \quad \sigma_1(x) = x - \beta_1 - \beta_2. \tag{4.1.20}$$

The polynomials $\{\rho_n(x)\}$ and $\{\sigma_n(x)\}$ play the role of $\{L_n^{(\pm 1/2)}(x)\}$ when $\{\mathcal{F}_n(x)\} = \{H_n(x)\}$.

4.2 Polynomials of Pollaczek Type

The Pollaczek-type polynomials $P_n^\lambda(x; a, b)$ are generated by

$$\begin{aligned}(n+1)P_{n+1}^\lambda(x; a, b) &= 2[(n + \lambda + a)x + b]P_n^\lambda(x; a, b) \\ &\quad - (n + 2\lambda - 1)P_{n-1}^\lambda(x; a, b), \quad n > 0,\end{aligned} \tag{4.2.1}$$

$$P_0^\lambda(x; a, b) = 1, \quad P_1^\lambda(x; a, b) = 2(\lambda + a)x + 2b. \tag{4.2.2}$$

Pollaczek (1949) introduced these polynomials when $\lambda = 1/2$ and $a \geq |b|$ and Szegő (1950a) studied them for general λ. It is clear that $C_n^{(\lambda)}(x) = P_n^\lambda(x; 0, 0)$. Special cases of the polynomials generated by (4.2.1)–(4.2.2) appeared through the application of the J-matrix method of quantum mechanics. Details and references are in Ismail (2005b, §5.8).

The monic polynomials are

$$Q_n^\lambda(x; a, b) := \frac{n!}{2^n(a+\lambda)_n} P_n^\lambda(x; a, b), \tag{4.2.3}$$

and the monic recurrence relation is

$$\begin{aligned}&Q_0^\lambda(x; a, b) = 1, \quad Q_1^\lambda(x; a, b) = x + b/(\lambda + a), \\ &Q_{n+1}^\lambda(x; a, b) = \left[x + \frac{b}{n + a + \lambda}\right] Q_n^\lambda(x; a, b) - \frac{n(n + 2\lambda - 1)}{(a + \lambda + n - 1)(a + \lambda + n)} Q_{n-1}^\lambda(x; a, b).\end{aligned} \tag{4.2.4}$$

There is no loss of generality in assuming $b \geq 0$, since

$$P_n^\lambda(-x; a, b) = (-1)^n P_n^\lambda(x; a, -b). \tag{4.2.5}$$

We have the **generating function** and the **hypergeometric representation**

$$\begin{aligned}&\sum_{n=0}^{\infty} P_n^\lambda(\cos\theta; a, b)t^n = \left(1 - te^{i\theta}\right)^{-\lambda + i\Phi(\theta)} \left(1 - te^{-i\theta}\right)^{-\lambda - i\Phi(\theta)}, \\ &\Phi(\theta) := [a\cos\theta + b]/\sin\theta,\end{aligned} \tag{4.2.6}$$

$$P_n^\lambda(\cos\theta; a, b) = e^{in\theta} \frac{(\lambda - i\Phi(\theta))_n}{n!} {}_2F_1\left(\left.\begin{matrix}-n, \lambda + i\Phi(\theta) \\ -n - \lambda + i\Phi(\theta)\end{matrix}\right| e^{-2i\theta}\right). \tag{4.2.7}$$

Szegő (1950a) proved that the **orthogonality relation** is

$$\int_{-1}^{1} P_m^\lambda(x; a, b)P_n^\lambda(x; a, b)w^\lambda(x; a, b)\, dx = \frac{2\pi\Gamma(n + 2\lambda)\,\delta_{m,n}}{2^{2\lambda}(n + \lambda + a)n!}, \tag{4.2.8}$$

when $a \geq |b|$, $\lambda > 0$, and w^λ is given by

$$w^\lambda(\cos\theta; a, b) = \left(1 - x^2\right)^{\lambda - 1/2} \exp((2\theta - \pi)\Phi(\theta)) \, |\Gamma(\lambda + i\Phi(\theta))|^2 . \tag{4.2.9}$$

This weight function is peculiar since

$$\int_0^\pi \log w^\lambda(\cos\theta; a, b)\, d\theta = -\infty,$$

so that it does not belong to the Szegő class and the asymptotic behavior cannot be obtained from the general theory for the Szegő class.

Let

$$1 > x_{n,1}(\lambda, a, b) > x_{n,2}(\lambda, a, b) > \cdots > x_{n,n}(\lambda, a, b) > -1 \tag{4.2.10}$$

be the zeros of $P_n^\lambda(x; a, b)$ and let

$$x_{n,k}(\lambda, a, b) = \cos\left(\theta_{n,k}(\lambda, a, b)\right). \tag{4.2.11}$$

Theorem 4.2.1 (Rui and Wong, 1996) *We have for* $a + b \neq 0$,

$$\theta_{n,k}\left(\frac{1}{2}; a, b\right) = \sqrt{\frac{a+b}{n}} + \frac{(a+b)^{1/6}}{2n^{5/6}}(-a_k) + O\left(n^{-7/6}\right), \tag{4.2.12}$$

where a_k *is the kth negative zero of the Airy function* $\mathrm{Ai}(x)$.

This should be contrasted with the case of ultraspherical polynomials where

$$\lim_{n\to\infty} n\theta_{n,k}(\nu, 0, 0) = j_{\nu-1/2,k}.$$

Rui and Wong proved an asymptotic formula for Pollaczek polynomials with $x = \cos(t/\sqrt{n}\,)$, which implies (4.2.12).

Let $\rho_1(z)$ and $\rho_2(z)$ be the roots of

$$\rho_1(x),\ \rho_2(x) = x \pm \sqrt{x^2 - 1}, \quad |\rho_1(x)| \leq |\rho_2(x)| . \tag{4.2.13}$$

In order for $\{P_m^\lambda(x; a, b)\}$ to be an orthogonal system it is necessary and sufficient that (i) or (ii) below hold:

$$\text{(i) } \lambda > 0 \text{ and } a + \lambda > 0, \quad \text{(ii) } -1/2 < \lambda < 0 \text{ and } -1 < a + \lambda < 0. \tag{4.2.14}$$

In all cases the orthogonality measure has an absolutely continuous part on $[-1, 1]$

$$\begin{aligned} \frac{d\mu^\lambda(x; a, b)}{dx} &= \frac{2^{2\lambda-1}(\lambda + a)}{\pi\Gamma(2\lambda)} \left(1 - x^2\right)^{\lambda - 1/2} \\ &\quad \times \exp\left((2\theta - \pi)\Phi(\theta)\right) |\Gamma(\lambda + i\Phi(\theta))|^2 . \end{aligned} \tag{4.2.15}$$

The measure μ^λ in (4.2.15) is normalized so that $\int_{\mathbb{R}} d\mu^\lambda(x; a, b) = 1$. There may be a discrete part outside $[-1, 1]$.

Let D be the set of points supporting point masses for μ^λ and set

$$\Delta_n = (n+\lambda)^2 + b^2 - a^2, \quad x_n = \frac{-ab + (n+\lambda)\sqrt{\Delta_n}}{a^2 - (n+\lambda)^2}, \quad y_n = \frac{-ab - (n+\lambda)\sqrt{\Delta_n}}{a^2 - (n+\lambda)^2}. \tag{4.2.16}$$

Theorem 4.2.2 (Charris and Ismail, 1987) *Let $a > |b|$. Then $D = \emptyset$ when $\lambda > 0$, but $D = \{x_0, y_0\}$, and $x_0 > 1, y_0 < -1$ if $\lambda < 0$.*

Theorem 4.2.3 (Charris and Ismail, 1987) *When $b \geqq 0$ and $a \leqq b$, the set D is as follows:*

Region I	(i)	$a < b$.	*Then* $D = \{x_n : n \geqq 0\}$.
	(ii)	$a = b$.	*Then* $D = \emptyset$.
Region II	(i)	$-b \leqq a < b$.	*Then* $D = \{x_n : n \geqq 0\}$.
	(ii)	$a < -b$.	*Then* $D = \{x_n : n \geqq 0\} \bigcup \{y_n : n \geqq 0\}$.
Region III	(i)	$-b < a < b$.	*Then* $D = \{x_n : n \geqq 0\}$, $x_0 > 1$.
	(ii)	$a = -b \neq 0$.	*Then* $D = \{x_n : n \geqq 1\}$.
	(iii)	$a < -b$.	*Then* $D = \{x_n : n > 1\} \bigcup \{y_n : n > 1\}$.
	(iv)	$a = b > 0\ (= 0)$.	*Then* $D = \{x_0\}\ (= \emptyset)$.
Region IV	(i)	$-b < a$.	*Then* $D = \{x_n : n \geqq 0\}$, $x_0 > 1$.
	(ii)	$b = -a$.	*Then* $D = \{x_n : n \geqq 1\}$.
	(iii)	$a < -b$.	*Then* $D = \{x_n : n \geqq 1\} \bigcup \{y_n : n \geqq 1\}$.

In all the regions, $x_n < -1$ and $y_n > 1$ for $n \geqq 1$. Also, $x_0 < -1$ and $y_0 > 1$ if $\lambda > 0$.

The symmetry relation (4.2.5) shows that the case $a \leqq -b$, $b \leqq 0$ can be obtained from Theorem 4.2.3 interchanging x_n and y_n, $n \geqq 0$.

Let J_ζ be the point mass at $x = \zeta$. If $n > 0$ then

$$\begin{aligned} J_{x_n} &= (\lambda + a)\rho_1^{2n}\left(1 - \rho_1^2\right)^{2\lambda} \frac{(2\lambda)_n \left[a\sqrt{\Delta_n} - b(n+\lambda)\right]}{n!\sqrt{\Delta_n}\,[a^2 - (n+\lambda)^2]}, \\ J_{y_n} &= (\lambda + a)\rho_1^{2n}\left(1 - \rho_1^2\right)^{2\lambda} \frac{(2\lambda)_n \left[a\sqrt{\Delta_n} + b(n+\lambda)\right]}{n!\sqrt{\Delta_n}\,[a^2 - (n+\lambda)^2]}. \end{aligned} \tag{4.2.17}$$

On the other hand,

$$J_{x_0} = -2(\lambda + a)\rho_1(x_0) \frac{\left[a\sqrt{\Delta_0} - b\lambda\right]^2}{\sqrt{\Delta_0}\,(a^2 - \lambda^2)^2}, \tag{4.2.18}$$

$$J_{y_0} = -2(\lambda + a)\rho_1(y_0) \frac{\left[a\sqrt{\Delta_0} + b\lambda\right]^2}{\sqrt{\Delta_0}\,(a^2 - \lambda^2)^2}. \tag{4.2.19}$$

The details are in Charris and Ismail (1987).

With w^λ defined in (4.2.9) we have the **orthogonality relation**

$$\int_{-1}^{1} w^\lambda(x;a,b)P_m^\lambda(x;a,b)P_n^\lambda(x;a,b)\,dx + \frac{\pi\Gamma(2\lambda)}{\lambda+a}2^{1-2\lambda}\sum_{\zeta\in D} P_n^\lambda(\zeta;a,b)P_m^\lambda(\zeta;a,b)J_\zeta \qquad (4.2.20)$$
$$= \frac{2\pi\Gamma(n+2\lambda)}{2^{2\lambda}(n+\lambda+a)n!}\delta_{m,n}.$$

Askey and Ismail (1984) studied the symmetric case $b = 0$. Their polynomials are random walk polynomials associated with

$$\lambda_n = an + b, \quad \mu_n = n, \qquad (4.2.21)$$

a problem originally proposed by Karlin and McGregor (1958), who considered only the case $a = 0$. Around the same time, Carlitz (1958), independently and using a completely different approach, studied the same polynomials ($a = 0$). Let

$$G_n(x;a,b) = r_n(x)a^n\frac{(b/a)_n}{n!}, \qquad (4.2.22)$$

with λ_n and μ_n as in (4.2.21). The **recurrence relation** satisfied by $\{G_n(x;a,b)\}$ is

$$[b+n(a+1)]xG_n(x;a,b) = (n+1)G_{n+1}(x;a,b) + (an+b-a)G_{n-1}(x;a,b). \qquad (4.2.23)$$

Set

$$\xi = \sqrt{(a+1)^2x^2-4a}, \qquad (4.2.24)$$
$$\alpha = \frac{x(a+1)}{2a} + \frac{\xi}{2a}, \quad \beta = x(a+1)2a - \frac{\xi}{2a},$$

and

$$A = -\frac{b}{2a} - \frac{x(1-a)b}{2a\xi}, \quad B = -\frac{b}{2a} - \frac{x(1-a)b}{2a\xi}. \qquad (4.2.25)$$

Then we have the **generating functions**

$$\sum_{n=0}^{\infty} G_n(x;a,b)t^n = (1-t/\alpha)^A(1-t/\beta)^B, \qquad (4.2.26)$$

$$\sum_{n=0}^{\infty}\frac{(\lambda)_n}{(b/a)_n}t^nG_n(x;a,b) = (1-t/\alpha)^{-\lambda}\,{}_2F_1\left(\left.\begin{matrix}\lambda, -B\\ b/a\end{matrix}\right|\frac{t\xi}{1-t/\alpha}\right), \qquad (4.2.27)$$

and the **explicit form**

$$G_n(x;a,b) = \frac{(-B)_n}{n!}\beta^{-n}\,{}_2F_1\left(\left.\begin{matrix}-n, -A\\ -n+B+1\end{matrix}\right|\beta/\alpha\right) = \frac{(b/a)_n}{n!}\alpha^{-n}\,{}_2F_1\left(\left.\begin{matrix}-n, -B\\ b/a\end{matrix}\right| -\xi\alpha\right). \qquad (4.2.28)$$

We let

$$x_k = (b + 2ak)[(b + k(a+1))(b + ka(a+1))]^{-1/2},$$
$$J_k = \frac{ba^k(b/a)_k}{2k!}\frac{[b(1-a)]^{1+b/a}[b + k(a+1)]^{k-1}}{[b + ka(a+1)]^{k+1+b/a}}, \tag{4.2.29}$$

$$w(x; a, b) = \frac{b2^{-1+b/a}}{\pi(a+1)\Gamma(b/a)} \exp\left(\frac{b(a-1)}{a(a+1)}(\theta - \pi/2)\cot\theta\right)$$
$$\times (\sin\theta)^{-1+b/a}\left|\Gamma\left(\frac{b}{2a} + i\frac{b(1-a)}{2a(a+1)}\cot\theta\right)\right|^2, \tag{4.2.30}$$

$$x := \frac{2\sqrt{a}}{1+a}\cos\theta, \quad 0 < \theta < \pi. \tag{4.2.31}$$

The polynomials $\{G_n\}$ are orthogonal with respect to a positive measure if and only if a and b belong to one of four regions. Set the following:

Region I $\quad a > 1, b > 0.$ $\quad$ Here, K is empty.

Region II $\quad 0 \le a < 1, b > 0.$ $\quad$ Here, $K = \{0, 1, \ldots\}$.

Region III $\quad a < 1, 0 < a + b < a.$ $\quad$ Here, $K = \{0\}$.

Region IV $\quad a > 1, 0 < a + b < a.$ $\quad$ Here, $K = \{1, 2, \ldots\}$.

The orthogonality relation is

$$\int_{-2\sqrt{a}/(1+a)}^{2\sqrt{a}/(1+a)} G_m(x; a, b)G_n(x; a, b)w(x; a, b)\,dx$$
$$+ \sum_{k\in K} J_k\{G_m(x_k; a, b)\,G_n(x_k, a, b) + G_m(-x_k, a, b)\,G_m(-x_k, a, b)\}$$
$$= \frac{ba^n(b/a)_n}{n![b + n(a+1)]}\delta_{m,n}. \tag{4.2.32}$$

When $a = 0$, the generating function becomes

$$\sum_{n=0}^{\infty} G_n(x; 0, b)t^n = e^{tb/x}(1 - xt)^{(1-x^2)b/x^2}, \tag{4.2.33}$$

and the explicit form is

$$G_n(x; 0, b) = \sum_{k=0}^{n} \frac{b^{n-k}}{(n-k)!}x^{2k-n}\frac{\left(b(1 - 1/x^2)\right)_k}{k!}. \tag{4.2.34}$$

Moreover,

$$(b + n)xG_n(x; 0, b) = (n+1)G_{n+1}(x; 0, b) + bG_{n-1}(x; 0, a). \tag{4.2.35}$$

The orthogonality relation is

$$\sum_{k=0}^{\infty} J_k \{G_m(x_k;0,b)\,G_n(x_k;0,b) + G_m(-x_k;0,b)\,G_n(-x_k;0,b)\} = \frac{b^{n+1}}{(n+1)!(b+n)}\,\delta_{m,n}, \tag{4.2.36}$$

with

$$x_k = \sqrt{\frac{b}{b+k}}, \qquad J_k = \frac{b(b+k)^{k-1}}{2(k!)}\exp(-k-b), \tag{4.2.37}$$

$k = 0, 1, \ldots$. Carlitz (1958) proved (4.2.36) using (1.1.1).

Pollaczek also considered a four-parameter family of orthogonal polynomials generated by

$$P_0^\lambda(x;a,b,c) = 1, \quad P_1^\lambda(x;a,b,c) = \frac{2[(\lambda+a+c)x+b]}{c+1}, \tag{4.2.38}$$

$$2[(n+\lambda+a+c)x+b]P_n^\lambda(x;a,b,c) = (n+1+c)P_{n+1}^\lambda(x;a,b,c) + (n+2\lambda+c-1)P_{n-1}^\lambda(x;a,b,c). \tag{4.2.39}$$

Clearly, $\{P_n^\lambda(x;a,b,c)\}$ are the associated polynomials of $\{P_n^\lambda(x;a,b)\}$ of (4.2.1)–(4.2.2) and c is the association parameter. One can show that

$$\sum_{n=0}^{\infty} P_n^\lambda(\cos\theta;a,b,c)t^n = t^{-c}\left(1-te^{i\theta}\right)^{-\lambda+i\Phi(\theta)}\left(1-te^{-i\theta}\right)^{-\lambda-i\Phi(\theta)} \times \int_0^t u^{c-1}\left(1-ue^{i\theta}\right)^{\lambda-i\Phi(\theta)}\left(1-ue^{-i\theta}\right)^{\lambda+i\Phi(\theta)}du, \tag{4.2.40}$$

where Φ is as in (4.2.6). The absolutely continuous component of the orthogonality measure is

$$w^\lambda(x;a,b,c) = \frac{(2\sin\theta)^{2\lambda-1}e^{(2\theta-\pi)\Phi(\theta)}}{\pi\Gamma(2\lambda+c)\Gamma(c+1)} \frac{|\Gamma(\lambda+c+i\Phi(\theta))|^2}{\left|{}_2F_1\left(\left.\begin{matrix}1-\lambda+i\Phi(\theta),c\\ c+\lambda+i\Phi(\theta)\end{matrix}\right|e^{2i\theta}\right)\right|^2}. \tag{4.2.41}$$

An **explicit formula** is

$$P_n^\lambda(\cos\theta;a,b,c) = \frac{A_{-1}B_n - A_nB_{-1}}{A_{-1}B_0 - A_0B_{-1}}, \tag{4.2.42}$$

where

$$\begin{aligned} A_n &:= \frac{\Gamma(2\lambda+c+n)}{\Gamma(c+n+1)\Gamma(2\lambda)}e^{i(c+n)\theta}\,{}_2F_1\left(\left.\begin{matrix}-c-n,\lambda+i\Phi(\theta)\\ 2\lambda\end{matrix}\right|1-e^{-2i\theta}\right),\\ B_n &:= \frac{\Gamma(1-\lambda+i\Phi(\theta)\Gamma(1-\lambda-i\Phi(\theta))}{\Gamma(2-2\lambda)}(2\sin\theta)^{1-2\lambda}e^{i(2\lambda+c+n-1)\theta}\\ &\quad\times {}_2F_1\left(\left.\begin{matrix}1-2\lambda-c-n,1-\lambda+i\Phi(\theta)\\ 2-2\lambda\end{matrix}\right|1-e^{-2i\theta}\right). \end{aligned} \tag{4.2.43}$$

When the orthogonality measure is absolutely continuous, the orthogonality relation becomes

$$\int_{-1}^{1} P_m^\lambda(x;a,b,c)P_n^\lambda(x;a,b,c)w^\lambda(x;a,b,c)\,dx = \frac{(2\lambda+c-1)_n}{(c+1)_n(n+\lambda+a+c)}\delta_{m,n}. \tag{4.2.44}$$

4.3 Associated Laguerre and Hermite Polynomials

The Laguerre polynomials are birth and death process polynomials with rates $\lambda_n = n + \alpha + 1$, $\mu_n = n$. According to Remark 4.1.1, when n is replaced by $n + c$ we will have two birth and death process models

$$\text{Model I}: \quad \lambda_n = n + c + \alpha + 1, \quad \mu_n = n + c, \qquad n \ge 0, \tag{4.3.1}$$

$$\text{Model II}: \quad \lambda_n = n + c + \alpha + 1, \quad \mu_{n+1} = n + c, \quad n \ge 0,\ \mu_0 = 0. \tag{4.3.2}$$

Let

$$\eta := 0 \text{ in Model I}, \quad \eta := 1 \text{ in Model II}. \tag{4.3.3}$$

Theorem 4.3.1 (Ismail, Letessier, and Valent, 1988; Askey and Wimp, 1984) *Let $\{L_n^{(\alpha)}(x;c)\}$ and $\{\mathcal{L}_n^{(\alpha)}(x;c)\}$ be the F_n in Models I and II, respectively, and let $\mu(0;x)$ and $\mu(1;x)$ be their spectral measures. Then*

$$\begin{gathered} L_0^{(\alpha)}(x;c) = 1, \quad L_1^{(\alpha)}(x;c) = \frac{2c+\alpha+1-x}{c+1}, \\ (2n+2c+\alpha+1-x)L_n^{(\alpha)}(x;c) \\ = (n+c+1)L_{n+1}^{(\alpha)}(x;c) + (n+c+\alpha)L_{n-1}^{(\alpha)}(x;c), \quad n > 0 \end{gathered} \tag{4.3.4}$$

and

$$\begin{gathered} \mathcal{L}_0^{(\alpha)}(x;c) = 1, \quad \mathcal{L}_1^{(\alpha)}(x;c) = \frac{c+\alpha+1-x}{c+1}, \\ (2n+2c+\alpha+1-x)\mathcal{L}_n^{(\alpha)}(x;c) \\ = (n+c+1)\mathcal{L}_{n+1}^{(\alpha)}(x;c) + (n+c+\alpha)\mathcal{L}_{n-1}^{(\alpha)}(x;c), \quad n > 0, \end{gathered} \tag{4.3.5}$$

$$\int_0^\infty \frac{d\mu(\eta;x)}{x+p} = \frac{\Psi(c+1,1-\alpha;p)}{\Psi(c,1-\alpha-\eta;p)}, \quad \eta = 0,1. \tag{4.3.6}$$

Moreover, the measures $\mu(\eta;x)$; $\eta = 0,1$ are absolutely continuous and

$$\begin{aligned} \mu'(0;x) &= x^\alpha e^{-x}\frac{\left|\Psi(c,1-\alpha;xe^{-i\pi})\right|^{-2}}{\Gamma(c+1)\Gamma(1+c+\alpha)}, \\ \mu'(1;x) &= x^\alpha e^{-x}\frac{\left|\Psi(c,-\alpha,xe^{-i\pi})\right|^{-2}}{\Gamma(c+1)\Gamma(1+c+\alpha)}. \end{aligned} \tag{4.3.7}$$

Furthermore, the polynomials $\{L_n^{(\alpha)}(x;c)\}$ *and* $\{\mathcal{L}_n^{(\alpha)}(x;c)\}$ *have the orthogonality relations*

$$\int_0^\infty p_{m,\eta}(x;\alpha,c)p_{n,\eta}(x;\alpha,c)\,d\mu(\eta;x) = \frac{(\alpha+c+1)_n}{(c+1)_n}\delta_{m,n}, \quad \eta = 0, 1, \tag{4.3.8}$$

where

$$p_{n,0}(x;\alpha,c) = L_n^{\alpha}(x;c), \quad p_{n,1}(x;\alpha,c) = \mathcal{L}_n^{\alpha}(x;c). \tag{4.3.9}$$

The polynomials have the explicit forms

$$\begin{aligned} p_{n,\eta}(x) = p_{n,\eta}(x;\alpha,c) &= \frac{(\alpha+1)_n}{n!} \\ &\times \sum_{m=0}^{n} \frac{(-n)_m x^m}{(c+1)_m(\alpha+1)_m}\, {}_3F_2\left(\left.\begin{matrix} m-n, m+1-\alpha-\eta, c \\ -\alpha-n, c+m+1 \end{matrix}\right| 1\right). \end{aligned} \tag{4.3.10}$$

In view of (3.8.10)–(3.8.11) we define associated Hermite polynomials by

$$\begin{aligned} H_{2n+1}(x;c) &= 2x(-4)^n(1+c/2)_n L_n^{1/2}\left(x^2;c/2\right), \\ H_{2n}(x;c) &= (-4)^n(1+c/2)_n \mathcal{L}_n^{-1/2}\left(x^2;c/2\right). \end{aligned} \tag{4.3.11}$$

Their orthogonality relation is

$$\int_{\mathbb{R}} \frac{H_m(x;c)H_n(x;c)}{\left|D_{-c}\left(xe^{i\pi/2}\sqrt{2}\right)\right|^2}\,dx = 2^n\sqrt{\pi}\,\Gamma(n+c+1)\,\delta_{m,n}. \tag{4.3.12}$$

The function D_{-c} in (4.3.12) is a parabolic cylinder function

$$D_{2\nu}(2x) = 2^{\nu} e^{-x^2}\Psi\left(-\nu, 1/2; 2x^2\right). \tag{4.3.13}$$

The polynomials $\{L_n^{(\alpha)}(x;c)\}$ and $\{H_n(x;c)\}$ were introduced in Askey and Wimp (1984), where their weight functions and explicit formulas were also found. Ismail, Letessier, and Valent (1988) realized that birth and death processes naturally give rise to two families of associated Laguerre polynomials and found an explicit representation and the weight function for the second family. The results on Model II are from Ismail, Letessier, and Valent (1988). It is then appropriate to call $\{L_n^{(\alpha)}(x;c)\}$ the Askey–Wimp polynomials and refer to $\{\mathcal{L}_n^{(\alpha)}(x;c)\}$ as the ILV polynomials, after the authors of Ismail, Letessier, and Valent (1988).

4.4 Associated Jacobi Polynomials

The Jacobi polynomials arise from a birth and death process with

$$\begin{aligned} \lambda_n &= \frac{2(n+\beta+1)(n+\alpha+\beta+1)}{(2n+\alpha+\beta+1)(2n+\alpha+\beta+2)}, \quad n \ge 0, \\ \mu_n &= \frac{2n(n+\alpha)}{(2n+\alpha+\beta)(2n+\alpha+\beta+1)}, \quad n \ge 0. \end{aligned} \tag{4.4.1}$$

The **three-term recurrence relation** for associated Jacobi polynomials is

$$\begin{aligned}2(n+c+1)(n+c+\gamma)(2n+2c+\gamma-1)y_{n+1}(x)\\ = (2n+2c+\gamma)\left[(2n+2c+\gamma-1)(2n+2c+\gamma+1)x+\left(\alpha^2-\beta^2\right)\right]\\ \times y_n(x) - 2(n+c+\alpha)(n+c+\beta)(2n+2c+\gamma+1)y_{n-1}(x),\end{aligned} \tag{4.4.2}$$

where

$$\gamma := \alpha+\beta+1. \tag{4.4.3}$$

As in Remark 4.1.1, we generate two systems of polynomial solutions, $\{P_n^{(\alpha,\beta)}(x;c)\}$ and $\{\mathcal{P}_n^{(\alpha,\beta)}(x;c)\}$, from the initial conditions

$$P_{-1}^{(\alpha,\beta)}(x;c) = 0, \quad P_0^{(\alpha,\beta)}(x;c) = 1, \tag{4.4.4}$$

$$\mathcal{P}_0^{(\alpha,\beta)}(x;c) = 1, \quad \mathcal{P}_1^{(\alpha,\beta)}(x;c) = \frac{(1+\gamma)(\gamma+2c)_2}{2(c+1)(\gamma+c)}x - \frac{\beta+c+1}{c+1}. \tag{4.4.5}$$

We shall refer to $\{P_n^{(\alpha,\beta)}(x;c)\}$ as the Wimp polynomials and to $\{\mathcal{P}_n^{(\alpha,\beta)}(x;c)\}$ as the Ismail–Masson polynomials. The Wimp polynomials have the property

$$P_n^{(\alpha,\beta)}(-x;c) = (-1)^n P_n^{(\beta,\alpha)}(x;c). \tag{4.4.6}$$

Theorem 4.4.1 (Wimp, 1987) *The polynomials $P_n^{(\alpha,\beta)}(x;c)$ have the* ***explicit form***

$$\begin{aligned}P_n^{(\alpha,\beta)}(x;c) = \frac{(\gamma+2c)_n(\alpha+c+1)_n}{(\gamma+c)_n n!}\sum_{k=0}^{n}\frac{(-n)_k(n+\gamma+2c)_k}{(c+1)_k(\alpha+c+1)_k}\left(\frac{1-x}{2}\right)^k\\ \times {}_4F_3\left(\left.\begin{matrix}k-n, n+\gamma+k+2c, \alpha+c, c\\ \alpha+k+c+1, k+c+1, \gamma+2c-1\end{matrix}\right|1\right),\end{aligned} \tag{4.4.7}$$

and satisfy the orthogonality relation

$$\int_{-1}^{1} P_m^{(\alpha,\beta)}(t;c)P_n^{(\alpha,\beta)}(t;c)\frac{(1-t)^\alpha(1+t)^\beta}{|F(t)|^2}\,dt = 0 \tag{4.4.8}$$

if $m \neq n$, where

$$F(t) := {}_2F_1\left(\left.\begin{matrix}c, 2-\gamma-c\\ 1-\beta\end{matrix}\right|\frac{1+t}{2}\right) + K(c)(1+t)\,{}_2F_1\left(\left.\begin{matrix}\beta+c, 1-\alpha-c\\ 1+\beta\end{matrix}\right|\frac{1+t}{2}\right), \tag{4.4.9}$$

with

$$K(c) = e^{i\pi\beta}\frac{\Gamma(-\beta)\Gamma(c+\beta)\Gamma(c+\gamma-1)}{2\Gamma(\beta)\Gamma(c+\gamma-\beta-1)\Gamma(c)}. \tag{4.4.10}$$

The polynomials $\{P_n^{(\alpha,\beta)}(x;c)\}$ satisfy the **differential equation**

$$A_0(x)y'''' + A_1(x)y''' + A_2(x)y'' + A_3(x)y' + A_4(x)y = 0, \tag{4.4.11}$$

with

$$A_0(x) = 1 - x^2, \quad A_1(x) = -10x\left(1 - x^2\right),$$

$$A_2(x) = -(1-x)^2\left(2K + 2C + \gamma^2 - 25\right),$$

$$+ 2(1-x)(2k + 2C + 2\alpha\gamma) + 2(\alpha+1) - 26, \tag{4.4.12}$$

$$A_3(x) = 3(1-x)\left(2K + 2C + \gamma^2 - 5\right) - 6(K + C + \alpha\gamma + \beta - 2),$$

$$A_4(x) = n(n+2)(n+\gamma+2c)(n+\gamma+2c-2),$$

where

$$K = (n+c)(n+\gamma+c), \quad C = (c-1)(c+\alpha+\beta). \tag{4.4.13}$$

Moreover, they have the explicit formula

$$\begin{aligned} P_n^{(\alpha,\beta)}(x;c) = {} & \frac{\Gamma(c+1)\Gamma(\gamma+c)}{\alpha\Gamma(\alpha+c)\Gamma(\beta+c)(\gamma+2c-1)} \\ & \times \left\{ \frac{\Gamma(c+\beta)\Gamma(n+\alpha+c+1)}{\Gamma(\gamma+c-1)\Gamma(n+c+1)} \, {}_2F_1\left(\left.\begin{matrix} c, 2-\gamma-c \\ 1-\alpha \end{matrix}\right| \frac{1-x}{2}\right)\right. \\ & \times {}_2F_1\left(\left.\begin{matrix} -n-c, n+\gamma+c \\ \alpha+1 \end{matrix}\right| \frac{1-x}{2}\right) \\ & \frac{\Gamma(\alpha+c)}{\Gamma(c)} \frac{\Gamma(n+\beta+1+c)}{\Gamma(n+c+\gamma)} \\ & \times {}_2F_1\left(\left.\begin{matrix} 1-c, \gamma+c-1 \\ \alpha+1 \end{matrix}\right| \frac{1-x}{2}\right) \\ & \left. \times {}_2F_1\left(\left.\begin{matrix} n+c+1, 1-n-\gamma-c \\ 1-\alpha \end{matrix}\right| \frac{1-x}{2}\right)\right\}. \end{aligned} \tag{4.4.14}$$

The polynomials $\{P_n^{(\alpha,\beta)}(x;c)\}$ have the asymptotic behavior

$$\begin{aligned} P_n^{(\alpha,\beta)}(x;c) \approx {} & \frac{\Gamma(c+1)\Gamma(\gamma+c)(2n\pi)^{-1/2}2^{(\beta-\alpha)/2}}{(\gamma+2c-1)\Gamma(\alpha+c)\Gamma(\beta+c)\left(1-x^2\right)^{1/4}} \\ & \times \left\{ 2^\alpha \frac{\Gamma(\beta+c)\Gamma(\alpha)}{\Gamma(\gamma+c-1)(1-x)^{\alpha/2}(1+x)^{\beta/2}} \right. \\ & \times {}_2F_1\left(\left.\begin{matrix} c, 2-\gamma-c \\ 1-\alpha \end{matrix}\right| \frac{1-x}{2}\right) \\ & \times \cos(n\theta + (c+\gamma/2)\theta - \pi(\alpha+1/2)/2) \\ & + \cos(n\theta + (c+\gamma/2)\theta + \pi(\alpha-1/2)/2) \\ & \left. \times \frac{\Gamma(\alpha+c)\Gamma(-\alpha)}{\Gamma(c)(1-x)^{-\alpha/2}(1+x)^{-\beta/2}} \, {}_2F_1\left(\left.\begin{matrix} 1-c, \gamma+c-1 \\ \alpha+1 \end{matrix}\right| \frac{1-x}{2}\right)\right\}, \end{aligned} \tag{4.4.15}$$

as $n \to \infty$, where $x = \cos\theta$, $0 < \theta < \pi$. With $R = \sqrt{1+t^2-2xt}$ we have the generating function

$$\begin{aligned}
&\sum_{n=0}^{\infty} \frac{(c+\gamma)_n(c+1)_n}{n!(\gamma+2c+1)_n} t^n P_n^{(\alpha,\beta)}(x;c) \\
&= \frac{1/\beta}{(\gamma+2c-1)}\left(\frac{2}{1+t+R}\right)^{\gamma+c}(\beta+c)(\gamma+c-1) \\
&\times {}_2F_1\left(\left.\begin{matrix} c, 2-\gamma-c \\ 1-\beta \end{matrix}\right| \frac{1+x}{2}\right) {}_2F_1\left(\left.\begin{matrix} -c, \gamma+c \\ \beta+1 \end{matrix}\right| \frac{1+t-R}{2t}\right) \\
&\times {}_2F_1\left(\left.\begin{matrix} \alpha+c+1, \gamma+c \\ \gamma+2c+1 \end{matrix}\right| \frac{2t}{1+t+R}\right) - c(\gamma+c-\beta-1) \\
&\times \left(\frac{1+t+R}{2}\right)^{\beta} {}_2F_1\left(\left.\begin{matrix} c+\beta, 1-c-\alpha \\ \beta+1 \end{matrix}\right| \frac{1+x}{2}\right) \\
&\times {}_2F_1\left(\left.\begin{matrix} c+\alpha+1, -c-\beta \\ 1-\beta \end{matrix}\right| \frac{1+t-R}{2t}\right) {}_2F_1\left(\left.\begin{matrix} \gamma+c+\beta, \gamma+c \\ \gamma+2c+1 \end{matrix}\right| \frac{2t}{1+t+R}\right).
\end{aligned} \tag{4.4.16}$$

When $c = 0$, (4.4.16) reduces to (3.2.7).

Ismail and Masson (1991) proved

$$\begin{aligned}
\mathcal{P}_n^{(\alpha,\beta)}(x;c) &= \frac{(-1)^n(\gamma+2c)_n(\beta+c+1)_n}{n!(\gamma+c)_n} \sum_{k=0}^{n} \frac{(-n)_k(\gamma+n+2c)_k}{(1+c)_k(c+1+\beta)_k} \\
&\times \left(\frac{1+x}{2}\right)^k {}_4F_3\left(\left.\begin{matrix} k-n, n+\gamma+k+2c, c+\beta+1, c \\ k+c+\beta+1, k+c+1, \gamma+2c \end{matrix}\right| 1\right),
\end{aligned} \tag{4.4.17}$$

$$\begin{aligned}
(-1)^n\mathcal{P}_n^{(\alpha,\beta)}(x;c) &= \frac{(c+\beta+1)_n}{(c+1)_n} {}_2F_1\left(\left.\begin{matrix} -n-c, n+c+\gamma \\ \beta+1 \end{matrix}\right| \frac{1+x}{2}\right) \\
&\times {}_2F_1\left(\left.\begin{matrix} c, 1-c-\gamma \\ -\beta \end{matrix}\right| \frac{1+x}{2}\right) - \frac{c(c+\alpha)_{n+1}(1+x)}{2\beta(\beta+1)(c+\gamma)_n} \\
&\times {}_2F_1\left(\left.\begin{matrix} n+c+1, 1-n-c-\gamma \\ 1-\beta \end{matrix}\right| \frac{1+x}{2}\right) {}_2F_1\left(\left.\begin{matrix} 1-c, c+\gamma \\ 2+\beta \end{matrix}\right| \frac{1+x}{2}\right).
\end{aligned} \tag{4.4.18}$$

Consequently,

$$\mathcal{P}_n^{(\alpha,\beta)}(-1;c) = \frac{(-1)^n(c+\beta+1)_n}{(c+1)_n}. \tag{4.4.19}$$

Ismail and Masson established the **asymptotic formula**

$$\begin{aligned}
\mathcal{P}_n^{(\alpha,\beta)}(x;c) &\approx \frac{\Gamma(\beta+1)\Gamma(c+1)}{\sqrt{n\pi}\,(c+\beta+1)}\left(\frac{1-x}{2}\right)^{-\alpha-1/2}\left(\frac{1+x}{2}\right)^{-\beta-1/2} \\
&\times W(x)\cos[(n+c+\gamma/2)\theta + c + (2\gamma-1)/4 - \eta],
\end{aligned} \tag{4.4.20}$$

with

$$W(x) = \left| {}_2F_1\left(\begin{matrix} c, -c-\beta-\alpha \\ -\beta \end{matrix} \middle| \frac{1+x}{2}\right) + \mathcal{K}(1+x)^{\beta+1}\, {}_2F_1\left(\begin{matrix} c+\beta+1, 1-c-\alpha \\ 2+\beta \end{matrix} \middle| \frac{1+x}{2}\right) \right|, \tag{4.4.21}$$

$x = \cos\theta$, $\theta \in (0, \pi)$, and

$$\mathcal{K} = \frac{\Gamma(c+\gamma)\Gamma(c+\beta+1)}{\Gamma(c)\Gamma(c+\alpha)\Gamma(2+\beta)} 2^{-\beta-1} e^{i\pi\beta}.$$

The phase shift η is derived from

$$\begin{aligned} W(x)\cos\eta &= \left[{}_2F_1\left(\begin{matrix} c, -c-\beta-\alpha \\ -\beta \end{matrix} \middle| \frac{1+x}{2}\right) \right. \\ &\left. + \mathcal{K}(1+x)^{\beta+1}\, {}_2F_1\left(\begin{matrix} c+\beta+1, 1-c-\alpha \\ 2+\beta \end{matrix} \middle| \frac{1+x}{2}\right) \right] \cos(\pi\beta/2), \\ W(x)\sin\eta &= \left[{}_2F_1\left(\begin{matrix} c, -c-\beta-\alpha \\ -\beta \end{matrix} \middle| \frac{1+x}{2}\right) \right. \\ &\left. - \mathcal{K}(1+x)^{\beta+1}\, {}_2F_1\left(\begin{matrix} c+\beta+1, 1-c-\alpha \\ 2+\beta \end{matrix} \middle| \frac{1+x}{2}\right) \right] \sin(\pi\beta/2). \end{aligned} \tag{4.4.22}$$

Ismail and Masson also gave the **generating function**

$$\begin{aligned} \sum_{n=0}^{\infty} \frac{(\gamma+c)_n (c+1)_n t^n}{n!(\gamma+2c+1)_n} \mathcal{P}_n^{\alpha,\beta}(x;c) &= \left\{\frac{2}{1+t+R}\right\}^{c+\gamma} {}_2F_1\left(\begin{matrix} c, 1-c-\gamma \\ -\beta \end{matrix} \middle| \frac{1+x}{2}\right) \\ &\times {}_2F_1\left(\begin{matrix} -c, c+\gamma \\ 1+\beta \end{matrix} \middle| \frac{1+t-R}{2t}\right) {}_2F_1\left(\begin{matrix} c+1+\alpha, \gamma \\ \gamma+2c+1 \end{matrix} \middle| \frac{2t}{1+t+R}\right) \\ &- \frac{c(c+\alpha)}{\beta(\beta+1)} \left\{\frac{2}{1+t+R}\right\}^{c+1} \left(\frac{1+x}{2}\right) {}_2F_1\left(\begin{matrix} 1-c, 2+\gamma \\ 2+\beta \end{matrix} \middle| \frac{1+x}{2}\right) \\ &\times {}_2F_1\left(\begin{matrix} 1-c-\gamma, c+1 \\ 1-\beta \end{matrix} \middle| \frac{1+t-R}{2t}\right) {}_2F_1\left(\begin{matrix} \beta+c+1, c+1 \\ \gamma+2c+1 \end{matrix} \middle| \frac{2t}{1+t+R}\right), \end{aligned}$$

where $R = \sqrt{1-2xt+t^2}$. When $c = 0$, the above generating function reduces to (3.2.7).

Finally, the orthogonality relation is

$$\int_{-1}^{1} \mathcal{P}_m^{(\alpha,\beta)}(x;c)\mathcal{P}_n^{(\alpha,\beta)}(x;c) \frac{(1-x)^\alpha(1+x)^\beta}{W^2(x)}\, dx = h_n^{(\alpha,\beta)}(c)\, \delta_{m,n}, \tag{4.4.23}$$

where

$$h_n^{(\alpha,\beta)}(c) = \frac{2^{\alpha+\beta+1}\Gamma(c+1)\Gamma^2(\beta+1)\Gamma(c+\alpha+n+1)\Gamma(c+\beta+n+1)}{(2n+2c+\gamma)\Gamma(c+\gamma+n)\Gamma^2(c+\beta+1)(c+1)_n}. \tag{4.4.24}$$

Analogous to (3.15.11), one can define two families of associated Bessel polynomials by

$$y_n(x;a,b;c) = \lim_{\lambda\to\infty} \frac{n!}{(\lambda+1)_n} P_n^{(\lambda,a-\lambda)}(1+2\lambda x/b;c), \tag{4.4.25}$$

$$\mathcal{Y}_n(x;a,b;c) = \lim_{\lambda\to\infty} \frac{n!}{(\lambda+1)_n} P_n^{(\lambda,a-\lambda)}(1+2\lambda x/b;c). \tag{4.4.26}$$

Therefore $\gamma = a+1$ and from (4.4.7) and (4.4.18) we find

$$\begin{aligned}
y_n(x;a,b;c) &= \frac{(a+1+2c)_n}{(a+1+c)_n} \sum_{k=0}^{n} \frac{(-n)_k(n+a+1+2c)_k}{(c+1)_k} \left(-\frac{x}{b}\right)^k \\
&\times {}_3F_2\left(\left.\begin{matrix} k-n, n+a+1+2c+k, c \\ k+c+1, a+2c \end{matrix}\right| 1\right), \\
\mathcal{Y}_n(x;a,b;c) &= \frac{(a+1+2c)_n}{(a+c+1)_n} \sum_{k=0}^{n} \frac{(-n)_k(a+1+n+2c)_k}{(c+1)_k(c+1+\beta)_k} \left(-\frac{x}{b}\right)^k \\
&\times {}_3F_2\left(\left.\begin{matrix} k-n, n+a+1+k+2c, c \\ k+c+1, a+1+2c \end{matrix}\right| 1\right).
\end{aligned} \tag{4.4.27}$$

Generating functions and asymptotics can be established by taking limits of the corresponding formulas for associated Jacobi polynomials.

4.5 Sieved Polynomials

The three-term recurrence relation of the q-ultraspherical polynomials $\{C_n(x;\beta|q)\}$ is (7.2.1). Let $k > 1$ be a positive integer and

$$\omega_k = \exp(2\pi i/k).$$

Al-Salam, Allaway, and Askey (1984) observed that the following limits exist:

$$\lim_{s\to1^-} \frac{\left(s^{\lambda k}; s^{\lambda k}\right)_n}{(s^{2\lambda k}; s^{\lambda k})_n} C_n\left(x; s^{\lambda k+1}\omega_k \mid s\omega_k\right) = c_n^\lambda(x;k), \tag{4.5.1}$$

$$\lim_{s\to1^-} C_n\left(x; s^{\lambda k+1}\omega_k \mid s\omega_k\right) = B_n^\lambda(x;k). \tag{4.5.2}$$

Their **three-term recurrence relations** are

$$\begin{aligned}
2xc_n^\lambda(x;k) &= c_{n+1}^\lambda(x;k) + c_{n-1}^\lambda(x;k), \quad k \nmid n, \\
2x(\lambda+m)c_{mk}^\lambda(x;k) &= (m+2\lambda)c_{mk+1}^\lambda(x;k) + mc_{mk-1}^\lambda(x;k), \quad m>0,
\end{aligned} \tag{4.5.3}$$

$$\begin{aligned}
2xB_n^\lambda(x;k) &= B_{n+1}^\lambda(x;k) + B_{n-1}^\lambda(x;k), \quad k \nmid n+1, \\
2x(\lambda+m)B_{mk-1}^\lambda(x;k) &= mB_{mk}^\lambda(x;k) + (2\lambda+m)B_{mk-2}^\lambda(x;k), \quad m>0,
\end{aligned} \tag{4.5.4}$$

with the initial conditions

$$\begin{aligned}
B_0^\lambda(x;k) = 1, &\quad B_1^\lambda(x;k) = 2x, \\
c_0^\lambda(x;k) = 1, &\quad c_1^\lambda(x;k) = x.
\end{aligned} \tag{4.5.5}$$

The polynomials $\{c_n^\lambda(x;k)\}$ are called **sieved ultraspherical polynomials of the first kind** while $\{B_n^\lambda(x;k)\}$ are **sieved ultraspherical polynomials of the second kind. Generating functions** are

$$\sum_{n=0}^{\infty} B_n^\lambda(x;k)t^n = \frac{(1-2xt+t^2)^{-1}}{[1-2T_k(x)t^k+t^{2k}]^\lambda}, \tag{4.5.6}$$

$$\sum_{n=0}^{\infty} b_n c_n^\lambda(x;k)t^n = \left(1-t^2\right)\frac{\left(1-2xt+t^2\right)^{-1}}{[1-2T_k(x)t^k+t^{2k}]^\lambda}, \tag{4.5.7}$$

where

$$b_n = \frac{(\lambda+1)_{\lfloor n/k\rfloor}(2\lambda)_{\lfloor n/k\rfloor}}{(1)_{\lfloor n/k\rfloor}(\lambda)_{\lfloor n/k\rfloor}}. \tag{4.5.8}$$

Clearly, (4.5.6) and (4.5.7) imply

$$b_n c_n^\lambda(x;k) = B_n^\lambda(x;k) - B_{n-2}^\lambda(x;k). \tag{4.5.9}$$

Moreover, Darboux's method implies (Ismail, 1985a)

$$B_n^\lambda(\cos\theta;k) = \left(\frac{n}{2k}\right)^\lambda \frac{\cos\left((n+k\lambda+1)\theta-\lambda\pi\ell-\pi(\lambda+1)/2\right)}{\Gamma(\lambda+1)\sin\theta|\sin(k\theta)|^\lambda}[1+o(1)], \tag{4.5.10}$$

$$c_n^\lambda(\cos\theta;k) = 2\left(\frac{n}{2k}\right)^\lambda \frac{\cos\left((n+k\lambda)\theta-\lambda\pi\ell-\pi\lambda/2\right)}{(\lfloor n/k\rfloor)^{2\lambda}\Gamma(\lambda+1)|\sin(k\theta)|^\lambda}[1+o(1)], \tag{4.5.11}$$

for $\ell\pi < k\theta < (\ell+1)\pi$ and $\ell = 0, 1, \ldots, k-1$.

The **orthogonality relations** are

$$\begin{aligned}\int_{-1}^{1} & B_m^\lambda(x;k)B_n^\lambda(x;k)\left(1-x^2\right)^{\lambda+1/2}|U_{k-1}(x)|^{2\lambda}\,dx \\ &= \frac{\sqrt{\pi}\,\Gamma(\lambda+1/2)}{\Gamma(\lambda+1)}\frac{(\lambda+1)_{\lfloor n/k\rfloor}(2\lambda+1)_{\lfloor (n+1)/k\rfloor}}{2(1)_{\lfloor n/k\rfloor}(\lambda+1)_{\lfloor (n+1)/k\rfloor}}\delta_{m,n},\end{aligned} \tag{4.5.12}$$

and

$$\begin{aligned}\int_{-1}^{1} & c_m^\lambda(x;k)c_n^\lambda(x;k)\left(1-x^2\right)^{\lambda-1/2}|U_{k-1}(x)|^{2\lambda}\,dx \\ &= \frac{\sqrt{\pi}\,\Gamma(\lambda+1/2)}{\Gamma(\lambda+1)}\frac{(1)_{\lfloor n/k\rfloor}(\lambda)_{\lfloor n/k\rfloor}}{(\lambda+1)_{\lfloor n/k\rfloor}(2\lambda)_{\lfloor n/k\rfloor}}\delta_{m,n},\end{aligned} \tag{4.5.13}$$

for $\lambda > 0$. Sieved versions of the Pollaczek polynomials were introduced in Ismail (1985a) and Charris and Ismail (1987).

The weight functions in (4.5.12) and (4.5.13) have $k-1$ zeros in $(-1,1)$. Thus we can think of the sieved ultraspherical polynomials as polynomials orthogonal on several adjacent intervals. Polynomials orthogonal on several disjoint intervals were introduced in Ismail (1986b) and Geronimo and Van Assche (1986). Let $c > 0$ and

$$I_0(x;k) = 1,\ I_1(x;k) = x, \quad J_0(x;k) = 1,\ J_1(x;k) = 2x, \tag{4.5.14}$$

$$\begin{aligned} (1+c)xI_{nk}(x;k) &= cI_{nk+1}(x;k) + I_{nk-1}(x;k), \quad n > 0, \\ 2xI_n(x;k) &= I_{n+1}(x;k) + I_{n-1}(x;k), \quad n > 0, \text{ if } k \nmid n, \end{aligned} \tag{4.5.15}$$

$$\begin{aligned} (1+c)xJ_{nk-1}(x;k) &= J_{nk}(x;k) + cJ_{nk-2}(x;k), \quad n > 0, \\ 2xJ_n(x;k) &= J_{n+1}(x;k) + J_{n-1}(x;k), \quad n > 0, \text{ if } k \nmid n+1. \end{aligned} \tag{4.5.16}$$

The polynomials $\{I_n(x;k)\}$ and $\{J_n(x;k)\}$ generalize the Chebyshev polynomials of the first and second kinds, respectfully, and were introduced by Ismail (1986b). Let

$$E(c) = \left\{x:\ -\frac{2\sqrt{c}}{1+c} < T_k(x) < \frac{2\sqrt{c}}{1+c}\right\}. \tag{4.5.17}$$

We have the **generating functions**

$$\sum_{n=0}^{\infty} J_n(x;k)t^n = \frac{1 - 2t^k T_k(x) + t^{2k}}{1 - 2xt + tr}\left[1 - t^k(1+c)T_k(x) + ct^{2k}\right]^{-1}, \tag{4.5.18}$$

$$\sum_{n=0}^{\infty} J_{nk+\ell}(x;k)t^n = \frac{U_\ell(x) + tU_{k-\ell-2}(x)}{1 - t(1+c)T_k(x) + ct^2}. \tag{4.5.19}$$

Moreover,

$$\begin{aligned} c^{-n/2}J_{nk+\ell} &= U_\ell(x)U_n(\cos\phi) + U_{k-\ell-2}(x)U_{n-1}(\cos\phi), \\ \cos\phi &= \frac{1+c}{2\sqrt{c}}T_k(x). \end{aligned} \tag{4.5.20}$$

Clearly, ϕ is real if and only if $x \in \overline{E(c)}$.

The **orthogonality relation** of $\{J_n(x;k)\}$ is

$$\int_{E(c)} J_m(x;k)J_n(x;k)w_J(x;k)\,dx = \tau_n\,\delta_{m,n} \quad \text{if } c > 1,$$

$$\begin{aligned} &\int_{E(c)} J_m(x;k)J_n(x;k)w_J(x;k)\,dx \\ &\quad + \frac{2(1-c)}{k}\sum_{j=1}^{k-1}\left(1 - x_j^2\right)J_m\left(x_j;k\right)J_n\left(x_j;k\right) = \tau_n\,\delta_{m,n} \quad \text{if } c < 1, \end{aligned} \tag{4.5.21}$$

$$w_J(x;k) := \frac{\sqrt{4c - (1+c)^2T_k^2(x)}}{\pi\,|U_{k-1}(x)|},$$

$$x_j := \cos(j\pi/k), \quad \tau_{nk+\ell} = c^n,\ 0 \le \ell < k-2, \quad \tau_{nk+k-1} = \frac{2c^{n+1}}{1+c}.$$

The **continued fraction** associated with $\{J_n(x;k)\}$ is

$$\cfrac{2}{2x - \cfrac{d_1}{(1+d_1)x - \cfrac{d_2}{(1+d_2)x - \cdots \cfrac{d_n}{(1+d_n)x - \ddots}}}} = \frac{(1+c)T_k(x) + 2U_{k-2}(x) - \sqrt{(1+c)^2T_k^2(x) - 4c}}{U_{k-1}(x)}, \tag{4.5.22}$$

valid for $x \notin [-1, 1]$, where

$$d_n = 1 \text{ if } k \nmid n+1, \quad \text{and} \quad d_{nk-1} = c. \tag{4.5.23}$$

Let

$$w_I(x;k) := \frac{\sqrt{4c - (1+c)^2T_k^2(x)}}{2\pi(1-x^2)\,|U_{k-1}(x)|}. \tag{4.5.24}$$

The **orthogonality relation** of $\{I_n(x;k)\}$ is

$$\begin{gathered} \int_{E(c)} I_m(x;k)I_n(x;k)w_I(x;k)\,dx = \sigma_n\,\delta_{m,n} \quad \text{if } c \geq 1, \\ \int_{E(c)} I_m(x;k)I_n(x;k)w_I(x;k)\,dx + \frac{(1-c)}{k}\sum_{j=1}^{k} I_m(x_j;k)I_n(x_j;k) = \sigma_n\,\delta_{m,n} \quad \text{if } c < 1, \end{gathered} \tag{4.5.25}$$

where $x_j := \cos(j\pi/k)$, as before, and

$$\sigma_0 = 1, \quad \sigma_{nk+\ell} = \frac{c^{-n}}{2}, \; 0 \leq \ell < k-1, \quad \sigma_{nk} = \frac{c^{1-n}}{1+c}, \; n > 0. \tag{4.5.26}$$

One can show that

$$\begin{gathered} (c+1)c^{n/2}I_{nk}(x;k) = cU_n\left(\frac{c+1}{2\sqrt{c}}T_k(x)\right) - U_{n_2}\left(\frac{c+1}{2\sqrt{c}}T_k(x)\right), \quad n > 0, \\ c^{(n+1)/2}I_{nk+\ell}(x;k) = \sqrt{c}\,T_\ell(x)\left(\frac{c+1}{2\sqrt{c}}T_k(x)\right) - T_{k-\ell}U_{n-1}\left(\frac{c+1}{2\sqrt{c}}T_k(x)\right), \\ n \geq 0, \; 0 < \ell < k. \end{gathered} \tag{4.5.27}$$

The **continued fraction** associated with $\{I_n(x;k)\}$ is

$$\cfrac{1}{2x - \cfrac{d_1}{(1+d_1)x - \cfrac{d_2}{(1+d_2)x - \cdots \cfrac{d_n}{(1+d_n)x - \ddots}}}} \tag{4.5.28}$$
$$= \frac{2cU_{k-1}(x)}{(c-1)T_k(x) + \sqrt{(1+c)^2T_k^2(x) - 4c}},$$

valid for $x \notin [-1, 1]$, where

$$d_n = 1 \text{ if } k \nmid n, \quad \text{and} \quad d_{nk} = 1/c. \tag{4.5.29}$$

If we let ρ and σ be the measures the $\{J_n(x;k)\}$ and $\{I_n(x;k)\}$ are orthogonal with respect to, respectively, then

$$\int_{-1}^{1} \frac{d\rho(t)}{x-t} = \frac{(1+c)T_k(x) + 2U_{k-2}(x) - \sqrt{(1+c)^2T_k^2(x) - 4c}}{U_{k-1}(x)}, \tag{4.5.30}$$

$$\int_{-1}^{1} \frac{d\sigma(t)}{x-t} = \frac{2cU_{k-1}(x)}{(c-1)T_k(x) + \sqrt{(1+c)^2T_k^2(x) - 4c}}. \tag{4.5.31}$$

There is extensive literature on sieved orthogonal polynomials. Van Assche and Magnus (1989) used them to construct discrete measures whose masses are dense in an interval. Geronimo and Van Assche (1986) showed how to use polynomial mappings to generate sieved polynomials from other orthogonal polynomials. When the polynomial mapping function is $T_k(x)$ one can generate the Al-Salam–Allaway–Askey polynomials from the ultraspherical polynomials. One can also derive some of the properties of $\{I_n(x;k)\}$ and $\{J_n(x;k)\}$, as well as the results in Charris and Ismail (1986), from polynomial mappings. Charris and Ismail (1987) introduced a sieved analogue of the Pollaczek polynomials and showed that it cannot be generated from a polynomial mapping technique. The symmetric case is in Ismail (1985a). In Charris, Ismail, and Monsalve (1994) the idea of block recurrences was introduced and applied to several classes of orthogonal polynomials.

5

Wilson and Related Polynomials

5.1 The Meixner–Pollaczek Polynomials

These polynomials appeared first in Meixner (1934) as orthogonal polynomials of Sheffer A-type zero relative to $\frac{d}{dx}$. This is equivalent to having a generating function of the form

$$\begin{aligned}
&\sum_{n=0}^{\infty} p_n(x) c_n t^n = A(t)\exp(xH(t)), \\
&A(t) = \sum_{n=0}^{\infty} a_n t^n, \quad H(t) = \sum_{n=0}^{\infty} h_{n+1} t^{n+1},
\end{aligned} \tag{5.1.1}$$

for some sequence $(c_n)_n$ with $a_0 c_n h_1 \neq 0$ for all n. Their **recurrence relation** is

$$\begin{aligned}
&(n+1)P_{n+1}^{(\lambda)}(x;\phi) - 2[x\sin\phi + (n+\lambda)\cos\phi]P_n^{(\lambda)}(x;\phi) \\
&\quad + (n+2\lambda-1)P_{n-1}^{(\lambda)}(x;\phi) = 0,
\end{aligned} \tag{5.1.2}$$

with the initial conditions

$$P_0^{(\lambda)}(x;\phi) = 1, \quad P_1^{(\lambda)}(x;\phi) = 2[x\sin\phi + \lambda\cos\phi]. \tag{5.1.3}$$

The **monic polynomials** are

$$P_n(x) = \frac{n!}{(2i\sin\phi)^n} P_n^{(\lambda)}(x;\phi) \tag{5.1.4}$$

and they satisfy

$$xP_n(x) = P_{n+1}(x) - \left(\frac{n+\lambda}{\tan\phi}\right)P_n(x) + \frac{n(n+2\lambda-1)}{4\sin^2\phi}P_{n-1}(x). \tag{5.1.5}$$

We shall assume

$$0 < \phi < \pi, \quad \lambda > 0$$

to ensure orthogonality with respect to a positive measure.

A **generating function** and two **explicit forms** are

$$\sum_{n=0}^{\infty} P_n^{(\lambda)}(x;\phi)t^n = \left(1-te^{i\phi}\right)^{-\lambda+ix}\left(1-te^{-i\phi}\right)^{-\lambda-ix}, \tag{5.1.6}$$

$$P_n^{(\lambda)}(x;\phi) = \frac{(\lambda+ix)_n}{n!}e^{-in\phi}\,{}_2F_1\left(\left.\begin{matrix}-n,\lambda-ix\\ -n-\lambda-ix+1\end{matrix}\right| e^{2i\phi}\right), \tag{5.1.7}$$

$$P_n^{(\lambda)}(x;\phi) = \frac{(2\lambda)_n}{n!}e^{in\phi}\,{}_2F_1\left(\left.\begin{matrix}-n,\lambda+ix\\ 2\lambda\end{matrix}\right| 1-e^{-2i\phi}\right). \tag{5.1.8}$$

Darboux's method leads to the asymptotic formulas

$$P_n^{(\lambda)}(x;\phi) \approx \begin{cases}\dfrac{(\lambda-ix)_n}{n!}\left(1-e^{-2i\phi}\right)^{-\lambda-ix}e^{in\phi}, & \operatorname{Im} x>0,\\ \dfrac{(\lambda+ix)_n}{n!}\left(1-e^{2i\phi}\right)^{-\lambda+ix}e^{-in\phi}, & \operatorname{Im} x<0.\end{cases} \tag{5.1.9}$$

When x is real, then Darboux's method gives

$$P_n^{(\lambda)}(x;\phi) \approx \frac{(\lambda-ix)_n}{n!}\left(1-e^{-2i\phi}\right)^{-\lambda-ix}e^{in\phi} + \frac{(\lambda+ix)_n}{n!}\left(1-e^{2i\phi}\right)^{-\lambda+ix}e^{-in\phi}.$$

The **orthogonality relation** for Meixner–Pollaczek polynomials is

$$\int_{\mathbb{R}} w_{\mathrm{MP}}^{(\lambda)}(x;\phi)P_m^{(\lambda)}(x;\phi)P_n^{(\lambda)}(x;\phi)\,dx = \frac{2\pi\Gamma(n+2\lambda)}{(2\sin\phi)^{2\lambda}n!}\delta_{m,n}, \tag{5.1.10}$$

where

$$w_{\mathrm{MP}}^{(\lambda)}(x;\phi) := e^{(2\phi-\pi)x}\Gamma(\lambda+ix)\Gamma(\lambda-ix). \tag{5.1.11}$$

The explicit formula (5.1.8) implies the **generating functions**

$$\sum_{n=0}^{\infty} P_n^{(\lambda)}(x;\phi)\frac{t^n}{(2\lambda)_n} = \exp\left(te^{i\phi}\right)\,{}_1F_1\left(\left.\begin{matrix}\lambda+ix\\ 2\lambda\end{matrix}\right| -2it\sin\varphi\right), \tag{5.1.12}$$

$$\sum_{n=0}^{\infty} \frac{(\gamma)_n}{(2\lambda)_n}P_n^{(\lambda)}(x;\phi)\frac{t^n}{e^{in\phi}} = (1-t)^{-\gamma}\,{}_2F_1\left(\left.\begin{matrix}\gamma,\lambda+ix\\ 2\lambda\end{matrix}\right| \frac{1-e^{-2i\phi}}{t-1}t\right). \tag{5.1.13}$$

The **lowering and raising** operators are

$$\tilde{\Delta}_i P_n^{(\lambda)}(x;\phi) = 2\sin\phi P_{n-1}^{(\lambda+1/2)}(x;\phi), \tag{5.1.14}$$

$$\frac{1}{w_{\mathrm{MP}}^{(\lambda)}(x;\phi)}\tilde{\Delta}_i w_{\mathrm{MP}}^{(\lambda+1/2)}(x;\phi)P_{n-1}^{(\lambda+1/2)}(x;\phi) = -nP_n^{(\lambda)}(x;\phi), \tag{5.1.15}$$

where $\tilde{\Delta}_h$ is defined in (1.2.4). Hence we have the **Rodrigues formula** and the second-order **divided difference equation**

$$w_{\mathrm{MP}}^{(\lambda)}(x;\phi)P_n^{(\lambda)}(x;\phi) = \frac{(-1)^n}{n!}\left(\tilde{\Delta}_i\right)^n w_{\mathrm{MP}}^{(\lambda+n/2)}(x;\phi), \tag{5.1.16}$$

$$\frac{1}{w_{\mathrm{MP}}^{(\lambda)}(x;\phi)}\tilde{\Delta}_i w_{\mathrm{MP}}^{(\lambda+1/2)}(x;\phi)\tilde{\Delta}_i P_n^{(\lambda)}(x;\phi) = -2n\sin\phi P_n^{(\lambda)}(x;\phi), \tag{5.1.17}$$

respectively. Equation (5.1.17) can be expanded out as

$$\begin{aligned}e^{i\phi}(\lambda - ix)y(x+i) + 2i[x\cos\phi - (n+\lambda)\sin\phi]y(x) &= e^{-i\phi}(\lambda + ix)y(x-i), \\ y(x) &= P_n^{(\lambda)}(x;\phi).\end{aligned} \tag{5.1.18}$$

We have the **bilinear generating function**

$$\begin{aligned}\sum_{n=0}^{\infty} \frac{n!t^n}{(2\lambda_1)_n} P_n^{(\lambda_1)}(\xi;\phi_1)\, P_n^{(\lambda)}(\xi;\phi) &= \left[1 - te^{i(\phi_1-\phi)}\right]^{-\lambda-i\eta}\left[1 - te^{i(\phi_1+\phi)}\right]^{-\lambda+i\eta} \\ &\times F_1\left(\lambda_1 + i\xi, \lambda + i\eta, \lambda - i\eta, 2\lambda_1; \frac{-2it\sin\phi_1}{e^{i\phi} - e^{i\phi_1}t}, \frac{-2it\sin\phi_1}{e^{-i\phi} - e^{i\phi_1}t}\right),\end{aligned} \tag{5.1.19}$$

a special case of which is the following generalization of the Poisson kernel:

$$\begin{aligned}\sum_{n=0}^{\infty} \frac{n!t^n}{(2\lambda)_n} P_n^{(\lambda)}(\xi;\phi_1)\, P_n^{(\lambda)}(\xi;\phi) &= \left[1 - te^{i(\phi_1-\phi)}\right]^{-\lambda-i\eta}\left[1 - te^{i(\phi-\phi_1)}\right]^{-\lambda+i\eta} \\ &\times \left[1 - te^{i(\phi_1+\phi)}\right]^{i\xi+i\eta} {}_2F_1\left(\left.\begin{matrix}\lambda + i\xi, \lambda + i\eta \\ 2\lambda\end{matrix}\right| \frac{-4t\sin\phi\sin\phi_1}{(1 - te^{i(\phi_1-\phi)})(1 - te^{i(\phi-\phi_1)})}\right).\end{aligned} \tag{5.1.20}$$

Formula (5.1.20) is due to Meixner (1942) in a different notation; other proofs are in Rahman (1988) and Ismail and Stanton (1997), where (5.1.20) was also noted. The Poisson kernel is the case $\phi = \phi_1$ of (5.1.20).

We note the **limiting relations**

$$\lim_{\phi\to 0} P_n^{(\alpha)}(-x/\phi;\phi) = L_n^{(2\alpha+1)}(2x), \tag{5.1.21}$$

$$\lim_{\lambda\to\infty} \lambda^{-n/2} P_n^{(\lambda)}\left(\frac{x\sqrt{\lambda} - \lambda\cos\phi}{\sin\phi};\phi\right) = \frac{H_n(x)}{n!}. \tag{5.1.22}$$

5.2 Wilson Polynomials

The Wilson polynomials are the most general orthogonal polynomials of hypergeometric type. They were introduced in Wilson (1982) and are related to the Wigner 6−j symbols; see Biedenharn and Louck (1981, volume 2). Their hypergeometric form is

$$W_n(x;\mathbf{t}) = \prod_{j=2}^{4}\left(t_1 + t_j\right)_n {}_4F_3\left(\left.\begin{matrix}-n, n + t_1 + t_2 + t_3 + t_4, t_1 + i\sqrt{x}, t_1 - i\sqrt{x} \\ t_1 + t_2, t_1 + t_3, t_1 + t_4,\end{matrix}\right| 1\right), \tag{5.2.1}$$

where $\mathbf{t}$ stands for the vector (t_1, t_2, t_3, t_4). We shall assume that the parameters t_j, $1 \le j \le 4$ are either real or if one is complex then another one is its complex conjugate. With

$$\begin{aligned}A_n &= \frac{\left(n - 1 + \sum_{j=1}^4 t_j\right)(n + t_1 + t_2)(n + t_1 + t_3)(n + t_1 + t_4)}{\left(2n - 1 + \sum_{j=1}^4 t_j\right)\left(2n + \sum_{j=1}^4 t_j\right)}, \\ C_n &= \frac{n\,(n + t_2 + t_3 - 1)(n + t_2 + t_4 - 1)(n + t_3 + t_4 - 1)}{\left(2n - 2 + \sum_{j=1}^4 t_j\right)\left(2n - 1 + \sum_{j=1}^4 t_j\right)},\end{aligned} \tag{5.2.2}$$

the three-term recurrence relation becomes

$$\left(x+t_1^2\right)\tilde{W}_n(x) = A_n\tilde{W}_{n+1}(x) - (A_n+C_n)\tilde{W}_n(x) + C_n\tilde{W}_{n-1}(x),$$
$$\tilde{W}_n(x) = \tilde{W}_n(x;\mathbf{t}) := \frac{W_n(x;\mathbf{t})}{\prod_{j=2}^4 \left(t_1+t_j\right)_n}, \tag{5.2.3}$$

and the monic polynomials $\{P_n(x)\}$ satisfy

$$\left(x+a^2\right)P_n(x) = P_{n+1}(x) - (A_n+C_n)P_n(x) + A_{n-1}C_nP_{n-1}(x),$$
$$P_n(x) := \frac{(-1)^n}{\left(n-1+\sum_{j=1}^4 t_j\right)_n} W_n(x;\mathbf{t}). \tag{5.2.4}$$

Let

$$w(x;\mathbf{t}) := \frac{\prod_{j=1}^4 \Gamma\left(t_j+i\sqrt{x}\right)\Gamma\left(t_j+i\sqrt{x}\right)}{2\sqrt{x}\,\Gamma\left(2i\sqrt{x}\right)\Gamma\left(-2i\sqrt{x}\right)}, \quad x>0. \tag{5.2.5}$$

The orthogonality relation for $\operatorname{Re} t_j > 0$, $1 \le j \le 4$ is

$$\frac{1}{2\pi}\int_0^\infty w(x;\mathbf{t})W_m(x;\mathbf{t})W_n(x;\mathbf{t})\,dx$$
$$= \frac{\prod_{1\le j<k\le 4}\Gamma\left(t_j+t_k\right)}{\Gamma(t_1+t_2+t_3+t_4+2n)} n!\,(n+t_1+t_2+t_3+t_4-1)_n\,\delta_{m,n}. \tag{5.2.6}$$

If $t_1 < 0$ the orthogonality relation (5.2.6) becomes

$$\frac{1}{2\pi}\int_0^\infty w(x;\mathbf{t})W_m(x;\mathbf{t})W_n(x;\mathbf{t})\,dx + \frac{\prod_{j=1}^3 \Gamma\left(t_j+t_1\right)\Gamma\left(t_j-t_1\right)}{\Gamma(-2t_1)}$$
$$\times \sum_{k,t_1+k<0} \frac{(2t_1)_k\,(t_1+1)_k\prod_{j=2}^4\left(t_1+t_j\right)_k}{k!\,(t_1)\,k\prod_{j=2}^4\left(t_1+1-t_j\right)_k} W_m(k-t_1;\mathbf{t})\,W_n(k-t_1;\mathbf{t}) \tag{5.2.7}$$
$$= \frac{n!\prod_{1\le j<k\le 4}\Gamma\left(t_j+t_k\right)}{\Gamma(t_1+t_2+t_3+t_4+2n)}(n+t_1+t_2+t_3+t_4-1)_n\,\delta_{m,n}.$$

The polynomials have the symmetry relation

$$W_n(x,t_1,t_2,t_3,t_4) = W_n\left(x,t_{\sigma(1)},t_{\sigma(2)},t_{\sigma(3)},t_{\sigma(4)}\right), \tag{5.2.8}$$

where σ is any permutation on $\{1,2,3,4\}$. This encodes the symmetries of the 6$-j$ symbols (Biedenharn and Louck, 1981). Formula (5.2.8) obviously holds when σ fixes 1 but gives the Whipple transformation.

The **raising and lowering operators** are

$$\mathcal{W}(W_n(x;\mathbf{t})) = -n(n+t_1+t_2+t_3+t_4-1)\,W_n(x;\mathbf{t}-1/2), \tag{5.2.9}$$

$$\frac{1}{w(x;\mathbf{t})}\mathcal{W}(w(x;\mathbf{t}+\mathbf{1/2})W_n(x;\mathbf{t}+1/2)) = W_{n+1}(x;\mathbf{t}), \tag{5.2.10}$$

respectively, where $w(x;\mathbf{t})$ is as in (5.2.5) and $\mathcal{W}$ is the Wilson operator defined through (1.2.6)–(1.2.5). The second-order difference equation is

$$\frac{1}{w(x;\mathbf{t})}\mathcal{W}\left(w(x;\mathbf{t}+1/2)\mathcal{W}W_n(x;\mathbf{t})\right) = -n\,(n+t_1+t_2+t_3+t_4-1)\,W_n(x;\mathbf{t}). \tag{5.2.11}$$

Equivalently, $z = W_n(x^2;\mathbf{t})$ solves

$$n\,(n+t_1+t_2+t_3+t_4-1)\,z(x) = B(x)z(x+i) - [B(x)+D(x)]z(x) + D(x)z(x-i), \tag{5.2.12}$$

with

$$B(x) = \frac{\prod_{j=1}^{4}\left(t_j - ix\right)}{2ix(2ix-1)}, \qquad D(x) = \frac{\prod_{j=1}^{4}\left(t_j + ix\right)}{2ix(2ix+1)}. \tag{5.2.13}$$

Most of the results stated so far are from Wilson (1980). A **generating function** is (Wilson, 1991)

$$\sum_{n=0}^{\infty}\frac{W_n(x;\mathbf{t})t^n}{(t_1+t_2)_n\,(t_3+t_4)_n} = {}_2F_1\left(\left.\begin{matrix} t_1+i\sqrt{x}\,, t_2+i\sqrt{x} \\ t_1+t_2 \end{matrix}\right| t\right) {}_2F_1\left(\left.\begin{matrix} t_3-i\sqrt{x}\,, t_4-i\sqrt{x} \\ t_3+t_4, \end{matrix}\right| t\right). \tag{5.2.14}$$

The following generating function follows from (3.2.4):

$$\begin{aligned}\sum_{n=0}^{\infty}&\frac{(s-1)_n W_n\,(x;\mathbf{t})}{n!\,(t_1+t_2)_n\,(t_1+t_3)_n\,(t_1+t_4)_n}t^n \\ &= (1-t)^{1-s}\,{}_4F_3\left(\left.\begin{matrix} s/2, (s-1)/2, t_1+i\sqrt{x}, t_1-i\sqrt{x} \\ t_1+t_2, t_1+t_3, t_1+t_4 \end{matrix}\right| \frac{-4t}{(1-t)^2}\right),\end{aligned} \tag{5.2.15}$$

where $s = t_1 + t_2 + t_3 + t_4$.

Theorem 5.2.1 (Wilson, 1991) *The large-degree asymptotics of the Wilson polynomials are given by the following expressions:*

(i) *If the parameters t_j, $1 \le j \le 4$ are positive except for complex conjugate pair(s) with positive real parts and $x > 0$ then*

$$W_n(x;\mathbf{t}) = C_n\left[2\left|A\left(i\sqrt{x}\right)\right|\cos\left(2\sqrt{x}\,\ln n - \arg\left(A\left(i\sqrt{x}\right)\right)\right) + O(1/n)\right], \tag{5.2.16}$$

where

$$C_n = (2\pi)^{3/2}e^{-3n}n^{3n+s-1}, \quad s := \sum_{j=1}^{4} t_j,$$

$$A(z) = \Gamma(2z)/\prod_{j=1}^{4}\Gamma\left(z+t_j\right).$$

(ii) *If* $\operatorname{Im} z > 0$ *and* $A(-iz) \neq 0$, *then*

$$W_n\left(z^2;\mathbf{t}\right) = C_n A(-iz) n^{-2iz}[1+O(1/n)], \tag{5.2.17}$$

even if $-4z^2$ is the square of an integer.

(iii) *If* $\operatorname{Im} z > 0$ *and* $A(-iz) = 0$ *(this case corresponds to having a mass point) then*

$$W_n\left(z^2;\mathbf{t}\right) = C_n A(iz) n^{2iz}[1 + O(1/n)], \tag{5.2.18}$$

except that the right-hand side must be doubled if $-4z^2$ *is the square of an integer.*

In the case $A(iz) = A(-iz) = 0$ we know only that $W_n(x;\mathbf{t}) \to 0$ as $n \to \infty$. This is the case when the measure has a point mass. Wilson (1991) contains complete asymptotic expansions in the cases covered by Theorem 5.2.1.

5.3 Continuous Dual Hahn Polynomials

The continuous dual Hahn polynomials arise as the limiting case $t_4 \to \infty$ of the Askey–Wilson polynomials. Their **hypergeometric representation** is

$$\frac{S_n(x;t_1,t_2,t_3)}{(t_1+t_2)_n(t_1+t_3)_n} = {}_3F_2\left(\left.\begin{matrix}-n, t_1+i\sqrt{x}, t_1-i\sqrt{x}\\ t_1+t_2, t_1+t_3\end{matrix}\right| 1\right). \tag{5.3.1}$$

Thus

$$S_n(x;t_1,t_2,t_3) = (-1)^n x^n + \text{ lower-order terms.}$$

It is clear that

$$S_n(x;t_1,t_2,t_3) = \lim_{t_4\to\infty} (t_4)^{-n} W_n(x;t_1,t_2,t_3,t_4). \tag{5.3.2}$$

The **weight function** for the continuous dual Hahn polynomials is

$$w_{\mathrm{CDH}}(x;t_1,t_2,t_3) := \frac{1}{2\sqrt{x}}\left|\frac{\Gamma\left(t_1+i\sqrt{x}\right)\Gamma\left(t_2+i\sqrt{x}\right)\Gamma\left(t_3+i\sqrt{x}\right)}{\Gamma\left(2i\sqrt{x}\right)}\right|^2. \tag{5.3.3}$$

Throughout the rest of this section we assume that either t_1, t_2, and t_3 are real or one is real and the other two are complex conjugates.

If $\operatorname{Re} t_j > 0$, $j = 1, 2, 3$, then we have the **orthogonality relation**

$$\begin{aligned}\frac{1}{2\pi}\int_0^\infty & w_{\mathrm{CDH}}(x;t_1,t_2,t_3) S_m(x;t_1,t_2,t_3) S_n(x;t_1,t_2,t_3)\,dx \\ &= n!\prod_{j=1}^{3}\Gamma\left(n+t_1+t_2+t_3-t_j\right)\delta_{mn}.\end{aligned} \tag{5.3.4}$$

On the other hand, if $t_1 < 0$ and $t_1 + t_2$, $t_1 + t_3$ are positive or a pair of complex conjugates with positive real parts, then the orthogonality relation becomes

$$\frac{1}{2\pi}\int_0^\infty w_{\mathrm{CDH}}(x;t_1,t_2,t_3)\,S_m(x;t_1,t_2,t_3)\,S_n(x;t_1,t_2,t_3)\,dx$$
$$+\frac{\Gamma(t_1+t_2)\Gamma(t_1+t_3)\Gamma(t_2-t_1)\Gamma(t_3-t_1)}{\Gamma(-2t_1)}$$
$$\times\sum_{\substack{k=0,1,2,\dots\\ t_1+k<0}}\frac{(2t_1)_k(t_1+1)_k(t_1+t_2)_k(t_1+t_3)_k}{(t_1)_k(t_1-t_2+1)_k(t_1-t_3+1)_k\,k!}(-1)^k \tag{5.3.5}$$
$$\times S_m\left(-(t_1+k)^2;t_1,t_2,t_3\right)S_n\left(-(t_1+k)^2;t_1,t_2,t_3\right)$$
$$=n!\prod_{j=1}^{3}\Gamma\left(n+t_1+t_2+t_3-t_j\right)\delta_{mn}.$$

The orthogonality relations (5.3.4)–(5.3.5) establish the symmetry relation

$$S_n(x;t_1,t_2,t_3)=S_n(x;t_{\sigma(1)},t_{\sigma(2)},t_{\sigma(3)}), \tag{5.3.6}$$

where σ is a permutation on $\{1,2,3\}$. The **three-term recurrence relation** is

$$\begin{aligned}-\left(t_1^2+x\right)S_n(x;t_1,t_2,t_3)&=S_{n+1}(x;t_1,t_2,t_3)-(A_n+C_n)\,S_n(x;t_1,t_2,t_3)\\&\quad+A_{n-1}C_nS_{n-1}(x;t_1,t_2,t_3),\end{aligned} \tag{5.3.7}$$

where

$$A_n=(n+t_1+t_2)(n+t_1+t_3),\qquad C_n=n(n+t_2+t_3-1).$$

The polynomials $\{S_n(x;t_1,t_2,t_3)\,/\,(t_1+t_2)_n(t_1+t_3)_n\}$ satisfy the birth and death recursion

$$-yQ_n(y)=A_nQ_{n+1}(y)-(A_n+C_n)\,Q_n(y)+C_nQ_{n-1}(y) \tag{5.3.8}$$

with $y=x+t_1^2$.

The function $y(x)=S_n(x^2;a,b,c)$ is a polynomial solution to the **difference equation**

$$ny(x)=B(x)y(x+i)-[B(x)+D(x)]\,y(x)+D(x)y(x-i), \tag{5.3.9}$$

where

$$B(x)=\frac{\left(t_1-i\sqrt{x}\right)\left(t_2-i\sqrt{x}\right)\left(t_3-i\sqrt{x}\right)}{2i\sqrt{x}\left(2i\sqrt{x}-1\right)},$$
$$D(x)=\frac{\left(t_1+i\sqrt{x}\right)\left(t_2+i\sqrt{x}\right)\left(t_3+i\sqrt{x}\right)}{2i\sqrt{x}\left(2i\sqrt{x}+1\right)}.$$

Recall the definition of the Wilson operator from (1.2.6)–(1.2.5). The **raising and lowering operators** are

$$\mathcal{W}S_n(x;t_1,t_2,t_3)=-nS_{n-1}(x;t_1+1/2,t_2+1/2,t_3+1/2), \tag{5.3.10}$$

$$\begin{aligned}&\mathcal{W}\left[w(x;t_1,t_2,t_3)\,S_n(x;t_1,t_2,t_3)\right]\\&\quad=w(x;t_1-1/2,t_2-1/2,t_3-1/2)\,S_{n+1}(x;t_1-1/2,t_2-1/2,t_3-1/2).\end{aligned} \tag{5.3.11}$$

Formula (5.3.11) implies the **Rodrigues-type formula**

$$w(x;t_1,t_2,t_3)\,S_n(x;t_1,t_2,t_3) = \mathcal{W}^n\left[w\left(x;t_1+\frac{n}{2},t_2+\frac{n}{2},t_3+\frac{n}{2}\right)\right]. \tag{5.3.12}$$

The continuous dual Hahn polynomials have the **generating functions**

$$(1-t)^{-t_3+i\sqrt{x}}\,{}_2F_1\left(\left.\begin{matrix} t_1+i\sqrt{x},\, t_2+i\sqrt{x} \\ t_1+t_2 \end{matrix}\right| t\right) = \sum_{n=0}^{\infty}\frac{S_n(x;t_1,t_2,c)}{(t_1+t_2)_n\, n!}t^n, \tag{5.3.13}$$

$$e^t\,{}_2F_2\left(\left.\begin{matrix} t_1+i\sqrt{x},\, t_1-i\sqrt{x} \\ t_1+t_2,\, t_1+t_3 \end{matrix}\right| -t\right) = \sum_{n=0}^{\infty}\frac{S_n(x;t_1,t_2,t_3)}{(t_1+t_2)_n(t_1+t_3)_n\, n!}t^n, \tag{5.3.14}$$

$$(1-t)^{-\gamma}\,{}_3F_2\left(\left.\begin{matrix} \gamma,\, t_1+i\sqrt{x},\, t_1-i\sqrt{x} \\ t_1+t_2,\, t_1+t_3 \end{matrix}\right| \frac{t}{t-1}\right) = \sum_{n=0}^{\infty}\frac{(\gamma)_n S_n(x;t_1,t_2,t_3)}{(t_1+t_2)_n(t_1+t_3)_n\, n!}t^n, \tag{5.3.15}$$

for arbitrary γ. Two more generating functions follow from (5.3.13) by permuting t_1, t_2, t_3, since $S_n(x;t_1,t_2,t_3)$ is invariant under such permutations.

5.4 Continuous Hahn Polynomials

The symmetric continuous Hahn polynomials were introduced in Askey and Wilson (1985) and then Atakishiyev and Suslov (1985) introduced the general case. Other proofs of the orthogonality are in Askey (1985), Kalnins and Miller (1988), and Koelink (1996). Special cases of the continuous Hahn polynomials appeared in the work of Bateman, Pasternack, and Touchard. Details and references are in Koelink (1996).

As in §5.2 we use the notation $\mathbf{t} = (t_1,t_2,t_3,t_4)$. Their hypergeometric representation is

$$\begin{aligned} p_n(x;\mathbf{t}) &= \frac{(t_1+t_3)_n(t_1+t_4)_n}{(-i)^n n!} \\ &\quad\times {}_3F_2\left(\left.\begin{matrix} -n,\, n+t_1+t_2+t_3+t_4-1,\, t_1+ix \\ t_1+t_3,\, t_1+t_4 \end{matrix}\right| 1\right). \end{aligned} \tag{5.4.1}$$

Let

$$w(x;\mathbf{t}) := \Gamma(t_1+ix)\Gamma(t_2+ix)\Gamma(t_3-ix)\Gamma(t_4-ix). \tag{5.4.2}$$

If $\mathrm{Re}\,(t_1,t_2,t_3,t_4) > 0$, $t_3=\bar{t}_1$, and $t_4=\bar{t}_2$, then the continuous Hahn polynomials satisfy the **orthogonality relation**

$$\begin{aligned} &\frac{1}{2\pi}\int_{\mathbb{R}} w(x;\mathbf{t})\,p_m(x;\mathbf{t})\,p_n(x;\mathbf{t})\,dx \\ &= \frac{\Gamma(n+t_1+t_3)\Gamma(n+t_1+t_4)\Gamma(n+t_2+t_3)\Gamma(n+t_2+t_4)}{(2n+t_1+t_2+t_3+t_4-1)\Gamma(n+t_1+t_2+t_3+t_4-1)\,n!}\delta_{mn}. \end{aligned} \tag{5.4.3}$$

The **three-term recurrence relation** of the continuous Hahn polynomials is

$$(a+ix)\,\tilde{p}_n(x) = A_n\tilde{p}_{n+1}(x) - (A_n+C_n)\,\tilde{p}_n(x) + C_n\tilde{p}_{n-1}(x), \tag{5.4.4}$$

where

$$\tilde{p}_n(x) := \tilde{p}_n(x;\mathbf{t}) = \frac{n!}{i^n (t_1+t_3)_n (t_1+t_4)_n} p_n(x;\mathbf{t})$$

and

$$\begin{aligned} A_n &= -\frac{(n+t_1+t_2+t_3+t_4-1)(n+t_1+t_3)(n+t_1+t_4)}{(2n+t_1+t_2+t_3+t_4-1)(2n+t_1+t_2+t_3+t_4)}, \\ C_n &= \frac{n(n+t_2+t_3-1)(n+t_2+t_4-1)}{(2n+t_1+t_2+t_3+t_4-2)(2n+t_1+t_2+t_3+t_4-1)}. \end{aligned} \tag{5.4.5}$$

On the other hand, the monic polynomials

$$P_n(x) = \frac{n!}{(n+t_1+t_2+t_3+t_4-1)_n} p_n(x;\mathbf{t}) \tag{5.4.6}$$

satisfy the recurrence relation

$$xP_n(x) = P_{n+1}(x) + i(A_n + C_n + t_1) P_n(x) - A_{n-1}C_n P_{n-1}(x). \tag{5.4.7}$$

The **difference equation** satisfied by the continuous Hahn polynomials is

$$\begin{aligned} &B(x)y(x+i) - [B(x)+D(x)]y(x) + D(x)y(x-i) \\ &= n(n+t_1+t_2+t_3+t_4-1)y(x), \end{aligned} \tag{5.4.8}$$

where $y(x) = p_n(x;\mathbf{t})$ and

$$B(x) = (t_3 - ix)(t_4 - ix), \quad D(x) = (t_1 + ix)(t_2 + ix). \tag{5.4.9}$$

The **lowering operator** is

$$\mathcal{W} p_n(x;\mathbf{t}) = (n+t_1+t_2+t_3+t_4-1)\, p_{n-1}\left(x;\mathbf{t}+\frac{1}{2}\right). \tag{5.4.10}$$

The **raising operator** is

$$\mathcal{W}\left[w(x;\mathbf{t})\, p_n(x;\mathbf{t})\right] = -(n+1) w\left(x;\mathbf{t}-\frac{1}{2}\right) \times p_{n+1}\left(x;\mathbf{t}-\frac{1}{2}\right). \tag{5.4.11}$$

A **Rodriguez-type formula** is

$$w(x;\mathbf{t})\, p_n(x;\mathbf{t}) = \frac{(-1)^n}{n!} \mathcal{W}^n \left[w\left(x;\mathbf{t}+\frac{n}{2}\right)\right]. \tag{5.4.12}$$

Koelink (1996) showed the following **connections with Jacobi polynomials**:

$$\int_{\mathbb{R}} e^{-2ixz}(1-\tan shx)^{\alpha}(1+\tan shx)^{\beta}P_n^{(\gamma,\delta)}(x)\,dx = 2^{\alpha+\beta-1}\frac{\Gamma(\alpha+iz)\Gamma(\beta-iz)}{i^n\Gamma(\alpha+\beta+n)}p_n(z;\alpha,\delta-\beta+1,\gamma-\alpha+1,\beta), \tag{5.4.13}$$

$$\int_0^{\infty}\frac{x^{\alpha}}{(1+x)^{\alpha+\beta}}P_n^{(\gamma,\delta)}\left(\frac{1-x}{1+x}\right)x^{-iz-1}\,dx = \frac{\Gamma(\alpha-iz)\Gamma(\beta-iz)}{i^n\Gamma(\alpha+\beta+n)}p_n(-z;\alpha,\delta-\beta+1,\gamma-\alpha+1,\beta). \tag{5.4.14}$$

Using (5.4.13) and contiguous relations for the ${}_2F_1$ function one can derive several recursion relations for continuous Hahn polynomials. Examples are

$$\begin{aligned} ix\,(n+t_1+t_4)\,p_n\,(x;t_1+1,t_2,t_3,t_4) &= (t_1+t_4)\,(t_1+ix)\,p_n\,(x;t_1+1,t_2,t_3-1,t_4) \\ &\quad + i\,(n+t_1+t_2+t_3+t_4-1)\,(t_1+ix)\,(t_4-ix)\,p_{n-1}\,(x;t_1+1,t_2,t_3,t_4+1), \end{aligned} \tag{5.4.15}$$

$$\begin{aligned} (2n+t_1+t_2+t_3+t_4)\,(t_1+ix)\,p_n\,(x;t_1+1,t_2,t_3,t_4) &= (t_1+t_4+n)\,(t_1+t_3+n)\,p_n\,(x;t_1,t_2,t_3,t_4) \\ &\quad + i(n+1)p_{n+1}\,(x;t_1+1,t_2,t_3,t_4). \end{aligned} \tag{5.4.16}$$

We have the **generating function**

$${}_1F_1\left(\left.\begin{matrix} t_1+ix \\ t_1+t_3 \end{matrix}\right| -it\right)\,{}_1F_1\left(\left.\begin{matrix} t_4-ix \\ t_2+t_4 \end{matrix}\right| it\right) = \sum_{n=0}^{\infty}\frac{p_n\,(x;\mathbf{t})}{(t_1+t_3)_n\,(t_2+t_4)_n}t^n, \tag{5.4.17}$$

with a similar generating function if we interchange t_3 and t_4. As Koelink (1996) observed, the generating function below follows from (5.4.13) and (3.2.1):

$$(1-t)^{1-s}\,{}_3F_2\left(\left.\begin{matrix} \frac{1}{2}(s-1),\frac{1}{2}(t_1+t_2+t_3+t_4),t_1+ix \\ t_1+t_3,t_1+t_4 \end{matrix}\right| -\frac{4t}{(1-t)^2}\right) = \sum_{n=0}^{\infty}\frac{(t_1+t_2+t_3+t_4-1)_n}{(t_1+t_3)_n\,(t_1+t_4)_n\,i^n}p_n\,(x;\mathbf{t})\,t^n, \tag{5.4.18}$$

where $s = t_1+t_2+t_3+t_4$. Note that (5.4.18) also follows from (3.2.4). Koelink proved

$$\sum_{n=0}^{\infty}\frac{(-it)^n p_n\,(x;\mathbf{t})}{(t_1+t_3)_n\,(t_1+t_4)_n\,(t_2+t_4)_n} = \sum_{k=0}^{\infty}\sum_{s=0}^{\infty}\frac{(-t)^s t^k\,(t_1+ix)_s\,(t_4-ix)_k}{s!\,(t_1+t_3)_s\,k!\,(t_2+t_4)_k\,(t_1+t_4)_{s+k}} \tag{5.4.19}$$

from (3.2.3) and (5.4.13).

6

Discrete Orthogonal Polynomials

6.1 Meixner and Charlier Polynomials

The **Meixner polynomials** $\{M_n(x;\beta,c)\}$ are orthogonal with respect to the negative binomial distribution. Let

$$w(x;\beta,c) = (\beta)_x c^x/x!, \quad x = 0,1,\ldots, \; c \in (0,1). \tag{6.1.1}$$

The attachment procedure of Section 1.6 leads to the **explicit form**

$$M_n(x;\beta,c) = {}_2F_1\left(\begin{matrix}-n,-x\\ \beta\end{matrix}\middle| 1-\frac{1}{c}\right) \tag{6.1.2}$$

and the **orthogonality relation**

$$\sum_{x=0}^{\infty} M_n(x;\beta,c)M_m(x;\beta,c)\frac{(\beta)_x}{x!}c^x = \frac{n!(1-c)^{-\beta}}{c^n(\beta)_n}\delta_{m,n}, \tag{6.1.3}$$

for $\beta > 0$, $0 < c < 1$. Their **three-term recurrence relation** is

$$\begin{aligned}-xM_n(x;\beta,c) &= c(\beta+n)(1-c)^{-1}M_{n+1}(x;\beta,c)\\ &\quad - [n+c(\beta+n)](1-c)^{-1}M_n(x;\beta,c) + n(1-c)^{-1}M_{n-1}(x;\beta,c),\end{aligned} \tag{6.1.4}$$

with the initial conditions

$$M_0(x;\beta,c) = 1, \quad M_1(x;\beta,c) = 1 + \frac{x(c-1)}{\beta c}. \tag{6.1.5}$$

Hence

$$M_n(x;\beta,c) = \frac{(1-1/c)^n}{(\beta)_n}x^n + \text{lower-order terms.} \tag{6.1.6}$$

The **generating functions** are

$$\sum_{n=0}^{\infty} \frac{(\beta)_n}{n!} M_n(x;\beta,c)t^n = (1-t/c)^x(1-t)^{-x-\beta}, \tag{6.1.7}$$

$$\sum_{n=0}^{\infty} \frac{t^n}{n!} M_n(x;\beta,c) = e^t \, {}_1F_1\left(\left.\begin{matrix}-x\\ \beta\end{matrix}\right| \frac{1-c}{c}t\right), \tag{6.1.8}$$

$$\sum_{n=0}^{\infty} \frac{(\gamma)_n}{n!} M_n(x;\beta,c)t^n = (1-t)^{-\gamma} \, {}_2F_1\left(\left.\begin{matrix}\gamma, -x\\ \beta\end{matrix}\right| \frac{(1-c)t}{c(1-t)}\right). \tag{6.1.9}$$

Recall the finite difference operators from (1.2.3). A **lowering operator** is

$$\Delta M_n(x;\beta,c) = \frac{n(c-1)}{\beta c} M_{n-1}(x;\beta+1,c). \tag{6.1.10}$$

An adjoint relation to (6.1.10) is

$$c(\beta+x)M_n(x;\beta+1,c) - xM_n(x-1;\beta+1,c) = c\beta M_{n+1}(x;\beta,c), \tag{6.1.11}$$

or equivalently,

$$\nabla\left[\frac{(\beta+1)_x}{x!}c^x M_n(x;\beta+1,c)\right] = \frac{(\beta)_x c^x}{x!} M_{n+1}(x;\beta,c). \tag{6.1.12}$$

Both (6.1.11) and (6.1.12) are **raising operators** for Meixner polynomials. This leads to the **difference equation** for Meixner polynomials,

$$\begin{aligned} &c(\beta+x)M_n(x+1;\beta,c) - [x+c(\beta+x)]M_n(x;\beta,c) + xM_n(x-1;\beta,c) \\ &\quad = n(c-1)M_n(x;\beta,c). \end{aligned} \tag{6.1.13}$$

It is important to note that the expression defining $M_n(x;\beta,c)$ in (6.1.2) is symmetric in x and n. Hence every formula we derive for $M_n(x;\beta,c)$ has a dual formula with x and n interchanged.

The discrete **Rodrigues formula** is

$$\frac{(\beta)_x c^x}{x!} M_n(x;\beta,c) = \nabla^n\left[\frac{(\beta+n)_x}{x!}c^x\right]. \tag{6.1.14}$$

More generally, we have

$$\frac{(\beta)_x c^x}{x!} M_{n+k}(x;\beta,c) = \nabla^k\left[\frac{(\beta+k)_x}{x!}c^x M_n(x;\beta+k,c)\right]. \tag{6.1.15}$$

The **relation to Jacobi polynomials** is

$$\frac{(\beta)_n}{n!} M_n(x;\beta,c) = P_n^{(\beta-1,-n-\beta-x)}((2-c)/c). \tag{6.1.16}$$

The limiting relation

$$\lim_{c\to 1^-} M_n(x/(1-c);\alpha+1,c) = \frac{n!}{(\alpha+1)_n} L_n^{(\alpha)}(x) \tag{6.1.17}$$

follows from (6.1.2) and (3.8.3). Another limiting case is

$$\lim_{\beta\to\infty} M_n(x;\beta, a/(\beta+a)) = C_n(x;a), \tag{6.1.18}$$

where $\{C_n(x;a)\}$ are the **Charlier polynomials**,

$$C_n(x;a) = {}_2F_0\,(-n,-x;—;-1/a)\,. \tag{6.1.19}$$

The orthogonality relation (6.1.3) and the generating function (6.1.7) imply

$$\sum_{x=0}^{\infty} C_m(x;a)C_n(x;a)\frac{a^x}{x!} = \frac{n!}{a^n}e^a\,\delta_{m,n}, \tag{6.1.20}$$

$$\sum_{n=0}^{\infty} C_n(x;a)\frac{t^n}{n!} = (1-t/a)^x e^t. \tag{6.1.21}$$

On the other hand, (6.1.11) and (6.1.10) establish the functional equations

$$\Delta C_n(x;a) = -\frac{n}{a}C_{n-1}(x;a), \tag{6.1.22}$$

$$aC_n(x;a) - xC_{n-1}(x-1;a) = aC_{n+1}(x;a). \tag{6.1.23}$$

The following **recurrence relation** follows from (6.1.4) and (6.1.18):

$$\begin{gathered} -xC_n(x;a) = aC_{n+1}(x;a) - (n+a)C_n(x;a) + nC_{n-1}(x;a), \\ C_0(x;a) = 1, \quad C_1(x;a) = (a-x)/a. \end{gathered} \tag{6.1.24}$$

We also have the binomial-type identity

$$C_n(x+y;a) = \sum_{k=0}^{n}\binom{n}{k}(-y)_{n-k}a^{-n+k}C_k(x;a). \tag{6.1.25}$$

The **relation to Laguerre polynomials** is

$$(-a)^n C_n(x;a) = n!L_n^{(x-n)}(a). \tag{6.1.26}$$

Rui and Wong (1994) derived uniform asymptotic developments for Charlier polynomials, which implies asymptotics of the kth largest zero of $C_n(x;a)$ as $n\to\infty$ and k is even allowed to depend on n.

6.2 Hahn, Dual Hahn, and Krawtchouk Polynomials

The weight function for the **Hahn polynomials** is

$$w(x;\alpha,\beta,N) := \frac{(\alpha+1)_x}{x!}\frac{(\beta+1)_{N-x}}{(N-x)!}, \qquad x = 0,1,\ldots,N. \tag{6.2.1}$$

An **explicit representation** is

$$Q_n(x) = Q_n(x;\alpha,\beta,N) = {}_3F_2\left(\left.\begin{matrix}-n, n+\alpha+\beta+1, -x\\ \alpha+1, -N\end{matrix}\right|1\right), \tag{6.2.2}$$

$n = 0, 1, \ldots, N$, while the **orthogonality relation** is

$$\sum_{x=0}^{N} Q_m(x;\alpha,\beta,N)Q_n(x;\alpha,\beta,N)w(x;\alpha,\beta,N) \\ = \frac{n!(N-n)!(\beta+1)_n(\alpha+\beta+n+1)_{N+1}}{(N!)^2(\alpha+\beta+2n+1)(\alpha+1)_n}\delta_{m,n}. \tag{6.2.3}$$

The **lowering operator** is

$$\Delta Q_n(x;\alpha,\beta,N) = -\frac{n(n+\alpha+\beta+1)}{N(\alpha+1)}Q_{n-1}(x;\alpha+1,\beta+1,N-1). \tag{6.2.4}$$

Moreover, (1.3.3) gives the endpoint evaluations

$$Q_n(0;\alpha,\beta,N) = 1, \quad Q_n(N;\alpha,\beta,N) = (-1)^n\frac{(\beta+1)_n}{(\alpha+1)_n}, \\ Q_n(-\alpha-1;\alpha,\beta,N) = (\alpha+\beta+N+2)_n\frac{(N-n)!}{N!}. \tag{6.2.5}$$

The **leading term** is

$$Q_n(x;\alpha,\beta,N) = \frac{(\alpha+\beta+n+1)_n}{(\alpha+1)_n(-N)_n}x^n + \frac{n(\alpha+\beta+1)_{n-1}}{2(\alpha+1)_n(-N)_n} \\ \times[(\alpha-\beta)(n-1)-2N(n+\alpha)]x^{n-1} + \text{lower-order terms}. \tag{6.2.6}$$

The **three-term recurrence relation** is

$$-xQ_n(x;\alpha,\beta,N) = \lambda_n Q_{n+1}(x;\alpha,\beta,N) + \mu_n Q_{n-1}(x;\alpha,\beta,N) \\ -[\lambda_n+\mu_n]Q_n(x;\alpha,\beta,N), \tag{6.2.7}$$

with

$$\lambda_n = \frac{(\alpha+\beta+n+1)(\alpha+n+1)(N-n)}{(\alpha+\beta+2n+1)(\alpha+\beta+2n+2)}, \\ \mu_n = \frac{n(n+\beta)(\alpha+\beta+n+N+1)}{(\alpha+\beta+2n)(\alpha+\beta+2n+1)}. \tag{6.2.8}$$

The limiting relations below follow from (6.2.2) and (6.1.2):

$$\lim_{N\to\infty} Q_n(Nx;\alpha,\beta,N) = P_n^{(\alpha,\beta)}(1-2x)n!/(\alpha+1)_n, \tag{6.2.9}$$

$$\lim_{N\to\infty} Q_n(x;\alpha,N((1-c)/c),N) = M_n(x;\alpha,c). \tag{6.2.10}$$

The adjoint relation to (6.2.4) is

$$(x+\alpha)(N+1-x)Q_n(x;\alpha,\beta,N) - x(\beta+N+1-x)Q_n(x-1;\alpha,\beta,N) \\ = \alpha(N+1)Q_{n+1}(x;\alpha-1,\beta-1,N-1) \tag{6.2.11}$$

or, equivalently,

$$\nabla[w(x;\alpha,\beta,N)Q_n(x;\alpha,\beta,N)] \\ = \frac{N+1}{\beta}w(x;\alpha-1,\beta-1,N+1)Q_{n+1}(x;\alpha-1,\beta-1,N+1). \tag{6.2.12}$$

This establishes the **second-order difference equation**

$$\frac{1}{w(x,\alpha,\beta,N)}\nabla(w(x;\alpha+1,\beta+1,N-1)\Delta Q_n(x;\alpha,\beta,N)) \\ = -\frac{n(n+\alpha+\beta+1)}{(\alpha+1)(\beta+1)}Q_n(x;\alpha,\beta,N). \tag{6.2.13}$$

Equation (6.2.13), when expanded out, reads

$$(x-N)(\alpha+x+1)\nabla\Delta y_n(x)+[x(\alpha+\beta+2)-N(\alpha+1)]\nabla y_n(x) \\ = n(n+\alpha+\beta+1)y_n(x), \tag{6.2.14}$$

or, equivalently,

$$(x-N)(\alpha+x+1)y_n(x+1)-[(x-N)(\alpha+x+1)+x(x-\beta-N-1)]y_n(x) \\ + x(x-\beta-N-1)y_n(x-1) = n(n+\alpha+\beta+1)y_n(x), \tag{6.2.15}$$

where $y_n(x) = Q_n(x;\alpha,\beta,N)$.

The orthogonality of the Hahn polynomials is equivalent to the orthogonality of the Clebsch–Gordon coefficients for SU(2), or $3-j$ symbols; see Koornwinder (1981).

The **dual Hahn polynomials** arise when we interchange n and x in (6.2.2). They are

$$R_n(\lambda(x)) = R_n(\lambda(x);\gamma,\delta,N) = {}_3F_2\left(\begin{matrix}-n,-x,x+\gamma+\delta+1\\ \gamma+1,-N\end{matrix}\,\middle|\,1\right), \\ \lambda(x) = x(x+\gamma+\delta+1), \tag{6.2.16}$$

for $n = 0,1,2,\ldots,N$. When $\gamma > -1$ and $\delta > -1$, or for $\gamma < -N$ and $\delta < -N$, the **orthogonality relation** dual to (6.2.3) is

$$\sum_{x=0}^{N}\frac{(2x+\gamma+\delta+1)(\gamma+1)_x(-N)_xN!}{(-1)^2(x+\gamma+\delta+1)_{N+1}(\delta+1)_xx!}R_m(\lambda(x);\gamma,\delta,N)R_n(\lambda(x);\gamma,\delta,N) \\ = \frac{\delta_{mn}}{\binom{\gamma+n}{n}\binom{\delta+N-n}{N-n}}. \tag{6.2.17}$$

The **three-term recurrence relation** for the dual Hahn polynomials is

$$\lambda(x)R_n(\lambda(x)) = A_nR_{n+1}(\lambda(x))-(A_n+C_n)R_n(\lambda(x))+C_nR_{n-1}(\lambda(x)), \tag{6.2.18}$$

where

$$A_n = (n+\gamma+1)(n-N), \quad C_n = n(n-\delta-N-1). \tag{6.2.19}$$

The **monic polynomials**

$$P_n(\lambda(x)) = (\gamma+1)_n(-N)_nR_n(\lambda(x);\gamma,\delta,N) \tag{6.2.20}$$

satisfy the **recurrence relation**

$$xP_n(x) = P_{n+1}(x)-(A_n+C_n)P_n(x)+A_{n-1}C_nP_{n-1}(x). \tag{6.2.21}$$

The dual Hahn polynomials satisfy the **difference equation**

$$-ny(x) = B(x)y(x+1) - [B(x) + D(x)]y(x) + D(x)y(x-1),$$
$$y(x) = R_n(\lambda(x); \gamma, \delta, N), \tag{6.2.22}$$

where

$$B(x) = \frac{(x+\gamma+1)(x+\gamma+\delta+1)(N-x)}{(2x+\gamma+\delta+1)(2x+\gamma+\delta+2)},$$
$$D(x) = \frac{x(x+\gamma+\delta+N+1)(x+\delta)}{(2x+\gamma+\delta)(2x+\gamma+\delta+1)}. \tag{6.2.23}$$

The **lowering operator** formula is

$$R_n(\lambda(x+1); \gamma, \delta, N) - R_n(\lambda(x); \gamma, \delta, N)$$
$$= -\frac{n(2x+\gamma+\delta+2)}{(\gamma+1)N} R_{n-1}(\lambda(x); \gamma+1, \delta, N-1), \tag{6.2.24}$$

or, equivalently,

$$\frac{\Delta R_n(\lambda(x); \gamma, \delta, N)}{\Delta\lambda(x)} = -\frac{n}{(\gamma+1)N} R_{n-1}(\lambda(x); \gamma+1, \delta, N-1). \tag{6.2.25}$$

The **raising operator** formula is

$$(x+\gamma)(x+\gamma+\delta)(N+1-x)R_n(\lambda(x); \gamma, \delta, N)$$
$$- x(x+\gamma+\delta+N+1)(x+\delta)R_n(\lambda(x-1); \gamma, \delta, N) \tag{6.2.26}$$
$$= \gamma(N+1)(2x+\gamma+\delta)R_{n+1}(\lambda(x); \gamma-1, \delta, N+1)$$

or, equivalently,

$$\frac{\nabla[\omega(x; \gamma, \delta, N)R_n(\lambda(x); \gamma, \delta, N)}{\nabla\lambda(x)}$$
$$= \frac{1}{\gamma+\delta}\omega(x; \gamma-1, \delta, N+1)R_{n+1}(\lambda(x); \gamma-1, \delta, N+1), \tag{6.2.27}$$

where

$$\omega(x; \gamma, \delta, N) = \frac{(-1)^x(\gamma+1)_x(\gamma+\delta+1)_x(-N)_x}{(\gamma+\delta+N+2)_x(\delta+1)_x x!}.$$

Iterating (6.2.27), we derive the **Rodrigues-type formula**

$$\omega(x; \gamma, \delta, N)R_n(\lambda(x); \gamma, \delta, N) = (\gamma+\delta+1)_n \left(\nabla_\lambda\right)^n [\omega(x; \gamma+n, \delta, N-n)], \tag{6.2.28}$$

where

$$\nabla_\lambda := \frac{\nabla}{\nabla\lambda(x)}.$$

The following **generating functions** hold for $x = 0, 1, 2, \ldots, N$:

$$(1-t)^{N-x} {}_2F_1\left(\left.\begin{matrix} -x, -x-\delta \\ \gamma+1 \end{matrix}\right| t\right) = \sum_{n=0}^{N} \frac{(-N)_n}{n!} R_n(\lambda(x); \gamma, \delta, N)t^n, \tag{6.2.29}$$

$$(1-t)^x\,{}_2F_1\left(\begin{matrix}x-N,x+\gamma+1\\-\delta-N\end{matrix}\,\middle|\,t\right)=\sum_{n=0}^{N}\frac{(\gamma+1)_n(-N)_n}{(-\delta-N)_n n!}R_n(\lambda(x);\gamma,\delta,N)t^n, \tag{6.2.30}$$

$$\left[e^t\,{}_2F_2\left(\begin{matrix}-x,x+\gamma+\delta+1\\\gamma+1,-N\end{matrix}\,\middle|\,-t\right)\right]_N=\sum_{n=0}^{N}\frac{R_n(\lambda(x);\gamma,\delta,N)}{n!}t^n, \tag{6.2.31}$$

$$\left[(1-t)^{-a}\,{}_3F_2\left(\begin{matrix}a,-x,x+\gamma+\delta+1\\\gamma+1,-N\end{matrix}\,\middle|\,\frac{t}{t-1}\right)\right]_N=\sum_{n=0}^{N}\frac{(a)_n}{n!}R_n(\lambda(x);\gamma,\delta,N)t^n, \tag{6.2.32}$$

where a is an arbitrary parameter. In the above we used the notation

$$\left[\sum_{m=k}^{\infty}a_n t^n\right]_N:=\sum_{n=0}^{N}a_n t^n, \quad \text{for } N\ge k. \tag{6.2.33}$$

The **Krawtchouk polynomials** are

$$K_n(x;p,N)={}_2F_1\left(-n,-x;-N;1/p\right), \quad n=0,1,2,\dots,N. \tag{6.2.34}$$

Formally they are Meixner polynomials with $\beta=-N$. The limiting relation

$$\lim_{t\to\infty}Q_n(x;pt,(1-p)t,N)=K_n(x;p,N)$$

enables us to derive many results for the Krawtchouk polynomials from the corresponding results for the Hahn polynomials. In particular, we establish the **orthogonality relation**

$$\sum_{x=0}^{N}\binom{N}{x}p^x(1-p)^{N-x}K_m(x,p,N)K_n(x;p,N)=\frac{(-1)^n n!}{(-N)_n}\left(\frac{1-p}{p}\right)^n\delta_{m,n}, \tag{6.2.35}$$

$0<p<1$, and the **recurrence relation**

$$\begin{aligned}-xK_n(x;p,N)&=p(N-n)K_{n+1}(x;p,N)\\&\quad-[p(N-n)+n(1-p)]K_n(x;p,N)+n(1-p)K_{n-1}(x;p,N).\end{aligned} \tag{6.2.36}$$

The **monic polynomials** $\{P_n(x)\}$ satisfy the **normalized recurrence relation**

$$xP_n(x)=P_{n+1}(x)+[p(N-n)+n(1-p)]P_n(x)+np(1-p)(N+1-n)P_{n-1}(x), \tag{6.2.37}$$

where

$$K_n(x;p,N)=\frac{1}{(-N)_n p^n}P_n(x).$$

The corresponding **difference equation** is

$$-ny(x)=p(N-x)y(x+1)-[p(N-x)+x(1-p)]y(x)+x(1-p)y(x-1), \tag{6.2.38}$$

where

$$y(x)=K_n(x;p,N).$$

The lowering operator is

$$\Delta K_n(x;p,N) = -\frac{n}{Np}K_{n-1}(x;p,N-1). \tag{6.2.39}$$

On the other hand, the raising operator is

$$(N+1-x)K_n(x;p,N) - x\left(\frac{1-p}{p}\right)K_n(x-1;p,N) = (N+1)K_{n+1}(x;p,N+1) \tag{6.2.40}$$

or, equivalently,

$$\nabla\left[\binom{N}{x}\left(\frac{p}{1-p}\right)^x K_n(x;p,N)\right] = \binom{N-n}{x}\left(\frac{p}{1-p}\right)^x K_{n+1}(x;p,N+1), \tag{6.2.41}$$

which leads to the Rodrigues-type formula

$$\binom{N}{x}\left(\frac{p}{1-p}\right)^x K_n(x;p,N) = \nabla^n\left[\binom{N-n}{x}\left(\frac{p}{1-p}\right)^x\right]. \tag{6.2.42}$$

The following generating functions hold for $x = 0, 1, 2, \ldots, N$:

$$\left(1 - \frac{(1-p)}{p}t\right)^x (1+t)^{N-x} = \sum_{n=0}^{N}\binom{N}{n}K_n(x;p,N)t^n, \tag{6.2.43}$$

$$\left[e^t\,{}_1F_1\left(\left.\begin{matrix}-x\\-N\end{matrix}\right|-\frac{t}{p}\right)\right]_N = \sum_{n=0}^{N}\frac{K_n(x;p,N)}{n!}t^n, \tag{6.2.44}$$

and

$$\left[(1-t)^{-\gamma}\,{}_2F_1\left(\left.\begin{matrix}\gamma,-x\\-N\end{matrix}\right|\frac{t}{p(t-1)}\right)\right]_N = \sum_{n=0}^{N}\frac{(\gamma)_n}{n!}K_n(x;p,N)t^n, \tag{6.2.45}$$

where γ is an arbitrary parameter.

The Krawtchouk polynomials are self-dual because they are symmetric in n and x. They are also the eigenmatrices of the Hamming scheme $H(n,q)$ (Bannai and Ito, 1984, Theorem 3.2.3). The orthogonality of the Krawtchouk polynomials is equivalent to the unitarity of unitary representations of SU(2) (Koornwinder, 1982).

Krawtchouk polynomials have been applied to many areas of mathematics. We shall briefly discuss their role in coding theory. The **Lloyd polynomials** $L_n(x;p,N)$ are

$$L_n(x;p,N) = \sum_{m=0}^{n}\binom{N}{m}\frac{p^m}{(1-p)^m}K_m(x;p,N).$$

The important cases are when $1/(1-p)$ is an integer. It turns out that

$$L_n(x;p,N) = \frac{p^n}{(1-p)^n}\binom{N-1}{n}K_n(x;p,N-1),$$

so the zeros of L_n are related to the zeros of K_n. One issue that arises in coding theory is to describe all integer zeros of K_n. In other words, for fixed p such that $1/(1-p)$ is an integer, describe all triples of positive integers (n, x, N) such that $K_n(x; p, N) = 0$ (Habsieger, 2001a). Habsieger and Stanton (1993) gave a complete list of solutions in the cases $N-2n \in \{1, 2, 3, 4, 5, 6\}$, $N - 2n = 8$, or x odd. Let $N(n, N)$ be the number of integer zeros of $K_n(x; 1/2, N)$. Two conjectures in this area are due to Krasikov and Litsyn (Habsieger, 2001a; Krasikov and Litsyn, 1996).

Conjecture 6.2.1 *For $2n - N < 0$, we have*

$$N(n, N) \leq \begin{cases} 3 & \text{if } n \text{ is odd,} \\ 4 & \text{if } n \text{ is even.} \end{cases}$$

Conjecture 6.2.2 *Let $n = \binom{m}{2}$. Then the only integer zeros of $K_n(x; 1/2, m^2)$ are 2, $m^2 - 2$, and $m^2/4$ for $m \equiv 2 \pmod 4$.*

Hong (1986) showed the existence of a noninteger zero for K_n when $1/p - 1$ is an integer greater than 2. For a survey of these results, see Habsieger (2001a); see also Habsieger (2001b).

The strong asymptotics of $K_n(x; p, N)$ when $n, N \to \infty$, $x > 0$ but n/N is fixed are in Ismail and Simeonov (1998), while a uniform asymptotic expansion is in Li and Wong (2000). Sharapudinov studied the asymptotic properties of $K_n(x; p, N)$ $n, N \to \infty$ with $n = O(N^{1/3})$. He also studied the asymptotics of the zeros of $K_n(x; p, N)$ when $n = o(N^{1/4})$. These results are in Sharapudinov (1988). More recently, Qiu and Wong (2004) gave an asymptotic expansion for the Krawtchouk polynomials and their zeros. The WKB technique was applied in Dominici (2008) to the study of the asymptotics of $K_n(x; p, N)$.

Let $q = 1/(1 - p)$ be a positive integer and denote the Hamming space $(\mathbb{Z}/q\mathbb{Z})^n$ by H, and O is the origin in H. For $X \subset H$, $X \neq \phi$ the Radon transform T_X is defined on functions $f\colon H \to \mathbb{C}$ by

$$T_X(f)(u) = \sum_{v \in u+X} f(v),$$

for $u \in H$. For $\mathbf{x} = (x_1, \ldots, x_N)$, $\mathbf{y} = (y_1, \ldots, y_N)$ in H, the Hamming distance between $\mathbf{x}$ and $\mathbf{y}$ is

$$d(\mathbf{x}, \mathbf{y}) = |\{i\colon 1 \leq i \leq N \text{ and } x_i \neq y_i\}|\,.$$

Let

$$S(\mathbf{x}, n) = \{\mathbf{y}\colon \mathbf{y} \in H, d(\mathbf{x}, \mathbf{y}) = n\}\,,$$
$$B(\mathbf{x}, n) = \{\mathbf{y}\colon \mathbf{y} \in H, d(\mathbf{x}, \mathbf{y}) \leq n\}\,.$$

Theorem 6.2.3 *The Radon transform $T_{S(O,n)}$ is invertible if and only if the polynomial $K_n(x, p, N)$, $q = 1/(1 - p)$ has no integer roots. The Radon transform $T_{B(O,n)}$ is invertible if and only if the (Lloyd) polynomial $K_n(x, p, N - 1)$ has no integer zeros.*

Theorem 6.2.3 is in Diaconis and Graham (1985) for $T_{S(O,n)}$, but Habsieger (2001a) pointed out that their proof method works for $T_{B(O,n)}$.

Another problem in graph theory, whose solution involves zeros of Krawtchouk polynomials, is a graph reconstruction problem. Let I be a subset of vertices of a graph G. Construct a new graph G_I by switching with respect to I. That is, if $u \in I$, $v \ni I$, then u and v are adjacent (nonadjacent) in G_I if and only if they are nonadjacent (adjacent) in G. Assume that G has N vertices. The n-switching deck is the multiset of unlabeled graphs $D_n(G) = \{G_I : |I| = n\}$. Stanley proved that G may be reconstructible from $D_n(G)$ if $K_n(x; 1/2, N)$ has no even zeros.

We have only mentioned samples of problems where an object has a certain property if the zeros of a Krawtchouk polynomial lie on the spectrum $\{0, 1, \ldots, N\}$.

6.3 Difference Equations

Let $\{p_n(x)\}$ be a polynomial system orthonormal with respect to a discrete measure supported on $\{s, s+1, \ldots, t\} \subset \mathbb{R}$, where s is finite but t is finite or infinite:

$$\sum_{\ell=s}^{t} p_m(\ell)p_n(\ell)w(\ell) = \delta_{m,n}, \quad \sum_{\ell=s}^{t} w(\ell) = 1. \tag{6.3.1}$$

We assume that

$$w(s-1) = 0, \quad w(t+1) = 0. \tag{6.3.2}$$

Define $u(x)$, the discrete analogue of the function $v(x)$ of Section 2.8, by

$$\Delta w(x) = w(x+1) - w(x) = -u(x+1)w(x+1). \tag{6.3.3}$$

Theorem 6.3.1 (Ismail, Nikolova, and Simeonov, 2004) *Let* $p_n(x) = \gamma_n x^n +$ lower-order terms, *satisfy* (6.3.1). *Then,*

$$\Delta p_n(x) = A_n(x)p_{n-1}(x) - B_n(x)p_n(x), \tag{6.3.4}$$

where $A_n(x)$ and $B_n(x)$ are given by

$$A_n(x) = \frac{a_n p_n(t+1)p_n(t)}{(t-x)} w(t) + a_n \sum_{\ell=s}^{t} p_n(\ell)p_n(\ell-1)\frac{u(x+1)-u(\ell)}{(x+1-\ell)} w(\ell), \tag{6.3.5}$$

$$B_n(x) = \frac{a_n p_n(t+1)p_{n-1}(t)}{(t-x)} w(t) + a_n \sum_{\ell=s}^{t} p_n(\ell)p_{n-1}(\ell-1)\frac{u(x+1)-u(\ell)}{(x+1-\ell)} w(\ell). \tag{6.3.6}$$

Relation (6.3.4) produces a **lowering operator**. A **raising operator** is

$$\frac{1}{A_n(x)} [\Delta + B_n(x)]\, p_n(x) = \left(\frac{x-b_n}{a_n}\right) p_n(x) - \frac{a_{n+1}}{a_n} p_{n+1}(x). \tag{6.3.7}$$

Set

$$L_{n,1} := \Delta + B_n(x), \quad L_{n+1,2} := -\Delta - B_n(x) + \frac{(x-b_n)}{a_n} A_n(x). \tag{6.3.8}$$

The operators $L_{n,1}$ and $L_{n,2}$ generate a second-order difference equation

$$L_{n,2}\left(\frac{1}{A_n(x)}L_{n,1}\right)p_n(x) = \frac{a_n A_{n-1}(x)}{a_{n-1}}p_n(x). \tag{6.3.9}$$

Equation (6.3.9) can be written in the form

$$\Delta^2 p_n(x) + R_n(x)\Delta p_n(x) + S_n(x)p_n(x) = 0, \tag{6.3.10}$$

where

$$\begin{aligned} R_n(x) = &-\frac{\Delta A_n(x)}{A_n(x)} + B_n(x+1) + \frac{B_{n-1}(x)A_n(x+1)}{A_n(x)} \\ &-\frac{(x-b_{n-1})}{a_{n-1}}\frac{A_{n-1}(x)A_n(x+1)}{A_n(x)}, \end{aligned} \tag{6.3.11}$$

$$\begin{aligned} S_n(x) = &\left(B_{n-1}(x) - 1 - \frac{(x-b_{n-1})}{a_{n-1}}A_{n-1}(x)\right)\frac{B_n(x)A_n(x+1)}{A_n(x)} \\ &+ B_n(x+1) + \frac{a_n A_{n-1}(x)A_n(x+1)}{a_{n-1}}. \end{aligned} \tag{6.3.12}$$

For applications it is convenient to have equation (6.3.10) in the form

$$\begin{aligned} &y(x+1) + (R_n(x-1) - 2)\,y(x) \\ &\quad + [S_n(x-1) - R_n(x-1) + 1]\,y(x-1) = 0. \end{aligned} \tag{6.3.13}$$

Analogously to (2.10.1) we define the function of the second kind by

$$q_n(x) := \frac{1}{w(x)}\sum_{\ell=s}^{t}\frac{p_n(\ell)}{x-\ell}w(\ell), \quad x \notin \{s, s+1, \ldots, t\}. \tag{6.3.14}$$

Theorem 6.3.2 (Ismail, Nikolova, and Simeonov, 2004) *Assume that* (6.3.2) *holds. Then the function* $q_n(x)$ *satisfies the three-term recurrence relation* (2.8.4), *and the lowering and raising relations* (6.3.4) *and* (6.3.7). *Moreover,* $q_n(x)$ *also satisfies* (6.3.10).

We also have the recurrences

$$B_{n+1}(x) - B_{n-1}(x) = \frac{A_n(x)}{a_n} + \frac{(x-b_n)}{a_n}A_n(x) - \frac{(x-b_{n-1})}{a_{n-1}}A_{n-1}(x), \tag{6.3.15}$$

$$\begin{aligned} &\left(B_{n-1}(x) - \frac{(x-b_{n-1})}{a_{n-1}}A_{n-1}(x)\right)B_n(x) - \left(B_n(x) - \frac{(x-b_n)}{a_n}A_n(x)\right)B_{n+1}(x) \\ &= -\frac{A_n(x)}{a_n} + \frac{a_{n+1}A_n(x)A_{n+1}(x)}{a_n} - \frac{a_n A_{n-1}(x)A_n(x)}{a_{n-1}}, \end{aligned} \tag{6.3.16}$$

$$\begin{aligned} \left(1 + \frac{1}{x-b_{n+1}}\right)B_{n+1}(x) - B_n(x) = &\frac{(x+1-b_{n+1})}{a_{n+1}}A_{n+1}(x) - \frac{a_{n+2}A_{n+2}(x)}{x-b_{n+1}} \\ &-\frac{(x-b_n)}{a_n}A_n(x) + \frac{a_{n+1}^2 A_n(x)}{a_n(x-b_{n+1})} + \frac{1}{x-b_{n+1}}. \end{aligned} \tag{6.3.17}$$

In (6.3.16) we substitute for $B_{n-1}(x) - B_{n+1}(x)$ using (6.3.15) to obtain the identity

$$B_{n+1}(x) - \left(1 + \frac{1}{x - b_n}\right) B_n(x) = \frac{a_{n+1}A_{n+1}(x)}{x - b_n} - \frac{a_n^2 A_{n-1}(x)}{a_{n-1}(x - b_n)} - \frac{1}{x - b_n}. \tag{6.3.18}$$

The above recurrences lead to the following theorem.

Theorem 6.3.3 *The functions $A_n(x)$ and $B_n(x)$ satisfy fifth-order nonhomogeneous recurrence relations.*

Recall the definition of the discrete discriminants in (2.9.6).

Theorem 6.3.4 (Ismail, Nikolova, and Simeonov, 2004) *Let $\{p_n(x)\}$ be a family of orthogonal polynomials generated by* (2.8.3)–(2.8.4). *Assume that $\{p_n(x)\}$ satisfy* (6.3.1). *Then the discrete discriminant is given by*

$$D(p_n; \Delta) = \prod_{j=1}^{n} \left\{ \frac{A_n\left(x_{n,j}\right)}{a_n} \right\} \prod_{k=1}^{n} a_n^{2k-2n+2}. \tag{6.3.19}$$

Meixner polynomials: The functions $A_n(x)$ and $B_n(x)$, as well as the discrete discriminant for Meixner polynomials, were computed in Ismail, Nikolova, and Simeonov (2004). The results are summarized in the following theorem.

Theorem 6.3.5 *The Meixner polynomials satisfy*

$$\Delta M_n(x;\beta,c) = \frac{n}{\beta + x} M_n(x;\beta,c) - \frac{n}{(\beta + x)c} M_{n-1}(x;\beta,c) \tag{6.3.20}$$

and their discrete discriminant is given by

$$D(M_n(x;\beta,c);\Delta) = \frac{(1 - 1/c)^{n^2-n}}{c^{n(n-1)/2}} \prod_{k=1}^{n} \frac{k^k}{(\beta + k - 1)^{2n-k-1}}. \tag{6.3.21}$$

Hahn polynomials: Ismail, Nikolova, and Simeonov (2004) also contains the following theorem.

Theorem 6.3.6 *The Hahn polynomials satisfy*

$$\Delta Q_n(x;\alpha,\beta,N) = A_n(x) Q_{n-1}(x;\alpha,\beta,N) - B_n(x) Q_n(x;\alpha,\beta,N), \tag{6.3.22}$$

where

$$A_n(x) = \frac{n(\alpha + \beta + n + N + 1)(\beta + n)}{(\alpha + \beta + 2n)(x + \alpha + 1)(x - N)} \tag{6.3.23}$$

and

$$B_n(x) = -\frac{n}{(\alpha + N + 1)(\alpha + \beta + 2n)} \times \left(\frac{(N - n + 1)(\beta + n)}{x + \alpha + 1} + \frac{(\alpha + n)(\alpha + \beta + n + N + 1)}{x - N} \right). \tag{6.3.24}$$

Moreover,

$$D\left(Q_n(x;\alpha,\beta,N);\Delta\right) = \prod_{k=1}^{n}\left(\frac{k^k(\alpha+\beta+n+k)^{n-k}(\alpha+\beta+N+k+1)^{k-1}}{(\beta+k)^{1-k}(\alpha+k)^{2n-k-1}(N-k+1)^{2n-k-1}}\right). \tag{6.3.25}$$

The discriminants of the Jacobi polynomials as in (2.9.14) can be obtained from the generalized discriminants of the Hahn polynomials through the limiting relation (6.2.9) while the discrete discriminant for Meixner polynomials could have been obtained from (6.3.25) through the limiting process in (6.2.10).

6.4 Lommel Polynomials and Related Polynomials

By iterating (1.2.19) we see that $J_{\nu+n}(z)$ is a linear combination of $J_\nu(z)$ and $J_{\nu-1}(z)$ with coefficients that are polynomials in $1/z$. Indeed, we have the following theorem.

Theorem 6.4.1 *Define polynomials* $\{R_{n,\nu}\}$ *by*

$$R_{0,\nu}(z) = 1, \quad R_{1,\nu}(z) = 2\nu/z, \tag{6.4.1}$$

$$R_{n+1,\nu}(z) = \frac{2(n+\nu)}{z}R_{n,\nu}(z) - R_{n-1,\nu}(z). \tag{6.4.2}$$

Then

$$J_{\nu+n}(z) = R_{n,\nu}(z)J_\nu(z) - R_{n-1,\nu+1}(z)J_{\nu-1}(z). \tag{6.4.3}$$

Moreover, with $h_{n,\nu}(z) := R_{n,\nu}(1/z)$ *the Lommel polynomials have the explicit form*

$$h_{n,\nu}(z) := R_{n,\nu}(1/z) = \sum_{r=0}^{\lfloor n/2\rfloor}\frac{(-1)^r(n-r)!(\nu)_{n-r}}{r!(n-2r)!(\nu)_r}\left(\frac{z}{2}\right)^{n-2r}. \tag{6.4.4}$$

Watson (1944) refers to $\{R_{nm\nu}(x)\}$ as the **Lommel polynomials** while he calls $\{h_{n,\nu}(x)\}$ the modified Lommel polynomials. It is clear from (6.4.1) and (6.4.2) that $\{h_{n,\nu}(x)\}$ is a system of orthogonal polynomials when $\nu > 0$. The large n behavior of $R_{n,\nu}(x)$, or $h_{n,\nu}(x)$, is given by Hurwitz's formula (Ismail, 2005b; Watson, 1944)

$$\lim_{n\to\infty}\frac{(z/2)^{n+\nu}R_{n,\nu+1}(z)}{\Gamma(n+\nu+1)} = J_\nu(z), \tag{6.4.5}$$

which holds uniformly on compact subsets of $\mathbb{C}$.

From (6.4.2) and (6.4.1) and using the notation in Theorem 2.4.2 we see that

$$D_n(z) = R_{n,\nu}(z), \quad N_n(z) = 2\nu R_{n-1,\nu+1}(z).$$

Therefore Hurwitz's theorem, (6.4.5), establishes the validity of

$$\frac{J_\nu(z)}{J_{\nu-1}(z)} = \cfrac{1}{2\nu/z - \cfrac{1}{2(\nu+1)/z - \cdots \cfrac{1}{2(\nu+n)/z - \ddots}}} \tag{6.4.6}$$

for all finite z when $J_{\nu-1}(z) \neq 0$, and the continued fraction converges uniformly over all compact subsets of $\mathbb{C}$ not containing $z = 0$ or any zero of $z^{1-\nu}J_{\nu-1}(z)$. The case $\nu = 1/2$ of formula (6.4.6) was known to Lambert in 1761 who used it to prove the irrationality of π because the continued fraction (6.4.6) becomes a continued fraction for $\tan z$; see (1.2.18). According to Wallisser (2000), Lambert gave explicit formulas for the polynomials $R_{n,1/2}(z)$ and $R_{n,3/2}(z)$, from which he established Hurwitz's theorem in the cases $\nu = -1/2, 1/2$, and then proved (6.4.6) for $\nu = 1/2$. This is remarkable since Lambert had polynomials with no free parameters and parameter-dependent explicit formulas are much easier to prove.

Rewrite (6.4.1)–(6.4.2) in terms of $\{h_{n,\nu}(x)\}$ as

$$h_{0,\nu}(x) = 1, \quad h_{1,\nu}(x) = 2\nu x, \tag{6.4.7}$$

$$2x(n+\nu)h_{n,\nu}(x) = h_{n+1,\nu}(x) + h_{n-1,\nu}(x). \tag{6.4.8}$$

Theorem 6.4.2 *For $\nu > 0$ the polynomials $\{h_{n,\nu}(x)\}$ are orthogonal with respect to a discrete measure α_ν normalized to have total mass 1, where*

$$\int_{\mathbb{R}} \frac{d\alpha_\nu(t)}{z-t} = 2\nu \frac{J_\nu(1/z)}{J_{\nu-1}(1/z)}. \tag{6.4.9}$$

Moreover, the orthogonality relation is

$$\begin{aligned}\sum_{k=1}^{\infty} \frac{1}{j_{\nu,k}^2} \{h_{n,\nu+1}(1/j_{\nu,k})\, h_{m,\nu+1}(1/j_{\nu,k}) + h_{n,\nu+1}(-1/j_{\nu,k})\, h_{m,\nu+1}(-1/j_{\nu,k})\} \\ = \frac{\delta_{m,n}}{2(\nu+n+1)}.\end{aligned} \tag{6.4.10}$$

Special values are

$$h_{2n+1,\nu}(0) = 0, \quad h_{2n}(0) = (-1)^n. \tag{6.4.11}$$

H. M. Schwartz (1940) gave a proof of (6.4.10) without justifying that $\alpha_\nu(\{0\}) = 0$. Later, Dickinson (1954) rediscovered (6.4.10) but made a numerical error and did not justify $\alpha_\nu(\{0\}) = 0$. A more general class of polynomials was considered in Dickinson, Pollack, and Wannier (1956), again without justifying that $x = 0$ does not support a mass. Goldberg corrected this slip and pointed out that in some cases of the class of polynomials considered by Dickinson, Pollack, and Wannier (1956), $\mu(\{0\})$ may indeed be positive; see Goldberg (1965).

The Lommel polynomials can be used to settle a generalization of the Bourget hypothesis (Bourget, 1866). Bourget conjectured that when ν is a nonnegative integer and m is a positive integer then $z^{-\nu}J_\nu(z)$ and $z^{-\nu-m}J_{\nu+m}(z)$ have no common zeros. Siegel (1929) proved that $J_\nu(z)$ is not an algebraic number when ν is a rational number and z, $z \neq 0$ is an algebraic number. If $J_\nu(z)$ and $J_{\nu+n}(z)$ have a common zero z_0, $z_0 \neq 0$, then (6.4.3) shows that $R_{n-1,\nu+1}(z_0) = 0$ since $z^{-\nu}J_\nu(z)$ and $z^{1-\nu}J_{\nu-1}(z)$ have no common zeros. Hence z_0 is an algebraic number. When ν is a rational number, this contradicts Siegel's theorem and then Bourget's conjecture follows not only for integer values of ν but also for any rational number ν.

The Bessel polynomials are related to special Lommel polynomials through

$$y_n(x) = i^{-n} h_{n,1/2}(iz) + i^{1-n} h_{n-1,3/2}(iz). \tag{6.4.12}$$

Wimp (1985) introduced a generalization of the Lommel polynomials. The **Wimp polynomials** arise when one iterates the three-term recurrence relation of the Coulomb wave functions (Abramowitz and Stegun, 1965) as in Theorem 6.4.1. The explicit definition is

$$W_n(x;\alpha,\gamma) = \sum_{k=0}^{n} \frac{i^k(-\gamma+i\alpha-n)_k}{k!(-2\gamma-2n)_k} x^{n-k} {}_3F_2\left(\left.\begin{matrix} -k, 2n+2\gamma+1-k, \gamma-i\alpha \\ n+\gamma-i\alpha+1-k, 2\gamma \end{matrix}\right| 1\right). \tag{6.4.13}$$

The corresponding three-term **recurrence relation** is

$$\begin{aligned} zy_n(z) = y_{n+1}(z) &+ \frac{\alpha}{2(n+\gamma+1)_2} y_n(z) \\ &+ \frac{(n+\gamma+1+i\alpha)(n+\gamma+1-i\alpha)}{(2n+2\gamma+1)_3(2n+2\gamma+2)} y_{n-1}(z). \end{aligned} \tag{6.4.14}$$

In this case $W_n^*(x;\alpha,\gamma) = W_{n-1}(x;\alpha,\gamma+1)$. From (6.4.13) it follows that the **asymptotic formula**

$$\lim_{n\to\infty} \frac{W_n(x;\alpha,\gamma)}{z^n} = \exp(i/2z)\, {}_1F_1\left(\left.\begin{matrix} \gamma-i\alpha \\ 2\gamma \end{matrix}\right| \frac{i}{z}\right) \tag{6.4.15}$$

holds uniformly in compact subsets of the complex z-plane. The corresponding measure μ_W is defined through

$$\frac{{}_1F_1(\gamma-i\alpha+1; 2\gamma+2; i/z)}{{}_1F_1(\gamma-i\alpha; 2\gamma; i/z)} = \int_{\mathbb{R}} \frac{1}{z-t}\, d\mu_W(t) \tag{6.4.16}$$

for all z such that ${}_1F_1(\gamma-i\alpha; 2\gamma; i/z) \neq 0$. The Coulomb wave function ${}_1F_1(\gamma-i\alpha; 2\gamma; iz)$ is known to have only real and simple zeros for $\gamma > -1$ and α real. The orthogonality measure is purely discrete with point masses at the reciprocals of the zeros of ${}_1F_1(\gamma-i\alpha; 2\gamma; iz)$ and the mass is equal to the residue of the left-hand side of (6.4.16).

One can iterate (1.4.44) and establish (Ismail, 1982)

$$q^{n\nu+n(n-1)/2} J_{\nu+n}^{(k)}(x;q) = R_{n,\nu}(x;q) J_\nu^{(k)}(x;q) - R_{n-1,\nu+1}(x;q) J_{\nu-1}^{(k)}(x;q), \tag{6.4.17}$$

where $k = 1, 2$, and $R_{n,\nu}(x;q)$ is a polynomial in $1/x$ of degree n. Ismail (1982) introduced the following q-analogue of the Lommel polynomials:

$$h_{n,\nu}(x;q) = R_{n,\nu}(1/x;q). \tag{6.4.18}$$

He proved the following results:

$$2x(1-q^{n+\nu})\, h_{n,\nu}(x;q) = h_{n+1,\nu}(x;q) + q^{n+\nu-1} h_{n-1,\nu}(x;q), \tag{6.4.19}$$

$$h_{0,\nu}(x;q) = 1, \quad h_{1,\nu}(x;q) = 2(1-q^{n+\nu})\, x, \tag{6.4.20}$$

$$h_n(x;q) = \sum_{j=0}^{\lfloor n/2 \rfloor} \frac{(-1)^j (q^\nu, q; q)_{n-j}}{(q, q^\nu; q)_j (q; q)_{n-2j}} (2x)^{n-2j} q^{j(j+\nu-1)}, \tag{6.4.21}$$

$$\sum_{n=0}^{\infty} h_{n,\nu}(x;q)t^n = \sum_{j=0}^{\infty} \frac{(-2xtq^\nu)^j \left(-\frac{1}{2}t/x;q\right)_j}{(2xt;q)_{j+1}} q^{j(j-1)/2}, \tag{6.4.22}$$

$$\lim_{n\to\infty} \frac{R_{n,\nu+1}(x;q)}{(x/2)^{n+\nu}(q;q)_\infty} = J_\nu^{(2)}(x;q), \tag{6.4.23}$$

and the polynomials of the second kind are

$$h_{n,\nu}^*(x;q) = 2\,(1-q^\nu)\,h_{n-1,\nu+1}(x;q). \tag{6.4.24}$$

Theorem 6.4.3 (Ismail, 1982) *The functions $z^{-\nu}J_\nu^{(2)}(z;q)$ and $z^{-\nu-1}J_{\nu+1}^{(2)}(z;q)$ have no common zeros for ν real. Moreover, for $\nu > 0$, $\{h_{n,\nu}(x;q)\}$ are orthogonal with respect to a purely discrete measure α_ν, with*

$$\int_{\mathbb{R}} \frac{d\alpha_\nu(t;q)}{z-t} = 2\,(1-q^\nu)\,\frac{J_\nu^{(2)}(1/z;q)}{J_{\nu-1}^{(2)}(1/z;q)}, \qquad z \notin \operatorname{supp}\mu. \tag{6.4.25}$$

Furthermore, for $\nu > -1$, $z^{-\nu}J_\nu^{(2)}(z;q)$ has only real and simple zeros. Let

$$0 < j_{\nu,1}(q) < j_{\nu,2}(q) < \cdots < j_{\nu,n}(q) < \cdots \tag{6.4.26}$$

be the positive zeros of $J_\nu^{(2)}(z;q)$. Then $\{h_{n,\nu}(x;q)\}$ satisfy the orthogonality relation

$$\int_{\mathcal{R}} h_{m,\nu}(x;q)h_{n,\nu}(x;q)\,d\alpha_\nu(x) = q^{n(2\nu+n-1)/2}\frac{1-q^\nu}{1-q^{\nu+n}}\,\delta_{m,n}, \tag{6.4.27}$$

where α_ν is supported on $\{\pm 1/j_{\nu,n}(q)\colon n = 1,2,\ldots\} \cup \{0\}$, but $x = 0$ does not support a positive mass.

Motivated by (3.15.6), Abdi (1965) defined q-**Bessel polynomials** by

$$Y_n(x,a) = \frac{\left(q^{a-1};q\right)_n}{(q;q)_n}\,{}_2\phi_1\left(q^{-n}, q^{n+a-1}; 0, q, x/2\right). \tag{6.4.28}$$

There is an alternate approach to discover the q-analogue of the Bessel polynomials. They should arise from the q-Bessel functions in the same way as the Bessel polynomials came from Bessel functions. This was done in Ismail (1981) and leads to a different polynomial sequence. Iterate the first equation in (1.4.54) to get

$$\begin{aligned} q^{\nu n+n(n-1)/2}K_{\nu+n}^{(j)}(z;q) &= i^n R_{n,\nu}(ix;q)K_\nu^{(j)}(z;q) \\ &\quad + i^{n-1}R_{n-1,\nu+1}(ix;q)K_{\nu-1}^{(j)}(z;q), \end{aligned} \tag{6.4.29}$$

which implies

$$q^{n^2/2}K_{n+1/2}^{(j)}(z;q) = \left[i^n R_{n,1/2}(ix) + i^{n-1}R_{n-1,3/2}(ix)\right] K_{1/2}^{(j)}(z;q). \tag{6.4.30}$$

In analogy with (6.4.12) we define

$$y_n(x \mid q) = i^{-n}h_{n,1/2}(ix) + i^{1-n}h_{n-1,3/2}(ix). \tag{6.4.31}$$

By considering the cases of odd and even n in (6.4.31) and applying (6.4.21) we derive the explicit representation

$$y_n\left(x \mid q^2\right) = q^{n(n-1)/2}\, {}_2\phi_1\left(q^{-n}, q^{n+1}, -q; q, -2qx\right). \tag{6.4.32}$$

The analogue of $y_n(x; a)$ $(= y_n(x; a, 2))$ is

$$y_n\left(x; a \mid q^2\right) = q^{n(n-1)/2}\, {}_2\phi_1\left(q^{-n}, q^{n+a-1}, -q; q, -2qx\right). \tag{6.4.33}$$

Clearly, $y_n\left(x/\left(1-q^2\right), a \mid q^2\right) \to {}_2F_0(-n, n+a-1, —; -x/2) = y_n(x; a)$, as $q \to 1$.

Theorem 6.4.4 *Set*

$$w_{\mathrm{QB}}(z; a) = \sum_{n=0}^{\infty} \frac{(-1; q)_n}{(q^{a-1}; q)_n} (-2z)^{-n}. \tag{6.4.34}$$

For $r > 1/2$ the polynomials $y_n(z; a \mid q)$ satisfy the orthogonality relation

$$\frac{1}{2\pi i} \oint_{|z|=r} y_n\left(z; a \mid q^2\right) y_n\left(z; a \mid q^2\right) w_{\mathrm{QB}}(z; a)\, dz = \frac{(-1)^{n+1} q^{n^2}\left(-q^{a-1}, q; q\right)_n}{(-q, q^{a-1}; q)_n \left(1 - q^{2n+a-1}\right)} \delta_{m,n}. \tag{6.4.35}$$

6.5 An Inverse Operator

Let $w_\nu(x)$ denote the weight function of the ultraspherical polynomials,

$$w_\nu(x) = \left(1 - x^2\right)^{\nu - 1/2}, \quad x \in (-1, 1). \tag{6.5.1}$$

Ismail and Zhang (1994) introduced a right inverse to $\frac{d}{dx}$ on $l^2[[-1,1], w_{\nu+1}]$ by

$$(T_\nu g)(x) = \int_{-1}^{1} \left(1 - t^2\right)^{\nu + 1/2} K_\nu(x, t) g(t)\, dt, \tag{6.5.2}$$

where

$$K_\nu(x, t) = \frac{\Gamma(\nu)\pi^{-1/2}}{\Gamma(\nu + 1/2)} \sum_{n=1}^{\infty} \frac{(n-1)!(n+\nu)}{(2\nu+1)_n} C_n^\nu(x) C_{n-1}^{\nu+1}(t). \tag{6.5.3}$$

The kernel $K_\nu(x, t)$ is a Hilbert–Schmidt kernel on $L^2[w_{\nu+1}] \times L^2[w_{\nu+1}]$, as can be seen from (3.5.8). The operator T_ν has the property

$$(T_\nu g)(x) \sim \sum_{n=1}^{\infty} \frac{g_{n-1}}{2\nu} C_n^\nu(x) \quad \text{if } g(x) \sim \sum_{n=0}^{\infty} g_n C_n^{\nu+1}(x), \tag{6.5.4}$$

where $\sim$ means "has the orthogonal expansion." We assumed that $\nu > 0$ and

$$g \in L^2[[-1,1], w_{\nu+1}], \quad \text{that is,} \sum_{n=0}^{\infty} |g_n|^2 \frac{(2\nu+2)_n}{n!(\nu+n+1)} < \infty. \tag{6.5.5}$$

Consider T_ν as a mapping,

$$T_\nu \colon L^2[w_{\nu+1}] \to L^2[w_\nu] \subset L^2[w_{\nu+1}].$$

Theorem 6.5.1 (Ismail and Zhang, 1994) *Let the positive zeros of $J_\nu(x)$ be as in* (1.2.27) *and let R_ν be the closure of the span of $\{C_n^\nu(x)\colon n = 1,2,\dots\}$ in $L^2[w_{\nu+1}]$. Then R_ν is an invariant subspace for T_ν in $L^2[w_{\nu+1}]$, and*

$$L^2[w_{\nu+1}] = R_\nu \oplus R_\nu^\perp,$$

where

$$R_\nu^\perp = \operatorname{span}\left\{\left(1-x^2\right)^{-1}\right\} \text{ for } \nu > 1/2 \quad \text{and} \quad R_\nu^\perp = \{0\} \text{ for } 1/2 \geq \nu > 0.$$

Moreover, the eigenvalues of the integral operator T_ν are $\{\pm i/j_{\nu,k}\colon k = 1,2,\dots\}$. The eigenfunctions have the ultraspherical series expansion

$$g(x; \pm i/j_{\nu,k}) \sim \sum_{n=1}^{\infty} (\mp i)^{n-1} \frac{(\nu+n)}{\nu+1} C_n^\nu(x) h_{n-1,\nu+1}(1/j_{\nu,k}). \tag{6.5.6}$$

A q-analogue of Theorem 6.5.1 is also in Ismail and Zhang (1994).
In the rest of this chapter we record properties of discrete q-orthogonal polynomials.

6.6 q-Sturm–Liouville Problems

There are two q-Sturm–Liouville equations associated with q-polynomials. They are

$$\frac{1}{w(x)} D_{q^{-1},x}\left(p(x) D_{q,x} Y(x,\lambda)\right) = \lambda Y(x,\lambda), \tag{6.6.1}$$

$$\frac{1}{W(x)} D_{q,x}\left(P(x) D_{q^{-1},x} Z(x,\lambda)\right) = \Lambda Z(x,\lambda). \tag{6.6.2}$$

We assume that

$$w(x) > 0 \text{ and } p(x) > 0, \quad \text{for } x = x_k,\ x = y_k, \tag{6.6.3}$$

$$W(x) > 0 \text{ and } P(x) > 0, \quad \text{for } x = r_k,\ x = s_k, \tag{6.6.4}$$

where $\{x_k\}$ and $\{y_k\}$ are as in (1.4.8), while $\{r_k\}$ and $\{s_k\}$ are as in (1.4.13).

The eigenfunctions are assumed to take finite values at x_{-1} and y_{-1}. Moreover, we assume $w(x_{-1}) - w(y_{-1}) = 0$. The eigenfunctions of (6.6.2) are similarly defined.

Theorem 6.6.1 *Under the above assumptions the operator*

$$T = \frac{1}{w} D_{q^{-1},x} p D_q$$

is symmetric, hence it has real eigenvalues, and the eigenfunctions corresponding to distinct eigenvalues are orthogonal.

A rigorous theory of q-Sturm is now available in Annaby and Mansour (2005). This paper corrects many of the results in Exton (1983), which also used inconsistent notation.

Let $\{p_n(x)\}$ satisfy the orthogonality relation

$$\int_a^b p_m(x)p_n(x)w(x)\,d_qx = \delta_{m,n}. \tag{6.6.5}$$

Theorem 6.6.2 (Ismail, 2003) *Let $\{p_n(x)\}$ be a sequence of discrete q-orthonormal polynomials. Then they have a lowering (annihilation) operator of the form*

$$D_q p_n(x) = A_n(x)p_{n-1}(x) - B_n(x)p_n(x),$$

where $A_n(x)$ and $B_n(x)$ are given by

$$A_n(x) = a_n \frac{w(y/q)p_n(y)p_n(y/q)}{x - y/q}\bigg|_a^b + a_n \int_a^b \frac{u(qx) - u(y)}{qx - y} p_n(y)p_n(y/q)w(y)\,d_qy, \tag{6.6.6}$$

$$B_n(x) = a_n \frac{w(y/q)p_n(y)p_{n-1}(y/q)}{x - y/q}\bigg|_a^b + a_n \int_a^b \frac{u(qx) - u(y)}{qx - y} p_n(y)p_{n-1}(y/q)w(y)\,d_qy, \tag{6.6.7}$$

where u is defined by

$$D_q w(x) = -u(qx)w(qx), \tag{6.6.8}$$

and $\{a_n\}$ are the recursion coefficients.

We set

$$L_{1,n} := B_n + D_q, \tag{6.6.9}$$

$$L_{2,n} := \frac{x - b_n - 1}{a_{n-1}} A_{n-1}(x) - B_{n-1}(x) - D_q. \tag{6.6.10}$$

One can prove the lowering and raising relations

$$L_{1,n}p_n(x) = A_n(x)p_{n-1}(x), \quad L_{2,n}p_{n-1}(x) = \frac{a_n}{a_{n-1}} A_{n-1}(x)p_n(x), \tag{6.6.11}$$

and the second-order q-difference equation

$$D_q^2 p_n(x) + R_n(x)D_q p_n(x) + S_n(x)p_n(x) = 0, \tag{6.6.12}$$

where

$$R_n(x) = B_n(qx) - \frac{D_qA_n(x)}{A_n(x)} + \frac{A_n(qx)}{A_n(x)}\left[B_{n-1}(x) - \frac{(x - b_n - 1)\,A_{n-1}(x)}{a_{n-1}}\right], \tag{6.6.13}$$

$$\begin{aligned} S_n(x) = {} & \frac{a_n}{a_{n-1}} A_n(qx)a_{n-1}(x) + D_qB_n(x) - \frac{B_n(x)}{A_n(x)} D_qA_n(x) \\ & + B_n(x)\frac{A_n(qx)}{A_n(x)}\left[B_{n-1}(x) - \frac{(x - b_n - 1)\,A_{n-1}(x)}{a_{n-1}}\right]. \end{aligned} \tag{6.6.14}$$

A more symmetric form of (6.6.12) is

$$\begin{aligned} &p_n(qx) - [1 + q + (1-q)xR_n(x/q)]\,p_n(x) \\ &\quad + \left[q + x^2(1-q)^2q^{-1}S_n(x/q) + (1-q)xR_n(x/q)\right]p_n(x/q) = 0. \end{aligned} \tag{6.6.15}$$

Theorem 6.6.3 (Ismail, 2003) *Let $\{p_n\}$ satisfy (6.6.5). The q-discriminant of p_n is given by*

$$D(p_n;q) = \left[\prod_{j=1}^{n} \frac{A_n\left(x_{nj}\right)}{a_n}\right]\left[\prod_{k=1}^{n} a_k^{2k-2n+2}\right],$$

where $\{a_n\}$ are the recursion coefficients and $\{x_{nj}\colon 1 \le j \le n\}$ are the zeros of $p_n(x)$.

6.7 The Al-Salam–Carlitz Polynomials

The **Al-Salam–Carlitz polynomials** $\{U_n^{(a)}(x;q)\}$ satisfy (Al-Salam and Carlitz, 1965; Chihara, 1978)

$$\sum_{n=0}^{\infty} U_n^{(a)}(x;q)\frac{t^n}{(q;q)_n} = \frac{(t,at;q)_\infty}{(tx;q)_\infty}, \tag{6.7.1}$$

$$U_0^{(a)}(x;q) := 1, \quad U_1^{(a)}(x;q) := x - (1+a), \tag{6.7.2}$$

$$U_{n+1}^{(a)}(x;q) = [x - (1+a)q^n]\,U_n^{(a)}(x;q) + aq^{n-1}(1-q^n)\,U_{n-1}^{(a)}(x;q), \tag{6.7.3}$$

$n > 0$. Note that $\{U_n^a(x;q)\}$ are essentially birth and death process polynomials with rates $\lambda_n = aq^n$ and $\mu_n = 1 - q^n$. The **orthogonality relation** is

$$\begin{aligned} &\sum_{k=0}^{\infty}\left[\frac{q^k U_m^{(a)}\left(q^k;q\right) U_n^{(a)}\left(q^k;q\right)}{(q,q/a;q)_k(a;q)_\infty} + \frac{q^k U_m^{(a)}\left(aq^k;q\right) U_n^{(a)}\left(aq^k;q\right)}{(q,aq;q)_k(1/a;q)_\infty}\right] \\ &\quad = (-a)^n q^{n(n-1)/2}(q;q)_n\,\delta_{m,n}, \quad a < 0. \end{aligned} \tag{6.7.4}$$

The above form of the orthogonality relation is from Ismail (1985b). The original form in Al-Salam and Carlitz (1965) involved a complicated sum.

The inner product associated with the Al-Salam–Carlitz polynomials correspond to $b = 1$ in (1.4.9). The orthogonality relation (6.7.4) can be written in the form

$$\int_a^1 \frac{(qx,qx/a;q)_\infty U_m^{(a)}(x;q)U_n^{(a)}(x;q)}{(q,a,q/a;q)_\infty(1-q)}\,d_qx = (-a)^n q^{n(n-1)/2}(q;q)_n\,\delta_{m,n}. \tag{6.7.5}$$

An **explicit formula** is

$$\begin{aligned} U_n^{(a)}(x;q) &= \sum_{k=0}^{n} \frac{(q;q)_n(-a)^{n-k}}{(q;q)_k(q;q)_{n-k}} q^{(n-k)(n-k-1)/2} x^k (1/x;q)_k \\ &= (-a)^n q^{n(n-1)/2}\,{}_2\phi_1\left(q^{-n}, 1/x; 0; q, qx/a\right). \end{aligned} \tag{6.7.6}$$

The polynomials $\{U_n^{(a)}(x;q)\}$ have the **lowering and raising operators**

$$D_{q,x}U_n^{(a)}(x;q) = \frac{1-q^n}{1-q}U_{n-1}^{(a)}(x;q), \tag{6.7.7}$$

$$\frac{D_{q^{-1},x}\left((qx,qx/a;q)_\infty U_n^{(a)}(x;q)\right)}{(qx,qx/a;q)_\infty} = \frac{q^{1-n}}{a(1-q)}U_{n+1}^{(a)}(x;q), \tag{6.7.8}$$

respectively. Hence we have the q-**Sturm–Liouville equation**

$$\frac{1}{(qx,qx/a;q)_\infty}D_{q^{-1},x}\left((qx,qx/a;q)_\infty D_{q,x}U_n^{(a)}(x;q)\right) = \frac{(1-q^n)q^{2-n}}{a(1-q)^2}U_n^{(a)}(x;q). \tag{6.7.9}$$

An equivalent form of (6.7.9) is

$$\begin{aligned}&\left[a+x^2-x(1+a)\right]D_{q^{-1},x}D_{q,x}U_n^{(a)}(x;q) + \frac{q(x-1-a)}{1-q}D_qU_n^{(a)}(x;q)\\ &\quad= \frac{(1-q^n)\,q^{2-n}}{(1-q)^2}U_n^{(a)}(x;q).\end{aligned} \tag{6.7.10}$$

The **Rodrigues-type formula** is

$$U_n^{(a)}(x;q) = \frac{(1-q)^n a^n q^{n(n-3)/2}}{(qx,qx/a;q)_\infty}D^n_{q^{-1},x}\left\{(qx,qx/a;q)_\infty\right\}. \tag{6.7.11}$$

The numerator polynomials $\{U_n^{(a)*}(x;q)\}$ satisfy (6.7.3) with the initial conditions

$$U_0^{(a)*}(x;q) = 0, \quad U_1^{(a)*}(x;q) = 1. \tag{6.7.12}$$

The numerator polynomials have the generating function

$$\sum_{n=0}^{\infty}\frac{U_n^{(a)*}(x;q)}{(q;q)_n}t^n = t\sum_{n=0}^{\infty}\frac{q^n(t,at;q)_n}{(xt;q)_{n+1}}. \tag{6.7.13}$$

For $x \neq 0$ the large n behaviors of $U_n^{(a)}(x;q)$ and $U_n^{(a)*}(x;q)$ are given by

$$U_n^{(a)}(x;q) = (1/x,a/x;q)_\infty x^n[1+o(1)], \tag{6.7.14}$$

$$U_n^{(a)*}(x;q) = (q;q)_\infty x^{n-1}\sum_{k=0}^{\infty}\frac{q^k(1/x,a/x;q)_k}{(q;q)_k}[1+o(1)], \tag{6.7.15}$$

as $n \to \infty$. Hence, for nonreal z we have

$$F(z) := \lim_{n\to\infty}\frac{U_n^{(a)*}(z;q)}{U_n^{(a)}(z;q)} = \frac{(q;q)_\infty}{z(1/z,a/z;q)_\infty}\,{}_2\phi_1(1/z,a/z;0;q,q). \tag{6.7.16}$$

The Perron–Stieltjes inversion formula implies the orthogonality relation (6.7.4).

The corresponding continued fraction is

$$\cfrac{1}{x-(1+a)-\cfrac{a(1-q)}{x-(1+a)q-\cdots\cfrac{aq^{n-1}(1-q^n)}{x-(1+a)q^n-\ddots}}} \tag{6.7.17}$$
$$= \frac{(q;q)_\infty}{z(1/z,a/z;q)_\infty}\,{}_2\phi_1(1/z,a/z;0;q,q).$$

A second family of Al-Salam–Carlitz polynomials $\{V_n^{(a)}(x;q)\}$ is generated by

$$V_0^{(a)}(x;q)=1,\quad V_1^{(a)}(x;q)=x-1-a, \tag{6.7.18}$$
$$V_{n+1}^{(a)}(x;q)=\left[x-(1+a)q^{-n}\right]V_n^{(a)}(x;q)-aq^{1-2n}\,(1-q^n)\,V_{n-1}^{(a)}(x;q), \tag{6.7.19}$$

$n>0$. The $V_n^{(a)}$ correspond to formally replacing q by $1/q$ in $U_n^{(a)}(x;q)$ and are orthogonal with respect to a positive measure if and only if $0<aq$, $-1<q<1$. They have the **generating function**

$$V^{(a)}(x;t):=\sum_{n=0}^{\infty}V_n^{(a)}(x;q)\frac{q^{\binom{n}{2}}}{(q;q)_n}(-t)^n=\frac{(xt;q)_\infty}{(t,at;q)_\infty},\quad |t|<\min(1,1/a), \tag{6.7.20}$$

which implies the **explicit form**

$$V_n^{(a)}(x;q)=(-1)^nq^{-n(n-1)/2}\sum_{k=0}^{n}\frac{(q;q)_na^{n-k}}{(q;q)_k(q;q)_{n-k}}(x;q)_k. \tag{6.7.21}$$

The **lowering and raising operators** for the $V_n^{(a)}$ are

$$D_{q^{-1},x}V_n^{(a)}(x;q)=q^{1-n}\frac{(1-q^n)}{(1-q)}V_{n-1}^{(a)}(x;q), \tag{6.7.22}$$
$$(x,x/a;q)_\infty D_{q,x}\frac{V_n^{(a)}(x;q)}{(x,x/a;q)_\infty}=\frac{q^n}{a(q-1)}V_{n+1}^{(a)}(x;q), \tag{6.7.23}$$

respectively. The corresponding q-**Sturm–Liouville equation** is

$$(x,x/a;q)_\infty D_{q,x}\frac{1}{(x,x/a;q)_\infty}D_{q^{-1},x}V_n^{(a)}(x;q)=-\frac{1-q^n}{a(1-q)^2}V_n^{(a)}(x;q), \tag{6.7.24}$$

that is,

$$\left[a-x(1+a)+x^2\right]D_{q,x}D_{q^{-1},x}V_n+\frac{1+a-x}{1-q}D_{q^{-1},x}V_n+\frac{1-q^n}{(1-q)^2}V_n=0. \tag{6.7.25}$$

Observe that the coefficients in (6.7.25) correspond to replacing q by $1/q$ in (6.7.10).

An **orthogonality relation** for $\{V_n^{(a)}(x;q)\}$ is

$$\sum_{k=0}^{\infty}\frac{a^kq^{k^2}}{(q,aq;q)_k}V_m^{(a)}\left(q^{-k};q\right)V_n^{(a)}\left(q^{-k};q\right)=\frac{(q;q)_na^n}{(qa;q)_nq^{n^2}}\delta_{m,n}. \tag{6.7.26}$$

The moment problem associated with $\{V_n^{(a)}(x;q)\}$ is determinate if and only if $0 < a \le q$ or $1/q \le a$. In the first case the unique solution is

$$m^{(a)} = (aq;q)_\infty \sum_{n=0}^{\infty} \frac{a^n q^{n^2}}{(q,aq;q)_n} \delta_{q^{-n}}, \tag{6.7.27}$$

and in the second case it is

$$\sigma^{(a)} = (q/a;q)_\infty \sum_{n=0}^{\infty} \frac{a^{-n} q^{n^2}}{(q,q/a;q)_n} \delta_{aq^{-n}}. \tag{6.7.28}$$

The total mass of these measures was evaluated to 1 in Ismail (1985b). Recall that δ_b is a unit measure supported at $x = b$.

If $q < a < 1/q$ the problem is indeterminate and both measures are solutions. In Berg and Valent (1994) the following one-parameter family of solutions with an analytic density was found:

$$\nu(x;a,q,\gamma) = \frac{\gamma|a-1|(q,aq,q/a;q)_\infty}{\pi a\left[(x/a;q)_\infty^2 + \gamma^2(x;q)_\infty^2\right]}, \quad \gamma > 0,\ a \ne 1. \tag{6.7.29}$$

For a similar formula when $a = 1$, see Berg and Valent (1994).

If μ is a solution to the moment problem, then the orthogonality relation becomes

$$\int_{\mathbb{R}} V_m^{(a)}(x;q) V_n^{(a)}(x;q)\, d\mu(x) = a^n q^{-n^2} (q;q)_n\, \delta_{m,n}. \tag{6.7.30}$$

6.8 q-Jacobi Polynomials

The orthogonality relation (6.7.5) implies the following integral evaluation, via the procedure in Section 1.6:

$$\frac{(at_1t_2;q)_\infty}{(t_1,t_2,at_1,at_2;q)_\infty} = \sum_{n=0}^{\infty} \frac{q^n/(q,q/a;q)_n}{(a,t_1q^n,t_2q^n;q)_\infty} + \sum_{n=0}^{\infty} \frac{q^n/(q,qa;q)_n}{(1/a,at_1q^n,at_2q^n;q)_\infty}, \tag{6.8.1}$$

which is the nonterminating analogue of the Chu–Vandermonde sum. This gives the total mass of a positive measure and the corresponding orthogonal polynomials have the **basic hypergeometric representation**

$$\varphi_n(x;a,t_1,t_2) = {}_3\phi_2\left(\left.\begin{matrix} q^{-n}, at_1t_2q^{n-1}, xt_1 \\ t_1, at_1 \end{matrix}\right| q,q\right). \tag{6.8.2}$$

Two **other representations** are

$$\varphi_n(x;a,t_1,t_2) = \frac{\left(q^{1-n}/t_2x;q\right)_n}{(at_1;q)_n} \left(at_1t_2xq^{n-1}\right)^n {}_3\phi_2\left(\left.\begin{matrix} q^{-n}, q^{1-n}/at_2, 1/x \\ t_1, q^{1-n}/t_2x \end{matrix}\right| q,q\right), \tag{6.8.3}$$

$$\varphi_n(x;t_1,t_2,a) = \frac{(q,at_2;q)_n}{(at_1;q)_n} \sum_{k=0}^{n} \frac{(1/x;q)_{n-k}(t_1x)^{n-k}}{(q,t_1;q)_{n-k}} \frac{(-at_1)^k (t_2x;q)_k q^{k(k-1)/2}}{(q,at_2;q)_k}. \tag{6.8.4}$$

Therefore (6.8.4) establishes the generating function (Ismail and Wilson, 1982)

$$\sum_{n=0}^{\infty} \varphi_n(x; t_1, t_2, a) \frac{(at_1; q)_n}{(q, at_2; q)_n} t^n = {}_2\phi_1(1/x, 0; t_1; q, t_1 x)\, {}_1\phi_1(t_2 x; at_2; q, at_1 t). \tag{6.8.5}$$

The **orthogonality relation** is

$$\begin{aligned}
&\int_a^1 \varphi_m(x; a, t_1, t_2)\, \varphi_n(x; a, t_1, t_2) \frac{(qx, qx/a; q)_\infty}{(xt_1, xt_2; q)_\infty} d_q x \\
&= \frac{(1-q)(q, a, q/a; q)_\infty \left(q, t_2, at_2, at_1t_2q^{n-1}; q\right)_n \left(at_1t_2q^{2n}; q\right)_\infty}{(t_1, at_1, t_2, at_2; q)_\infty (t_1, at_1; q)_n} \left(-at_1^2\right)^n q^{n(n-1)} \delta_{m,n}.
\end{aligned} \tag{6.8.6}$$

The polynomials $\{\varphi_n(x; a, t_1, t_2)\}$ are the **big q-Jacobi polynomials** of Andrews and Askey (1985) in a different normalization. The **little q-Jacobi polynomials** arise as the limiting case

$$p_n(x; \alpha, \beta) = \lim_{a\to\infty} \varphi_n(ax; a, \alpha q, \alpha q/a) = {}_2\phi_1\left(q^{-n}, \alpha\beta q^{n+1}; q\alpha; q, qx\right). \tag{6.8.7}$$

They contain the Jacobi polynomials as the limiting case

$$\lim_{q\to 1} p_n\left(x; q^\alpha, q^\beta\right) = \frac{n!}{(\alpha+1)_n} P_n^{(\alpha,\beta)}(1-2x). \tag{6.8.8}$$

The **orthogonality relation** of the little q-Jacobi polynomials is

$$\begin{aligned}
&\sum_{k=0}^{\infty} \frac{(\beta q; q)_k}{(q; q)_k} (aq)^k p_m\left(q^k; \alpha, \beta\right) p_n\left(q^k; \alpha, \beta\right) \\
&= \frac{\left(\alpha\beta q^2\right)_\infty}{(\alpha q; q)_\infty} \frac{(1-\alpha\beta q)}{(1-\alpha\beta q^{2n+1})} \frac{(q; \beta q; q)_n}{(\alpha q; \alpha\beta q; q)_n} \delta_{m,n}.
\end{aligned} \tag{6.8.9}$$

This follows from (6.8.6) and (6.8.7).

The **three-term recurrence relation** is

$$(xt_1 - 1)\, y_n(x) = A_n y_{n+1}(x) - (A_n + C_n)\, y_n(x) + C_n y_{n-1}(x), \tag{6.8.10}$$

where

$$A_n = \frac{(1-t_1q^n)(1-at_1q^n)\left(1-at_1t_2q^{n-1}\right)}{(1-at_1t_2q^{2n})(1-at_1t_2q^{2n-1})}, \tag{6.8.11}$$

$$C_n = -\frac{at_1^2q^n(1-q^n)\left(1-t_2q^{n-1}\right)\left(1-at_1q^{n-1}\right)}{(1-at_1t_2q^{2n-1})(1-at_1t_2q^{2n-2})}. \tag{6.8.12}$$

Up to the scaling $x \to (x+1)/t_1$, $\{\varphi_n\}$ becomes a family of birth and death process polynomials with birth rates $\{A_n\}$ and death rates $\{C_n\}$. The **lowering and raising operators** are

$$D_{q,x}\varphi_n(x; a, t_1, t_2) = \frac{t_1q^{1-n}(1-q^n)\left(1-at_1t_2q^{n-1}\right)}{(1-q)(1-t_1)(1-at_1)} \varphi_{n-1}(x; a, qt_1, qt_2) \tag{6.8.13}$$

and

$$\begin{aligned}
&\frac{(t_1x/q, t_2x/q; q)_\infty}{(qx, qx/a; q)_\infty} D_{q^{-1},x}\left(\frac{(qx, qx/a; q)_\infty}{(t_1x, t_2x; q)_\infty} \varphi_n(x; a, t_1, t_2)\right) \\
&\quad = -\frac{(1-t_1/q)(1-at_1/q)q^2}{at_1(1-q)} \varphi_{n+1}(x; a, t_1/q, t_2/q),
\end{aligned} \tag{6.8.14}$$

respectively.

An immediate consequence of (6.8.14) is the **Rodrigues-type** formula

$$\varphi_n(x; a, t_1, t_2) = \frac{a^n t_1^n (1-q)^n}{(t_1, at_1; q)_n q^n} \frac{(t_1x, t_2x; q)_\infty}{(qx, qx/a; q)_\infty} D^n_{q^{-1},x}\left(\frac{(qx, qx/a; q)_\infty}{(q^n t_1 x, q^n t_2 x; q)_\infty}\right). \tag{6.8.15}$$

Hence the big q-Jacobi polynomials are solutions to the q-Sturm–Liouville problem

$$\begin{aligned}
&\frac{(t_1x, t_2x; q)_\infty}{(qx, qx/a; q)_\infty} D_{q^{-1},x}\left(\frac{(qx, qx/a; q)_\infty}{(qt_1x, qt_2x; q)_\infty} D_{q,x}\varphi_n(x; a, t_1, t_2)\right) \\
&\quad = \frac{(1-q^n)\left(1-at_1t_2q^{n-1}\right)q^{2-n}}{a(1-q)^2} \varphi_n(x; a, t_1, t_2)
\end{aligned} \tag{6.8.16}$$

or, equivalently,

$$\begin{aligned}
&\left[x^2 - x(1+a) + a\right] D_{q^{-1},x} D_{q,x} y + q\frac{(1-at_1t_2)x + a(t_1+t_2) - a - 1}{(1-q)} D_{q,x} y \\
&\quad = \frac{(1-q^n)\left(1-at_1t_2q^{n-1}\right)}{(1-q)^2} q^{2-n} y.
\end{aligned} \tag{6.8.17}$$

Thus $y = \varphi_n(x; a, t_1, t_2)$ is a solution to (6.8.17). Thus, the little q-Jacobi polynomials satisfy

$$x(x-1)D_{q^{-1},x}D_{q,x}y + q\frac{\left(1-q^2\alpha\beta\right)x + q\alpha - 1}{1-q} D_{q,x}y = \frac{(1-q^n)\left(1-\alpha\beta q^{n+1}\right)}{(1-q)^2} y. \tag{6.8.18}$$

Note that the q-Sturm–Liouville property (6.8.16) implies the orthogonality relation (6.8.6).

The large n asymptotics of the big and little q-Jacobi polynomials were developed in Ismail and Wilson (1982) through the application of Darboux's method to generating functions.

Theorem 6.8.1 *We have the asymptotic formulas*

$$\lim_{n\to\infty} (t_1x)^{-n} \varphi_n(x; a, t_1, t_2) = \frac{(1/x, a/x; q)_\infty}{(t_1, at_1; q)_\infty}, \tag{6.8.19}$$

for $x \neq 0, q^m, aq^m, m = 0, 1, \dots,$ and

$$\begin{aligned}
\lim_{n\to\infty} \frac{q^{nm-\binom{n}{2}}}{(-at_1)^n} \varphi_n(q^m; a, t_1, t_2) &= \frac{(t_2; q)_\infty q^{m^2}}{(at; q)_\infty (t_1, t_2; q)_m}, \\
\lim_{n\to\infty} \frac{q^{nm-\binom{n}{2}}}{(-t_1)^n} \varphi_n(aq^m; a, t_1, t_2) &= \frac{(at_2; q)_\infty}{(t_1; q)_\infty} \frac{1}{(t_1; q)_m}.
\end{aligned} \tag{6.8.20}$$

The little and big q-Jacobi functions can be defined as

$$\frac{1}{w(x)}\int_{\alpha}^{\beta}\frac{w(t)u_n(t)}{x-t}w(t)\,dt,$$

where $u_n(x)$ is a little (big) q-Jacobi polynomial, $(\alpha,\beta)=(0,1)$ $((\alpha,\beta)=(a,1))$, respectively, and w is the corresponding weight function. Kadell (2005) introduced a different type of little q-Jacobi function and used it to give new derivations of several summation theorems for q-series.

Theorem 6.8.2 (Ismail, 2003) *The q-discriminant $\Delta_n(a,b)$ of the little q-Jacobi polynomials in the normalization*

$$\frac{(aq;q)_n}{(q;q)_n}\,{}_2\phi_1\left(\left.\begin{matrix}q^{-n},abq^{n+1}\\ aq\end{matrix}\right|q,qx\right)$$

is given by

$$\Delta_n(a,b)=a^{n(n-1)/2}q^{-n(n-1)(n+1)/3}\prod_{j=1}^{n}\left(\frac{1-q^j}{1-q}\right)^{j+2-2n}$$
$$\times\prod_{k=1}^{n}\left(\frac{1-aq^{k-1}}{1-q}\right)^{k-1}\left(\frac{1-bq^k}{1-q}\right)^{k-1}\left(\frac{1-abq^{n+k}}{1-q}\right)^{n-k}.$$

6.9 q-Hahn Polynomials

The q**-Hahn polynomials** are

$$Q_n(x;\alpha,\beta,N)=Q_n(x;\alpha,\beta,N;q)={}_3\phi_2\left(\left.\begin{matrix}q^{-n},\alpha\beta q^{n+1},x\\ \alpha q,q^{-N}\end{matrix}\right|q,q\right),\tag{6.9.1}$$

$n=0,1,\ldots,N$. Their orthogonality relation is

$$\begin{aligned}\sum_{j=0}^{N}&\frac{\left(\alpha q,q^{-N};q\right)_j}{\left(q,q^{-N}/\beta;q\right)_j}(\alpha\beta q)^{-j}Q_m\left(q^{-j};\alpha,\beta,N\right)Q_n\left(q^{-j};\alpha,\beta,N\right)\\ &=\frac{\left(\alpha\beta q^2;q\right)_N}{(\beta q;q)_N(\alpha q)^N}\frac{\left(q,\alpha\beta q^{N+2},\beta q;q\right)_n}{\left(\alpha q,\alpha\beta q,q^{-N};q\right)_n}\frac{(1-\alpha\beta q)(-\alpha q)^n}{(1-\alpha\beta q^{2n+1})}q^{\binom{n}{2}-Nn}\delta_{m,n}.\end{aligned}\tag{6.9.2}$$

The three-term recurrence relation is

$$\begin{aligned}(1-x)Q_n(x;\alpha,\beta,N)&=A_nQ_{n+1}(x;\alpha,\beta,N)\\ &\quad-(A_n+C_n)\,Q_n(x;\alpha,\beta,N)+C_nQ_{n-1}(x;\alpha,\beta,N),\end{aligned}\tag{6.9.3}$$

where

$$
\begin{aligned}
A_n &= -\frac{\left(1-q^{n-N}\right)\left(1-\alpha q^{n+1}\right)\left(1-\alpha\beta q^{n+1}\right)}{\left(1-\alpha\beta q^{2n+1}\right)\left(1-\alpha\beta q^{2n+2}\right)},\\
C_n &= \frac{\alpha q^{n-N}\left(1-q^{n}\right)\left(1-\alpha\beta q^{n+N+1}\right)\left(1-\beta q^{n}\right)}{\left(1-\alpha\beta q^{2n}\right)\left(1-\alpha\beta q^{2n+1}\right)}.
\end{aligned}
\tag{6.9.4}
$$

The lowering operator for Q_n is

$$
D_{q^{-1}} Q_n(x;\alpha,\beta,N) = \frac{q^{1-n}\left(1-q^{n}\right)\left(1-\alpha\beta q^{n+1}\right)}{(1-q)(1-\alpha q)\left(1-q^{-N}\right)} Q_{n-1}(x;\alpha q,\beta q,N-1). \tag{6.9.5}
$$

With

$$
w(x,\alpha,\beta,N) = \frac{\left(\alpha q, q^{-N};q\right)_u}{\left(q,q^{-N}/\beta;q\right)_u}\frac{1}{(\alpha\beta)^u}, \tag{6.9.6}
$$

where $x = q^{-u}$, $u = 0,1,\ldots,N$, the raising operator is

$$
D_q\left(w(x;\alpha q,\beta q,N-1)Q_n(x;\alpha q,\beta q,N-1)\right) = \frac{w(x,\alpha,\beta,N)}{1-q} Q_{n+1}(x,\alpha,\beta,N). \tag{6.9.7}
$$

The Rodrigues formula is

$$
Q_n(x;\alpha,\beta,N) = \frac{(1-q)^k}{w(x;\alpha,\beta,N)} D_q^k\left(w\left(x;\alpha q^k,\beta q^k,N-k\right)Q_{n-k}\left(x;\alpha q^k,\beta q^k,N-k\right)\right). \tag{6.9.8}
$$

In particular,

$$
Q_n(x;\alpha,\beta,N) = \frac{(1-q)^n}{w(x;\alpha,\beta,N)} D_q^n\left(w\left(x,\alpha q^n,\beta q^n,N-n\right)\right). \tag{6.9.9}
$$

The second-order operator equation is

$$
\begin{aligned}
&\frac{1}{w(x;\alpha,\beta,N)} D_q\left(w(x;\alpha q,\beta q,N-1)D_{q^{-1}}Q_n(x;\alpha,\beta,N)\right)\\
&\quad = \frac{q^{1-n}\left(1-q^{n}\right)\left(1-\alpha\beta q^{n+1}\right)}{(1-q)^2(1-\alpha q)\left(1-q^{-N}\right)} Q_n(x;\alpha,\beta,N).
\end{aligned}
$$

The generating functions

$$
\sum_{n=0}^{N} \frac{\left(q^{-N};q\right)_n}{(q,\beta q;q)_n} Q_n(x;\alpha,\beta,N)t^n = {}_1\phi_1\left(\left.\begin{matrix} x \\ \alpha q\end{matrix}\right| q,\alpha q t\right) {}_2\phi_1\left(\left.\begin{matrix} xq^{-N},0 \\ \beta q\end{matrix}\right| q,xt\right) \tag{6.9.10}
$$

and

$$
\begin{aligned}
&\sum_{n=0}^{N} \frac{\left(\alpha q, q^{-N};q\right)_n}{(q;q)_n} q^{-\binom{n}{2}} Q_n(x;\alpha,\beta,N)t^n\\
&\quad = {}_2\phi_1\left(\left.\begin{matrix} x,\beta q^{N+1}x \\ 0\end{matrix}\right| q,-\alpha t q^{1-N}/x\right) {}_2\phi_0\left(q^{-N}/x,\alpha q/x \,\middle|\, q,-tx\right)
\end{aligned}
\tag{6.9.11}
$$

hold when $x = 1, q^{-1}, \ldots, q^{-N}$.

6.10 A Family of Biorthogonal Rational Functions

Through generating functions, the orthogonality relation (6.7.30) is equivalent to the integral evaluation

$$\int_{\mathbb{R}} \frac{(xt_1, xt_2; q)_\infty \, d\mu(x)}{(t_1, at_1, t_2, at_2; q)_\infty} = \frac{1}{(at_1t_2/q; q)_\infty}, \quad |t_1|, |t_2| < \sqrt{q/a}.$$

The bootstrap method suggests seeking functions orthogonal with respect to a measure ν defined by

$$d\nu(x) = (xt_1, xt_2; q)_\infty \, d\mu(x), \tag{6.10.1}$$

where the measure μ satisfies (6.7.30). The result is that the rational functions

$$\psi_n(x; a, t_1, t_2) = {}_3\phi_2\left(\begin{matrix} q^{-n}, t_1, at_2 \\ xt_1, at_1t_2 \end{matrix} \,\middle|\, q, q \right) \tag{6.10.2}$$

satisfy the biorthogonality relation

$$\begin{aligned} &\int_{\mathbb{R}} \psi_m(x; a, t_2, t_1) \psi_n(x; a, t_1, t_2) \, d\nu(x) \\ &= \frac{(t_1, at_1, t_2, at_2; q)_\infty \, (q; q)_n}{(at_1t_2/q; q)_\infty \, (at_1t_2/q; q)_n} (at_1t_2/q)^n \, \delta_{m,n}. \end{aligned} \tag{6.10.3}$$

The ψ_n are essentially the rational functions studied in Al-Salam and Verma (1983). This treatment is from Berg and Ismail (1996) and is adopted in Ismail (2005b).

7

Some q-Orthogonal Polynomials

The continuous q-ultraspherical and continuous q-Hermite polynomials first appeared in Rogers' work on the Rogers–Ramanujan identities in 1893–95 (Askey and Ismail, 1983). They belong to the Fejér class of polynomials having a generating function of the form

$$\sum_{n=0}^{\infty} \phi_n(\cos\theta)t^n = \left|F\left(re^{i\theta}\right)\right|^2, \tag{7.0.1}$$

where $F(z)$ is analytic in a neighborhood of $z = 0$. Feldheim (1941) and Lanzewizky (1941) independently proved that the only orthogonal generalized polynomials in the Fejér class are either the ultraspherical polynomials or the q-ultraspherical polynomials or special cases of them. They proved that F has to be F_1 or F_2, or some limiting cases of them, where

$$F_1(z) = (1-z)^{-\nu}, \quad F_2(z) = \frac{(\beta z; q)_\infty}{(z; q)_\infty}. \tag{7.0.2}$$

The continuous q-Hermite polynomials correspond to $\beta = 0$. We shall always assume

$$0 < q < 1. \tag{7.0.3}$$

7.1 q-Hermite Polynomials

The **continuous q-Hermite polynomials** $\{H_n(x \mid q)\}$ are generated by

$$H_0(x \mid q) = 1, \quad H_1(x \mid q) = 2x, \tag{7.1.1}$$

$$2xH_n(x \mid q) = H_{n+1}(x \mid q) + (1-q^n)\, H_{n-1}(x \mid q). \tag{7.1.2}$$

They are q-analogues of the Hermite polynomials because

$$\lim_{q\to 1^-} \left(\frac{2}{1-q}\right)^{n/2} H_n\left(x\sqrt{(1-q)/2} \mid q\right) = H_n(x). \tag{7.1.3}$$

The continuous q-Hermite polynomials have the properties

$$\sum_{n=0}^{\infty} H_n(\cos\theta \mid q) \frac{t^n}{(q;q)_n} = \frac{1}{(te^{i\theta}, te^{-i\theta}; q)_\infty}, \tag{7.1.4}$$

$$H_n(\cos\theta \mid q) = \sum_{k=0}^{n} \frac{(q;q)_n}{(q;q)_k (q;q)_{n-k}} e^{i(n-2k)\theta}, \tag{7.1.5}$$

$$H_n(\cos\theta \mid q) = \sum_{k=0}^{n} \frac{(q;q)_n}{(q;q)_k (q;q)_{n-k}} \cos(|n-2k|\theta), \tag{7.1.6}$$

$$H_n(-x \mid q) = (-1)^n H_n(x \mid q), \tag{7.1.7}$$

$$\max\{|H_n(x \mid q)| : -1 \le x \le 1\} = H_n(1 \mid q) = (-1)^n H_n(-1 \mid q), \tag{7.1.8}$$

and the maximum in (7.1.8) is attained only at $x = \pm 1$. The **orthogonality relation** is

$$\int_{-1}^{1} H_m(x \mid q) H_n(x \mid q) w(x \mid q)\, dx = \frac{2\pi (q;q)_n}{(q;q)_\infty} \delta_{m,n}, \tag{7.1.9}$$

$$w(x \mid q) = \frac{\left(e^{2i\theta}, e^{-2i\theta}; q\right)_\infty}{\sqrt{1-x^2}}, \quad x = \cos\theta,\ 0 \le \theta \le \pi. \tag{7.1.10}$$

The special case $\lambda = 0$ of (8.6.5) yields

$$\frac{H_n(\cos\theta \mid q^2)}{(q^2;q^2)_n} = \sum_{k=0}^{n} \frac{\left(qe^{2i\theta}; q^2\right)_k}{(q^2;q^2)_k (q;q)_{n-k}} e^{i(n-2k)\theta}. \tag{7.1.11}$$

The **linearization of products** and its inverses are

$$H_m(x \mid q) H_n(x \mid q) = \sum_{k=0}^{m\wedge n} \frac{(q;q)_m (q;q)_n}{(q;q)_k (q;q)_{m-k} (q;q)_{n-k}} H_{m+n-2k}(x \mid q), \tag{7.1.12}$$

$$\frac{H_{n+m}(x \mid q)}{(q;q)_m (q;q)_n} = \sum_{k=0}^{m\wedge n} \frac{(-1)^k q^{k(k-1)/2}}{(q;q)_k} \frac{H_{n-k}(x \mid q)}{(q;q)_{n-k}} \frac{H_{m-k}(x \mid q)}{(q;q)_{m-k}}. \tag{7.1.13}$$

Let $V_n(q)$ denote an n-dimensional vector space over a field with q elements. The q-binomial coefficient $\left[{n \atop k}\right]_q$ counts the number of $V_k(q)$ such that $V_k(q)$ is a subspace of a fixed $V_n(q)$. One can view $H_n(\cos\theta \mid q)$ as a generating function for $\left[{n \atop k}\right]_q$, $k = 0, 1, \ldots,$ since

$$z^{-n} H_n(\cos\theta \mid q) = \sum_{k=0}^{n} \begin{bmatrix} n \\ k \end{bmatrix}_q z^{-2k}, \quad z = e^{i\theta}.$$

One can prove (7.1.12) by classifying the subspaces of a $V_{n+m}(q)$ according to the dimensions of their intersections with $V_n(q)$ and $V_m(q)$; see Ismail, Stanton, and Viennot (1987).

The **Poisson kernel** of the H_n is

$$\sum_{n=0}^{\infty} \frac{H_n(\cos\theta \mid q) H_n(\cos\phi \mid q)}{(q;q)_n} t^n = \frac{\left(t^2; q\right)_\infty}{(te^{i(\theta+\phi)}, te^{i(\theta-\phi)}, te^{-i(\theta+\phi)}, te^{-i(\theta-\phi)}; q)_\infty}. \tag{7.1.14}$$

In fact, the evaluation of the Poisson kernel is equivalent to the linearization formula (7.1.12), which is equivalent to the evaluation of the integral

$$\int_0^{\pi} \frac{\left(e^{2i\theta}, e^{-2i\theta}; q\right)_\infty \, d\theta}{\prod_{j=1}^{3}\left(1 - t_j e^{i\theta}, t_j e^{-i\theta}; q\right)_\infty} = \frac{2\pi}{(q;q)_\infty} \frac{1}{\prod_{1\le j<k\le 3}\left(t_j t_k; q\right)_\infty} \tag{7.1.15}$$

(Ismail and Stanton, 1988). A q-analogue of the integrals (3.8.85) is

$$I(t_1, t_2, \dots, t_k) := \frac{(q;q)_\infty}{2\pi} \int_0^{\pi} \frac{\left(e^{2i\theta}, e^{-2i\theta}; q\right)_\infty}{\prod_{j=1}^{k}\left(t_j e^{i\theta}, t_j e^{-i\theta}; q\right)_\infty} \, d\theta. \tag{7.1.16}$$

The case $k = 4$ is the Askey–Wilson integral evaluation (8.2.1).

Theorem 7.1.1 (Ismail, Stanton, and Viennot, 1987) *Let*

$$I(t_1, t_2, \dots, t_k) = \sum_{n_1,\dots n_k=0}^{\infty} g(n_1, n_2, \dots, n_k) \prod_{j=1}^{k} \frac{t_j^{n_j}}{(q;q)_{n_j}}. \tag{7.1.17}$$

Then

$$g(n_1, n_2, \dots, n_k) = \sum_{n_{ij}} \prod_{\ell=1}^{k} \begin{bmatrix} n_\ell \\ n_{\ell 1}, \dots n_{\ell k} \end{bmatrix}_q \prod_{1\le i<j\le k} (q;q)_{n_{ij}} q^B, \tag{7.1.18}$$

where the summation is over all nonnegative integral symmetric matrices (n_{ij}) such that $n_{ii} = 0$ and $\sum_{i=1}^{k} n_{ij} = n_j$ for $1 \le j \le k$. Furthermore,

$$B = \sum_{1\le i<j<m<\ell\le k}' n_{im} n_{j\ell}. \tag{7.1.19}$$

The q-multinomial coefficient in (7.1.18) is

$$\begin{bmatrix} n \\ n_1, \dots n_k \end{bmatrix}_q := (q;q)_n \Big/ \prod_{j=1}^{k} (q;q)_{n_j}, \quad \sum_{j=1}^{k} n_j = n. \tag{7.1.20}$$

In addition to Theorem 7.1.1, Ismail, Stanton, and Viennot (1987) contains combinatorial interpretations of the moments and the polynomials as generating functions of certain statistics.

When $k = 5$, $I(t_1, \dots, t_5)$ is a multiple of ${}_3\phi_2$. The invariance of $I(t_1, \dots, t_5)$ under $t_j \leftrightarrow t_j$ gives all the known transformation formulas for ${}_3\phi_2$. Details are in Ismail, Stanton, and Viennot (1987).

Special values: We have

$$H_{2n+1}(0 \mid q) = 0 \quad \text{and} \quad H_{2n}(0 \mid q) = (-1)^n(-q;q)_n, \tag{7.1.21}$$

$$H_n\left(\pm\left(q^{1/4} + q^{-1/4}\right)/2 \mid q\right) = (\pm 1)^n q^{-n/4}\left(-q^{1/2}; q^{1/2}\right)_n. \tag{7.1.22}$$

The polynomials $\{H_n(x \mid q)\}$ have the ladder operators

$$\mathcal{D}_q H_n(x \mid q) = \frac{2(1-q^n)}{1-q} q^{(1-n)/2} H_{n-1}(x \mid q), \tag{7.1.23}$$

$$\frac{1}{w(x \mid q)} \mathcal{D}_q \{w(x \mid q) H_n(x \mid q)\} = -\frac{2q^{-n/2}}{1-q} H_{n+1}(x \mid q), \tag{7.1.24}$$

where $w(x \mid q)$ is as defined in (7.1.10).

The polynomials $\{H_n(\cos\theta \mid q^{-1})\}$ satisfy

$$H_n\left(\cos\theta \mid q^{-1}\right) = \sum_{k=0}^{n} \frac{(q;q)_n}{(q;q)_k (q;q)_{n-k}} q^{k(k-n)} e^{i(n-2k)\theta}, \tag{7.1.25}$$

$$\sum_{n=0}^{\infty} \frac{H_n\left(\cos\theta \mid q^{-1}\right)}{(q;q)_n} (-1)^n t^n q^{\binom{n}{2}} = \left(te^{i\theta}, te^{-i\theta}; q\right)_\infty . \tag{7.1.26}$$

Using the approach outlined in Section 2.12 one can prove the following theorem (Ismail and Stanton, 1997, 2002).

Theorem 7.1.2 *The continuous q-Hermite polynomials have the q-integral representations*

$$H_n(\cos\theta \mid q) = \frac{1}{(1-q)e^{i\theta}\left(q, qe^{2i\theta}, e^{-2i\theta}; q\right)_\infty} \times \int_{e^{-i\theta}}^{e^{i\theta}} y^n \left(qye^{i\theta}, qye^{-i\theta}; q\right)_\infty d_q y, \tag{7.1.27}$$

$$\frac{H_n(\cos\theta \mid q)}{(q;q)_n} = \frac{\left(\lambda e^{i\theta}, \lambda e^{-i\theta}, qe^{i\theta}/\lambda, qe^{-i\theta}/\lambda; q\right)_\infty}{2(1-q)i\sin\theta\left(q, q, qe^{2i\theta}, qe^{-2i\theta}; q\right)_\infty} \times \int_{e^{-i\theta}}^{e^{i\theta}} y^n \frac{\left(qye^{i\theta}, qye^{-i\theta}; q\right)_\infty}{(\lambda y, qy/\lambda, \lambda/y, q/(\lambda y); q)_\infty} d_q y, \tag{7.1.28}$$

$$\frac{H_n\left(\cos\theta \mid q^2\right)}{(q;q)_n} = \frac{\left(\sqrt{q}\,e^{i\theta}, \sqrt{q}\,e^{i\theta}, \sqrt{q}\,e^{-i\theta}, \sqrt{q}\,e^{-i\theta}; q\right)_\infty}{2(1-q)i\sin\theta\left(q, q, qe^{2i\theta}, qe^{-2i\theta}; q\right)_\infty} \times \int_{e^{-i\theta}}^{e^{i\theta}} y^n \frac{\left(qye^{i\theta}, qye^{-i\theta}, -\sqrt{q}/y; q\right)_\infty}{\left(\sqrt{q}\,y, \sqrt{q}\,y, \sqrt{q}/y; q\right)_\infty} d_q y, \tag{7.1.29}$$

$$\frac{H_n(\cos\theta \mid q^2)}{(-q;q)_n} = \frac{\left(qe^{2i\theta}, qe^{-2i\theta}; q^2\right)_\infty}{2(1-q)i\sin\theta\left(q, -q, qe^{2i\theta}, qe^{-2i\theta}; q\right)_\infty} \times \int_{e^{-i\theta}}^{e^{i\theta}} y^n \frac{\left(qye^{i\theta}, qye^{-i\theta}; q\right)_\infty}{(qy^2; q^2)_\infty} d_q y. \tag{7.1.30}$$

Additional **moment representations** follow from Theorem 8.1.1 since $H_n(x \mid q) = \lim_{t_1 \to 0} p_n(x; t_1, 0) / t_1^n$.

The **generating functions**

$$\sum_{n=0}^{\infty} \frac{H_{2n}(\cos\theta \mid q)}{(q^2; q^2)_n} t^n = \frac{(-t; q)_\infty}{(te^{2i\theta}, te^{-2i\theta}; q^2)_\infty}, \tag{7.1.31}$$

$$\sum_{n=0}^{\infty} \frac{H_n\left(\cos\theta \mid q^2\right)}{(q; q)_n} t^n = \frac{\left(qt^2; q^2\right)_\infty}{(te^{i\theta}, te^{-i\theta}; q)_\infty} \tag{7.1.32}$$

follow from the orthogonality relation (7.1.9) (Ismail and Stanton, 2003b).

Bryc, Matysiak, and Szabłowski (2005) proved that

$$\begin{vmatrix} H_0(x \mid q) & H_1(x \mid q) & \cdots & H_{n-1}(x \mid q) \\ H_1(x \mid q) & H_2(x \mid q) & \cdots & H_n(x \mid q) \\ \vdots & \vdots & & \vdots \\ H_{n-1}(x \mid q) & H_n(x \mid q) & \cdots & H_{2n-2}(x \mid q) \end{vmatrix} = (-1)^n q^{\binom{n}{3}} \prod_{k=1}^{n-1} (q; q)_j. \tag{7.1.33}$$

Theorem 7.1.3 (Bryc, Matysiak, and Szabłowski, 2005) *Let μ be the probability measure*

$$d\mu\,(x, t_1, t_2) = \frac{(t_1 t_2, q; q)_\infty}{2\pi} w\left(x; te^{i\phi}, te^{-i\phi}\right) dx, \tag{7.1.34}$$

where $w\,(x; t_1, t_2 \mid q)$ is the weight function of the Al-Salam–Chihara polynomials as in (8.1.3). *Then*

$$\int_{-1}^{1} H_n(x \mid q)\, d\mu\left(x; te^{i\phi}, te^{-i\phi}\right) = t^n H_n(\cos\phi \mid q). \tag{7.1.35}$$

Conversely, if (7.1.35) *holds for a probability measure μ and $t > 0$, $\phi \in [-\pi, \pi]$, then μ must be as given above.*

One can show that the eigenvalues of the integral operator

$$(Tf)(\cos\phi) = \int_{-1}^{1} f(x)\, d\mu\left(x; te^{i\phi}, te^{-i\phi}\right) \tag{7.1.36}$$

are $\{t^n : n = 0, 1, \dots\}$ and the corresponding eigenfunctions are $\{H_n(\cos\phi \mid q)\}$.

7.2 q-Ultraspherical Polynomials

The **continuous q-ultraspherical polynomials** are the solution to

$$2x\,(1 - \beta q^n)\, C_n(x; \beta \mid q) = \left(1 - q^{n+1}\right) C_{n+1}(x; \beta \mid q) + \left(1 - \beta^2 q^{n-1}\right) C_{n-1}(x; \beta \mid q), \tag{7.2.1}$$

$n > 0$ with the initial values

$$C_0(x; \beta \mid q) = 1, \quad C_1(x; \beta \mid q) = 2x(1 - \beta)/(1 - q). \tag{7.2.2}$$

It is easy to see that

$$C_n(\cos\theta;\beta \mid q) = \sum_{k=0}^{n} \frac{(\beta;q)_k(\beta;q)_{n-k}}{(q;q)_k(q;q)_{n-k}} e^{i(n-2k)\theta}, \tag{7.2.3}$$

$$C_n(\cos\theta;\beta \mid q) = \frac{(\beta;q)_n e^{in\theta}}{(q;q)_n} {}_2\phi_1\left(\begin{matrix} q^{-n}, \beta \\ q^{1-n}/\beta \end{matrix} \middle| q, qe^{-2i\theta}/\beta\right). \tag{7.2.4}$$

It is clear that

$$\begin{aligned} C_n(x;0 \mid q) &= H_n(x \mid q)/(q;q)_n, \\ C_n(-x;\beta \mid q) &= (-1)^n C_n(x;\beta \mid q), \\ C_n(x;\beta \mid q) &= \frac{2^n(\beta;q)_n}{(q;q)_n} x^n + \text{ lower-order terms.} \end{aligned} \tag{7.2.5}$$

The **orthogonality relation**

$$\int_{-1}^{1} C_m(x;\beta \mid q) C_n(x;\beta \mid q) w(x \mid \beta)\, dx = \frac{2\pi(\beta, q\beta;q)_\infty}{(q,\beta^2;q)_\infty} \frac{(1-\beta)\left(\beta^2;q\right)_n}{(1-\beta q^n)\,(q;q)_n} \delta_{m,n} \tag{7.2.6}$$

holds for $|\beta| < 1$, with

$$w(\cos\theta \mid \beta) = \frac{\left(e^{2i\theta}, e^{-2i\theta};q\right)_\infty}{(\beta e^{2i\theta}, \beta e^{-2i\theta};q)_\infty} (\sin\theta)^{-1}. \tag{7.2.7}$$

We have the **generating functions**

$$\sum_{n=0}^{\infty} C_n(\cos\theta;\beta \mid q) t^n = \frac{\left(t\beta e^{i\theta}, t\beta e^{-i\theta};q\right)_\infty}{(te^{i\theta}, te^{-i\theta};q)_\infty}, \tag{7.2.8}$$

$$\sum_{n=1}^{\infty} \frac{C_n(\cos\theta;\beta \mid q)}{(\beta^2;q)_n} q^{\binom{n}{2}} t^n = \left(-te^{-i\theta};q\right)_\infty {}_2\phi_1\left(\begin{matrix} \beta, \beta e^{2i\theta} \\ \beta^2 \end{matrix} \middle| q, -te^{-i\theta}\right). \tag{7.2.9}$$

Moreover, one can establish

$$\frac{\left(\gamma te^{i\theta}, \gamma te^{-i\theta};q\right)_\infty}{(te^{i\theta}, te^{-i\theta};q)_\infty} = \sum_{n=0}^{\infty} \frac{1-\beta q^n}{1-\beta} C_n(\cos\theta;\beta \mid q) F_n(t), \tag{7.2.10}$$

where

$$F_n(t) = \frac{t^n(\gamma;q)_n}{(q\beta;q)_n} {}_2\phi_1\left(\gamma/\beta, \gamma q^n; \gamma q^{n+1}; q, t^2\right). \tag{7.2.11}$$

In particular, this implies another plane wave expansion (Koornwinder, 2005, (2.20))

$$\begin{aligned} \left(-isq^{\frac{1}{2}}e^{i\theta}, -isq^{\frac{1}{2}}e^{-i\theta};q\right)_\infty &= \frac{q^{\frac{1}{2}\alpha^2}}{s^\alpha} \frac{(q;q)_\infty}{(q^{\alpha+1};q)_\infty} \sum_{k=0}^{\infty} i^k q^{\frac{1}{2}k^2 + \frac{1}{2}k\alpha} \frac{1-q^{\alpha+k}}{1-q^\alpha} \\ &\quad \times J_{\alpha+k}^{(2)}\left(2sq^{-\frac{1}{2}\alpha};q\right) C_k\left(\cos\theta;q^\alpha \mid q\right). \end{aligned} \tag{7.2.12}$$

Other **basic hypergeometric representations** are

$$C_n(\cos\theta;\beta \mid q) = \frac{\left(\beta^2;q\right)_n e^{-in\theta}}{(q;q)_n\beta^n}\, {}_3\phi_2\left(\left.\begin{matrix} q^{-n},\beta,\beta e^{2i\theta} \\ \beta^2,0 \end{matrix}\,\right|\, q,q\right), \tag{7.2.13}$$

$$C_n(\cos\theta;\beta \mid q) = \frac{\left(\beta^2;q\right)_n}{\beta^{n/2}(q;q)_n}\, {}_4\phi_3\left(\left.\begin{matrix} q^{-n},q^n\beta^2,\sqrt{\beta}\,e^{i\theta},\sqrt{\beta}\,e^{-i\theta} \\ \beta q^{1/2},-\beta q^{1/2},-\beta \end{matrix}\,\right|\, q,q\right). \tag{7.2.14}$$

Formula (7.2.14) follows from identifying the continuous q-ultraspherical polynomials as special Askey–Wilson polynomials. As $q \to 1$, the representation (7.2.14) with $\beta = q^\nu$ reduces to the first line in (3.5.1). Comparing (8.1.8) and (7.2.8) we relate the Al-Salam–Chihara polynomials (of Section 9.1) to the continuous q-ultraspherical polynomials through

$$C_n(\cos\theta;\beta \mid q) = \frac{\left(\beta^2;q\right)_n}{(q;q)_n}\,\frac{p_n\left(\cos\theta;\beta e^{i\theta},\beta e^{-i\theta} \mid q\right)}{\beta^n e^{in\theta}}. \tag{7.2.15}$$

Theorem 8.1.1 is transformed to

$$\begin{aligned} C_n(\cos\theta;\beta \mid q) &= \frac{\left(\beta e^{2i\theta},\beta e^{-2i\theta},\beta,\beta;q\right)_\infty}{(1-q)e^{i\theta}\left(q,\beta^2,qe^{2i\theta},qe^{-2i\theta};q\right)_\infty} \\ &\quad \times \frac{\left(\beta^2;q\right)_n}{(q;q)_n}\int_{e^{-i\theta}}^{e^{i\theta}} y^n \frac{\left(qye^{i\theta},qye^{-i\theta};q\right)_\infty}{\left(\beta e^{i\theta}y,\beta e^{-i\theta}y;q\right)_\infty}\, d_qy \end{aligned} \tag{7.2.16}$$

and

$$\begin{aligned} C_n(\cos\theta;\beta \mid q) &= \frac{\left(\beta e^{2i\theta},qe^{-2i\theta}/\beta,\beta,q/\beta;q\right)_\infty}{2(1-q)i\sin\theta\left(q,q,qe^{2i\theta},qe^{-2i\theta};q\right)_\infty} \\ &\quad \times \int_{e^{-i\theta}}^{e^{i\theta}} y^n \frac{\left(qye^{i\theta},qye^{-i\theta},\beta e^{-i\theta}/y;q\right)_\infty}{\left(qye^{-i\theta}/\beta,\beta ye^{i\theta},qe^{-i\theta}/(\beta y);q\right)_\infty}\, d_qy. \end{aligned} \tag{7.2.17}$$

Special cases:

$$C_n(x;q \mid q) = U_n(x), \quad n \ge 0, \tag{7.2.18}$$

$$\lim_{\beta\to 1}\frac{\left(1-\beta^2q^n\right)}{\left(1-\beta^2\right)} C_n(x;\beta \mid q) = T_n(x), \quad \lim_{q\to 1} C_n\left(x;q^\nu \mid q\right) = C_n^\nu(x), \tag{7.2.19}$$

for $n \ge 0$, $\{C_n^\nu(x)\}$ being the ultraspherical polynomials of Section 3.4.

It is clear from (7.2.3) that

$$\max\left\{C_n(x;\beta \mid q):\; -1 \le x \le 1\right\} = C_n(1;\beta \mid q). \tag{7.2.20}$$

Unlike the ultraspherical polynomials, $C_n(1;\beta \mid q)$ for general β does not have a closed form expression. However, we have the **special values**

$$C_n\left(\left(\beta^{1/2}+\beta^{-1/2}\right)/2;\beta \,\middle|\, q\right) = \beta^{-n/2}\frac{\left(\beta^2;q\right)_n}{(q;q)_n} = (-1)^n C_n\left(-\left(\beta^{1/2}+\beta^{-1/2}\right)/2;\beta \,\middle|\, q\right). \tag{7.2.21}$$

Moreover,

$$C_{2n+1}(0;\beta \mid q) = 0, \quad C_{2n}(0;\beta \mid q) = \frac{(-1)^n \left(\beta^2;q^2\right)_n}{(q^2;q^2)_n}. \tag{7.2.22}$$

Furthermore,

$$\max\left\{C_n(x;\beta \mid q) : |x| \le \left(\beta^{1/2}+\beta^{-1/2}\right)/2, x \text{ real}\right\} = \beta^{-n/2}\frac{\left(\beta^2;q\right)_n}{(q;q)_n}. \tag{7.2.23}$$

An important special case of the C_n is

$$\lim_{\beta\to\infty} \beta^{-n} C_n(x;\beta \mid q) = \frac{q^{n(n-1)/2}(-1)^n}{(q;q)_n} H_n\left(x \mid q^{-1}\right) \tag{7.2.24}$$

where $H_n(x \mid q^{-1})$ is as in (7.1.25).

The orthogonality relation (7.2.6) is equivalent to the evaluation of the q-beta integral

$$\begin{aligned}&\int_0^\pi \frac{\left(t_1\beta e^{i\theta}, t_1\beta e^{-i\theta}, t_2\beta e^{i\theta}, t_2\beta e^{-i\theta}, e^{2i\theta}, e^{-2i\theta};q\right)_\infty}{(t_1e^{i\theta}, t_1e^{-i\theta}, t_2e^{i\theta}, t_2e^{-i\theta}, \beta e^{2i\theta}, \beta e^{-2i\theta};q)_\infty}\,d\theta \\ &\quad = \frac{(\beta, q\beta;q)_\infty}{(q,\beta^2;q)_\infty}\,{}_2\phi_1\left(\beta^2,\beta;q\beta;q,t_1t_2\right), \quad |t_1|<1,\ |t_2|<1.\end{aligned} \tag{7.2.25}$$

The **lowering operator** is

$$\mathcal{D}_q C_n(x;\beta \mid q) = \frac{2(1-\beta)}{1-q} q^{(1-n)/2} C_{n-1}(x;q\beta \mid q), \tag{7.2.26}$$

while the **raising operator** is

$$\begin{aligned}&\mathcal{D}_q[w(x \mid \beta) C_n(x;\beta \mid q)] \\ &\quad = -\frac{2q^{-\frac{1}{2}n}\left(1-q^{n+1}\right)\left(1-\beta^2q^{n-1}\right)}{(1-q)(1-\beta q^{-1})} w\left(x \mid \beta q^{-1}\right) C_{n+1}\left(x;\beta q^{-1} \mid q\right).\end{aligned} \tag{7.2.27}$$

We also have

$$\begin{aligned}\left(1-2x\beta+\beta^2\right) C_{n-1}(x;q\beta \mid q) &= \frac{\left(1-\beta^2q^n\right)\left(1-\beta^2q^{n-1}\right)}{(1-\beta)(1-\beta q^n)} C_{n-1}(x;\beta \mid q) \\ &\quad - \frac{\beta(1-q^n)\left(1-q^{n+1}\right)}{(1-\beta)(1-\beta q^n)} C_{n+1}(x;\beta \mid q),\end{aligned} \tag{7.2.28}$$

$$\mathcal{D}_q C_n(x;\beta \mid q) = A_n(x) C_{n-1}(x;\beta \mid q) - B_n(x) C_n(x;\beta \mid q), \tag{7.2.29}$$

where

$$\begin{aligned}A_n(x) &= \frac{2q^{(1-n)/2}(1+\beta)\left(1-\beta^2q^{n-1}\right)}{(1-q)(1-2x\beta+\beta^2)}, \\ B_n(x) &= \frac{4\beta x q^{(1-n)/2}(1-q^n)}{(1-q)(1-2x\beta+\beta^2)}.\end{aligned} \tag{7.2.30}$$

Rogers' **connection coefficient** formula for $\{C_n(x;\beta \mid q)\}$ is (Rogers, 1894)

$$C_n(x;\gamma \mid q) = \sum_{k=0}^{\lfloor n/2 \rfloor} \frac{\beta^k(\gamma/\beta;q)_k(\gamma;q)_{n-k}}{(q;q)_k(q\beta;q)_{n-k}} \frac{\left(1-\beta q^{n-2k}\right)}{(1-\beta)} C_{n-2k}(x;\beta \mid q). \tag{7.2.31}$$

Important special and limiting cases are (cf. (7.2.5))

$$C_n(x;\gamma \mid q) = \sum_{k=0}^{\lfloor n/2 \rfloor} \frac{(-\gamma)^k(\gamma;q)_{n-k}}{(q;q)_k} q^{\binom{k}{2}} \frac{H_{n-2k}(x \mid q)}{(q;q)_{n-2k}}, \tag{7.2.32}$$

$$\frac{H_n(x \mid q)}{(q;q)_n} = \sum_{k=0}^{\lfloor n/2 \rfloor} \frac{\beta^k}{(q;q)_k(q\beta;q)_{n-k}} \frac{\left(1-\beta q^{n-2k}\right)}{(1-\beta)} C_{n-2k}(x;\beta \mid q), \tag{7.2.33}$$

$$\frac{H_n(x \mid q)}{(q;q)_n} = \sum_{k=0}^{\lfloor n/2 \rfloor} \frac{(-1)^k q^{k(3k-2n-1)/2}}{(q;q)_k(q;q)_{n-2k}} H_{n-2k}\left(x \mid q^{-1}\right), \tag{7.2.34}$$

$$H_n\left(x \mid q^{-1}\right) = \sum_{s=0}^{\lfloor n/2 \rfloor} \frac{q^{-s(n-s)}(q;q)_n}{(q;q)_s(q;q)_{n-2s}} H_{n-2s}(x \mid q). \tag{7.2.35}$$

In view of the orthogonality relation (7.2.3) the connection coefficient formula (7.2.31) is equivalent to the integral evaluation

$$\begin{aligned} &\int_0^\pi \frac{\left(t\gamma e^{i\theta}, t\gamma e^{-i\theta}, e^{2i\theta}, e^{-2i\theta};q\right)_\infty}{\left(te^{i\theta}, te^{-i\theta}, \beta e^{2i\theta}, \beta e^{-2i\theta};q\right)_\infty} C_m(\cos\theta;\beta \mid q)\, d\theta \\ &\quad = \frac{(\beta, q\beta;q)_\infty(\gamma;q)_m t^m}{(q,\beta^2;q)_\infty\,(q\beta;q)_m}\, {}_2\phi_1\left(\gamma/\beta, \gamma q^m; q^{m+1}\beta; q, \beta t^2\right). \end{aligned} \tag{7.2.36}$$

Since $C_n(x;q \mid q) = U_n(x)$ is independent of q we can use (7.2.32) and (7.2.33) to establish the change of basis formula (Bressoud, 1981)

$$\frac{H_n(x \mid q)}{(q;q)_n} = \sum_{k=0}^{\lfloor n/2 \rfloor} \frac{q^k\left(1-q^{n-2k+1}\right)}{(q;q)_k(q;q)_{n-k+1}} \sum_{j=0}^{\lfloor n/2 \rfloor - k} \frac{(-1)^j p^{\binom{j+1}{2}}(p;p)_{n-2k-j}}{(p;p)_j} \frac{H_{n-2k-2j}(x \mid p)}{(p;p)_{n-2k-2j}}. \tag{7.2.37}$$

Similarly, we get the more general connection formula and its inverse

$$\begin{aligned} C_n(x;\gamma \mid q) = &\sum_{k=0}^{\lfloor n/2 \rfloor} \frac{q^k(\gamma/q;q)_k(\gamma;q)_{n-k}\left(1-q^{n-2k+1}\right)}{(q;q)_k\left(q^2;q\right)_{n-k}(1-q)} \sum_{j=0}^{\lfloor n/2 \rfloor - k} \frac{(\beta)^j(p/\beta;q)_j(p;p)_{n-2k-j}}{(p;p)_j(p\beta;q)_{n-2k-j}} \\ &\times \frac{\left(1-\beta p^{n-2k-2j}\right)}{(1-\beta)} C_{n-2k-2j}(x;\beta \mid p), \end{aligned} \tag{7.2.38}$$

$$\frac{\left(\beta e^{2i\theta},\beta e^{-2i\theta};q\right)_\infty}{\left(\gamma e^{2i\theta},\gamma e^{-2i\theta};q\right)_\infty}C_n(\cos\theta;\gamma\mid q)=\sum_{k=0}^{\infty}d(k,n)C_{n+2k}(\cos\theta;\beta\mid q),$$
$$d(k,n)=\beta^k\left(1-\gamma q^{n+2k}\right)\frac{(\gamma/\beta;q)_k(q^{n+1};q)_{2k}\left(\gamma^2q^{n+2k},\beta q^{n+k+1},\beta;q\right)_\infty}{(q;q)_k\left(\gamma q^{n+k},\beta^2q^n,\gamma;q\right)_\infty}; \tag{7.2.39}$$

see Askey and Ismail (1983).

Denote the coefficient of $H_{n-2k-2j}(x\mid p)$ in (7.2.37) by $c_{n,n-2k}(q,p)$. One can show the special evaluations (Ismail and Stanton, 2003b)

$$\begin{aligned}
c_{2n,0}(p,q)&=\sum_{j=-n}^{n}(-1)^jq^{n-j}q^{j(j+1)/2}\begin{bmatrix}2n\\ n-1\end{bmatrix}_p,\\
c_{2n,0}\left(q^2,q\right)&=(-1)^nq^{n^2}\left(q;q^2\right)_n,\\
c_{2n,0}(-q,q)&=(-q)^n\left(-1;q^2\right)_n,\\
c_{2n,0}\left(q^{1/2},q\right)&=q^{n/2}\left(q^{1/2};q\right)_n,\\
c_{2n,0}\left(q^{1/3},q\right)&=q^{n/3}\left(q^{2n/3};q^{-1/3}\right)_n,\\
c_{2n,0}\left(q^{2/3},q\right)&=q^{2n/3}\left(q^{1/3};q^{2/3}\right)_n.
\end{aligned} \tag{7.2.40}$$

Each case in (7.2.40) leads to a Rogers–Ramanujan type identity; see Ismail and Stanton (2003c).

L. J. Rogers (1894) found the **linearization coefficients for continuous q-ultraspherical polynomials**. He proved

$$\begin{aligned}
C_m(x;\beta\mid q)C_n(x;\beta\mid q)&=\sum_{k=0}^{m\wedge n}\frac{(q;q)_{m+n-2k}(\beta;q)_{m-k}(\beta;q)_{n-k}(\beta;q)_k\left(\beta^2;q\right)_{m+n-k}}{\left(\beta^2;q\right)_{m+n-2k}(q;q)_{m-k}(q;q)_{n-k}(q;q)_k(\beta q;q)_{m+n-k}}\\
&\quad\times\frac{\left(1-\beta q^{m+n-2k}\right)}{(1-\beta)}C_{m+n-2k}(x;\beta\mid q).
\end{aligned} \tag{7.2.41}$$

An inverse is (Askey and Ismail, 1983)

$$\frac{(\beta;q)_m(\beta;q)_n}{(q;q)_m(q;q)_n}C_{m+n}(x;\beta\mid q)=\frac{(\beta;q)_{m+n}}{(q;q)_{m+n}}\sum_{k=0}^{m\wedge n}b(k,m,n)C_{m-k}(x;\beta\mid q)C_{n-k}(x;\beta\mid q),$$
$$b(k,m,n)=\frac{\left(1-q^{2k-m-n}/\beta^2\right)\left(q^{-m-n}/\beta^2,1/\beta;q\right)_k}{\left(1-q^{-m-n}/\beta^2\right)\left(q;q^{1-m-n}/\beta;q\right)_k}. \tag{7.2.42}$$

The linearization formula (7.2.41) is equivalent to

$$\begin{aligned}
\frac{\left(\beta re^{i\theta},\beta re^{-i\theta},\beta se^{i\theta},\beta se^{-i\theta};q\right)_\infty}{\left(re^{i\theta},re^{-i\theta},se^{i\theta},se^{-i\theta};q\right)_\infty}&=\sum_{m,n=0}^{\infty}\frac{\left(q,\beta^2;q\right)_{m+n}}{\left(\beta^2,\beta q;q\right)_{m+n}}\frac{(\beta;q)_m(\beta;q)_n}{(q;q)_m(q;q)_n}\frac{1-\beta q^{m+n}}{1-\beta}\\
&\quad\times C_{m+n}(\cos\theta;\beta\mid q)r^ms^n\,{}_2\phi_1\left(\left.\begin{matrix}\beta,\beta^2q^{m+n}\\ \beta q^{m+n+1}\end{matrix}\right|q,rs\right).
\end{aligned}$$

Theorem 7.2.1 (Ismail and Stanton, 1988) *We have the bilinear generating function*

$$\begin{aligned}
&\frac{\left(\beta te^{i(\theta+\phi)}, \beta te^{i(\theta-\phi)}, \beta te^{i(\phi-\theta)}, \beta te^{-i(\theta+\phi)}; q\right)_\infty}{\left(te^{i(\theta+\phi)}, te^{i(\theta-\phi)}, te^{i(\phi-\theta)}, te^{-i(\theta+\phi)}; q\right)_\infty} \\
&\quad = \sum_{n=0}^{\infty} \frac{\left(q, \beta^2; q\right)_n}{(\beta^2, \beta q; q)_n} \frac{1-\beta q^n}{1-\beta} t^n \, {}_2\phi_1\left(\left.\begin{matrix}\beta, \beta^2 q^n \\ \beta q^{n+1}\end{matrix}\right| q, t^2\right) \\
&\quad \times C_n(\cos\theta; \beta \mid q) C_n(\cos\phi; \beta \mid q).
\end{aligned} \tag{7.2.43}$$

Theorem 7.2.1 implies that the left-hand side of (7.2.43) is a symmetric Hilbert–Schmidt kernel and (7.2.43) is the expansion guaranteed by Mercer's theorem (Tricomi, 1957) for $\beta \in (-1, 1)$.

The special case of the q-ultraspherical polynomials of Theorem 8.2.3 was proved by Rahman and Verma (1986b) and takes the form

$$\begin{aligned}
C_n(\cos\theta; \beta \mid q) &= \frac{\left(\beta, \beta e^{2i\theta}; q\right)_\infty}{(q, e^{2i\theta}; q)_\infty} e^{-in\theta} \, {}_2\phi_1\left(\left.\begin{matrix}q/\beta, qe^{-2i\theta}/\beta \\ qe^{-2i\theta}\end{matrix}\right| q, \beta^2 q^n\right) \\
&+ \text{ a similar term with } \theta \text{ replaced by } -\theta.
\end{aligned} \tag{7.2.44}$$

The q-integral form of (7.2.44) is (7.2.16). One application of (7.2.44) is

$$\begin{aligned}
\sum_{n=0}^{\infty} C_n(\cos\theta; \beta \mid q) \frac{(\lambda; q)_n t^n}{(\beta^2; q)_n} &= \frac{2i \sin\theta \left(\beta, \beta, \beta e^{2i\theta}, \beta e^{-2i\theta}; q\right)_\infty}{(1-q)\left(q, \beta^2, e^{2i\theta}, e^{-2i\theta}; q\right)_\infty} \\
&\times \int_{e^{i\theta}}^{e^{-i\theta}} \frac{\left(que^{i\theta}, que^{-i\theta}, \lambda ut; q\right)_\infty}{(\beta ue^{i\theta}, \beta ue^{-i\theta}, ut; q)_\infty} \, d_q u.
\end{aligned} \tag{7.2.45}$$

The following theorem gives a closed form expression for the Turánian (see (2.12.3)) of continuous q-ultraspherical polynomials.

Theorem 7.2.2 (Ismail, 2005c) *Let* $\zeta = \left(\beta^{1/2} + \beta^{-1/2}\right)/2$ *and*

$$\mathcal{C}_n(x; \beta \mid q) := \frac{C_n(x; \beta \mid q)}{C_n(\zeta; \beta \mid q)} = \frac{(q; q)_n}{(\beta^2; q)_n} C_n(x; \beta \mid q). \tag{7.2.46}$$

Then the Turánian of $\{\mathcal{C}_n(x; \beta \mid q)\}$ *is given by*

$$\begin{aligned}
T\left(\mathcal{C}_0(\cos\theta; \beta \mid q), \ldots, \mathcal{C}_n(\cos\theta; \beta \mid q)\right) &= (-1)^{n(n+1)/2} q^{(n+1)n(n-1)/6} \\
&\times \prod_{j=1}^{n} \left(\beta e^{2i\theta}, \beta e^{-2i\theta}; q\right)_j \frac{\left(q, \beta, \beta, \beta^2/q; q\right)_j}{(\beta^2, \beta^2/q; q)_{2j}}.
\end{aligned} \tag{7.2.47}$$

The case $\beta = 0$ of Theorem 7.2.2 is in Bryc, Matysiak, and Szabłowski (2005).

7.3 Asymptotics

For x in the complex plane set

$$x = \cos\theta \quad \text{and} \quad e^{\pm i\theta} = x \pm \sqrt{x^2 - 1}. \tag{7.3.1}$$

We choose the branch of the square root that makes $\sqrt{x^2-1}/x \to 1$ as $x \to \infty$. This makes

$$\left|e^{-i\theta}\right| \le \left|e^{i\theta}\right|, \quad \text{with} = \text{if and only if } x \in [-1,1]. \tag{7.3.2}$$

Darboux's method shows that as $n \to \infty$ we have

$$C_n(x;\beta \mid q) = \frac{e^{-in\theta}\left(\beta, \beta e^{-2i\theta}; q\right)_\infty}{(q, e^{-2i\theta}; q)_\infty}[1 + o(1)], \quad x = \cos\theta \in \mathbf{C} \setminus [-1,1], \tag{7.3.3}$$

$$C_n(\cos\theta;\beta \mid q) = 2\sqrt{\frac{(\beta, \beta, \beta e^{2i\theta}, \beta e^{-2i\theta}; q)_\infty}{(q, q, e^{2i\theta}, e^{-2i\theta}; q)_\infty}}\cos(n\theta + \phi)[1 + o(1)],$$
$$x = \cos\theta \in (-1,1), \quad \text{as } n \to \infty, \tag{7.3.4}$$

with

$$\phi = \arg\left[\frac{\left(\beta e^{2i\theta}; q\right)_\infty}{(e^{2i\theta}; q)_\infty}\right]. \tag{7.3.5}$$

Moreover,

$$(-1)^n C_n(-1;\beta \mid q) = C_n(1;\beta \mid q) = n\frac{(\beta;q)_\infty^2}{(q;q)_\infty^2}[1 + o(1)], \quad \text{as } n \to \infty. \tag{7.3.6}$$

In the normalization of Theorem 2.3.5, one has

$$\alpha_n = 0, \quad \beta_n = \frac{\left(1 - \beta^2 q^{n-1}\right)(1 - q^n)}{4(1 - \beta q^n)(1 - \beta q^{n-1})}, \quad \zeta_n = \frac{1-\beta}{1-\beta q^n}\frac{\left(\beta^2;q\right)_n}{(q;q)_n},$$
$$\frac{P_n(x)}{\sqrt{\zeta_n}} = \frac{C_n(x;\beta \mid q)}{\sqrt{u_n}}, \quad u_n = \frac{1-\beta}{1-\beta q^n}\frac{\left(\beta^2;q\right)_n}{(q;q)_n}.$$

7.4 Integrals and the Rogers–Ramanujan Identities

The **Rogers–Ramanujan identities** are

$$\sum_{n=0}^{\infty} \frac{q^{n^2}}{(q;q)_n} = \frac{1}{(q, q^4; q^5)_\infty}, \tag{7.4.1}$$

$$\sum_{n=0}^{\infty} \frac{q^{n^2+n}}{(q;q)_n} = \frac{1}{(q^2, q^3; q^5)_\infty}. \tag{7.4.2}$$

The Rogers–Ramanujan identities and their generalizations play a central role in the theory of partitions (Andrews, 1976, 1986).

It was realized in Garrett, Ismail, and Stanton (1999) that the Rogers–Ramanujan identities result from the evaluation of an integral in two different ways. By considering the integral

$$I_m(t) = \frac{(q;q)_\infty}{2\pi} \int_0^\pi H_m(\cos\theta \mid q) \left(te^{i\theta}, te^{-i\theta}, e^{2i\theta}, e^{-2i\theta}; q\right)_\infty d\theta, \tag{7.4.3}$$

Garrett, Ismail, and Stanton (1999) proved that

$$\sum_{n=0}^\infty \frac{q^{n^2+mn}}{(q;q)_n} = \frac{1}{(q;q)_\infty} \sum_{s=0}^m \begin{bmatrix} m \\ s \end{bmatrix}_q q^{2s(s-m)} \left(q^5, q^{3+4s-2m}, q^{2-4s+2m}; q^5\right)_\infty \tag{7.4.4}$$

and its inverse

$$\frac{\left(q^{3-2m}, q^{2+2m}; q^5\right)_\infty}{(q,q^2,q^3,q^4;q^5)_\infty} = \sum_{j=0}^{[m/2]} (-1)^j q^{2j(j-m)+j(j+1)/2} \begin{bmatrix} m-j \\ j \end{bmatrix}_q \sum_{s=0}^\infty \frac{q^{s^2+s(m-2j)}}{(q;q)_s} \tag{7.4.5}$$

holds for $m = 0, 1, \ldots$. In the process of proving (7.4.4), Garrett, Ismail, and Stanton established

$$I_m(t) = (-t)^m q^{\binom{m}{2}} \sum_{n=0}^\infty \frac{q^{n^2-n}}{(q;q)_n} \left(t^2 q^m\right)^n. \tag{7.4.6}$$

Theorem 7.4.1 (Garrett, Ismail, and Stanton, 1999) *The following generalization of the Rogers–Ramanujan identities holds for* $m = 0, 1, \ldots$ *:*

$$\sum_{n=0}^\infty \frac{q^{n^2+mn}}{(q;q)_n} = \frac{(-1)^m q^{-\binom{m}{2}} a_m(q)}{(q,q^4;q^5)_\infty} + \frac{(-1)^{m+1} q^{-\binom{m}{2}} b_m(q)}{(q^2,q^3;q^5)_\infty}, \tag{7.4.7}$$

where

$$\begin{aligned} a_m(q) &= \sum_j q^{j^2+j} \begin{bmatrix} m-j-2 \\ j \end{bmatrix}_q, \\ b_m(q) &= \sum_j q^{j^2} \begin{bmatrix} m-j-1 \\ j \end{bmatrix}_q. \end{aligned} \tag{7.4.8}$$

The polynomials $\{a_m(q)\}$ and $\{b_m(q)\}$ were considered by Schur in conjunction with his proof of the Rogers–Ramanujan identities. We shall refer to them as the Schur polynomials. They are solutions of the three-term recurrence relation

$$y_{m+2} = y_{m+1} + q^m y_m, \tag{7.4.9}$$

with the initial conditions

$$a_0(q) = 1, \ a_1(q) = 0, \quad \text{and} \quad b_0(q) = 0, \ b_1(q) = 1. \tag{7.4.10}$$

The **Schur polynomials** have the generating functions

$$\sum_{n=0}^\infty a_n(q) t^n = \sum_{n=0}^\infty \frac{t^{2n} q^{n(n-1)}}{(t;q)_n}, \quad \sum_{n=0}^\infty b_n(q) t^n = \sum_{n=0}^\infty \frac{t^{2n+1} q^{n^2}}{(t;q)_{n+1}}. \tag{7.4.11}$$

We can then prove (7.4.7) from knowledge of the Rogers–Ramanujan identities (Ismail and Stanton, 2003c). Moreover, (7.4.9) defines y_m for $m > 0$ from y_0 and y_1. Indeed, one can show that

$$\begin{aligned} b_{1-m}(q) &= (-1)^m q^{-\binom{m}{2}} a_m(q), \quad m \geq 1, \\ a_{1-m}(q) &= (-1)^{m+1} q^{-\binom{m}{2}} b_m(q), \quad m \geq 1. \end{aligned} \tag{7.4.12}$$

Theorem 7.4.2 *The generalized Rogers–Ramanujan identities (7.4.7) hold for all integers m, where $a_m(q)$ and $b_m(q)$ are given by (7.4.8) for $m \geq 0$ and by (7.4.12) when $m < 0$.*

Carlitz (1959) proved the case $m \leq 0$ of Theorem 7.4.2.

Theorem 7.4.3 (Garrett, Ismail, and Stanton, 1999) *The following quintic transformations hold:*

$$\begin{aligned} \sum_{n=0}^{\infty} \frac{q^{n^2}(qf)^{2n}}{(q;q)_n} &= \frac{\left(f^4q^5;q\right)_\infty}{(f^4q^5, f^6q^{10};q^5)_\infty (f^2q^3;q)_\infty} \\ &\times \sum_{n=0}^{\infty} \frac{1-f^6q^{10n+5}}{1-f^6q^5} \frac{\left(f^6q^5, f^4q^{10};q^5\right)_n}{(q^5, f^2;q^5)_n} \frac{\left(f^2;q\right)_{5n}}{(f^4q^6;q)_{5n}} q^{5\binom{n}{2}} \left(-f^4q^{10}\right)^n \\ &= \frac{\left(f^4q^9, f^2q^5, f^4q^6;q^5\right)_\infty}{(f^2q^3;q)_\infty} {}_3\phi_2\left(\left.\begin{matrix} f^2q^2, f^2q^3, f^2q^5 \\ f^4q^9, f^4q^6 \end{matrix}\right| q^5, f^2q^5\right) \\ &= \frac{\left(f^4q^8, f^2q^6, f^4q^6;q^5\right)_\infty}{(f^2q^3;q)_\infty} {}_3\phi_2\left(\left.\begin{matrix} f^2q, f^2q^3, f^2q^4 \\ f^4q^8, f^4q^6 \end{matrix}\right| q^5, f^2q^6\right). \end{aligned} \tag{7.4.13}$$

Observe that the Rogers–Ramanujan identities (7.4.1) and (7.4.2) correspond to the special cases $f = q^{-1}$ and $f = q^{-1/2}$ in the last two forms of (7.4.13).

7.5 A Generalization of the Schur Polynomials

Al-Salam and Ismail (1983) found a common generalization of $\{a_n(q)\}$ and $\{b_n(q)\}$. Their polynomials are generated by

$$\begin{aligned} U_0(x;a,b) &:= 1, \quad U_1(x;a,b) := x(1+a), \\ x(1+aq^n)\,U_n(x;a,b) &= U_{n+1}(x;a,b) + bq^{n-1}U_{n-1}(x;a,b), \end{aligned} \tag{7.5.1}$$

for $q \in (0,1)$, $b > 0$, $a > -1$. Set

$$F(x;a) := \sum_{k=0}^{\infty} \frac{(-1)^k x^k q^{k(k-1)}}{(q,-a;q)_k}. \tag{7.5.2}$$

The function F is essentially a $J_\nu^{(2)}$ q-Bessel function. One can show that the numerator polynomial is

$$U_n^*(x;a,b) = (1+a)U_{n-1}(x;qa,qb), \tag{7.5.3}$$

and

$$\sum_{n=0}^{\infty} U_n(x;a,b)t^n = \sum_{m=0}^{\infty} \frac{(bt/(ax);q)_m}{(xt;q)_{m+1}}(axt)^m q^{m(m-1)/2}, \tag{7.5.4}$$

$$U_n(x;a,b) = \sum_{k=0}^{\lfloor n/2 \rfloor} \frac{(-a,q;q)_{n-k}(-b)^k x^{n-2k}}{(-a,q;q)_k(q;q)_{n-2k}} q^{k(k-1)}. \tag{7.5.5}$$

Moreover,

$$\lim_{n\to\infty} x^{-n} U_n(x;a,b) = (-a;q)_\infty F\left(b/x^2;a\right), \tag{7.5.6}$$

and the functions $F(z;a)$ satisfy

$$\begin{gathered}(z-a)F(zq;a) + (a-q)F(z;a) + qF(z/q;a) = 0,\\ [(1+a)qF(z;a) - F(z/q;a)] = zF(qz;qa).\end{gathered} \tag{7.5.7}$$

Theorem 7.5.1 *The functions $F(z;a)$ and $F(qz;qa)$ have no common zeros. Let $\mu^{(a)}$ be the normalized orthogonality measure of $\{U_n(x;a,b)\}$. Then*

$$\int_{\mathbb{R}} \frac{d\mu^{(a)}(y)}{z-y} = \frac{F\left(qbz^{-2};qa\right)}{zF\left(bz^{-2};a\right)}, \tag{7.5.8}$$

where $F(bz^{-2};a) \neq 0$ and $z \neq 0$.

Theorem 7.5.2 *For $a > -1$, $q \in (0,1)$, the function $F(z;a)$ has only positive simple zeros. The zeros of $F(z;a)$ and $F(zq;aq)$ interlace. The measure $\mu^{(a)}$ is supported at $\{\pm\sqrt{b/x_n(a)}\}$ where $\{x_n(a)\}$ are the zeros of $F(z;a)$ arranged in increasing order.*

Observe that

$$a_{n+2}(q) = U_n\left(1;0,-q^2\right), \quad b_{n+2}(q) = U_{n+1}(1;0,-q). \tag{7.5.9}$$

The q-Lommel polynomials of Section 6.4 are $\{U_n(2x;-q^\nu,q^\nu)\}$.

It is true that

$$(-a;q)_\infty F(x;a) = G(x/a;a), \tag{7.5.10}$$

where

$$G(x;a) = \sum_{m=0}^{\infty} \frac{(x;q)_m}{(q;q)_m} a^m q^{\binom{m}{2}}. \tag{7.5.11}$$

The continued fraction associated with (7.5.1) is (Al-Salam and Ismail, 1983)

$$\cfrac{1+a}{x(1+a) - \cfrac{b}{x(1+aq) - \cdots \cfrac{bq^{n-2}}{x(1+aq^n) - \ddots}}} \tag{7.5.12}$$

$$= \frac{F\left(bq/x^2;aq\right)}{xF\left(b/x^2;a\right)} = \frac{(1+a)}{x}\frac{G\left(b/ax^2;aq\right)}{G\left(b/ax^2;a\right)}.$$

Formula (7.5.12) appeared in Ramanujan's lost notebook. George Andrews gave another proof of it in Andrews (1981) without identifying its partial numerators or denominators.

7.6 Associated q-Ultraspherical Polynomials

The **associated continuous q-ultraspherical polynomials** $\{C_n^{(\alpha)}(x;\beta \mid q)\}$ (Bustoz and Ismail, 1982) are recursively defined by

$$C_0^{(\alpha)}(x;\beta \mid q) = 1, \quad C_1^{(\alpha)}(x;\beta \mid q) = \frac{2(1-\alpha\beta)}{(1-\alpha q)} x, \tag{7.6.1}$$

$$\begin{aligned} 2x(1-\alpha\beta q^n) C_n^{(\alpha)}(x;\beta \mid q) &= \left(1-\alpha q^{n+1}\right) C_{n+1}^{(\alpha)}(x;\beta \mid q) \\ &+ \left(1-\alpha\beta^2 q^{n-1}\right) C_{n-1}^{(\alpha)}(x;\beta \mid q), \quad n > 0. \end{aligned} \tag{7.6.2}$$

A generating function is (Bustoz and Ismail, 1982)

$$\sum_{n=0}^{\infty} C_n^{(\alpha)}(x;\beta \mid q) t^n = \frac{1-\alpha}{1-2xt+t^2} \, {}_2\phi_1\left(\left. \begin{matrix} q, \beta t e^{i\theta}, \beta t e^{-i\theta} \\ q t e^{i\theta}, q t e^{-i\theta} \end{matrix} \right| q, \alpha \right). \tag{7.6.3}$$

Let $\mu(.; \alpha, \beta)$ be the orthogonality measure of $\{C_n^{(\alpha)}(x;\beta \mid q)\}$. Then

$$\int_{\mathbb{R}} \frac{d\mu(t;\alpha,\beta)}{x-t} = \frac{2(1-\alpha\beta)\, {}_2\phi_1(\beta, \beta\rho_1^2; q\rho_1^2; q, q, \alpha)}{(1-\alpha)\rho_2 \, {}_2\phi_1(\beta, \beta\rho_1^2; q\rho_1^2; q, q\alpha)}, \quad x \notin \mathbb{R} \tag{7.6.4}$$

where ρ_1 and ρ_2 are as in (4.2.13), that is, $\rho_1, \rho_2 = x \pm \sqrt{x^2-1}$, $|\rho_1| \le |\rho_2|$. As $n \to \infty$ we have

$$\begin{aligned} C_n^{(\alpha)}(\cos\theta;\beta \mid q) &\approx \frac{(1-\alpha)i}{2\sin\theta} e^{-i(n+1)\theta} \, {}_2\phi_1\left(\left. \begin{matrix} \beta e^{2i\theta}, \beta \\ q e^{2i\theta} \end{matrix} \right| q, \alpha \right) \\ &+ \text{ a similar term with } \theta \text{ replaced by } -\theta, \quad 0 < \theta < \pi. \end{aligned} \tag{7.6.5}$$

The orthogonality measure has no singular part if the denominator in (7.6.4) has no zeros. The orthogonality measure is absolutely continuous if $0 < q < 1$ and

$$0 < \beta < 1, \; 0 < \alpha < 1, \quad \text{or} \quad q^2 \le \beta < 1, \; -1 < \alpha < 0 \tag{7.6.6}$$

(Bustoz and Ismail, 1982). When the measure is purely absolutely continuous then

$$\begin{aligned} &\int_0^{\pi} C_m^{(\alpha)}(\cos\theta;\beta \mid q) C_n^{(\alpha)}(\cos\theta;\beta \mid q) w(\cos\theta;\alpha,\beta) \sin\theta \, d\theta \\ &= \frac{(1-\alpha\beta)\left(\alpha\beta^2;q\right)_n}{(1-\alpha\beta q^n)(\alpha q;q)_n} \delta_{m,n}, \end{aligned} \tag{7.6.7}$$

$$w(\cos\theta;\alpha,\beta) = \frac{2}{\pi} \frac{(1-\alpha\beta)\left(\alpha\beta^2;q\right)_\infty}{(1-\alpha)(\alpha;q)_\infty} \left| {}_2\phi_1\left(\left. \begin{matrix} \beta e^{2i\theta}, \beta \\ q e^{2i\theta} \end{matrix} \right| q, \alpha \right) \right|^{-2}. \tag{7.6.8}$$

The continued fraction associated with (7.6.2) is

$$\cfrac{1}{x-\cfrac{\beta_1}{x-\cfrac{\beta_2}{x-\ddots}}} = \frac{2(1-\alpha\beta)\,{}_2\phi_1(\beta,\beta\rho_1^2;q\rho_1^2;q,q,\alpha)}{(1-\alpha)\rho_2\,{}_2\phi_1(\beta,\beta\rho_1^2;q\rho_1^2;q,q\alpha)}, \quad x\notin\mathbb{R}, \tag{7.6.9}$$

$$\beta_n = \frac{1}{4}\frac{\left(1-\alpha\beta^2q^{n-1}\right)(1-\alpha q^n)}{(1-\alpha\beta q^n)(1-\alpha\beta q^{n-1})}.$$

Consider the convergents $\{C_n\}$ of (7.6.9). When $x = 1/2$, $C_{3n+\epsilon}$ converges for $\epsilon = 0, \pm 1$ and the limits are known (Ismail and Stanton, 2006).

We consider the function

$$\Phi_n(\theta;\beta,\alpha) = \int_{e^{-i\theta}}^{e^{i\theta}} \frac{y^n}{1-q}\frac{\left(qye^{i\theta},qye^{\ i\theta},\beta e^{-i\theta}/y;q\right)_\infty}{(\alpha\beta e^{i\theta}y,qe^{-i\theta}/(\alpha\beta y),qye^{-i\theta}/\beta;q)_\infty}\,d_qy. \tag{7.6.10}$$

Theorem 7.6.1 *The function $\Phi_n(\theta,\beta,\alpha)$ has the hypergeometric representation*

$$\begin{aligned}\Phi_n(\theta,\beta,\alpha) &= e^{i(n+1)\theta}\frac{\left(q,\alpha q^{n+1},qe^{2i\theta},e^{-2i\theta};q\right)_\infty}{(q/\beta,\alpha\beta q^n,\alpha\beta e^{2i\theta},qe^{2i\theta}/(\alpha\beta);q)_\infty}\\ &\times {}_2\phi_1\left(\left.\begin{matrix}q^{-n}/\alpha,\beta\\ q^{1-n}/(\alpha\beta)\end{matrix}\right| q,\frac{q}{\beta}e^{-2i\theta}\right), \quad n\geq 0.\end{aligned} \tag{7.6.11}$$

Corollary 7.6.2 *The function $v_n(\theta;\beta,\alpha)$ defined by*

$$v_n(\theta;\beta,\alpha) = \frac{\Phi_n(\theta;\beta,\alpha)}{\Phi_0(\theta;\beta,\alpha)} = e^{in\theta}\frac{(\alpha\beta;q)_n}{(q\alpha;q)_n}\,{}_2\phi_1\left(\left.\begin{matrix}q^{-n}/\alpha,\beta\\ q^{1-n}/(\alpha\beta)\end{matrix}\right| q,\frac{q}{\beta}e^{-2i\theta}\right) \tag{7.6.12}$$

satisfies the three-term recurrence relation (7.6.2).

When $\alpha = 1$ the extreme right-hand side of (7.6.12) reduces to the q-ultraspherical polynomial $C_n(\cos\theta;\beta \mid q)$. For $\alpha \neq 1$ it may not be a polynomial but, nevertheless, is a solution to (7.6.2).

The following two solutions of (7.6.2) are defined for a wider range of β:

$$v_n^{\pm}(\theta;\alpha,\beta) := e^{\pm(n+1)i\theta}\,{}_2\phi_1\left(\left.\begin{matrix}q/\beta,qe^{\pm 2i\theta}/\beta\\ qe^{\pm 2i\theta}\end{matrix}\right| q,\alpha\beta^2q^n\right), \tag{7.6.13}$$

provided that

$$-1 < \alpha\beta^2/q < 1. \tag{7.6.14}$$

The Casorati determinant of v_n^+ and v_n^-,

$$\Delta_n = v_{n+1}^+(\theta;\beta,\alpha)v_n^-(\theta;\beta,\alpha) - v_n^+(\theta;\beta,\alpha)v_{n+1}^-(\theta;\beta,\alpha),$$

is given by

$$\Delta_n = \frac{\left(\alpha q^{n+2};q\right)_\infty}{(\alpha\beta^2 q^n;q)_\infty} 2i\sin\theta. \tag{7.6.15}$$

Since $\{v_n^\pm\}$ is a basis of solutions of (7.6.2) we can show that

$$\begin{aligned} C_n^{(\alpha)}(\cos\theta;\beta \mid q) &= \frac{\left(\alpha\beta^2/q;q\right)_\infty}{2i\sin\theta(\alpha q;q)_\infty} \\ &\times \left[v_{-1}^-(\theta;\beta,\alpha)v_n^+(\theta;\beta,\alpha) - v_{-1}^+(\theta;\beta,\alpha)v_n^-(\theta;\alpha,\beta)\right]. \end{aligned} \tag{7.6.16}$$

Formula (7.6.16) is Rahman and Tariq (1997, (3.4)) and another proof is in Ismail and Stanton (2002).

Theorem 7.6.3 (Ismail and Stanton, 2002) *We have*

$$\begin{aligned} \sum_{n=0}^\infty \frac{(\lambda;q)_n}{(q;q)_n} C_{n+k}^{(\alpha)}(\cos\theta;\beta \mid q)t^n &= e^{ik\theta}\frac{\left(\lambda te^{i\theta},\alpha\beta^2/q;q\right)_\infty}{(1-e^{2i\theta})(te^{i\theta},\alpha q;q)_\infty} \\ &\times {}_2\phi_1\left(\left.\begin{matrix} q/\beta, qe^{-2i\theta}/\beta \\ qe^{-2i\theta}\end{matrix}\right| q, \frac{\alpha\beta^2}{q}\right) {}_3\phi_2\left(\left.\begin{matrix} q/\beta, qe^{2i\theta}/\beta, te^{i\theta} \\ qe^{2i\theta}, \lambda te^{i\theta}\end{matrix}\right| q, \alpha\beta^2 q^k\right) \\ &+ \text{ a similar term with } \theta \text{ replaced by } -\theta, \end{aligned} \tag{7.6.17}$$

$$\begin{aligned} \sum_{n=0}^\infty C_n(\cos\phi;\beta_1 \mid q)\, C_n^{(\alpha)}(\cos\theta;\beta \mid q)t^n &= \frac{\left(\alpha\beta^2/q,\beta_1 te^{i(\theta+\phi)},\beta_1 te^{i(\theta-\phi)};q\right)_\infty}{(1-e^{-2i\theta})(\alpha q, te^{i(\theta+\phi)}, te^{i(\theta-\phi)};q)_\infty} \\ &\times {}_2\phi_1\left(\left.\begin{matrix} q/\beta, qe^{-2i\theta}/\beta \\ qe^{-2i\theta}\end{matrix}\right| q, \frac{\alpha\beta^2}{q}\right) {}_4\phi_3\left(\left.\begin{matrix} q/\beta, qe^{2i\theta}/\beta, te^{i(\theta+\phi)}, te^{i(\theta-\phi)} \\ qe^{2i\theta}, \beta_1 te^{i(\theta+\phi)}, \beta_1 te^{i(\theta-\phi)}\end{matrix}\right| q, \alpha\beta^2\right) \\ &+ \text{ a similar term with } \theta \text{ replaced by } -\theta. \end{aligned} \tag{7.6.18}$$

The cases $\lambda = q$ or $k = 0$ of (7.6.17) are in Rahman and Tariq (1997).

We now give a Poisson-type kernel for the polynomials under consideration.

Theorem 7.6.4 (Ismail and Stanton, 2002) *A bilinear generating function for the associated continuous q-ultraspherical polynomials is given by*

$$\begin{aligned} &\sum_{n=0}^\infty C_n^{(\alpha_1)}(\cos\phi;\beta_1 \mid q)\, C_n^{(\alpha)}(\cos\theta;\beta \mid q)t^n \\ &= \frac{(1-\alpha_1)\left(\alpha\beta^2/q;q\right)_\infty}{(1-e^{-2i\theta})(\alpha q;q)_\infty} {}_2\phi_1\left(\left.\begin{matrix} q/\beta, qe^{-2i\theta}/\beta \\ qe^{-2i\theta}\end{matrix}\right| q, \frac{\alpha\beta^2}{q}\right) \\ &\times \sum_{k=0}^\infty \frac{\left(q/\beta, qe^{2i\theta}/\beta;q\right)_k \alpha^k\beta^{2k}}{[1-2\cos\phi\, te^{i\theta}q^k + t^2q^{2k}e^{2i\theta}](q, qe^{2i\theta};q)_k} \\ &\times {}_3\phi_2\left(\left.\begin{matrix} q, \beta_1 tq^k e^{i(\theta+\phi)}\beta_1 tq^k e^{i(\theta-\phi)} \\ 4q^{k+1}te^{i(\theta+\phi)}, q^{k+1}te^{i(\theta-\phi)}\end{matrix}\right| q, \alpha_1\right) \\ &+ \text{ a similar term with } \theta \text{ replaced by } -\theta. \end{aligned} \tag{7.6.19}$$

The case $\alpha = \alpha_1$ of (7.6.18) is in Rahman and Tariq (1997).

7.7 Two Systems of q-Orthogonal Polynomials

A **q-analogue of the Pollaczek polynomials** was introduced in Charris and Ismail (1987). They also appeared later in Al-Salam and Chihara (1987), which classifies all families of orthogonal polynomials having generating functions of the form

$$A(t)\prod_{n=0}^{\infty}\frac{1-\delta xH(q^m t)}{1-\theta xK(q^m t)}=\sum_{n=0}^{\infty}P_n(x)t^n, \quad m\geq 0,\ n\geq 0.$$

Following Charris and Ismail (1987) we denote the polynomials by $\{F_n(x;U,\Delta,V)\}$, or $\{F_n(x)\}$ for short. They are **generated** by

$$F_0(x)=1, \quad F_{-1}(x)=0, \tag{7.7.1}$$

$$2\left[(1-U\Delta q^n)x+Vq^n\right]F_n(x)=\left(1-q^{n+1}\right)F_{n+1}(x)+\left(1-\Delta^2q^{n-1}\right)F_{n-1}(x), \tag{7.7.2}$$

$n\geq 0$. One can show that

$$\sum_{n=0}^{\infty}F_n(\cos\theta)t^n=\frac{(t/\xi,t/\eta;q)_\infty}{(te^{i\theta},te^{-i\theta};q)_\infty}, \tag{7.7.3}$$

$$F_n(\cos\theta)=e^{in\theta}\frac{\left(e^{-i\theta}/\xi;q\right)_n}{(q;q)_n}\,{}_2\phi_1\left(\left.\begin{matrix}q^{-n},e^{i\theta}/\eta\\ q^{1-n}e^{i\theta}\xi\end{matrix}\right|q,qe^{-i\theta}\xi\right), \tag{7.7.4}$$

where

$$1+2q(V-x\Delta U)\Delta^{-2}y+q^2\Delta^{-2}y^2=(1-q\xi y)(1-q\eta y), \tag{7.7.5}$$

and ξ and η depend on x.

The numerators $\{F_n^*(x)\}$ have the generating function

$$\sum_{n=0}^{\infty}F_n^*(\cos\theta)t^n=2t(1-U\Delta)\sum_{n=0}^{\infty}\frac{(t/\xi,t/\eta;q)_nq^n}{(te^{i\theta},te^{-i\theta};q)_{n+1}}. \tag{7.7.6}$$

If $q,U,\Delta\in[0,1)$ and $1-U^2\pm 2V>0$ the orthogonality relation of the F_n is (Charris and Ismail, 1987)

$$\begin{aligned}\int_0^{\pi}&\frac{\left(e^{2i\theta},e^{-2i\theta};q\right)_\infty}{(e^{i\theta}/\xi,e^{-i\theta}/\xi,e^{i\theta}/\eta,e^{-i\theta}/\eta;q)_\infty}F_m(\cos\theta;U,\Delta,V)F_n(\cos\theta;U,\Delta,V)\,d\theta\\ &=\frac{2\pi}{(q,\Delta^2;q)_\infty}\frac{\left(\Delta^2;q\right)_n}{(1-U\Delta q^n)(q;q)_n}\delta_{m,n}.\end{aligned} \tag{7.7.7}$$

From (7.7.4) and (8.1.9) it follows that

$$F_n(x;U,\Delta,V)=\frac{(1/(\xi\eta);q)_n}{(q;q)_n}\eta^np_n\left(x;1/\eta,1/\xi\right), \tag{7.7.8}$$

where $\{p_n(x; t_1, t_2)\}$ are the Al-Salam–Chihara polynomials. We use Theorem 8.1.1 to state moment representations for the q-Pollaczek polynomials.

Theorem 7.7.1 *The q-Pollaczek polynomials have the q-integral representations*

$$\begin{aligned}\frac{(q;q)_n}{(\Delta^2;q)_n}F_n(x;U,\Delta,V) &= \frac{(e^{i\theta}/\eta, e^{-i\theta}/\eta, e^{i\theta}/\xi, e^{-i\theta}/\xi;q)_\infty}{(1-q)e^{i\theta}(q, qe^{2i\theta}, qe^{-2i\theta};q)_\infty} \\ &\times \int_{e^{-i\theta}}^{e^{i\theta}} y^n \frac{\left(qye^{i\theta}, qye^{-i\theta};q\right)_\infty}{(y/\eta, y/\xi;q)_\infty}\, d_qy,\end{aligned} \tag{7.7.9}$$

$$\begin{aligned}F_n(x;U,\Delta,V) &= \frac{\left(q\eta e^{i\theta}, q\eta e^{-i\theta}, e^{i\theta}/\eta, e^{-i\theta}/\eta;q\right)_\infty}{2(1-q)i\sin\theta\,(q,q,qe^{2i\theta},qe^{-2i\theta};q)_\infty} \\ &\times \int_{e^{-i\theta}}^{e^{i\theta}} y^n \frac{\left(qye^{i\theta}, qye^{-i\theta}, 1/(\xi y);q\right)_\infty}{(qy\eta, y/\eta, q\eta/y;q)_\infty}\, d_qy.\end{aligned} \tag{7.7.10}$$

Ismail and Mulla (1987) generalized the Chebyshev polynomials to polynomials generated by

$$\begin{aligned}\theta_0^{(a)}(x;q) = 1, \quad \theta_1^{(a)}(x;q) = 2x - a,& \\ 2x\theta_n^{(a)}(x;q) = \theta_{n+1}^{(a)}(x;q) + aq^n\theta_n^{(a)}(x;q) + \theta_{n-1}^{(a)}(x;q).&\end{aligned} \tag{7.7.11}$$

One can show that $\theta_n^{(a)}(-x;q) = (-1)^n\theta_n^{(-a)}(x;q)$, $(\theta_n^{(a)}(x;q))^* = 2\theta_{n-1}^{(a)}(x;q)$, $n \ge 0$, and

$$\sum_{n=0}^\infty \theta_n^{(a)}(x;q)t^n = \sum_{n=0}^\infty \frac{(-at)^k q^{\binom{k}{2}}}{(t/\rho_2(x), t/\rho_1(x);q)_{k+1}}, \tag{7.7.12}$$

where $\rho_1(x)$ and $\rho_2(x)$ are as in (4.2.13). The corresponding continued J-fraction is

$$\cfrac{2}{2x - a - \cfrac{1}{2x - aq - \cfrac{1}{2x - aq^2 - \cdots \cfrac{1}{2x - aq^n - \ddots}}}} \tag{7.7.13}$$

$$= 2\rho_1(x)\frac{M(x;aq,q)}{M(x;aq)} = \int_{\mathbb{R}} \frac{d\psi(t;a,q)}{x-t}, \qquad \operatorname{Im} x \neq 0,$$

where $\psi(x;a,q)$ is the orthogonality measure and

$$M(x;a,q) := \sum_{k=0}^\infty \frac{(-a\rho_1(x))^k q^{\binom{k}{2}}}{(q;q)_k\left(\rho_1^2(x);q\right)_{k+1}}. \tag{7.7.14}$$

The orthogonality relation is

$$\int_{\mathbb{R}} \theta_m^{(a)}(x;q)\theta_n^{(a)}(x;q)\, d\psi(x;a,q) = \delta_{m,n}, \tag{7.7.15}$$

$$\frac{d\psi(x;a,q)}{dx} = \frac{2}{\pi}\sqrt{1-x^2}\left|\sum_{k=0}^{\infty} \frac{\left(-ae^{i\theta}\right)^k q^{\binom{k}{2}}}{(q, qe^{2i\theta};q)_k}\right|^{-2}. \tag{7.7.16}$$

Moreover, for $x = \cos\theta$, $0 < \theta < \pi$, Darboux's method yields

$$\theta_n^{(a)}(\cos\theta;q) \approx 2\left|\sum_{k=0}^{\infty} \frac{\left(-ae^{i\theta}\right)^k q^{\binom{k}{2}}}{(q;q)_k \left(e^{2i\theta};q\right)_{k+1}}\right| \cos(n\theta + \varphi), \tag{7.7.17}$$

where

$$\varphi = \arg\left(\sum_{k=0}^{\infty} \frac{\left(-ae^{i\theta}\right)^k q^{\binom{k}{2}}}{(q;q)_k \left(e^{2i\theta};q\right)_{k+1}}\right). \tag{7.7.18}$$

One can prove that ψ is absolutely continuous if $q \in (0,1)$ and

$$|a|q < 1 + q - \sqrt{1+q^2}, \quad \text{or} \quad |a| \le (1-q)^2; \tag{7.7.19}$$

see Ismail and Mulla (1987). Darboux's method also shows that

$$\theta_n^{(a)}(1;q) \approx (n+1)\sum_{k=0}^{\infty} \frac{(-a)^k q^{\binom{k}{2}}}{(q;q)_\infty^2}. \tag{7.7.20}$$

It turned out that special cases of the continued fraction in (7.7.13) are related to continued fractions in Ramanujan's notes, which became known as the "lost notebook." For details see Ismail and Stanton (2006). One special case is when $x = 1/2 = \cos(\pi/3)$. At this point the continued fraction does not converge, but convergents of order $3k+s$, $s = -1, 0, 1$ converge. This follows from (7.7.17). The result is

$$\lim_{k\to\infty} \cfrac{1}{1 - \cfrac{1}{1+q - \cfrac{1}{1+q^2 - \cdots \cfrac{1}{1+q^{3k+s}}}}} \tag{7.7.21}$$

$$= -\omega^2 \frac{\left(q^2;q^3\right)_\infty}{(q;q^3)_\infty} \frac{\omega^{s+1} - \left(\omega^2 q;q\right)_\infty / (\omega q;q)_\infty}{\omega^{s-1} - (\omega^2 q;q)_\infty / (\omega q;q)_\infty},$$

where $\omega = e^{2\pi i/3}$. This was proved in Andrews et al. (2003, 2005). Ismail and Stanton (2006) gave a proof using the polynomials $\theta_n^{(a)}(x;q)$. Ismail and Stanton also extended (7.7.21) to any kth root of unity by letting $x = \cos(\pi/k)$. Many of the results in Andrews (1990) and Berndt and Sohn (2002) become transparent through the use of orthogonal polynomials.

8

The Askey–Wilson Family of Polynomials

8.1 Al-Salam–Chihara Polynomials

The **Al-Salam–Chihara polynomials** appeared in a characterization problem regarding convolutions of orthogonal polynomials. Al-Salam and Chihara (1976) only recorded the three-term recurrence relation and a generating function. The weight function was first found by Askey and Ismail (1983, 1984), who also named the polynomials after the ones who first identified them.

A **basic hypergeometric representation** is

$$p_n(x;t_1,t_2 \mid q) = {}_3\phi_2\left(\begin{matrix} q^{-n}, t_1 e^{i\theta}, t_1 e^{-i\theta} \\ t_1 t_2, 0 \end{matrix} \,\middle|\, q, q \right), \tag{8.1.1}$$

and the **orthogonality relation** is

$$\int_{-1}^{1} p_m(x;t_1,t_2 \mid q)\, p_n(x;t_1,t_2 \mid q)\, w_1(x;t_1,t_2 \mid q)\, dx = \frac{2\pi (q;q)_n t_1^{2n}}{(q, t_1 t_2; q)_\infty (t_1 t_2; q)_n} \delta_{m,n}, \tag{8.1.2}$$

where

$$w_1(x;t_1,t_2 \mid q) := \frac{\left(e^{2i\theta}, e^{-2i\theta}; q\right)_\infty}{(t_1 e^{i\theta}, t_1 e^{-i\theta}, t_2 e^{i\theta}, t_2 e^{-i\theta}; q)_\infty} \frac{1}{\sqrt{1-x^2}}, \quad x = \cos\theta. \tag{8.1.3}$$

The generating **three-term recurrence relation** is

$$\begin{aligned} &[2x - (t_1 + t_2) q^n]\, t_1 p_n(x;t_1,t_2 \mid q) \\ &\quad = (1 - t_1 t_2 q^n)\, p_{n+1}(x;t_1,t_2 \mid q) + t_1^2 (1 - q^n)\, p_{n-1}(x;t_1,t_2 \mid q), \end{aligned} \tag{8.1.4}$$

$$p_0(x;t_1,t_2 \mid q) = 1, \quad p_1(x;t_1,t_2 \mid q) = t_1 (2x - t_1 - t_2). \tag{8.1.5}$$

If $\{P_n(x;t_1,t_2|q)\}$ are the corresponding monic polynomials then

$$p_n(x;t_1,t_2 \mid q) = \frac{(2t_1)^n}{(t_1 t_2; q)_n} P_n(x;t_1,t_2 \mid q) \tag{8.1.6}$$

and

$$\begin{aligned}&[x-(t_1+t_2)q^n/2]\,t_1P_n(x;t_1,t_2\mid q)\\&\quad=P_{n+1}(x;t_1,t_2\mid q)+\frac{(1-q^n)\left(1-t_1t_2q^{n-1}\right)}{4}P_{n-1}(x;t_1,t_2\mid q).\end{aligned} \tag{8.1.7}$$

A **generating function** is

$$\sum_{n=0}^{\infty}\frac{(t_1t_2;q)_n}{(q;q)_n}p_n(\cos\theta;t_1,t_2\mid q)(t/t_1)^n=\frac{(tt_1,tt_2;q)_\infty}{(te^{-i\theta},te^{i\theta};q)_\infty}, \tag{8.1.8}$$

and it leads to the **alternative representation**

$$p_n(x;t_1,t_2\mid q)=\frac{\left(t_1e^{-i\theta};q\right)_n t_1^n e^{in\theta}}{(t_1t_2;q)_n}\,{}_2\phi_1\left(\left.\begin{matrix}q^{-n},t_2e^{i\theta}\\ q^{1-n}e^{i\theta}/t_1\end{matrix}\right|q,qe^{-i\theta}/t_1\right). \tag{8.1.9}$$

We can express a multiple of p_n as a Cauchy product in the form

$$p_n(\cos\theta;t_1,t_2\mid q)=\frac{(q;q)_nt_1^n}{(t_1t_2;q)_n}\sum_{k=0}^{n}\frac{\left(t_2e^{i\theta};q\right)_k}{(q;q)_k}e^{-ik\theta}\frac{\left(t_1e^{-i\theta};q\right)_{n-k}}{(q;q)_{n-k}}e^{i(n-k)\theta}. \tag{8.1.10}$$

When $x=\cos\theta\in[-1,1]$ and with $\left|e^{-i\theta}\right|=\left|e^{i\theta}\right|$, formula (8.1.10) leads to the **asymptotic formula**

$$\lim_{n\to\infty}\frac{p_n(\cos\theta;t_1,t_2\mid q)}{t_1^ne^{in\theta}}=\frac{\left(t_1e^{-i\theta},t_2e^{-i\theta};q\right)_\infty}{(t_1t_2,e^{-2i\theta};q)_\infty}. \tag{8.1.11}$$

It follows from (8.1.10) that

$$\max\{|p_n(x;t_1,t_2\mid q)|:\ -1\le x\le 1\}=|p_n(1;t_1,t_2\mid q)|\le Cn|t_1|^n, \tag{8.1.12}$$

for some constant C which depends only on t_1 and t_2.

The **lowering operator** is

$$\mathcal{D}_qp_n(x;t_1,t_2\mid q)=\frac{(1-q^n)\,t_1q^{n-1}}{(1-t_1t_2)(1-q)}p_{n-1}\left(x;q^{1/2}t_1,q^{1/2}t_2\mid q\right), \tag{8.1.13}$$

while the raising operator is

$$\begin{aligned}&\frac{2(1-t_1t_2)}{(q-1)t_1}w_1(x;t_1,t_2\mid q)\,p_n(x;t_1,t_2\mid q)\\&\quad=\mathcal{D}_qw_1\left(x;t_1,q^{1/2},t_2q^{1/2}\mid q\right)p_{n-1}\left(x;t_1q^{1/2},t_2q^{1/2}\mid q\right).\end{aligned}$$

When $q>1$, we can replace q by $1/q$ and realize that the polynomials involve two new parameters t_1 and t_2, and (8.1.4) can be normalized to become

$$\begin{aligned}&[2xq^n+t_1+t_2]\,r_n(x;t_1,t_2)\\&\quad=(t_1t_2+q^n)\,r_{n+1}(x;t_1,t_2)+(1-q^n)\,r_{n-1}(x;t_1,t_2),\end{aligned} \tag{8.1.14}$$

with

$$r_0(x;t_1,t_2):=1,\quad r_1(x;t_1,t_2)=\frac{(2x+t_1+t_2)}{(1+t_1t_2)}. \tag{8.1.15}$$

We have

$$\sum_{n=0}^{\infty} r_n(\sinh\xi;t_1,t_2)\frac{(1/t_1t_2;q)_n}{(q;q)_n}(t_1t_2t)^n = \frac{\left(-te^{\xi},te^{-\xi};q\right)_\infty}{(tt_1,tt_2;q)_\infty}, \tag{8.1.16}$$

$$r_n(\sinh\xi;t_1,t_2) = \frac{(q;q)_n}{t_1^n(1/t_1t_2;q)_n}\sum_{k=0}^{n}\frac{\left(e^{-\xi}/t_2;q\right)_k}{(q;q)_k}\frac{\left(-e^{\xi}/t_1;q\right)_{n-k}}{(q;q)_{n-k}}\left(\frac{t_1}{t_2}\right)^k. \tag{8.1.17}$$

It must be emphasized that the Al-Salam–Chihara polynomials are q-analogues of Laguerre polynomials as can be seen from (8.1.4),

$$\lim_{q\to 1^-} p_n\left(1-x(1-q)/2;q^{(\alpha+1)/2},q^{(\alpha+1)/2}\right) = \frac{n!}{(\alpha+1)_n}L_n^{(\alpha)}(x). \tag{8.1.18}$$

Theorem 8.1.1 *The Al-Salam–Chihara polynomials have the q-integral moment representations*

$$\begin{aligned}\frac{p_n(\cos\theta;t_1,t_2)}{t_1^n} &= \frac{\left(t_1e^{i\theta},t_1e^{-i\theta},t_2e^{i\theta},t_2e^{-i\theta};q\right)_\infty}{(1-q)e^{i\theta}\left(q,t_1t_2,qe^{2i\theta},e^{-2i\theta};q\right)_\infty}\\ &\quad\times\int_{e^{-i\theta}}^{e^{i\theta}} y^n\frac{\left(qye^{i\theta},qye^{-i\theta};q\right)_\infty}{(t_1y,t_2y;q)_\infty}\,d_qy,\end{aligned} \tag{8.1.19}$$

$$\begin{aligned}\frac{(t_1t_2;q)_n}{(q;q)_n}\frac{p_n(\cos\theta;t_1,t_2)}{t_1^n} &= \frac{\left(t_1e^{i\theta},t_1e^{-i\theta},qe^{i\theta}/t_1,qe^{-i\theta}/t_1;q\right)_\infty}{2(1-q)i\sin\theta\left(q,q,qe^{2i\theta},qe^{-2i\theta};q\right)_\infty}\\ &\quad\times\int_{e^{-i\theta}}^{e^{i\theta}} y^n\frac{\left(qye^{i\theta},qye^{-i\theta},t_2/y;q\right)_\infty}{\left(qy/t_1,t_1y,q/(yt_1);q\right)_\infty}\,d_qy.\end{aligned} \tag{8.1.20}$$

These integral representations lead to the following linear and bilinear generating functions:

$$\begin{aligned}&\sum_{n=0}^{\infty}\frac{(t_1t_2,\lambda/\mu;q)_n}{(q,q;q)_n}p_n(\cos\theta;t_1,t_2)\mu^n\\ &= \frac{e^{i\theta}\left(t_2e^{-i\theta},t_1e^{-i\theta},t_1\lambda e^{i\theta};q\right)_\infty}{2i\sin\theta\left(q,qe^{-2i\theta},t_1\mu e^{i\theta};q\right)_\infty}\,{}_3\phi_2\left(\left.\begin{matrix}qe^{i\theta}/t_1,qe^{i\theta}/t_2,t_1\mu e^{i\theta}\\ qe^{2i\theta},t_1\lambda e^{i\theta}\end{matrix}\right| q,t_1t_2\right)\\ &\quad-\text{ a similar term with }\theta\text{ replaced by }-\theta,\end{aligned} \tag{8.1.21}$$

$$\begin{aligned}&\sum_{n=0}^{\infty}\frac{(t_1t_2,s_1s_2;q)_n}{(q,q;q)_n}p_n(\cos\theta;t_1,t_2)\,p_n(\cos\phi;s_1,s_2)\left(\frac{t}{t_1s_1}\right)^n\\ &= \frac{\left(t_1e^{-i\theta},t_2e^{-i\theta},ts_1e^{i\theta},ts_2e^{i\theta};q\right)_\infty}{\left(q,e^{-2i\theta},te^{i(\theta+\phi)},te^{i(\theta-\phi)};q\right)_\infty}\,{}_4\phi_3\left(\left.\begin{matrix}te^{i(\theta+\phi)},te^{i(\theta-\phi)},qe^{i\theta}/t_1,qe^{i\theta}/t_2\\ ts_1e^{i\theta},ts_2e^{i\theta},qe^{2i\theta}\end{matrix}\right| q,t_1t_2\right)\\ &\quad+\text{ a similar term with }\theta\text{ replaced by }-\theta,\end{aligned} \tag{8.1.22}$$

and

$$\sum_{n=0}^{\infty} \frac{(t_1t_2;q)_n}{(q;q)_n t_1^n} t^n p_n(\cos\theta;t_1,t_2) p_n(\cos\phi;s_1,s_2)$$
$$= \frac{(s_1e^{-i\phi}, s_2e^{-i\phi}, tt_1s_1e^{i\phi}, tt_2s_1e^{i\phi};q)_\infty}{(s_1s_2, e^{-2i\phi}, ts_1e^{i(\theta+\phi)}, ts_1e^{i(\phi-\theta)};q)_\infty} \tag{8.1.23}$$
$$\times {}_4\phi_3\left(\begin{matrix} s_1e^{i\phi}, s_2e^{i\phi}, ts_1e^{i(\theta+\phi)}, ts_1e^{i(\phi-\theta)} \\ qe^{2i\phi}, tt_1s_1e^{i\phi}, tt_2s_1e^{i\phi} \end{matrix} \middle| q,q\right)$$
$$+ \text{ a similar term with } \theta \text{ and } \phi \text{ replaced by } -\theta \text{ and } -\phi.$$

The special case $s_1s_2 = t_1t_2$ of (8.1.23) is the Poisson kernel for the Al-Salam–Chihara polynomials (Askey, Rahman, and Suslov, 1996; Ismail and Stanton, 1997). It is

$$\sum_{n=0}^{\infty} \frac{(t_1t_2;q)_n}{(q;q)_n t_1^n} t^n p_n(\cos\theta;t_1,t_2)\, p_n(\cos\phi;s_1,s_2)$$
$$= \frac{\left(tt_1s_1e^{i\phi}, tt_2s_1e^{i\phi}, ts_1s_2e^{i\theta}, ts_1^2e^{i\theta};q\right)_\infty}{(tt_1t_2s_1e^{i(\theta+\phi)}, ts_1e^{i(\theta-\phi)}, ts_1e^{i(\theta+\phi)}, ts_1e^{i(\phi-\theta)};q)_\infty} \tag{8.1.24}$$
$$\times {}_8\phi_7\left(\begin{matrix} s_1^2ts_2e^{i(\theta+\phi)}/q,\ s_1\sqrt{qts_2}\,e^{i(\theta+\phi)/2}, -s_1\sqrt{qts_2}\,e^{i(\theta+\phi)/2}, t_2e^{i\theta}, t_1e^{i\theta}, \\ s_1\sqrt{ts_2/q}\,e^{i(\theta+\phi)/2}, -s_1\sqrt{ts_2/q}\,e^{i(\theta+\phi)/2}, tt_1s_1e^{i\phi}, tt_2s_1e^{i\phi}, \end{matrix}\right.$$
$$\left.\begin{matrix} s_1e^{i\phi}, s_2e^{i\phi}, ts_1e^{i(\theta+\phi)} \\ ts_1s_2e^{i\theta}, s_1^2te^{i\theta}, s_1s_2 \end{matrix} \middle| q, ts_1e^{-i(\theta+\phi)}\right).$$

Theorem 8.1.2 (Ismail, 2005c) *The Turánian (see* (2.12.3)*) of the Al-Salam–Chihara polynomials is given by*

$$T\left(p_0(\cos\theta,t_1,t_2),\ldots,p_n(\cos\theta,t_1,t_2)\right) = \left(-t_1^2\right)^{n(n+1)/2} q^{(n+1)n(n-1)/6}$$
$$\times \prod_{j=1}^{n} \left(t_1e^{i\theta}, t_1e^{-i\theta}, t_2e^{i\theta}, t_2e^{-i\theta};q\right)_j \frac{(q,t_1t_2/q;q)_j}{(t_1t_2,t_1t_2/q;q)_{2j}}. \tag{8.1.25}$$

8.2 The Askey–Wilson Polynomials

The orthogonality relation (8.1.2), the bound (8.1.12), and the generating function (8.1.8) imply the Askey–Wilson q-beta integral evaluation

$$\int_0^\pi \frac{\left(e^{2i\theta}, e^{-2i\theta};q\right)_\infty}{\prod_{j=1}^4 \left(t_je^{i\theta}, t_je^{-i\theta};q\right)_\infty}\, d\theta = \frac{2\pi\,(t_1t_2t_3t_4;q)_\infty}{(q;q)_\infty \prod_{1\le j<k\le 4}\left(t_jt_k;q\right)_\infty}, \tag{8.2.1}$$

for $|t_1|, |t_2| < 1$. Proofs are in Askey and Wilson (1985), Ismail and Stanton (1988), Rahman (1984), and Askey (1983). The polynomials orthogonal with respect to the weight function in the integral whose total mass is given by (8.2.1) are the **Askey–Wilson polynomials**. To save space we shall use the vector notation $\mathbf{t}$ to denote the ordered tuple (t_1, t_2, t_3, t_4). Their weight function is

$$w(x; \mathbf{t} \mid q) = w(x; t_1, t_2, t_3, t_4 \mid q) = \frac{\left(e^{2i\theta}, e^{-2i\theta}; q\right)_\infty}{\prod_{j=1}^{4}\left(t_j e^{i\theta}, t_j e^{-i\theta}; q\right)_\infty} \frac{1}{\sqrt{1-x^2}}, \tag{8.2.2}$$

$x = \cos\theta$. The polynomials have the **basic hypergeometric representation**

$$p_n(x; \mathbf{t} \mid q) = t_1^{-n} (t_1t_2, t_1t_3, t_1t_4; q)_n \, {}_4\phi_3\left(\left.\begin{matrix} q^{-n}, t_1t_2t_3t_4q^{n-1}, t_1e^{i\theta}, t_1e^{-i\theta} \\ t_1t_2, t_1t_3, t_1t_4 \end{matrix}\right| q, q\right) \tag{8.2.3}$$

and satisfy the orthogonality relation

$$\begin{aligned} &\int_{-1}^{1} p_m(x; \mathbf{t} \mid q) p_n(x; \mathbf{t} \mid q) w(x; \mathbf{t} \mid q)\, dx \\ &= \frac{2\pi \left(t_1t_2t_3t_4q^{2n}; q\right)_\infty \left(t_1t_2t_3t_4q^{n-1}; q\right)_n}{(q^{n+1}; q)_\infty \prod_{1\le j<k\le 4}\left(t_jt_kq^n; q\right)_\infty} \delta_{m,n}, \end{aligned} \tag{8.2.4}$$

for $\max\{|t_1|, |t_2|, |t_3|, |t_4|\} < 1$.

The following **generating function** was established in Ismail and Wilson (1982) through the use of the Sears transformation (1.4.23):

$$\begin{aligned} &\sum_{n=0}^{\infty} \frac{p_n(\cos\theta; \mathbf{t} \mid q)}{(q, t_1t_2, t_3t_4; q)_n} t^n \\ &= {}_2\phi_1\left(\left.\begin{matrix} t_1e^{i\theta}, t_2e^{i\theta} \\ t_1t_2 \end{matrix}\right| q, te^{-i\theta}\right) {}_2\phi_1\left(\left.\begin{matrix} t_3e^{-i\theta}, t_4e^{-i\theta} \\ t_3t_4 \end{matrix}\right| q, te^{i\theta}\right). \end{aligned} \tag{8.2.5}$$

The **three-term recurrence relation** is

$$2xp_n(x; \mathbf{t} \mid q) = A_n p_{n+1}(x; \mathbf{t} \mid q) + B_n p_n(x; \mathbf{t} \mid q) + C_n p_{n-1}(x; \mathbf{t} \mid q), \tag{8.2.6}$$

and the coefficients are given by

$$A_n = \frac{1 - t_1t_2t_3t_4q^{n-1}}{(1 - t_1t_2t_3t_4q^{2n-1})(1 - t_1t_2t_3t_4q^{2n})}, \tag{8.2.7}$$

$$C_n = \frac{(1-q^n)\prod_{1\le j<k\le 4}\left(1 - t_jt_kq^{n-1}\right)}{(1 - t_1t_2t_3t_4q^{2n-2})(1 - t_1t_2t_3t_4q^{2n-1})}, \tag{8.2.8}$$

$$B_n = t_1 + t_1^{-1} - A_nt_1^{-1}\prod_{j=2}^{4}\left(1 - t_1t_jq^n\right) - \frac{t_1C_n}{\prod_{2\le k\le 4}(1 - t_1t_kq^{n-1})}. \tag{8.2.9}$$

Denote the corresponding monic polynomials by $\{P_n(x; \mathbf{t})\}$, thus

$$p_n(x; \mathbf{t}) = 2^n \left(t_1t_2t_3t_4q^{n-1}; q\right)_n P_n(x; \mathbf{t}), \tag{8.2.10}$$

then we have

$$xP_n(x;\mathbf{t}\mid q) = P_{n+1}(x;\mathbf{t}\mid q) + B_nP_n(x;\mathbf{t}\mid q) + \frac{A_{n-1}C_n}{4}P_{n-1}(x;\mathbf{t}\mid q). \tag{8.2.11}$$

One can use (1.4.31) and (8.2.3) to find the **lowering operator**

$$\mathcal{D}_q p_n(x;\mathbf{t}\mid q) = 2\frac{(1-q^n)\left(1-t_1t_2t_3t_4q^{n-1}\right)}{(1-q)q^{(n-1)/2}}p_{n-1}\left(x;q^{1/2}\mathbf{t}\mid q\right). \tag{8.2.12}$$

The **raising relation** is

$$\frac{2q^{(1-n)/2}}{q-1}w(x;\mathbf{t}\mid q)p_n(x;\mathbf{t}\mid q) = \mathcal{D}_q\left(w\left(x;q^{1/2}\mathbf{t}\mid q\right)p_{n-1}\left(x;q^{1/2}\mathbf{t}\mid q\right)\right), \tag{8.2.13}$$

where $w(x;\mathbf{t}\mid q)$ is defined in (8.2.2). Iterating (8.2.13) gives the **Rodrigues-type formula**

$$w(x;\mathbf{t}\mid q)p_n(x;\mathbf{t}\mid q) = \left(\frac{q-1}{2}\right)^n q^{n(n-1)/4}\mathcal{D}_q^n\left[w\left(x;q^{n/2}\mathbf{t}\mid q\right)\right]. \tag{8.2.14}$$

Theorem 8.2.1 (Askey and Wilson, 1985; Ismail and Zhang, 2005) *The Askey–Wilson polynomials have the* ***connection relation***

$$p_n(x;\mathbf{b}) = \sum_{k=0}^{n} c_{n,k}(\mathbf{a},\mathbf{b})p_k(x;\mathbf{a}), \tag{8.2.15}$$

where

$$\begin{aligned} c_{n,k}(\mathbf{b},\mathbf{a}) &= \frac{(q;q)_n b_4^{k-n}\left(b_1b_2b_3b_4q^{n-1};q\right)_k(b_1b_4,b_2b_4,b_3b_4;q)_n}{(q;q)_{n-k}\left(q,a_1a_2a_3a_4q^{k-1};q\right)_k(b_1b_4,b_2b_4,b_3b_4;q)_k} \\ &\times q^{k(k-n)}\sum_{j,l\geq 0}\frac{\left(q^{k-n},b_1b_2b_3b4q^{n+k-1},a_4b_4q^k;q\right)_{j+l}q^{j+l}}{(b_1b_4q^k,b_2b_4q^k,b_3b_4q^k;q)_{j+l}(q;q)_j(q;q)_l} \\ &\times\frac{\left(a_1a_4q^k,a_2a_4q^k,a_3a_4q^k;q\right)_l(b_4/a_4;q)_j}{(a_4b_4q^k,a_1a_2a_3a_4q^{2k};q)_l}\left(\frac{b_4}{a_4}\right)^l. \end{aligned} \tag{8.2.16}$$

The proof in Ismail and Zhang (2005) uses (8.2.15) and Theorem 1.4.1. One very special case of their connection coefficients formula is (Askey and Wilson, 1985, (6.4)–(6.5))

$$p_n\left(x;\alpha,a_2,a_3,a_4\mid q\right) = \sum_{k=0}^{n} c_{k,n}p_n\left(x;a,a_2,a_3,a_4\mid q\right), \tag{8.2.17}$$

where

$$c_{k,n} = \frac{a^{n-k}(q;q)_n\left(\alpha a_2a_3a_4q^{n-1};q\right)_k(\alpha/a;q)_{n-k}}{(q,aa_2a_3a_4q^{k-1};q)_k(q,aa_2a_3a_4q^{2k};q)_{n-k}}\prod_{2\leq j<m\leq 4}\left(a_ja_mq^k;q\right)_{n-k}. \tag{8.2.18}$$

Rahman and Verma proved the following addition theorem which reduces to the Gegenbauer addition theorem as $q\to 1$.

Theorem 8.2.2 (Rahman and Verma, 1986a) *We have*

$$\begin{aligned} &p_n\left(z; a, aq^{1/2}, -a, -aq^{1/2} \mid q\right) \\ &= \sum_{k=0}^{n} \frac{(q;q)_n \left(a^4q^n, a^4/q, a^2q^{1/2}, -a^2q^{1/2}, -a^2; q\right)_k a^{n-k}}{(q;q)_k (q;q)_{n-k} (a^4/q;q)_{2k} (a^2q^{1/2}, -a^2q^{1/2}, -a^2;q)_n} \\ &\times p_{n-k}\left(x; aq^{k/2}, aq^{(k+1)/2}, -aq^{k/2}, -aq^{(k+1)/2} \mid q\right) \\ &\times p_{n-k}\left(y; aq^{k/2}, aq^{(k+1)/2}, -aq^{k/2}, -aq^{(k+1)/2} \mid q\right) \\ &\times p_k\left(z; ae^{i(\theta+\phi)}, ae^{-i(\theta+\phi)}, ae^{i(\theta-\phi)}, ae^{i(\phi-\theta)} \mid q\right), \end{aligned} \tag{8.2.19}$$

where $x = \cos\theta$, $y = \cos\phi$.

The q-ultraspherical polynomials are special cases of the Askey–Wilson polynomials,

$$p_n\left(x; \sqrt{\beta}, -\sqrt{\beta}, \sqrt{\beta q}, -\sqrt{\beta q} \mid q\right) = (q, -\beta; q)_n \frac{\left(q\beta^2; q^2\right)_n}{(\beta^2; q)_n} C_n(x; \beta \mid q). \tag{8.2.20}$$

Using (8.2.20) we let $q \to 1$ in (8.2.19) and see that it reduces to the Gegenbauer addition theorem, Theorem 3.5.1.

Koornwinder (2005) established a second addition theorem for the continuous q-ultraspherical polynomials. His result is

$$\begin{aligned} &p_n\left(\cos\theta, aq^{1/2}s/t, aq^{1/2}t/s, -aq^{1/2}st, -aq^{1/2}/st \mid q\right) \\ &= (-1)^n q^{n^2/2} \sum_{k=0}^{n} a^{n-k} \left(a^2q^{k+1}; q\right)_{n-k} q^{k/2} \\ &\times \frac{\left(q^{-n}, a^4q^{n+1}; q\right)_k}{(q, a^4q^k; q)_k} \frac{\left(-a^2s^2q^{k+1}, -a^2q^{k+1}/t^2; q\right)_{n-k}}{(s/t)^{n-k}} \\ &\times {}_2\phi_2\left(\begin{matrix} q^{k-n}, a^4q^{n+k+1} \\ a^2q^{k+1}, -a^2s^2q^{k+1} \end{matrix} \,\middle|\, q, -s^2q\right) {}_2\phi_2\left(\begin{matrix} q^{k-n}, a^4q^{n+k+1} \\ a^2q^{k+1}, -a^2q^{k+1}/t^2 \end{matrix} \,\middle|\, q, -q/t^2\right) \\ &\times p_k\left(\cos\theta; a, -a, aq^{1/2}, -aq^{1/2} \mid q\right). \end{aligned} \tag{8.2.21}$$

The special case $a = 1$ is in Koornwinder (1993).

Ismail and Stanton (1988) introduced the following generalization of the Askey–Wilson integral:

$$\begin{aligned} \mathcal{J}(t_1, t_2, t_3, t_4) &:= \frac{\left(q, \beta^2; q\right)_\infty}{2\pi(\beta, \beta; q)_\infty} \\ &\times \int_0^\pi \prod_{j=1}^4 \frac{\left(\beta t_j e^{i\theta}, \beta t_j e^{-i\theta}; q\right)_\infty}{\left(t_j e^{i\theta}, t_j e^{-i\theta}; q\right)_\infty} \frac{\left(e^{2i\theta}, e^{-2i\theta}; q\right)_\infty}{\left(\beta e^{2i\theta}, \beta e^{-2i\theta}; q\right)_\infty} \, d\theta. \end{aligned} \tag{8.2.22}$$

They proved that

$$\begin{aligned}
&\mathcal{J}\left(\rho e^{i\phi}, \rho e^{-i\phi}, \sigma e^{i\psi}, \sigma e^{-i\psi}\right) \\
&= \sum_{n=0}^{\infty} \frac{\left(q, \beta^2; q\right)_n}{(\beta; q)_{n+1}(\beta; q)_n} (\rho\sigma)^n C_n(\cos\phi; \beta \mid q) C_n(\cos\psi; \beta \mid q) \\
&\times {}_2\phi_1\left(\left.\begin{matrix} \beta, \beta^2 q^n \\ \beta q^{n+1} \end{matrix}\right| q, \rho^2\right) {}_2\phi_1\left(\left.\begin{matrix} \beta, \beta^2 q^n \\ \beta q^{n+1} \end{matrix}\right| q, \sigma^2\right),
\end{aligned} \tag{8.2.23}$$

for $|\rho| < 1$, $|\sigma| < 1$, $\beta \in (-1, 1)$, $|t_j| < 1$, $1 \le j \le 4$. The evaluation (8.2.23) generalizes (8.2.1) and reduces to it when $\beta = 0$. In fact, $\mathcal{J}$ is a Hilbert–Schmidt kernel (Tricomi, 1957) for the integral operator

$$(Tf)(x) = \int_0^{\pi} \mathcal{J}\left(\rho e^{i\theta}, \rho e^{-i\theta}, \sigma e^{i\phi}, \sigma e^{-i\phi}\right) w(\cos\phi \mid \beta) f(\cos\phi)\, d\phi,$$

$x = \cos\theta \in (-1, 1)$, $f \in L^2(w(x \mid \beta), -1, 1)$, and

$$w(\cos\phi \mid \beta) = \frac{\left(e^{2i\phi}, e^{-2i\phi}; q\right)_\infty}{(\beta e^{2i\phi}, \beta e^{-2i\phi}; q)_\infty}.$$

The eigenfunctions of T are q-ultraspherical polynomials and the eigenvalues can be found from (8.2.23). The details are in Ismail and Stanton (1988).

We now give an asymptotic development for $\{p_n(x, \mathbf{t} \mid q)\}$.

Theorem 8.2.3 (Ismail, 1986a) *With $z = e^{i\theta}$ and $x = \cos\theta$ the Askey–Wilson polynomials have the form*

$$\begin{aligned}
\frac{p_n(\cos\theta; \mathbf{t} \mid q)}{(q, t_1 t_2, t_3 t_4; q)_n} &= z^n \frac{(t_1/z, t_3/z; q)_\infty}{(z^{-2}, q; q)_\infty} \sum_{m=0}^{\infty} \frac{(qz/t_1, qz/t_3; q)_m}{(q, qz^2; q)_m} (t_1 t_3)^m q^{mn} \\
&\times {}_2\phi_2\left(\left.\begin{matrix} t_1 z, t_1/z \\ t_1 t_2, t_1 q^{-m}/z \end{matrix}\right| q, t_2 q^{-m}/z\right) {}_2\phi_2\left(\left.\begin{matrix} t_3 z, t_3/z \\ t_3 t_4, t_3 q^{-m}/z \end{matrix}\right| q, t_4 q^{-m}/z\right)
\end{aligned} \tag{8.2.24}$$

+ a similar term with z and $1/z$ interchanged.

The series (8.2.24) is both an explicit formula and an asymptotic series.

8.3 The Askey–Wilson Equation

By combining (8.2.12) and (8.2.12) we see that the Askey–Wilson polynomials satisfy the eigenvalue equation

$$\frac{1}{w(x; \mathbf{t})} \mathcal{D}_q \left(w\left(x; q^{1/2}\mathbf{t} \mid q\right) \mathcal{D}_q p_n(x; \mathbf{t})\right) = \lambda_n p_n(x; \mathbf{t}), \tag{8.3.1}$$

whose eigenvalues $\{\lambda_n\}$ are

$$\lambda_n := \frac{4q}{(1-q)^2}\left(1-q^{-n}\right)\left(1-\sigma_4 q^{n-1}\right), \quad n = 1, 2, \ldots, \tag{8.3.2}$$

$\sigma_4 = \sum_{j=1}^{4} t_j$. Note that the orthogonality relation (8.2.4) is implied by the eigenvalue equation (8.3.1).

Theorem 8.3.1 *If f is a polynomial solution of degree n of*

$$\frac{1}{w(x;\mathbf{t})}\mathcal{D}_q\left[w\left(x;q^{1/2}\mathbf{t}\right)\mathcal{D}_q f(x)\right] = \lambda f(x), \tag{8.3.3}$$

then $\lambda = \lambda_n$ and f is a constant multiple of $p_n(x;\mathbf{t}\mid q)$.

It is useful to write (8.3.1) in expanded form. Using (1.4.31)–(1.4.33) and the facts

$$\begin{aligned}\pi_2(x) &:= \frac{\mathcal{D}_q w\left(x, q^{1/2}\mathbf{t}\right)}{w(x,\mathbf{t})} \\ &= -q^{-1/2}\left[2(1+\sigma_4)x^2 - (\sigma_1+\sigma_3)x - 1 + \sigma_2 - \sigma_4\right],\end{aligned} \tag{8.3.4}$$

$$\pi_1(x) := \frac{\mathcal{A}_q w\left(x, q^{1/2}\mathbf{t}\right)}{w(x,\mathbf{t})} = \frac{2\left[2(\sigma_4-1)x + \sigma_1 - \sigma_3\right]}{(1-q)} \tag{8.3.5}$$

transforms (8.3.1) to

$$\pi_2(x)\mathcal{D}_q^2 y(x) + \pi_1(x)\mathcal{A}_q\mathcal{D}_q y(x) = \lambda_n y(x), \tag{8.3.6}$$

where λ_n is given by (8.3.2).

Let $\mathcal{H}_w$ be the weighted space $L^2([-1,1], w)$ with inner product

$$(f,g)_w := \int_{-1}^{1} f(x)\overline{g(x)}w(x)\,dx, \quad \|f\|_w := (f,f)_w^{1/2}. \tag{8.3.7}$$

Define an operator T by

$$Tf(x) := Mf(x) \tag{8.3.8}$$

for f in H_1, where

$$(Mf)(x) = -\frac{1}{w(x)}\mathcal{D}_q\left(p\mathcal{D}_q f\right)(x). \tag{8.3.9}$$

The functions p and w are assumed to be positive on $(-1,1)$ and to satisfy

$$\begin{aligned}&\text{(i)}\;\; p(x)/\sqrt{1-x^2} \in H_{1/2},\; 1/p \in L([-1,1]),\\ &\text{(ii)}\;\; w(x) \in L([-1,1]),\; 1/w \in L\left([-1,1]; \frac{1}{1-x^2}\right).\end{aligned} \tag{8.3.10}$$

The expression Mf is therefore defined for $f \in H_1$, and the operator T acts in $\mathcal{H}_w$. The domain H_1 of T is dense in $\mathcal{H}_w$ since it contains all polynomials.

Theorem 8.3.2 *The operator T is symmetric in $\mathcal{H}_w$ and positive.*

Consequently, the eigenfunctions corresponding to different eigenvalues are orthogonal.

The form domain of T, its self-adjoint closure, and Friedrichs extension are described in Brown, Evans, and Ismail (1996).

8.4 Continuous q-Jacobi Polynomials and Discriminants

The **continuous q-Jacobi polynomials** are multiples of Askey–Wilson polynomials with $t_1 = q^{(2\alpha+1)/4}$, $t_2 = q^{(2\alpha+3)/4}$, $t_3 = -q^{(2\beta+1)/4}$, $t_4 = -q^{(2\beta+3)/4}$. They are

$$P_n^{(\alpha,\beta)}(\cos\theta \mid q) = \frac{\left(q^{\alpha+1};q\right)_n}{(q;q)_n}\, {}_4\phi_3\left(\left.\begin{matrix} q^{-n}, q^{n+\alpha+\beta+1}, q^{(2\alpha+1)/4}e^{i\theta}, q^{(2\alpha+1)/4}e^{-i\theta} \\ q^{\alpha+1}, -q^{(\alpha+\beta+1)/2}, -q^{(\alpha+\beta+2)/2} \end{matrix}\,\right|\, q;q\right). \tag{8.4.1}$$

For $\alpha > -1/2$ and $\beta > -1/2$ their **orthogonality relation** is

$$\begin{aligned}\frac{1}{2\pi}\int_{-1}^{1} w\left(x \mid q^\alpha, q^\beta\right) P_m^{(\alpha,\beta)}(x \mid q) P_n^{(\alpha,\beta)}(x \mid q)\, dx \\ = \frac{\left(q^{(\alpha+\beta+2)/2}, q^{(\alpha+\beta+3)/2}; q\right)_\infty}{\left(q, q^{\alpha+1}, q^{\beta+1}, -q^{(\alpha+\beta+1)/2}, -q^{(\alpha+\beta+2)/2}; q\right)_\infty} \\ \times \frac{\left(1-q^{\alpha+\beta+1}\right)\left(q^{\alpha+1}, q^{\beta+1}, -q^{(\alpha+\beta+3)/2}; q\right)_n}{\left(1-q^{2n+\alpha+\beta+1}\right)\left(q, q^{\alpha+\beta+1}, -q^{(\alpha+\beta+1)/2}; q\right)_n} q^{(2\alpha+1)n/2}\delta_{mn},\end{aligned} \tag{8.4.2}$$

where

$$\begin{aligned}\sin\theta\, w\left(\cos\theta \mid q^\alpha, q^\beta\right) &= \left|\frac{\left(e^{2i\theta};q\right)_\infty}{\left(q^{(2\alpha+1)/4}e^{i\theta}, q^{(2\alpha+3)/4}e^{i\theta}, -q^{(2\beta+1)/4}e^{i\theta}, -q^{(2\beta+3)/4}e^{i\theta}; q\right)_\infty}\right|^2 \\ &= \left|\frac{\left(e^{i\theta}, -e^{i\theta}; q^{1/2}\right)_\infty}{\left(q^{(2\alpha+1)/4}e^{i\theta}, -q^{(2\beta+1)/4}e^{i\theta}; q^{1/2}\right)_\infty}\right|^2.\end{aligned} \tag{8.4.3}$$

The **recurrence relation** for the polynomials $\{\phi_n(x)\}$,

$$\phi_n(x \mid q) := \frac{(q;q)_n}{\left(q^{\alpha+1};q\right)_n} P_n^{(\alpha,\beta)}(x \mid q), \tag{8.4.4}$$

is

$$\begin{aligned}2x\phi_n(x \mid q) &= A_n\phi_{n+1}(x \mid q) \\ &+ \left[q^{(2\alpha+1)/4} + q^{-(2\alpha-1)/4} - (A_n + C_n)\right]\phi_n(x \mid q) + C_n\phi_{n-1}(x \mid q),\end{aligned} \tag{8.4.5}$$

where

$$A_n = \frac{\left(1 - q^{n+\alpha+1}\right)\left(1 - q^{n+\alpha+\beta+1}\right)\left(1 + q^{n+(\alpha+\beta+1)/2}\right)\left(1 + q^{n+(\alpha+\beta+2)/2}\right)}{q^{(2\alpha+1)/4}\left(1 - q^{2n+\alpha+\beta+1}\right)\left(1 - q^{2n+\alpha+\beta+2}\right)},$$

$$C_n = \frac{q^{(2\alpha+1)/4}\left(1 - q^{n}\right)\left(1 - q^{n+\beta}\right)\left(1 + q^{n+(\alpha+\beta)/2}\right)\left(1 + q^{n+(\alpha+\beta+1)/2}\right)}{\left(1 - q^{2n+\alpha+\beta}\right)\left(1 - q^{2n+\alpha+\beta+1}\right)}.$$

The **monic polynomials** satisfy the recurrence relation

$$\begin{aligned} xP_n(x) = P_{n+1}(x) &+ \frac{1}{2}\left[q^{(2\alpha+1)/4} + q^{-(2\alpha-1)/4} - (A_n + C_n)\right] P_n(x) \\ &+ \frac{1}{4} A_{n-1} C_n P_{n-1}(x), \end{aligned} \tag{8.4.6}$$

where

$$P_n^{(\alpha,\beta)}(x \mid q) = \frac{2^n q^{(2\alpha+1)n/4}\left(q^{n+\alpha+\beta+1}; q\right)_n}{\left(q, -q^{(\alpha+\beta+1)/2}, -q^{(\alpha+\beta+2)/2}; q\right)_n} P_n(x).$$

A **lowering operator** is

$$\mathcal{D}_q P_n^{(\alpha,\beta)}(x \mid q) = \frac{2q^{-n+(2\alpha+5)/4}\left(1 - q^{n+\alpha+\beta+1}\right)}{(1-q)\left(1 + q^{(\alpha+\beta+1)/2}\right)\left(1 + q^{(\alpha+\beta+2)/2}\right)} P_{n-1}^{(\alpha+1,\beta+1)}(x \mid q), \tag{8.4.7}$$

while a **raising operator** is

$$\begin{aligned} &\mathcal{D}_q \left[w\left(x \mid q^{\alpha}, q^{\beta}\right) P_n^{(\alpha,\beta)}(x \mid q) \right] \\ &\quad = -2q^{-(2\alpha+1)/4} \frac{\left(1 - q^{n+1}\right)\left(1 + q^{(\alpha+\beta-1)/2}\right)\left(1 + q^{(\alpha+\beta)/2}\right)}{1-q} \\ &\quad\quad \times w\left(x \mid q^{\alpha-1}, q^{\beta-1}\right) P_{n+1}^{(\alpha-1,\beta-1)}(x \mid q). \end{aligned} \tag{8.4.8}$$

The **Rodrigues-type formula** is

$$\begin{aligned} &w\left(x \mid q^{\alpha}, q^{\beta}\right) P_n^{(\alpha,\beta)}(x \mid q) \\ &\quad = \left(\frac{q-1}{2}\right)^n \frac{q^{n(n+2\alpha)/4}}{\left(q, -q^{(\alpha+\beta+1)/2}, -q^{(\alpha+\beta+2)/2}; q\right)_n} \mathcal{D}_q^n \left[w\left(x \mid q^{\alpha+n}, q^{\beta+n}\right) \right]. \end{aligned} \tag{8.4.9}$$

We have the **generating functions**

$$\begin{aligned} &{}_2\phi_1\left(\left. \begin{matrix} q^{(2\alpha+1)/4} e^{i\theta}, q^{(2\alpha+3)/4} e^{i\theta} \\ q^{\alpha+1} \end{matrix} \right| q; e^{-i\theta} t \right) {}_2\phi_1\left(\left. \begin{matrix} -q^{(2\beta+1)/4} e^{-i\theta}, -q^{(2\beta+3)/4} e^{-i\theta} \\ q^{\beta+1} \end{matrix} \right| q; e^{i\theta} t \right) \\ &\quad = \sum_{n=0}^{\infty} \frac{\left(-q^{(\alpha+\beta+1)/2}, -q^{(\alpha+\beta+2)}; q\right)_n}{\left(q^{\alpha+1}, q^{\beta+1}; q\right)_n} \frac{P_n^{(\alpha,\beta)}(x \mid q)}{q^{(2\alpha+1)n/4}} t^n, \end{aligned} \tag{8.4.10}$$

$$
\begin{aligned}
&{}_2\phi_1\left(\begin{matrix} q^{(2\alpha+1)/4}e^{i\theta}, -q^{(2\beta+1)/4}e^{i\theta} \\ -q^{(\alpha+\beta+1)/2} \end{matrix}\,\middle|\, q; e^{-i\theta}t\right) {}_2\phi_1\left(\begin{matrix} q^{(2\alpha+3)/4}e^{-i\theta}, -q^{(2\beta+3)/4}e^{-i\theta} \\ -q^{(\alpha+\beta+3)/2} \end{matrix}\,\middle|\, q; e^{i\theta}t\right) \\
&\quad = \sum_{n=0}^{\infty} \frac{\left(-q^{(\alpha+\beta+2)/2}; q\right)_n}{\left(-q^{(\alpha+\beta+3)/2}; q\right)_n} \frac{P_n^{(\alpha,\beta)}(x \mid q)}{q^{(2\alpha+1)n/4}} t^n,
\end{aligned}
\tag{8.4.11}
$$

$$
\begin{aligned}
&{}_2\phi_1\left(\begin{matrix} q^{(2\alpha+1)/4}e^{i\theta}, -q^{(2\beta+3)/4}e^{i\theta} \\ -q^{(\alpha+\beta+2)/2} \end{matrix}\,\middle|\, q; e^{-i\theta}t\right) {}_2\phi_1\left(\begin{matrix} q^{(2\alpha+3)/4}e^{-i\theta}, -q^{(2\beta+1)/4}e^{-i\theta} \\ -q^{(\alpha+\beta+2)/2} \end{matrix}\,\middle|\, q; e^{i\theta}t\right) \\
&\quad = \sum_{n=0}^{\infty} \frac{\left(-q^{(\alpha+\beta+1)/2}; q\right)_n}{\left(-q^{(\alpha+\beta+2)/2}; q\right)_n} \frac{P_n^{(\alpha,\beta)}(x \mid q)}{q^{((2\alpha+1)/4)n}} t^n.
\end{aligned}
\tag{8.4.12}
$$

M. Rahman (1981) takes $t_1 = q^{1/2}$, $t_2 = q^{\alpha+1/2}$, $t_3 = -q^{\beta+1/2}$, and $t_4 = -q^{1/2}$, and uses the normalization

$$
P_n^{(\alpha,\beta)}(x; q) = \frac{\left(q^{\alpha+1}, -q^{\beta+1}; q\right)_n}{(q, -q; q)_n} {}_4\phi_3\left(\begin{matrix} q^{-n}, q^{n+\alpha+\beta+1}, q^{1/2}e^{i\theta}, q^{1/2}e^{-i\theta} \\ q^{\alpha+1}, -q^{\beta+1}, -q \end{matrix}\,\middle|\, q; q\right). \tag{8.4.13}
$$

These two q-analogues of the Jacobi polynomials are connected by

$$
P_n^{(\alpha,\beta)}\left(x \mid q^2\right) = \frac{(-q; q)_n}{\left(-q^{\alpha+\beta+1}; q\right)_n} q^{n\alpha} P_n^{(\alpha,\beta)}(x; q). \tag{8.4.14}
$$

Special points: We have the special evaluations

$$
\begin{aligned}
P_n^{(\alpha,\beta)}(x_1 \mid q) &= \frac{\left(q^{\alpha+1}; q\right)_n}{(q; q)_n}, \\
P_n^{(\alpha,\beta)}(x_2 \mid q) &= \frac{\left(q^{\beta+1}\right)_n}{(q; q)_n}(-1)^n q^{(\alpha-\beta)n/2},
\end{aligned}
\tag{8.4.15}
$$

where

$$
x_1 = \frac{1}{2}\left(q^{(2\alpha+1)/4} + q^{-(2\alpha+1)/4}\right), \quad x_2 = -\frac{1}{2}\left(q^{(2\beta+1)/4} + q^{-(2\beta+1)/4}\right). \tag{8.4.16}
$$

The evaluations at x_1 and x_2 follow from the Pfaff–Saalschütz theorem (1.3.5).

The continuous q-Jacobi polynomials given by (8.4.11) and the continuous q-ultraspherical polynomials are connected by the quadratic transformations

$$
\begin{aligned}
C_{2n}\left(x; q^\lambda \mid q\right) &= \frac{\left(q^\lambda, -q; q\right)_n}{\left(q^{1/2}, -q^{1/2}\right)_n} q^{-\frac{1}{2}n} P_n^{(\lambda-\frac{1}{2},-\frac{1}{2})}\left(2x^2 - 1; q\right), \\
C_{2n+1}\left(x; q^\lambda \mid q\right) &= \frac{\left(q^\lambda, -q; q\right)_{n+1}}{\left(q^{1/2}, -q^{1/2}\right)_{n+1}} q^{-n/2} x P_n^{(\lambda-\frac{1}{2},\frac{1}{2})}\left(2x^2 - 1; q\right).
\end{aligned}
\tag{8.4.17}
$$

The continuous q-Jacobi polynomials are essentially invariant under $q \to q^{-1}$. Indeed,

$$
\begin{aligned}
P_n^{(\alpha,\beta)}\left(x \mid q^{-1}\right) &= q^{-n\alpha} P_n^{(\alpha,\beta)}(x \mid q), \\
P_n^{(\alpha,\beta)}\left(x; q^{-1}\right) &= q^{-n(\alpha+\beta)} P_n^{(\alpha,\beta)}(x; q).
\end{aligned}
\tag{8.4.18}
$$

Another form of the **lowering operator** is

$$\begin{aligned}&\left(1-2xq^{(2\alpha+1)/4}+q^{\alpha+1/2}\right)\left(1+2xq^{(2\beta+1)/4}+q^{\beta+1/2}\right)\mathcal{D}_q P_n^{(\alpha,\beta)}(x\mid q)\\&\quad=A_n(x)P_{n-1}^{(\alpha,\beta)}(x\mid q)+B_n(x)P_n^{(\alpha,\beta)}(x\mid q),\end{aligned}\tag{8.4.19}$$

where

$$\begin{aligned}A_n(x)&=2\frac{\left(1-q^{\alpha+n}\right)\left(1-q^{\beta+n}\right)}{(1-q)\left(1-q^{n+(\alpha+\beta)/2}\right)}\left(1+q^{(\alpha+\beta+1)/2}\right)q^{(\alpha-n)/2+3/4},\\B_n(x)&=2\frac{\left(1-q^{n}\right)\left(1-q^{(\alpha-\beta)/2}\right)}{(1-q)\left(1-q^{n+(\alpha+\beta)/2}\right)}\left(1+q^{n+\alpha+\beta+1/2}\right)q^{(\beta-n)/2+3/4}\\&\quad-4q^{(\alpha+\beta+2-n)/2}\frac{(1-q^n)}{1-q}x.\end{aligned}\tag{8.4.20}$$

Gasper characterized the domain of positivity of the linearization coefficients of the continuous q-Jacobi polynomials.

Theorem 8.4.1 (Gasper, 1983) *Let $0<q<1$, $\alpha,\beta>-1$. The linearization coefficients $c_{k,m,n}^{(\alpha,\beta)}(q)$ in*

$$P_m^{(\alpha,\beta)}(x)P_n^{(\alpha,\beta)}(x)=\sum_{k=|m-n|}^{m+n}c_{k,m,n}^{(\alpha,\beta)}(q)P_k^{(\alpha,\beta)}(x)\tag{8.4.21}$$

are nonnegative for $k,m,n=0,1,\dots$ if and only if either $\alpha\ge\beta$, $\alpha+\beta+1\ge 0$ or $-1/2>\beta>-1$, $c_{2,2,2}^{(\alpha,\beta)}(q)\ge 0$.

The relevant discriminant in the notation of (2.9.6) is (Ismail, 2005c)

$$\begin{aligned}D\left(P_n^{(\alpha,\beta)}(x\mid q);\mathcal{D}_q\right)&=\frac{2^{n^2-2n}q^{n(n-1)\alpha}}{(1-q)^n}\prod_{j=1}^{n}\left(1-q^{\alpha+\beta+n+j}\right)^{n-j}\\&\times\prod_{j=1}^{n}\left(1-q^j\right)^{j-2n+2}\left(1-q^{\alpha+j}\right)^{j-1}\left(1-q^{\beta+j}\right)^{j-1}\prod_{j=1}^{2n}\left(1+q^{(\alpha+\beta+j)/2}\right)^{j-2n}.\end{aligned}\tag{8.4.22}$$

8.5 q-Racah Polynomials

The q-**Racah polynomials** were introduced in Askey and Wilson (1979). They are defined by

$$R_n(\mu(x);\alpha,\beta;\gamma,\delta):={}_4\phi_3\left(\left.\begin{matrix}q^{-n},\alpha\beta q^{n+1},q^{-x},\gamma\delta q^{x+1}\\\alpha q,\beta\delta q,\gamma q\end{matrix}\right|q,q\right),\tag{8.5.1}$$

where

$$\mu(x)=q^{-x}+\gamma\delta q^{x+1},\tag{8.5.2}$$

and $\alpha q = q^{-N}$, for some positive integer N. Clearly, $(q^{-x}, \gamma\delta q^{x+1}; q)_m$ is a polynomial of degree m in $\mu(x)$, hence $R_n(\mu(x))$ is a polynomial of exact degree n in $\mu(x)$. Indeed,

$$R_n(\mu(x); \alpha, \beta, \gamma, \delta) = \frac{\left(\alpha\beta a^{n+1}; q\right)_n (\mu(x))^n}{(\alpha q, \beta\delta q, \gamma q; q)_n} + \text{lower-order terms.} \tag{8.5.3}$$

The case $q \to 1$ gives the Racah polynomials of the Racah–Wigner algebra of quantum mechanics (Biedenharn and Louck, 1981). See also Askey and Wilson (1982).

Let

$$w(x; \alpha, \beta, \gamma, \delta) := \frac{(\alpha q, \beta\delta q, \gamma q, \gamma\delta q; q)_x}{(q, \gamma\delta q/\alpha, \gamma q/\beta, \delta q; q)_x} \frac{\left(1 - \gamma\delta q^{2x+1}\right)}{(\alpha\beta q)^x (1 - \gamma\delta q)}. \tag{8.5.4}$$

Theorem 8.5.1 *The q-Racah polynomials satisfy the orthogonality relation*

$$\sum_{x=0}^{N} w(x; \alpha, \beta, \gamma, \delta) R_m(\mu(x); \alpha, \beta; \gamma, \delta) R_n(\mu(x); \alpha, \beta; \gamma, \delta) = h_n\, \delta_{m,n}, \tag{8.5.5}$$

where $\alpha q = q^{-N}$, $n = 0, 1, \ldots, N$ *and*

$$h_n = \frac{\left(\gamma/\alpha\beta, \delta/\alpha, 1/\beta, \gamma\delta q^2; q\right)_\infty}{(1/\alpha\beta q, \gamma\delta q/\alpha, \gamma q/\beta, \delta q; q)_\infty} \frac{(1 - \alpha\beta q)(\gamma\delta q)^n}{(1 - \alpha\beta q^{2n+1})} \frac{(q, \alpha\beta q/\gamma, \alpha q/\delta, \beta q; q)_n}{(\alpha q, \alpha\beta q, \beta\delta q, \gamma q; q)_n}. \tag{8.5.6}$$

The case $m = n = 0$ of (8.5.5) is a discrete analogue of the Askey–Wilson integral (8.2.1).

By reparametrizing the parameters of the Askey–Wilson polynomials we can prove that the R_n satisfy

$$\begin{aligned}(q^{-x} - 1)\left(1 - \gamma\delta q^{x+1}\right) & R_n(\mu(x)) \\ &= A_n R_{n+1}(\mu(x)) - (A_n + C_n) R_n(\mu(x)) + C_n R_{n-1}(\mu(x)),\end{aligned} \tag{8.5.7}$$

where

$$\begin{aligned} A_n &= \frac{\left(1 - \alpha q^{n+1}\right)\left(1 - \alpha\beta q^{n+1}\right)\left(1 - \beta\delta q^{n+1}\right)\left(1 - \gamma q^{n+1}\right)}{(1 - \alpha\beta q^{2n+1})(1 - \alpha\beta q^{2n+2})}, \\ C_n &= \frac{q(1 - q^n)(1 - \beta q^n)(\gamma - \alpha\beta q^n)(\gamma - \alpha q^n)}{(1 - \alpha\beta q^{2n})(1 - \alpha\beta q^{2n+1})}, \end{aligned} \tag{8.5.8}$$

with $R_0(\mu(x)) = 1$, $R_{-1}(\mu(x)) = 0$. Here we used $R_n(\mu(x))$ for $R_n(\mu(x); \alpha, \beta, \gamma, \delta)$. It is clear from (8.5.1) that $R_n(\mu(x))$ is symmetric under $x \leftrightarrow n$. Hence, (8.5.7) shows that $R_n(\mu(n))$ solves the difference equation

$$(q^{-n} - 1)\left(1 - \gamma\delta q^{n+1}\right) y(x) = A_x y(x+1) - (A_x + C_x) y(x) + C_x y(x-1). \tag{8.5.9}$$

In fact, (8.5.9) can be factored as a product of two first-order operators. Indeed,

$$\begin{aligned}
&\frac{\Delta R_n(\mu(x);\alpha,\beta,\gamma,\delta)}{\Delta\mu(x)}\\
&\quad=\frac{q^{1-n}(1-q^n)\left(1-\alpha\beta q^{n+1}\right)}{(1-q)(1-\alpha q)(1-\beta\delta q)(1-\gamma q)}R_{n-1}(\mu(x);\alpha q,\beta q,\gamma q,\delta),
\end{aligned}\tag{8.5.10}$$

$$\begin{aligned}
&\frac{\nabla\left(\bar{w}(x;\alpha,\beta,\gamma,\delta)R_n(\mu(x);\alpha,\beta,\gamma,\delta)\right)}{\nabla\mu(x)}\\
&\quad=\frac{\bar{w}(x;\alpha/q,\beta/q,\gamma/q,\delta)}{(1-q)(1-\gamma\delta)}R_{n+1}(\mu(x);\alpha/q,\beta/q,\gamma/q,\delta),
\end{aligned}\tag{8.5.11}$$

where

$$\bar{w}(x;\alpha,\beta,\gamma,\delta)=\frac{(\alpha q,\beta\delta q,\gamma q,\gamma\delta q;q)_x}{(q,q\gamma\delta/\alpha,\gamma q/\beta,\delta q;q)_x}(\alpha\beta)^{-x}.\tag{8.5.12}$$

Repeated applications of (8.5.11) gives the **Rodrigues formula**

$$\bar{w}(x;\alpha,\beta,\gamma,\delta)R_n(x;\alpha,\beta,\gamma,\delta)=(1-q)^n\left(\frac{\nabla}{\nabla\mu(x)}\right)^n\bar{w}\left(x;\alpha q^n,\beta q^n,\gamma q^n,\delta\right).\tag{8.5.13}$$

One can prove the following **generating function** using the Sears transformation (1.4.23). For $x=0,1,2,\ldots,N$ we have

$$\begin{aligned}
&{}_2\phi_1\left(\left.\begin{matrix}q^{-x},\beta\gamma^{-1}q^{-x}\\ \beta\delta q\end{matrix}\,\right|q;\gamma\delta q^{x+1}t\right){}_2\phi_1\left(\left.\begin{matrix}\alpha q,\gamma q^{x+1}\\ \alpha\delta^{-1}q\end{matrix}\,\right|q;q^{-x}t\right)\\
&\quad=\sum_{n=0}^{N}\frac{(\alpha q,\gamma q;q)_n}{(\alpha\delta^{-1}q,q;q)_n}R_n(\mu(x);\alpha,\beta,\gamma,\delta\mid q)t^n,
\end{aligned}\tag{8.5.14}$$

if $\alpha q=q^{-N}$ or $\gamma q=q^{-N}$.

8.6 Linear and Multilinear Generating Functions

In view of (7.2.15), every linear or multilinear generating function for the polynomials $\{p_n(x;t_1,t_2\mid q)\}$ leads to a similar result for $\{C_n(x;\beta\mid q)\}$. We shall not record these results here.

Theorem 8.6.1 *We have the Poisson-type kernel*

$$\begin{aligned}
&\sum_{n=0}^{\infty}\frac{(q;q)_n}{(\beta^2;q)_n}C_n(\cos\theta;\beta\mid q)C_n(\cos\phi;\beta\mid q)t^n\\
&\quad=\frac{\left(\beta te^{i(\theta+\phi)},\beta te^{i(\phi-\theta)},\beta te^{i(\theta-\phi)},\beta te^{i(\theta+\phi)};q\right)_\infty}{\left(\beta^2te^{i(\theta+\phi)},te^{i(\theta-\phi)},te^{i(\theta+\phi)},te^{i(\phi-\theta)};q\right)_\infty}\\
&\quad\quad\times{}_8W_7\left(\beta^2te^{i(\theta+\phi)}/q;\beta,\beta e^{2i\theta},\beta e^{2i\phi},\beta,e^{i(\theta+\phi)};te^{-i(\theta+\phi)}\right).
\end{aligned}\tag{8.6.1}$$

The following more symmetric form of (8.6.1) is due to Gasper and Rahman (1983):

$$\begin{aligned}\sum_{n=0}^{\infty} \frac{(q;q)_n}{(\beta^2;q)_n} & C_n(\cos\theta;\beta \mid q)C_n(\cos\phi;\beta \mid q)t^n \\ &= \frac{\left(\beta, t^2, \beta te^{i(\theta+\phi)}, \beta te^{i(\theta-\phi)}, \beta te^{i(\phi-\theta)}, \beta te^{-i(\theta+\phi)}; q\right)_\infty}{(\beta^2, \beta t^2, te^{i(\theta+\phi)}, te^{i(\theta-\phi)}, te^{i(\phi-\theta)}, te^{-i(\theta+\phi)}; q)_\infty} \\ &\times {}_8W_7\left(\beta t^2/q; te^{i(\theta+\phi)}, te^{i(\theta-\phi)}, te^{i(\phi-\theta)}, te^{-i(\theta+\phi)}, \beta; \beta\right).\end{aligned} \tag{8.6.2}$$

The moment representations not only give an integral representation for the Al-Salam–Chihara polynomials but also they give q-integral representations for other solutions to the same three-term recurrence relation. Details are in Ismail and Stanton (2002).

We next state a bibasic version of (8.1.22)–(8.1.23).

Theorem 8.6.2 *Let $p_n(x; t_1, t_2 \mid q)$ denote the Al-Salam–Chihara polynomials with base q. Then*

$$\begin{aligned}\sum_{n=0}^{\infty} & p_n(\cos\theta; t_1, t_2 \mid q)\, p_n(\cos\phi; s_1, s_2 \mid p) \frac{(t_1t_2;q)_n (s_1s_2;p)_n}{(q;q)_n(p;p)_n} \left(\frac{t}{t_1s_1}\right)^n \\ &= \frac{\left(t_1e^{-i\theta}, t_2e^{-i\theta}; q\right)_\infty}{(q, e^{-2i\theta}; q)_\infty} \sum_{k=0}^{\infty} \frac{\left(qe^{i\theta}/t_1, t_1e^{i\theta}, qe^{i\theta}/t_2; q\right)_k}{(q, qe^{2i\theta}, t_1e^{i\theta}; q)_k} (t_1t_2)^k \\ &\times \frac{\left(ts_1q^ke^{i\theta}, ts_2q^ke^{i\theta}; p\right)_\infty}{(tq^ke^{i(\theta+\phi)}, tq^ke^{i(\theta-\phi)}; p)_\infty} + \text{a similar term with } \theta \text{ replaced by } -\theta.\end{aligned} \tag{8.6.3}$$

Theorem 8.6.2 is in Van der Jeugt and Jagannathan (1998) with a quantum group derivation and in Ismail and Stanton (2002). The moment technique also gives

$$\begin{aligned}\sum_{n=0}^{\infty} \frac{H_{n+k}(\cos\theta \mid q)}{(q;q)_{n+k}} \frac{(\lambda;q)_n t^n}{(q;q)_n} &= \frac{\left(\lambda te^{i\theta};q\right)_\infty e^{ik\theta}}{(te^{i\theta}, e^{-2i\theta};q)_\infty}\, {}_1\phi_2\left(\left.\begin{matrix} te^{i\theta} \\ qe^{2i\theta}, \lambda te^{i\theta}\end{matrix}\,\right|\, q, q^{k+2}e^{2i\theta}\right) \\ &+ \text{a similar term with } \theta \text{ replaced by } -\theta.\end{aligned} \tag{8.6.4}$$

Theorem 8.6.3 *A generating function for the q-Hermite polynomials is*

$$\sum_{n=0}^{\infty} \frac{(\lambda;q)_n}{(q^2;q^2)_n} H_n\left(x \mid q^2\right) t^n = \frac{\left(\lambda te^{i\theta};q\right)_\infty}{(te^{i\theta};q)_\infty}\, {}_3\phi_2\left(\left.\begin{matrix}\lambda, \sqrt{q}\,e^{i\theta}, -\sqrt{q}\,e^{i\theta} \\ \lambda te^{i\theta}, -q\end{matrix}\,\right|\, q, te^{-i\theta}\right). \tag{8.6.5}$$

The ${}_3\phi_2$ in Theorem 8.6.3 is essentially bibasic on base q and q^2. If $\lambda = 0$ or $\lambda = -q$ the ${}_3\phi_2$ may be summed and the result is (7.1.4).

Ismail and Stanton used (7.1.27) and (7.1.14) to establish the generating functions

$$\begin{aligned}\sum_{n=0}^{\infty} & H_{n+k}(\cos\psi \mid q)H_n(\cos\theta \mid q)H_n(\cos\phi \mid q)\frac{t^n}{(q;q)_n} \\ &= \frac{e^{ik\psi}\left(t^2e^{2i\psi};q\right)_\infty}{(e^{-2i\psi}, te^{i(\psi+\theta+\phi)}, te^{i(\psi+\theta-\phi)}, te^{i(\psi+\phi-\theta)}, te^{i(\psi-\theta-\phi)};q)_\infty} \\ &\times {}_6\phi_5\left(\begin{matrix} te^{i(\psi+\theta+\phi)}, te^{i(\psi+\theta-\phi)}, te^{i(\psi+\phi-\theta)}, te^{i(\psi-\theta-\phi)}, 0, 0 \\ qe^{2i\psi}, te^{i\psi}, -te^{i\psi}, \sqrt{q}\,te^{i\psi}, -\sqrt{q}\,te^{i\psi}\end{matrix}\,\middle|\, q, q^{k+1}e^{i\psi}\right) \\ &+ \text{ a similar term with } \psi \text{ replaced by } -\psi,\end{aligned} \tag{8.6.6}$$

and

$$\begin{aligned}\sum_{n=0}^{\infty} & \frac{H_{n+k}(\cos\psi \mid q)}{(q;q)_{n+k}(q;q)_n}H_n(\cos\theta \mid q)H_n(\cos\phi \mid q)t^n \\ &= \frac{e^{ik\psi}\left(t^2e^{2i\psi};q\right)_\infty}{(e^{-2i\psi}, te^{i(\psi+\theta+\phi)}, te^{i(\psi+\theta-\phi)}, te^{i(\psi+\phi-\theta)}, te^{i(\psi-\theta-\phi)};q)_\infty} \\ &\times {}_4\phi_5\left(\begin{matrix} te^{i(\psi+\theta+\phi)}, te^{i(\psi+\theta-\phi)}, te^{i(\psi+\phi-\theta)}, te^{i(\psi-\theta-\phi)} \\ qe^{2i\psi}, te^{i\psi}, -te^{i\psi}, \sqrt{q}\,te^{i\psi}, -\sqrt{q}\,te^{i\psi}\end{matrix}\,\middle|\, q, q^{k+2}e^{3i\psi}\right) \\ &+ \text{ a similar term with } \psi \text{ replaced by } -\psi.\end{aligned} \tag{8.6.7}$$

Note that the right-hand sides of (8.6.6) and (8.6.7) are symmetric under any permutation of θ, ϕ, and ψ.

The trilinear generating function (8.6.6) contains two important product formulas for the continuous q-Hermite polynomials which will be stated in the next theorem.

Theorem 8.6.4 *With $K(\cos\theta, \cos\phi, \cos\psi)$ denoting the right-hand side of* (8.6.6), *we have the* ***product formulas***

$$\begin{aligned}H_n(\cos\theta \mid q)H_n(\cos\phi \mid q) &= \frac{(q;q)_\infty(q;q)_n}{2\pi t^n(q;q)_{n+k}}\int_0^\pi K(\cos\theta,\cos\phi,\cos\psi) \\ &\quad\times H_{n+k}(\cos\psi \mid q)\left(e^{2i\psi}, e^{-2i\psi};q\right)_\infty d\psi,\end{aligned} \tag{8.6.8}$$

and the more general formula

$$\begin{aligned}H_n(\cos\theta \mid q)H_{n+k}(\cos\psi \mid q) &= \frac{(q;q)_\infty}{2\pi t^n}\int_0^\pi K(\cos\theta,\cos\phi,\cos\psi) \\ &\quad\times H_n(\cos\phi \mid q)\left(e^{2i\phi}, e^{-2i\phi};q\right)_\infty d\phi.\end{aligned} \tag{8.6.9}$$

Note that (7.1.2), (7.1.1), and the initial conditions of $p_n(x; t_1, t_2)$ imply

$$(-1)^n \frac{\left(q;q^2\right)_N}{(q^2;q^2)_n} p_n\left(\cos 2\theta; -1, -q \mid q^2\right) = \frac{H_{2n}(\cos\theta \mid q)}{(q^2;q^2)_n},$$

$$\frac{2\left(q^3;q^2\right)_n}{(-q^2)^n\,(q^2;q^2)_n} \cos\theta\, p_n\left(\cos 2\theta; -q^2, -q \mid q^2\right) = \frac{H_{2n+1}(\cos\theta \mid q)}{(q^2;q^2)_n}.$$

Thus Theorem 8.1.1 gives q-integral moment representations for the following functions:

$$\frac{H_{2n}(x \mid q)}{(q^2;q^2)_n}, \quad \frac{H_{2n}(x \mid q)}{(-q;q^2)_n}, \quad \frac{H_{2n+1}(x \mid q)}{(q^2;q^2)_n}, \quad \frac{H_{2n+1}(x \mid q)}{(-q^3;q^2)_n}.$$

One can also derive several generating functions involving $H_{2n}(x \mid q)$ and $H_{2n+1}(x \mid q)$ from the corresponding results in Section 8.1.

8.7 Associated Askey–Wilson Polynomials

All the results in this section are from Ismail and Rahman (1991) and concern **associated Askey–Wilson polynomials**.

Consider the three-term recurrence relation

$$2xy_\alpha(x) = A_\alpha y_{\alpha+1}(x) + B_\alpha y_\alpha(x) + C_\alpha y_{\alpha-1}(x), \tag{8.7.1}$$

where

$$A_\alpha = \frac{(1 - t_1t_2t_3t_4q^{\alpha-1}) \prod_{k=2}^4 (1 - t_1t_kq^\alpha)}{t_1\,(1 - t_1t_2t_3t_4q^{2\alpha-1})(1 - t_1t_2t_3t_4q^{2\alpha})}, \tag{8.7.2}$$

$$C_\alpha = \frac{t_1\,(1 - q^\alpha) \prod_{2\le j<k\le 4} \left(1 - t_jt_kq^{\alpha-1}\right)}{(1 - t_1t_2t_3t_4q^{2\alpha-2})(1 - t_1t_2t_3t_4q^{2\alpha-1})}, \tag{8.7.3}$$

$$B_\alpha = t_1 + t_1^{-1} - A_\alpha - C_\alpha. \tag{8.7.4}$$

In this notation, when $x \to -x - t_1 - 1/t_1$ and $\alpha \to n + \alpha$, (8.7.1) becomes a three-term recurrence relation for birth and death process polynomials.

Throughout this section we choose z such that

$$x := \frac{1}{2}[z + 1/z], \quad |z| < 1. \tag{8.7.5}$$

Theorem 8.7.1 *Let $x \in \mathbb{C} \setminus [-1, 1]$. Then the functions*

$$r_\alpha(z) := \frac{(t_2t_3t_4q^\alpha/z;q)_\infty \prod_{k=2}^4 (t_1t_kq^\alpha;q)_\infty}{(t_1zq^\alpha;q)_\infty \prod_{2\le j<k\le 4} \left(t_jt_kq^\alpha;q\right)_\infty} \left(\frac{a}{z}\right)^\alpha$$
$$\times\, {}_8W_7\left(t_2t_3t_4/qz; t_2/z, t_3/z, t_4/z, t_1t_2t_3t_4q^{\alpha-1}, q^{-\alpha}; q, qz/t_1\right) \tag{8.7.6}$$

and

$$s_\alpha(z) := \frac{\left(t_1t_2t_3t_4q^{2\alpha}, t_2t_3t_4zq^{\alpha}; q\right)_\infty \prod_{k=2}^{4}\left(t_kzq^{\alpha+1}; q\right)_\infty}{\left(t_2t_2t_4zq^{2\alpha+1}, q^{\alpha+1}; q\right)_\infty \prod_{2\le j<k\le 4}\left(t_jt_kq^{\alpha}; q\right)_\infty}(az)^\alpha \times {}_8W_7\left(t_2t_3t_4zq^{2\alpha}; t_2t_3q^{\alpha}, t_2t_4q^{\alpha}, t_3t_4q^{\alpha}, q^{\alpha+1}, zq/t_1; q, t_1z\right) \tag{8.7.7}$$

are linearly independent solutions of (8.7.1).

The functions r_α and s_α have become known as the Ismail–Rahman functions. Let

$$z_n^\alpha(x) = t_1^{-n}\prod_{k=2}^{4}(1 - t_1t_kq^\alpha)_n\, y_{n+\alpha}(x). \tag{8.7.8}$$

The functional relation (8.7.1) becomes

$$2xz_n^{(\alpha)}(x) = A_n^{(\alpha)}z_{n+1}^{(\alpha)}(x) + B_n^{(\alpha)}z_n^{(\alpha)}(x) + C_n^{(\alpha)}z_{n-1}^{(\alpha)}(x), \tag{8.7.9}$$

with

$$A_n^{(\alpha)} = \frac{1 - t_1t_2t_3t_4q^{n+\alpha-1}}{\left(1 - t_1t_2t_3t_4q^{2n+2\alpha-1}\right)\left(1 - t_1t_2t_3t_4q^{2n+2\alpha}\right)}, \tag{8.7.10}$$

$$C_n^{(\alpha)} = \frac{\left(1 - q^{n+\alpha}\right)\prod_{1\le j<k\le 4}\left(1 - t_jt_kq^{n+\alpha-1}\right)}{\left(1 - t_1t_2t_3t_4q^{2n+2\alpha-2}\right)\left(1 - t_1t_2t_3t_4q^{2n+2\alpha-1}\right)}, \tag{8.7.11}$$

and

$$B_n^{(\alpha)} = t_1 + t_1^{-1} - A_n^{(\alpha)}t_1^{-1}\prod_{j=2}^{4}\left(1 - t_1t_jq^{n+\alpha}\right) - \frac{t_1C_n^{(\alpha)}}{\prod_{2\le j<k\le 4}\left(1 - t_jt_kq^{n+\alpha-1}\right)}. \tag{8.7.12}$$

Remark 4.1.1 suggests two polynomial solutions, say $\{p_n^{(\alpha)}\}$ and $\{q_n^{(\alpha)}\}$, of (8.7.9) having the initial conditions

$$\begin{aligned} p_0^{(\alpha)}(x; \mathbf{t} \mid q) &= 1, \quad p_1^{(\alpha)}(x; \mathbf{t} \mid q) = \left[2x - B_0^{(\alpha)}\right]/A_0^{(\alpha)}, \\ q_0^{(\alpha)}(x; \mathbf{t} \mid q) &= 1, \quad q_1^{(\alpha)}(x; \mathbf{t} \mid q) = \left[2x - \tilde{B}_0^{(\alpha)}\right]/A_0^{(\alpha)}, \end{aligned} \tag{8.7.13}$$

with

$$\tilde{B}_0^{(\alpha)} := t_1 + t_1^{-1} - A_0^{(\alpha)}t_1^{-1}\prod_{j=2}^{4}\left(1 - t_1t_jq^\alpha\right). \tag{8.7.14}$$

To simplify the writing we used the simplified notation $p_n^{(\alpha)}(x; \mathbf{t} \mid q)$ and $q_n^{(\alpha)}(x; \mathbf{t} \mid q)$ unless there is a need to exhibit the dependence on the parameters. Ismail and Rahman established the representations

$$p_n^{(\alpha)}(x; \mathbf{t} \mid q) = zt_1^{1-2\alpha-n}\frac{\prod_{2\le j<k\le 4}\left(t_jt_k; q\right)_\infty}{\prod_{k=2}^{4}\left(t_jt_kq^{\alpha+n}, t_kz; q\right)_\infty}\,\frac{\left(q^\alpha, t_1z; q\right)_\infty\left(t_2t_4q^{\alpha-1}, t_3t_4q^{\alpha-1}; q\right)_\infty}{\left(1 - t_1t_2t_3t_4q^{2\alpha-2}\right)\left(t_2t_3t_4/z; q\right)_\infty} \times \left\{s_{\alpha-1}(z)r_{n+\alpha}(z) - r_{\alpha-1}(z)s_{n+\alpha}(z)\right\} \tag{8.7.15}$$

and

$$q_n^{(\alpha)}(x;\mathbf{t}\mid q) = z t_1^{1-2\alpha-n} \frac{\prod_{2\le j<k\le 4}\left(t_jt_k;q\right)_\infty}{\prod_{k=2}^4\left(t_jt_kq^{\alpha+n},t_kz;q\right)_\infty} \frac{(q^\alpha,t_1z;q)_\infty\left(t_2t_4q^{\alpha-1},t_3t_4q^{\alpha-1};q\right)_\infty}{(1-t_1t_2t_3t_4q^{2\alpha-2})(t_2t_3t_4/z;q)_\infty} \times\{(s_{\alpha-1}(z)-s_\alpha(z))\,r_{n+\alpha}(z)-(r_{\alpha-1}(z)-r_\alpha(z))\,s_{n+\alpha}(z)\}. \tag{8.7.16}$$

Using transformation theory of basic hypergeometric series, Ismail and Rahman found the following compact representations:

$$\begin{aligned} p_n^{(\alpha)}(\cos\theta,\mathbf{t}\mid q) &= (t_1t_2q^\alpha,t_1t_3q^\alpha,t_1t_4q^\alpha;q)_n\,t_1^{-n} \\ &\times\sum_{k=0}^n \frac{\left(q^{-n},t_1t_2t_3t_4q^{2\alpha+n-1},t_1t_2t_3t_4q^{2\alpha-1},t_1e^{i\theta},t_2e^{-i\theta};q\right)_k}{(q,t_1t_2q^\alpha,t_1t_3q^\alpha,t_1t_4q^\alpha,t_1t_2t_3t_4q^{\alpha-1};q)_\infty}q^k \\ &\times {}_{10}W_9\left(t_1t_2t_3t_4q^{2\alpha+k-2};q^\alpha,t_2t_3q^{\alpha-1},t_2t_4q^{\alpha-1},t_3t_4q^{\alpha-1},\right. \\ &\qquad\left. q^{k+1},q^{k-n},t_1t_2t_3t_4q^{2\alpha+n+k-1};q,t_1^2\right) \end{aligned} \tag{8.7.17}$$

and

$$\begin{aligned} q_n^{(\alpha)}(\cos\theta,\mathbf{t}\mid q) &= (t_1t_2q^\alpha,t_1t_3q^\alpha,t_1t_4q^\alpha;q)_n\,t_1^{-n} \\ &\times\sum_{k=0}^n \frac{\left(q^{-n},t_1t_2t_3t_4q^{2\alpha+n-1},t_1t_2t_3t_4q^{2\alpha-1},t_1e^{i\theta},t_2e^{-i\theta};q\right)_k}{(q,t_1t_2q^\alpha,t_1t_3q^\alpha,t_1t_4q^\alpha,t_1t_2t_3t_4q^{\alpha-1};q)_\infty}q^k \\ &\times {}_{10}W_9\left(t_1t_2t_3t_4q^{2\alpha+k-2};q^\alpha,t_2t_3q^{\alpha-1},t_2t_4q^{\alpha-1},t_3t_4q^{\alpha-1},\right. \\ &\qquad\left. q^{k},q^{k-n},t_1t_2t_3t_4q^{2\alpha+n+k-1};q,qt_1^2\right). \end{aligned} \tag{8.7.18}$$

Formula (8.7.7) and a Wronskian-type formula can be used to establish the limiting relations

$$\lim_{n\to\infty} z^np_n^{(\alpha)}(x;\mathbf{t}\mid q) = (t_1z)^{1-\alpha}\,s_{\alpha-1}(z)\frac{(q^\alpha,t_1z;q)_\infty\prod_{2\le i<j\le 4}\left(t_it_jq^{\alpha-1};q\right)_\infty}{(1-t_1t_2t_3t_4q^{2\alpha-2})(t_1t_2t_3t_4q^{\alpha-1},z^2;q)_\infty} \tag{8.7.19}$$

and

$$\lim_{n\to\infty} z^nq_n^{(\alpha)}(x;\mathbf{t}\mid q) = (t_1z)^{1-\alpha}\left[s_{\alpha-1}(z)-s_\alpha(z)\right] \times\frac{(q^\alpha,t_1z;q)_\infty\prod_{2\le i<j\le 4}\left(t_it_jq^{\alpha-1};q\right)_\infty}{(1-t_1t_2t_3t_4q^{2\alpha-2})(t_1t_2t_3t_4q^{\alpha-1},z^2;q)_\infty}, \tag{8.7.20}$$

for $|t_1z|<1$.

Theorem 8.7.2 *Let $\mu^{(1)}(x;\mathbf{t},\alpha)$ and $\mu^{(2)}(x;\mathbf{t},\alpha)$ be the probability measures with respect to which $p_n^{(\alpha)}$ and $q_n^{(\alpha)}$ are orthogonal. Then*

$$\int_{\mathbb{R}}\frac{d\mu^{(1)}(y;\mathbf{t},\alpha)}{x-y} = \frac{2z\left(t_2t_3t_4zq^{2\alpha-1};q\right)_2}{(1-t_2t_3t_4zq^{\alpha-1})\prod_{k=2}^4(1-t_kzq^\alpha)} \times\frac{{}_8W_7\left(t_2t_3t_4zq^{2\alpha};t_2t_3q^\alpha,t_2t_4q^\alpha,t_3t_4q^\alpha,q^{\alpha+1},zq/t_1;q,t_1z\right)}{{}_8W_7\left(t_2t_3t_4zq^{2\alpha-2};t_2t_3q^{\alpha-1},t_2t_4q^{\alpha-1},t_3t_4q^{\alpha-1},q^\alpha,qz/t_1;q,t_1z\right)} \tag{8.7.21}$$

and

$$\int_{\mathbb{R}} \frac{d\mu^{(2)}(y;\mathbf{t},\alpha)}{x-y} = \frac{2z\left(t_2t_3t_4zq^{2\alpha-1};q\right)_2}{(1-t_2t_3t_4zq^{\alpha-1})\prod_{k=2}^{4}(1-t_kzq^{\alpha})} \times \frac{{}_8W_7\left(t_2t_3t_4zq^{2\alpha};t_2t_3q^{\alpha},t_2t_4q^{\alpha},t_3t_4q^{\alpha},q^{\alpha+1},zq/t_1;q,t_1z\right)}{{}_8W_7\left(t_2t_3t_4zq^{2\alpha-2};t_2t_3q^{\alpha-1},t_2t_4q^{\alpha-1},t_3t_4q^{\alpha-1},q^{\alpha},z/t_1;q,qt_1z\right)}, \tag{8.7.22}$$

which are valid in the complex x-plane cut along $[-1,1]$.

Theorem 8.7.3 (Ismail and Rahman, 1991) *The absolutely continuous components of* $\mu^{(1)}$ *and* $\mu^{(2)}$ *are given by*

$$\begin{aligned}\frac{d\mu^{(1)}(\cos\theta;\mathbf{t},\alpha)}{d\theta} &= \left(1-t_1t_2t_3t_4q^{2\alpha-2}\right)\left(t_1t_2t_3t_4q^{2\alpha-2};q\right)_\infty \\ &\times \frac{\left(q^{\alpha+1};q\right)_\infty\prod_{1\le j<k\le 4}\left(t_jt_kq^{\alpha};q\right)_\infty}{2\pi\left(1-t_1t_2t_3t_4q^{\alpha-2}\right)\left(t_1t_2t_3t_4q^{\alpha-2},t_1t_2t_3t_4q^{2\alpha};q\right)_\infty} \\ &\times \frac{\left(e^{2i\theta},e^{-2i\theta},q^{\alpha+1}e^{2i\theta},q^{\alpha+1}e^{-2i\theta};q\right)_\infty}{\left(qe^{2i\theta},qe^{-2i\theta};q\right)_\infty\prod_{k=1}^{4}\left(t_ke^{i\theta},t_ke^{-i\theta};q\right)_\infty} \\ &\times \left|{}_8W_7(q^{\alpha}e^{2i\theta};qe^{i\theta}/t_1,qe^{i\theta}/t_2,qe^{i\theta}/t_3,qe^{i\theta}/t_4,q^{\alpha};q,t_1t_2t_3t_4q^{\alpha-2})\right|^2,\end{aligned} \tag{8.7.23}$$

and

$$\begin{aligned}\frac{d\mu^{(2)}(\cos\theta;\mathbf{t},\alpha)}{d\theta} &= \frac{\left(q^{\alpha+1};q\right)_\infty\prod_{1\le j<k\le 4}\left(t_jt_kq^{\alpha};q\right)_\infty}{2\pi\left(t_1t_2t_3t_4q^{2\alpha};q\right)_\infty}\frac{\left(t_1t_2t_3t_4q^{2\alpha-1};q\right)_\infty}{\left(t_1t_2t_3t_4q^{\alpha-1};q\right)_\infty} \\ &\times \frac{1-2t_1xq^{\alpha}+t_1^2q^{2\alpha}}{1-2t_1x+t_1^2}\frac{\left(e^{2i\theta},e^{-2i\theta},q^{\alpha+1}e^{2i\theta},q^{\alpha+1}e^{-2i\theta};q\right)_\infty}{\left(qe^{2i\theta},qe^{-2i\theta};q\right)_\infty\prod_{k=1}^{4}\left(t_ke^{i\theta},t_ke^{-i\theta};q\right)_\infty} \\ &\times \left|{}_8W_7(q^{\alpha}e^{2i\theta};e^{i\theta}/t_1,qe^{i\theta}/t_2,qe^{i\theta}/t_3,qe^{i\theta}/t_4,q^{\alpha};q,t_1t_2t_3t_4q^{\alpha-1})\right|^2,\end{aligned} \tag{8.7.24}$$

respectively.

9

Orthogonal Polynomials on the Unit Circle

Leonid Golinskii

One way to generalize orthogonal polynomials on subsets of $\mathbb{R}$ is to consider orthogonality on curves in the complex plane. Among these generalizations, the most developed theory is the general theory of orthogonal polynomial on the unit circle $\mathbb{T}$. The basic sources for this chapter are Grenander and Szegő (1958), Szegő ([1939] 1975), Geronimus (1961, 1962), Simon (2004a,b), Ismail (2005b, Chapters 8 and 17), and recent papers which will be cited in the appropriate places.

In what follows we shall use Simon's abbreviation OPUC for orthogonal polynomials on the unit circle.

9.1 Definitions and Basic Properties

The **unit circle** $\mathbb{T}$ is by far the simplest closed curve in the complex plane with a number of additional properties, so polynomials orthogonal with respect to measures on $\mathbb{T}$ are of specific interest.

Consider the class $\mathcal{P}$ of all nontrivial probability measures μ on $[-\pi, \pi]$ (that is, not supported on a finite set, positive Borel measures with $\mu[-\pi, \pi] = 1$). The Lebesgue decomposition of μ is the decomposition

$$\mu = \mu_{\text{ac}} + \mu_s = \mu'(\theta)\frac{d\theta}{2\pi} + \mu_s \tag{9.1.1}$$

where $\mu' \in L^1([-\pi, \pi])$ is the Radon–Nikodym derivative of μ with respect to the Lebesgue measure and μ_s is the singular part of μ.

The **moments** (Fourier coefficients) of μ are defined by

$$\mu_k = \int_{-\pi}^{\pi} e^{-ik\theta}\, d\mu(\theta), \qquad k \in \mathbb{Z} = \{0, \pm 1, \pm 2, \dots\}, \tag{9.1.2}$$

and form a bounded sequence. The moments of μ generate the **Toeplitz determinants**

$$D_n = D_n(\mu) = \det(\mu_{i-k})_{i,k=0}^{n} > 0. \tag{9.1.3}$$

The theory of quadratic forms shows that D_n is strictly positive for all $n \in \mathbb{Z}_+ = \{0, 1, 2, \dots\}$.

The orthogonal polynomials with respect to μ arise as an outcome of the standard Gram–Schmidt procedure applied to the system of monomials $\{\zeta^n\}_{n\ge 0}$, $\zeta = e^{i\theta}$, in the Hilbert space $L^2_\mu([-\pi,\pi])$ of square-summable measurable functions on $\mathbb{T}$ with the inner product

$$(f,g)_\mu = \int_{-\pi}^{\pi} f(\zeta)\overline{g(\zeta)}\,d\mu(\theta), \quad \zeta = e^{i\theta}, \quad \|f\|^2_\mu = (f,f)_\mu.$$

There are two natural ways of normalization: the **orthonormal polynomials**

$$\phi_n(z) = \phi_n(z;\mu) = \kappa_n z^n + \text{ lower-order terms}, \quad (\phi_n,\phi_m)_\mu = \delta_{nm}, \tag{9.1.4}$$

$n, m \in \mathbb{Z}_+$, and the **monic orthogonal polynomials**

$$\Phi_n(z) = \Phi_n(z;\mu) = \kappa_n^{-1}\phi_n(z) = z^n + \ell_{n,n-1}z^{n-1} + \text{ lower-order terms}. \tag{9.1.5}$$

Both systems are uniquely determined when we require that $\kappa_n > 0$. The monic orthogonal polynomials are characterized by the property

$$\deg(P) = n, \quad (P,\zeta)_\mu = 0, \quad 0 \le j < n \quad \text{imply} \quad P = c_n\Phi_n. \tag{9.1.6}$$

The following expressions for monic orthogonal polynomials are similar to (2.1.4) and (2.1.6):

$$\Phi_n(z) = \frac{1}{D_{n-1}}\begin{vmatrix} \mu_0 & \mu_{-1} & \dots & \mu_{-n} \\ \mu_1 & \mu_0 & \dots & \mu_{-n+1} \\ \vdots & \vdots & & \vdots \\ \mu_{n-1} & \mu_{n-2} & \dots & \mu_{-1} \\ 1 & z & \dots & z^n \end{vmatrix} \tag{9.1.7}$$

and

$$\Phi_n(z) = \frac{1}{n!D_{n-1}}\int_{\mathbb{T}^n}\prod_{j=1}^n (z-\zeta_j)\prod_{1\le j<k\le n}|\zeta_j-\zeta_k|^2\prod_{j=1}^n d\mu(\zeta_j). \tag{9.1.8}$$

Equation (9.1.7) implies an important relation

$$(\Phi_n, z^n)_\mu = \|\Phi_n\|^2_\mu = \frac{D_n}{D_{n-1}}. \tag{9.1.9}$$

Let z_0 be a zero of Φ_n. Following an elegant argument due to H. Landau (Landau, 1987), we write $\Phi_n(z) = (z - z_0)P(z)$, $\deg P = n - 1$, so $\Phi_n \perp P$ and

$$zP(z) = \Phi_n(z) + z_0P(z), \quad \|zP\|^2_\mu = \|P\|^2_\mu = \|\Phi_n\|^2_\mu + |z_0|^2\|P\|^2_\mu,$$

hence $(1 - |z_0|^2)\|P\|^2_\mu = \|\Phi_n\|^2_\mu$ and so $|z_0| < 1$. In other words, all zeros of all orthogonal polynomials lie in the open unit disk $\mathbb{D} = \{|z| < 1\}$.

The following **extremal property** of monic orthogonal polynomials is one of the highlights of OPUC theory.

Theorem 9.1.1 *The minimum of the integral*

$$\int_{-\pi}^{\pi} |P(\zeta)|^2 \, d\mu(\theta) \tag{9.1.10}$$

taken over all monic polynomials P of degree n is attained when $P = \Phi_n$. The minimum value of the integral is κ_n^{-2}.

As a straightforward consequence we have Simon's variational principle (Simon, 2007b), which proved useful in the study of Schur and related flows. Note that one can define monic OPUC for any finite positive measure, even if not normalized, and of course $\Phi_n(z; c\mu) = \Phi_n(z; \mu)$ for any positive constant c.

Theorem 9.1.2 *Let μ be a nontrivial probability measure on $[-\pi, \pi]$, and $\{z_j\}_{j=1}^k$ be among the zeros of $\Phi_n(\mu)$. Then*

$$\Phi_n(z; \mu) = \prod_{j=1}^{k} (z - z_j) \Phi_{n-k}\left(z; \prod_{j=1}^{k} |z - z_j|^2 \, d\mu \right).$$

The **reverse polynomial** f^* of a polynomial f of degree n is $f^*(z) = z^n \overline{f}(1/z)$, that is,

$$f^*(z) = \sum_{k=0}^{n} \bar{f}_{n-k} z^k \quad \text{if} \quad f(z) = \sum_{k=0}^{n} f_k z^k. \tag{9.1.11}$$

Equation (9.1.6) now shows that

$$\deg(P) \le n, \quad P \perp \zeta^j, \quad j = 1, \ldots, n \quad \text{imply} \quad P = c_n \Phi_n^*. \tag{9.1.12}$$

The polynomials Φ_n^* are called the ***-reverse polynomials**. Clearly, $\Phi_n^*(0) = 1$.

The next result shows how systems of orthogonal polynomials on $\mathbb{T}$ are in one-to-one correspondence with pairs of special systems of polynomials orthogonal on $[-1, 1]$. The model is $\{z^n\}$ on $\mathbb{T}$ and the Chebyshev polynomials $\{\operatorname{Re} z^n = \cos n\theta\}$ and $\{\operatorname{Im} z^{n+1} / \operatorname{Im} z = \sin(n+1)\theta / \sin\theta\}$ on $[-1, 1]$.

Theorem 9.1.3 (Szegő's mapping theorem) *Let μ be a probability measure on $[-1, 1]$ and let ϕ_n be the polynomials orthonormal with respect to $d\mu(\cos\theta)$ on the unit circle. Assume further that $\{t_n(x)\}$ and $\{u_n(x)\}$ are orthonormal sequences of polynomials whose measures of orthogonality are $d\mu(x)$ and $c_2\left(1 - x^2\right) d\mu(x)$, respectively. With $z \in \mathbb{T}$ and $x = (z + 1/z)/2$ we have*

$$\begin{aligned} t_n(x) &= \left[1 + \phi_{2n}(0)/\kappa_{2n}\right]^{-1/2} \left[z^{-n} \phi_{2n}(z) + z^n \phi_{2n}(1/z)\right] \\ &= \left[1 - \phi_{2n}(0)/\kappa_{2n}\right]^{-1/2} \left[z^{-n+1} \phi_{2n-1}(z) + z^{n-1} \phi_{2n-1}(1/z)\right] \end{aligned} \tag{9.1.13}$$

and

$$\begin{aligned} u_n(x) &= \frac{z^{-n-1} \phi_{2n+2}(z) + z^{n+1} \phi_{2n+2}(1/z)}{\sqrt{1 - \phi_{2n+2}(0)/\kappa_{2n+2}} \, (z - 1/z)} \\ &= \frac{z^{-n} \phi_{2n+1}(z) + z^n \phi_{2n+1}(1/z)}{\sqrt{1 + \phi_{2n+2}(0)/\kappa_{2n+2}} \, (z - 1/z)}. \end{aligned} \tag{9.1.14}$$

9.2 Szegő Recurrence Relations and Verblunsky Coefficients

A key feature of the unit circle is that the multiplication operator $Uf = zf$ in $L^2_\mu(\mathbb{T})$ is unitary. So the difference $\Phi_{n+1}(z)-z\Phi_n(z)$ is of degree at most n and orthogonal to z^j for $j = 1, 2, \dots, n$, and by (9.1.12),

$$\Phi_{n+1}(z) = z\Phi_n(z) - \bar{\alpha}_n\Phi^*_n(z) \tag{9.2.1}$$

with some complex numbers α_n, known as the **Verblunsky coefficients**,

$$\alpha_n = -\overline{\Phi_{n+1}(0)} = (-1)^n \prod_{j=1}^{n+1} \bar{z}_{j,n+1}, \quad |\alpha_n| = \prod_{j=1}^{n+1} |z_{j,n+1}| < 1, \tag{9.2.2}$$

where $\{z_{j,n+1}\}$ are zeros of Φ_{n+1}. Applying (9.1.11) to (9.2.1) yields

$$\Phi^*_{n+1}(z) = \Phi^*_n(z) - \alpha_n z\Phi_n(z). \tag{9.2.3}$$

The recurrence relations (9.2.1) and (9.2.3) are the **Szegő recurrences**.

It follows from the unitarity of U and $\Phi^*_n \perp \Phi_{n+1}$, that

$$\|\Phi_{n+1}\|^2 = (1-|\alpha_n|^2)\|\Phi_n\|^2, \quad \|\Phi_n\|^2 = \kappa_n^{-2} = \prod_{j=0}^{n-1}(1-|\alpha_j|^2). \tag{9.2.4}$$

We set

$$\rho_j := \sqrt{1-|\alpha_j|^2}, \quad \text{so that } 0 < \rho_j \le 1. \tag{9.2.5}$$

Thus the leading coefficients κ_n satisfy $\kappa_{n+1}^{-1} = \rho_n\kappa_n^{-1}$, hence are given by

$$\kappa_n = \prod_{j=0}^{n-1} \frac{1}{\rho_j}.$$

Combining (9.2.1) and (9.2.3) we obtain the Szegő recurrence relations in matrix form:

$$\begin{bmatrix}\Phi_{n+1}(z)\\ \Phi^*_{n+1}(z)\end{bmatrix} = A(z,\alpha_n)\begin{bmatrix}\Phi_n(z)\\ \Phi^*_n(z)\end{bmatrix}, \quad A(z,\alpha) = \begin{bmatrix} z & -\bar{\alpha}\\ -z\alpha & 1\end{bmatrix}. \tag{9.2.6}$$

In other words,

$$\begin{bmatrix}\Phi_{n+1}(z)\\ \Phi^*_{n+1}(z)\end{bmatrix} = T_{n+1}(z)\begin{bmatrix}1\\ 1\end{bmatrix}, \quad T_p(z) := A(z,\alpha_{p-1})\dots A(z,\alpha_0). \tag{9.2.7}$$

The matrix $T_p(z) =$ is called a **transfer matrix**. This leads to the **inverse Szegő recurrences**

$$z\Phi_n(z) = \rho_n^{-2}\left[\Phi_{n+1}(z) + \bar{\alpha}_n\Phi^*_{n+1}(z)\right],$$
$$\Phi^*_n(z) = \rho_n^{-2}\left[\Phi^*_{n+1}(z) + \alpha_n\Phi_{n+1}(z)\right].$$

By eliminating Φ^*_n between the direct and inverse Szegő recurrences we get the three-term recurrence relation (see Geronimus, 1962)

$$\bar{\alpha}_{n-1}\Phi_{n+1}(z) = (\bar{\alpha}_n + \bar{\alpha}_{n-1}z)\Phi_n(z) - \bar{\alpha}_n\rho^2_{n-1}z\Phi_n(z) \tag{9.2.8}$$

for the Φ_j without any Φ_j^*, which is helpful for the study of the ratio asymptotics of orthogonal polynomials. Equation (9.2.8) has the defect that α_{n-1} can vanish.

The Szegő recurrence relations for orthonormal polynomials are

$$\begin{bmatrix} \phi_{n+1}(z) \\ \phi_{n+1}^*(z) \end{bmatrix} = \frac{1}{\rho_n} A(z,\alpha_n) \begin{bmatrix} \phi_n(z) \\ \phi_n^*(z) \end{bmatrix} = \prod_{j=0}^{n} \frac{1}{\rho_j} A(z,\alpha_j) \begin{bmatrix} 1 \\ 1 \end{bmatrix}. \tag{9.2.9}$$

The following fundamental result is proved in Verblunsky (1935).

Theorem 9.2.1 (Verblunsky's theorem) *Let $\mathbb{D}^\infty$ be the set of complex sequences $\{\alpha_j\}_{j=0}^\infty$ with $|\alpha_j| < 1$. Let S be the mapping from the set of all nontrivial probability measures on $[-\pi,\pi]$ to $\mathbb{D}^\infty$ defined by $S(\mu) = \{\alpha_j(\mu)\}_{j=0}^\infty$. Then S is one-to-one. Moreover, S is a homeomorphism if the space of measures and $\mathbb{D}^\infty$ are equipped with the weak* topology and the componentwise convergence topology, respectively.*

For a detailed discussion and several proofs see Simon (2004a). In fact, the analysis in Verblunsky (1935) shows that the moments μ_n of every such measure can be parametrized by elements of $\mathbb{D}^\infty$ via

$$\mu_{n+1} = \text{polynomial in } \{\alpha_0, \bar{\alpha}_0, \ldots, \alpha_{n-1}, \bar{\alpha}_{n-1}\} + \alpha_n \prod_{j=0}^{n-1} \rho_j^2.$$

Theorem 9.2.2 (Bernstein–Szegő approximation) *Let μ be a nontrivial probability measure on $[-\pi,\pi]$ with orthonormal polynomials ϕ_n. Let $dm = d\theta/2\pi$ and*

$$\mu^{(n)} = |\phi_n(\zeta;\mu)|^{-2}\, dm. \tag{9.2.10}$$

Then $\mu^{(n)}$ belongs to the same class of measures, with

$$\phi_k(z;\mu^{(n)}) = \phi_k(z;\mu), \quad k = 0,1,\ldots,n; \qquad \phi_k(z;\mu^{(n)}) = z^{k-n}\phi_n(z;\mu) \tag{9.2.11}$$

for $k \ge n$, so

$$\alpha_j(\mu^{(n)}) = \alpha_j(\mu), \quad j = 0,1,\ldots,n-1; \qquad \alpha_j(\mu^{(n)}) = 0, \quad j \ge n. \tag{9.2.12}$$

*Moreover, $\mu^{(n)} \to \mu$ as $n \to \infty$ in the *-weak topology.*

In fact, the measures with finite sequences of Verblunsky coefficients are exactly those of the form $\mu = c|P(\zeta)|^{-2}\, d\theta$, where c is picked to make μ a probability measure, and P is a monic polynomial of degree n with all zeros in $\mathbb{D}$. In this case $\Phi_k(z;\mu) = z^{k-n}P(z)$ for $k \ge n$.

The relation between measures μ and their Verblunsky coefficients in Theorem 9.2.1 is quite intricate, and very little can be said in the general setting. But there is an important situation – rotation of α – when some information about a corresponding family of measures is available. Specifically, let $\lambda \in \mathbb{T}$ and put $\alpha_{n,\lambda} = \lambda\alpha_n$, $n \in \mathbb{Z}_+$. The measures μ_λ with $\alpha_n(\mu_\lambda) = \alpha_{n,\lambda}$ (which exist by Theorem 9.2.1) are known as the **Aleksandrov measures** (or Aleksandrov–Clark measures). A representative with $\lambda = -1$ is of particular interest. The measure μ_{-1}

is called a **measure of the second kind**, and the corresponding orthogonal polynomials are called **polynomials of the second kind**. Special notation,

$$\Phi_n(z;\mu_{-1}) = \Psi_n(z), \quad \phi_n(z;\mu_{-1}) = \psi_n(z), \tag{9.2.13}$$

is standard for the monic orthogonal (orthonormal) polynomials of the second kind, respectively. The explicit formulas for Ψ_n are due to Geronimus (1961):

$$\Psi_n(z) = \int_{-\pi}^{\pi} \frac{\zeta+z}{\zeta-z} [\Phi_n(\zeta) - \Phi_n(z)]\, d\mu(\theta), \quad \zeta = e^{i\theta}, \tag{9.2.14}$$

$$\Psi_n^*(z) = z^n \int_{-\pi}^{\pi} \frac{\zeta+z}{\zeta-z} \left[\overline{\Phi_n(z^{-1})} - \overline{\Phi_n(\zeta)}\right] d\mu(\theta), \tag{9.2.15}$$

for $n \ge 1$. Clearly, both relations hold for orthonormal polynomials as well. There is another simple relation between Φ and Ψ:

$$\Psi_n^*(z)\Phi_n(z) + \Psi_n(z)\Phi_n^*(z) = 2z^n \prod_{j=0}^{n-1} \rho_j^2, \quad n \ge 1. \tag{9.2.16}$$

An important consequence is the following theorem.

Theorem 9.2.3 *Let μ be a nontrivial probability measure on $[-\pi,\pi]$ and $\mu^{(n)}$ be its Bernstein–Szegő approximations* (9.2.10). *Then for $z \in \mathbb{D}$,*

$$\frac{\Psi_n^*(z)}{\Phi_n^*(z)} = F(z,\mu^{(n)}) = \int_{-\pi}^{\pi} \frac{\zeta+z}{\zeta-z}\, d\mu^{(n)}(\theta), \quad \zeta = e^{i\theta}, \tag{9.2.17}$$

so

$$\lim_{n\to\infty} \frac{\Psi_n^*(z)}{\Phi_n^*(z)} = F(z,\mu) = \int_{-\pi}^{\pi} \frac{\zeta+z}{\zeta-z}\, d\mu(\theta) \tag{9.2.18}$$

uniformly on compact subsets of $\mathbb{D}$ and

$$\left| F(z,\mu) - \frac{\Psi_n^*(z)}{\Phi_n^*(z)} \right| = O(z^{n+1}), \quad z \to 0. \tag{9.2.19}$$

The function $F(\mu)$ in (9.2.18) is a **Carathéodory function** of μ. An explicit formula for the error in (9.2.19) is available:

$$F(z,\mu)\Phi_n^*(z) - \Psi_n^*(z) = z^n \int_{-\pi}^{\pi} \frac{\zeta+z}{\zeta-z} \overline{\Phi_n(\zeta)}\, d\mu(\theta). \tag{9.2.20}$$

Similarly to (9.2.19) we also have

$$F(z,\mu)\Phi_n(z) + \Psi_n(z) = 2\kappa_n^{-2} z^n + O(z^{n+1}), \quad z \to 0. \tag{9.2.21}$$

There is a converse to (9.2.19)/(9.2.21) due to Peherstorfer and Steinbauer (1995).

Theorem 9.2.4 *Given a nontrivial probability measure μ on $[-\pi, \pi]$ with the Carathéodory function $F(\mu)$, let p and q be polynomials of degree at most n with*

$$p(z)F(z,\mu) + q(z) = O(z^n), \quad p^*(z)F(z,\mu) - q^*(z) = O(z^{n+1}),$$

as $z \to 0$, where p^*, q^* are the reverse polynomials for degree n. Then $p = c\Phi_n(\mu)$, $q = c\Psi_n(\mu)$ for some constant c.

It turns out that the vector

$$\begin{pmatrix} \psi_n(z) \\ -\psi_n^*(z) \end{pmatrix}$$

provides a second linearly independent solution of the Szegő recurrence (9.2.9), and the Carathéodory function $F(\mu)$ has a property analogous to a defining property of the Weyl m-function in the case of differential equations (see Geronimo, 1992; Golinskii and Nevai, 2001).

Theorem 9.2.5 *For fixed $z \in \mathbb{D}$ a number $r = F(z,\mu)$ is the unique complex number so that*

$$\begin{pmatrix} \psi_n(z) \\ -\psi_n^*(z) \end{pmatrix} + r \begin{pmatrix} \phi_n(z) \\ \phi_n^*(z) \end{pmatrix} \in \ell^2\left(\mathbb{Z}_+, \mathbb{C}^2\right). \tag{9.2.22}$$

There is another important property of $F(\mu)$ related to mass points of the orthogonality measure, which follows directly from the definition (9.2.18):

$$\lim_{r \to 1-0} (1-r)F(r\zeta,\mu) = 2\mu\{\theta\} \quad \text{for all } \zeta = e^{i\theta} \in \mathbb{T}. \tag{9.2.23}$$

The **Christoffel kernels** (reproducing kernels)

$$K_n(z,w) = \sum_{j=0}^{n} \phi_j(z)\overline{\phi_j(w)} \tag{9.2.24}$$

arise with regard to the following extremal problem, cf. Theorem 9.1.1.

Theorem 9.2.6 *Let μ be a nontrivial probability measure on $[-\pi, \pi]$, and $\Pi_n(w)$ be a set of all polynomials P of degree at most n subject to $P(w) = 1$. Then*

$$\lambda_n(w) = \min_{P \in \Pi_n(w)} \int_{-\pi}^{\pi} |P(\zeta)|^2 \, d\mu(\theta) = \frac{1}{K_n(w,w)}. \tag{9.2.25}$$

The minimum is attained when $P(z) = K_n(z,w)/K_n(w,w)$.

Theorem 9.2.7 (Christoffel–Darboux formula) *For any $n \in \mathbb{Z}_+$ and $z, w \in \mathbb{C}$ with $\bar{w}z \neq 1$,*

$$\begin{aligned} K_n(z,w) &= \frac{\phi_{n+1}^*(z)\overline{\phi_{n+1}^*(w)} - \phi_{n+1}(z)\overline{\phi_{n+1}(w)}}{1 - \bar{w}z} \\ &= \frac{\phi_n^*(z)\overline{\phi_n^*(w)} - z\bar{w}\phi_n(z)\overline{\phi_n(w)}}{1 - \bar{w}z}. \end{aligned} \tag{9.2.26}$$

Here are some consequences of the Christoffel–Darboux formula. Setting $z = w$ we have

$$(1-|z|^2)\sum_{j=0}^{n}|\phi_j(z)|^2 = |\phi_{n+1}^*(z)|^2 - |\phi_{n+1}(z)|^2. \tag{9.2.27}$$

Setting $w = 0$ we come to

$$K_n(z,0) = \sum_{j=0}^{n}\phi_j(z)\overline{\phi_j(0)} = \phi_n^*(z)\overline{\phi_n^*(0)} = \kappa_n\phi_n^*(z). \tag{9.2.28}$$

The reproducing property of the Christoffel kernels is

$$\int_{-\pi}^{\pi} P(\zeta)\overline{K_n(\zeta,w)}\,d\mu(\theta) = P(w), \tag{9.2.29}$$

which holds for an arbitrary polynomial P of degree at most n and all complex w and follows directly from the definition. As a simple consequence of (9.2.29) one has a unit circle analogue of Theorem 2.1.8.

Theorem 9.2.8 *Let $M_n(\mu) = \|\mu_{i-j}\|_{i,j=0}^n$ be the Toeplitz matrix of the moments of μ. Let*

$$K_n(z,w) = \sum_{i,j=0}^{n} a_{ij}z^i\bar{w}^j$$

be the Taylor series expansion of the Christoffel kernel about the origin. Then

$$M_n^{-1}(\mu) = A^*, \quad A = \|a_{ij}\|_{i,j=0}^n.$$

One result connected to the Christoffel circle of ideas in the case of orthogonal polynomials on the real line is the Gauss–Jacobi quadrature formula. The following is its partial analogue for the unit circle case.

Theorem 9.2.9 *Suppose that the monic orthogonal polynomial $\Phi_n(\mu)$ has n distinct roots $z_1,\dots,z_n$. Then there exist complex numbers $\beta_1,\dots,\beta_n$ so that for each Laurent polynomial π of the form*

$$\pi(z) = \sum_{j=-n+1}^{n}\pi_j z^j$$

one has the equality

$$\int_{-\pi}^{\pi}\pi(\zeta)\,d\mu(\theta) = \sum_{j=1}^{n}\beta_j\pi(z_j). \tag{9.2.30}$$

9.3 Szegő's Theory and Its Extensions

Szegő's theorems may well be considered the most celebrated in OPUC theory. They have repeatedly served as a source for further development. For historical reasons one should state them in terms of Toeplitz determinants, $D_n(\mu)$ (cf. (9.1.3)). It follows from (9.1.9) and (9.2.4) that

$$D_n(\mu) = \prod_{j=0}^{n} \|\Phi_j\|^2 = \prod_{j=0}^{n-1} (1 - |\alpha_j|^2)^{n-j},$$

so $D_n^{1/n}(\mu)$ is monotone decreasing and

$$S(\mu) = \lim_{n\to\infty} (D_n(\mu))^{1/n} = \lim_{n\to\infty} \frac{D_n(\mu)}{D_{n-1}(\mu)} \tag{9.3.1}$$

exists and is a nonnegative number. Suppose $S(\mu) > 0$. Then $D_n(\mu)/S^{n+1}(\mu)$ is monotone increasing and

$$G(\mu) = \lim_{n\to\infty} \frac{D_n(\mu)}{S^{n+1}(\mu)} \tag{9.3.2}$$

exists and may be equal $+\infty$. Also, $G(\mu) < +\infty$ if and only if $\sum_{j=0}^{\infty} j|\alpha_j|^2 < \infty$ (Ibragimov's condition).

Szegő's theorems express S and G in terms of the absolutely continuous and singular components of μ (cf. (9.1.1)).

Theorem 9.3.1 (Szegő's theorem)

$$S(\mu) = \prod_{j=0}^{\infty} (1 - |\alpha_j|^2) = \exp\left(\frac{1}{2\pi} \int_{-\pi}^{\pi} \log \mu'(\theta)\, d\theta \right). \tag{9.3.3}$$

A striking feature of this result is that the α depend heavily on the singular component μ_s, whereas the product in (9.3.3) does not!

Szegő (1915) proved this for the case $\mu_s = 0$ in 1915 (in his very first paper!). The result does not depend on μ_s – this was shown by Verblunsky (1936).

It is immediate from Szegő's theorem that

$$\sum_{j=0}^{\infty} |\alpha_j|^2 < \infty \quad \text{if and only if} \quad \log \mu' \in L^1. \tag{9.3.4}$$

The equivalent conditions (9.3.4) are known as the **Szegő condition**, and the corresponding class of measures the **Szegő class**. Within this class, the **Szegő function**

$$D(z) = D(z,\mu) = \exp\left(\frac{1}{4\pi} \int_{-\pi}^{\pi} \frac{\zeta + z}{\zeta - z} \log \mu'(\theta)\, d\theta \right), \quad \zeta = e^{i\theta}, \quad |z| < 1, \tag{9.3.5}$$

which depends only on the absolutely continuous component of the orthogonality measure, is well defined. It is clear from the definition that $D(\mu)$ lies in the Hardy space $H^2(\mathbb{D})$, and the standard boundary value theory implies $D(\zeta) = \lim_{r\uparrow 1} D(rz)$ exists almost everywhere and $|D(\zeta)|^2 = \mu'(\theta)$ a.e., so $\|D\|_{H^2} \le 1$.

Theorem 9.3.2 (Strong Szegő theorem in Ibragimov's version) *If $\mu_s = 0$ and the Szegő condition holds, then*

$$G(\mu) = \prod_{j=0}^{\infty}(1-|\alpha_j|^2)^{-j-1} = \exp\left(\sum_{n=0}^{\infty} n|w_n|^2\right) = \exp\left(\int_{\mathbb{D}} \left|\frac{D'(z)}{D(z)}\right|^2 d^2z\right),$$

where w_n are the Fourier coefficients of $\log\mu'$, *and d^2z the normalized Lebesgue measure of* $\mathbb{D}$. *All the values may equal $+\infty$.*

For the modern approach to this result see Simon (2004a, Chapter 6).

Simon (2004a, Section 2.8) came up with the idea of extending Szegő's theorem by allowing "Pollaczek singularities" (so all quantities in (9.3.3) may be infinite). His result can be viewed as the first-order Szegő theorem: for any $\zeta_0 \in \mathbb{T}$,

$$|\zeta-\zeta_0|^2 \log\mu' \in L^1 \quad \text{if and only if} \quad \sum_{j=0}^{\infty} |\alpha_{j+1} - \bar\zeta_0\alpha_j|^2 + |\alpha_j|^4 < \infty.$$

Moreover, there is a precise formula for this case similar to the second equality in (9.3.3). The second-order Szegő theorem appeared in Simon and Zlatoš (2005).

Theorem 9.3.3 *Let $\zeta_k \in \mathbb{T}$, $k = 1, 2$. Then for $\zeta_1 \neq \zeta_2$,*

$$|\zeta-\zeta_1|^2|\zeta-\zeta_2|^2 \log\mu' \in L^1 \quad \textit{if and only if}$$

$$\sum_{j=0}^{\infty} |\alpha_{j+2} - (\bar\zeta_1+\bar\zeta_2)\alpha_{j+1} + \overline{\zeta_1\zeta_2}\alpha_j|^2 + |\alpha_j|^4 < \infty,$$

and for $\zeta_1 = \zeta_2$,

$$|\zeta-\zeta_1|^4 \log\mu' \in L^1 \textit{ if and only if}$$

$$\sum_{j=0}^{\infty} |\alpha_{j+2} - 2\bar\zeta_1\alpha_{j+1} + \bar\zeta_1^2\alpha_j|^2 + |\alpha_j|^6 < \infty.$$

The general conjecture called the **higher-order Szegő theorem** was formulated in Simon (2004a, Section 2.8). Given $\zeta_k \in \mathbb{T}$, $k = 1, \dots, n$ and $\zeta_p \neq \zeta_q$, $p \neq q$, define a polynomial

$$P(\zeta) := \prod_{k=1}^{n}(\zeta-\zeta_k)^{m_k}, \quad m_k \in \mathbb{N} = \{1, 2, \dots\}, \quad \overline{P}(\zeta) := \prod_{k=1}^{n}(\zeta-\bar\zeta_k)^{m_k},$$

and put $m = 1 + \max_k m_k$. Simon conjectures that

$$|P(\zeta)|^2 \log w \in L^1 \quad \text{if and only if} \quad \left(\overline{P}(S)\right)\{\alpha_j\} \in \ell^2 \text{ and } \{\alpha_j\} \in \ell^{2m},$$

where S is the shift operator: $S(\alpha_0, \alpha_1, \dots) = (\alpha_1, \alpha_2, \dots)$.

The following particular case of Simon's conjecture, which can be called the **higher-order Szegő theorem in ℓ^4**, is proved in Golinskii and Zlatoš (2007).

Theorem 9.3.4 *Assume that* $\{\alpha_j\} \in \ell^4$. *Then*

$$|P(\zeta)|^2 \log \mu' \in L^1 \quad \textit{if and only if} \quad \big(\overline{P}(S)\big)\{\alpha_j\} \in \ell^2.$$

The further advances concerning the polynomial Szegő class, that is, the class of measures μ with $|P(\zeta)|^2 \log \mu' \in L^1$, in particular, the asymptotics inside the disk and L^2 asymptotics on the circle, are in Denisov and Kupin (2006).

The celebrated Szegő asymptotic formula is one of the cornerstones of OPUC theory.

Theorem 9.3.5 (Szegő's limit theorem) *Suppose the Szegő condition* (9.3.4) *holds. Then*

$$\lim_{n\to\infty} \phi_n(z) = 0, \qquad \lim_{n\to\infty} \phi_n^*(z) = D^{-1}(z,\mu) \tag{9.3.6}$$

uniformly on compact sets of the open unit disk.

The result appeared in Szegő's pioneering 1920 paper (Szegő, 1920). Another closely related result concerns the asymptotics of the Christoffel kernels (see, for example, Grenander and Szegő, 1958, Chapter 3.4)

Theorem 9.3.6 *Under the Szegő condition* (9.3.4),

$$\lim_{n\to\infty} K_n(z,w) = \sum_{j=0}^{\infty} \phi_j(z)\overline{\phi_j(w)} = \frac{1}{1-\bar{w}z}\,\frac{1}{\overline{D(w,\mu)}}\,\frac{1}{D(z,\mu)} \tag{9.3.7}$$

uniformly on compact sets of the bidisk $(|z| < 1, |w| < 1)$.

In particular,

$$\lambda_\infty(w) := \min\left(\int_{-\pi}^{\pi} |P(\zeta)|^2\, d\mu(\theta) \colon P(w) = 1\right) = (1 - |w|^2)|D(w,\mu)|^2, \tag{9.3.8}$$

where the minimum is taken over the set of all polynomials P.

As for the asymptotics on the unit circle, we begin with L^2-convergence. The argument here uses a simple equality

$$\begin{aligned} &\frac{1}{2\pi}\int_{-\pi}^{\pi} |\phi_n^*(\zeta) - D^{-1}(\zeta)|^2 \mu'(\theta)\, d\theta + \int_{-\pi}^{\pi} |\phi_n^*(\zeta)|^2\, d\mu_s \\ &\quad = \int_{-\pi}^{\pi} |\phi_n^*(\zeta)|^2\, d\mu + \frac{1}{2\pi}\int_{-\pi}^{\pi} \frac{\mu'(\theta)}{|D(\zeta)|^2}\, d\theta - 2\mathrm{Re}\, D(0)\phi^*(0) \end{aligned}$$

(cf. Simon, 2004a, Section 2.4). Since the first two terms on the right-hand side are $1+1$, and from the Szegő limit theorem

$$\phi_n^*(0) = \kappa_n = \prod_{j=0}^{n-1}(1 - |\alpha_j|^2), \quad D^{-1}(0) = \prod_{j=0}^{n-1}(1 - |\alpha_j|^2),$$

it is not hard to obtain the following bound:

$$\frac{1}{2\pi}\int_{-\pi}^{\pi}|\phi_n^*(\zeta)-D^{-1}(\zeta)|^2\mu'(\theta)\,d\theta+\int_{-\pi}^{\pi}|\phi_n^*(\zeta)|^2\,d\mu_s\le 2\sum_{j=n}^{\infty}|\alpha_j|^2.$$

In particular, we have the following theorem.

Theorem 9.3.7 *Under the Szegő condition* (9.3.4),

$$\lim_{n\to\infty}\frac{1}{2\pi}\int_{-\pi}^{\pi}|\phi_n^*(\zeta)-D^{-1}(\zeta)|^2\mu'(\theta)\,d\theta=\lim_{n\to\infty}\int_{-\pi}^{\pi}|\phi_n^*(\zeta)|^2\,d\mu_s=0. \tag{9.3.9}$$

There are L^2-convergence results of a slightly different type, which deal with the limit relation

$$\lim_{n\to\infty}\frac{1}{\phi_n^*(\zeta)}=D(\zeta),\quad \zeta\in\mathbb{T}. \tag{9.3.10}$$

Theorem 9.3.8 *Under the Szegő condition, the convergence in* (9.3.10) *holds in the weak topology of $L^2(\mathbb{T})$, that is,*

$$\lim_{n\to\infty}\frac{1}{2\pi}\int_{-\pi}^{\pi}\overline{\frac{g(\zeta)}{\phi_n^*(\zeta)}}\,d\theta=\frac{1}{2\pi}\int_{-\pi}^{\pi}D(\zeta)\overline{g(\zeta)}\,d\theta,\quad g\in L^2.$$

Furthermore,

$$\lim_{n\to\infty}\left\|D-1/\phi_n^*\right\|_{L^2}^2=\mu_s([-\pi,\pi]).$$

In particular, the convergence in (9.3.10) *is in the L^2-norm if and only if μ is absolutely continuous* ($\mu_s=0$).

Khruschev (2001) proved the following nice limit relation that characterizes the Szegő class

$$\lim_{n\to\infty}\int_{-\pi}^{\pi}\left|\log|\phi_n(\zeta)|^{-2}-\log\mu'(\zeta)\right|\,d\theta=0.$$

As far as the Christoffel function on the circle goes, one can easily prove that for an arbitrary measure μ,

$$\lim_{n\to\infty}\lambda_n(\zeta)=\lim_{n\to\infty}\frac{1}{K_n(\zeta,\zeta)}=\left(\sum_{n=0}^{\infty}|\phi_n(\zeta)|^2\right)^{-1}=\mu\{\theta\} \tag{9.3.11}$$

for all $\zeta=e^{i\theta}\in\mathbb{T}$. A much more delicate result is due to Máté, Nevai, and Totik (1991).

Theorem 9.3.9 *For an arbitrary measure μ from the Szegő class one has*

$$\lim_{n\to\infty}\frac{n}{K_n(\zeta,\zeta)}=\mu'(\theta),\qquad \zeta=e^{i\theta},$$

a.e. on $[-\pi,\pi]$.

There are two natural ways to proceed from the Szegő theory. The first one is to consider proper subclasses of the Szegő class, and refine the above asymptotic results by imposing additional assumptions on the orthogonality measure, Verblunsky coefficients, ..., etc. For instance, one may ask whether the basic formula (9.3.6) holds on the unit circle (uniformly, pointwise, almost everywhere, etc.). Here is the classical result due to Szegő (see Grenander and Szegő, 1958, Theorem 3.5), which gives a rate of convergence in (9.3.6) on $\mathbb{T}$.

Given a continuous function $g \in C(\mathbb{T})$, define its modulus of continuity by

$$\omega(\delta; g) = \sup\{|g(h\zeta) - g(\zeta)| : |h - 1| \le \delta,\ h \in \mathbb{T}\}.$$

Theorem 9.3.10 *Let μ be absolutely continuous, $\mu = \frac{1}{2\pi}\mu'(\theta)\,d\theta$, with a positive and continuous density μ'. Assume also that*

$$\omega(\delta; \mu') \le C\left(\log\frac{1}{\delta}\right)^{-(1+\varepsilon)}, \quad \varepsilon > 0. \tag{9.3.12}$$

Then

$$\sup_{\mathbb{T}} |\phi_n^*(\zeta) - D^{-1}(\zeta)| \le C_1 (\log n)^{-\varepsilon}.$$

Assumption (9.3.12) was relaxed by B. Golinskii (Golinskii, 1967) to

$$\int_0^a \frac{\omega(t; \mu')}{t}\, dt < \infty,$$

and some rate of convergence was found in this case.

The uniform convergence in (9.3.6) in the closed unit disk can be guaranteed by certain assumptions on Verblunsky coefficients $\alpha_n(\mu)$ (see Geronimus, 1961, Chapter 8).

Theorem 9.3.11 *Let the Verblunsky coefficients α_n of the orthogonality measure μ satisfy*

$$\sum_{n=0}^{\infty} |\alpha_n(\mu)| < \infty. \tag{9.3.13}$$

Then

$$\max_{|z|\le 1} |\phi_n^*(z) - D^{-1}(z, \mu)| \le C \sum_{k=n}^{\infty} |\alpha_k(\mu)|, \quad n \in \mathbb{N}. \tag{9.3.14}$$

In the latter two results the property $\mu' > 0$ is crucial. Indeed, the asymptotic formula (9.3.6) cannot hold uniformly on $\mathbb{T}$ as long as μ' has zeros on $[-\pi, \pi]$. Badkov (1985) suggested a modified asymptotic formula, wherein $\phi_n^*(\zeta)$ is compared with $D^{-1}(r_n\zeta)$ with $r_n \uparrow 1$ as $n \to \infty$. More precisely, he proved that in a variety of situations, when μ' has algebraic zeros on $\mathbb{T}$, the limit relation

$$\phi_n^*(\zeta) = D^{-1}((1 - cn^{-1})\zeta)[1 + o(1)]$$

holds uniformly on $\mathbb{T}$.

Denote by $E(\mu)$ the subset of $[-\pi,\pi]$ on which $D^{-1}(\zeta)=\lim_{r\to 1-0}D^{-1}(r\zeta)$, $\zeta=e^{i\theta}$. Clearly, the normalized Lebesgue measure of this set is 1. The existence of the limit in (9.3.6) at ζ can be viewed as a **Tauberian problem**: under what conditions does

$$\lim_{n\to\infty}\phi_n^*(\zeta)=\lim_{n\to\infty}\lim_{r\to 1-0}\phi_n^*(r\zeta)=\lim_{r\to 1-0}\lim_{n\to\infty}\phi_n^*(r\zeta)=D^{-1}(\zeta)? \tag{9.3.15}$$

The following result is due to Geronimus (1961, Theorem 5.1). Put

$$\delta_n(\mu)=\sum_{k=n}^{\infty}|\alpha_k(\mu)|^2. \tag{9.3.16}$$

Theorem 9.3.12 *Assume that $\mu'\ge c>0$ a.e., and $\delta_n(\mu)=o(n^{-1})$. Then on the whole unit circle,*

$$|\phi_n^*(\zeta)-D^{-1}(\zeta)|\le|D^{-1}(\zeta)-D^{-1}(r_n\zeta)|+C(n\delta_n)^{1/3},\quad r_n=1-\left(\frac{\delta_n}{n}\right)^{2/3}.$$

In particular, (9.3.15) *holds on $E(\mu)$.*

Let us now turn to the boundedness of OPUC. The simplest bound comes out of the Szegő recurrences (9.2.1)/(9.2.3).

Theorem 9.3.13 *For $\zeta\in\mathbb{T}$ one has*

$$\prod_{j=0}^{n-1}(1-|\alpha_j|)\le|\Phi_n(\zeta)|\le\prod_{j=0}^{n-1}(1+|\alpha_j|). \tag{9.3.17}$$

Moreover,

$$\sup_{|z|\le 1}|\Phi_n(z)|\le\exp\left(\sum_{j=0}^{n-1}|\alpha_j|\right), \tag{9.3.18}$$

so under assumption (9.3.13),

$$\sup_n\sup_{|z|\le 1}|\Phi_n(z)|<\infty, \tag{9.3.19}$$

that is, the system Φ_n is uniformly bounded in the closed unit disk. If $B=\sup_j|\alpha_j|<1$, then

$$\inf_{\zeta\in\mathbb{T}}|\Phi_n(\zeta)|\ge\exp\left(\sum_{j=0}^{n-1}|\alpha_j|-\frac{1}{2(1-B)^2}\sum_{j=0}^{n-1}|\alpha_j|^2\right). \tag{9.3.20}$$

The uniform boundedness (9.3.19) holds under certain assumptions on the measure μ.

Theorem 9.3.14 *Let μ be absolutely continuous, and one of the following holds:*

(i) $0<c\le\mu'\le C<\infty$ *a.e., and the moments* $\mu_n=O(1/n)$, $n\to\infty$;
(ii) $0<c\le\mu'$ *a.e., and μ' is of bounded variation.*

Then Φ_n is uniformly bounded in the closed unit disk.

The study of the uniform boundedness (9.3.19) has a long history going back at least to Geronimus (1961). The uniform boundedness and uniform asymptotic representation for orthogonal polynomials is discussed in Golinskii and Golinskii (1998).

It was a conjecture of Steklov that the condition $\mu' \geq c > 0$ is sufficient for the uniform boundedness of orthogonal polynomials. This was proven false by Rakhmanov (1980, 1982a), who showed that for any $\delta < \frac{1}{2}$, there are examples with $\limsup_{n\to\infty} |\phi_n(1)| n^{-\delta} = \infty$. On the other hand,

$$\mu' \geq c > 0 \text{ a.e.} \quad \text{implies} \quad \|\phi_n\|_\infty \leq c^{-1/2}\sqrt{n+1}. \tag{9.3.21}$$

Statement (9.3.21) is sometimes called the Szegő estimate.

One can relax (9.3.13) and get some unbounded (but still useful) estimates. For instance,

$$\sum_{n=0}^{\infty} n|\alpha_n|^2 < \infty \quad \text{implies} \quad \left\|(\Phi_n^*)^{\pm 1}\right\|_\infty \leq C\exp(D\sqrt{\log n}),$$

where $\|\cdot\|_\infty$ is the L^∞-norm on $\mathbb{T}$, and C and D are suitable constants.

The second way to proceed from the Szegő theory is to go beyond the Szegő class and extend the above results to the case when the Szegő condition fails. The following three notable classes of measures, each of which contains the Szegő class as a proper subclass, come in naturally. These are

the **Erdős** class $\mathcal{E}$ of measures μ with $\mu' > 0$ a.e.;

the **Nevai** class $\mathcal{N}$ of measures μ with $\lim_{n\to\infty} \alpha_n(\mu) = 0$;

the **Rakhmanov** class $\mathcal{R}$ of measures μ with

$$\lim_{n\to\infty} \int_{-\pi}^{\pi} f(\zeta)|\phi_n(\zeta)|^2\, d\mu(\theta) = \frac{1}{2\pi}\int_{-\pi}^{\pi} f(\zeta)\, d\theta, \quad \zeta = e^{i\theta}, \quad \text{for all } f \in C(\mathbb{T}); \tag{9.3.22}$$

in other words, $|\phi_n|^2\, d\mu \to dm$ in the *-weak topology of the space of measures.

As it turns out, each class contains the former one as a proper subclass.

The first two classes were characterized by Máté, Nevai, and Totik (1985, 1987b) (see also Nevai, 1991) by means of the quantity

$$b_{n,k} = \frac{1}{2\pi}\int_{-\pi}^{\pi} \left| \left|\frac{\phi_n(\zeta)}{\phi_{n+k}(\zeta)}\right|^2 - 1\right| d\theta.$$

Their argument is based on the relation, which holds for arbitrary nontrivial probability measure μ,

$$\lim_{n\to\infty} \frac{1}{2\pi}\int_{-\pi}^{\pi} \left||\phi_n(\zeta)|^2\mu'(\theta) - 1\right|^2 d\theta \leq \limsup_{k\to\infty} b_{n,k}.$$

Theorem 9.3.15 *The following are equivalent:*

(i) $\mu \in \mathcal{E}$;

(ii) $\lim_{n\to\infty} \sup_{k\geq 1} b_{n,k} = 0$.

Theorem 9.3.16 *The following are equivalent:*

(i) $\mu \in \mathcal{N}$;
(ii) $\lim_{n\to\infty} \inf_{k\geq 1} b_{n,k} = 0$.

As a simple consequence one has $\mathcal{E} \subset \mathcal{N}$ (Rakhmanov's theorem). Moreover, a quantitative version of this relation was proved by Denisov (2004):

$$\limsup_{n\to\infty} |\alpha_n(\mu)| \leq 2\sqrt{2}\left[1 - m^2(\Omega)\right]^{1/2}, \tag{9.3.23}$$

where $\Omega = \{\theta \in [-\pi, \pi] : \mu'(\theta) > 0\}$, and $m(\Omega)$ is its normalized Lebesgue measure.

The next result is also due to Máté, Nevai, and Totik (1987b).

Theorem 9.3.17 *Let* $\mu \in \mathcal{E}$. *Then*

$$\lim_{n\to\infty} \int_{-\pi}^{\pi} \left| |\phi_n(\zeta)|^2 \mu'(\zeta) - 1 \right| d\theta = 0. \tag{9.3.24}$$

Moreover,

$$\lim_{n\to\infty} \int_{-\pi}^{\pi} \left(|\phi_n(\zeta)|^{-1} - \sqrt{\mu'(\theta)} \right)^2 d\theta = 0$$

if and only if μ *is absolutely continuous.*

Later on, Khruschev (2001) showed that (9.3.24) in fact characterizes the class $\mathcal{E}$, and suggested another characteristic property, namely

$$\lim_{n\to\infty} \frac{1}{2\pi} \int_{-\pi}^{\pi} \left[|\phi_n(\zeta)|^2 \mu'(\theta) \right]^a d\theta = 1 \quad \text{for all } 0 < a < 1.$$

There is another characterization of the Nevai class (Máté, Nevai, and Totik, 1987a).

Theorem 9.3.18 *Let* $\mu \in \mathcal{N}$. *Then*

$$\lim_{n\to\infty} \max_{|z|\leq 1} \frac{|\phi_n(z)|^2}{\sum_{k=0}^{n} |\phi_k(z)|^2} = 0.$$

Moreover, if the latter relation holds at least at one point $z_0 \in \mathbb{D}$ *then* $\mu \in \mathcal{N}$.

One of the most important results due to Máté, Nevai, and Totik (1987a) is the so-called **comparative** asymptotics outside the Szegő class.

Theorem 9.3.19 *Let* $\mu \in \mathcal{E}$. *Suppose*

$$\nu = g\mu, \quad g \geq 0, \quad \int g\, d\mu = 1,$$

and there is a polynomial Q *so that* $Qg^{\pm 1} \in L^\infty(\mu)$. *Then uniformly for* z, w *in compact subsets of* $\mathbb{D}$ *we have*

(1) $\lim_{n\to\infty} \frac{\phi_n^*(z,\nu)}{\phi_n^*(z,\mu)} = \exp\left(-\frac{1}{4\pi}\int_{-\pi}^{\pi} \frac{\zeta+z}{\zeta-z}\log g(\zeta)\,d\theta\right) = D(z,g^{-1});$

(2) $\lim_{n\to\infty} \frac{K_n^*(z,w;\nu)}{K_n^*(z,w;\mu)} = \overline{D(w,g^{-1})}D(z,g^{-1});$

(3) $\lim_{n\to\infty} \frac{\kappa_n(\nu)}{\kappa_n(\mu)} = \exp\left(-\frac{1}{4\pi}\int_{-\pi}^{\pi} \log g(\zeta)\,d\theta\right).$

Moreover,

$$\lim_{n\to\infty} \int_{-\pi}^{\pi} \left|\phi_n(\zeta;\nu)\overline{D(\zeta,g)} - \phi_n(\zeta;\mu)\right|^2 \mu'(\theta)\,d\theta = 0.$$

We now come to Rakhmanov's class, and give a characterization due to Khruschev (2001). We say that a sequence of Verblunsky coefficients obeys the **Máté–Nevai condition** if for each fixed $k \in \mathbb{N}$,

$$\lim_{n\to\infty} \alpha_n(\mu)\alpha_{n+k}(\mu) = 0. \tag{9.3.25}$$

Let us also introduce the probability measures

$$d\nu_{n,k} = \frac{1}{2\pi}\left[\frac{|\phi_n(\zeta)|^2}{|\phi_{n+k}(\zeta)|^2}\right] d\theta. \tag{9.3.26}$$

Theorem 9.3.20 *The following are equivalent:*

(i) $\mu \in \mathcal{R}$;
(ii) *the Máté–Nevai condition holds for* $\alpha_n(\mu)$;
(iii) $\nu_{n,k}$ *converges weakly to the normalized Lebesgue measure as* $n \to \infty$ *for all* $k \in \mathbb{N}$;
(iv) *uniformly on compact subsets of* $\mathbb{D}$, *we have*

$$\lim_{n\to\infty} \frac{\Phi_{n+1}^*(z)}{\Phi_n^*(z)} = 1. \tag{9.3.27}$$

It is obvious from this theorem that $\mathcal{N} \subset \mathcal{R}$. It is also easy to manufacture examples of measures outside the Nevai class which obey (9.3.25). Indeed, these are measures with sparse Verblunsky coefficients. Furthermore, the Rakhmanov measures which do not belong to the Nevai class are necessarily singular (Khruschev, 2001, Corollary 2.6).

Relation (9.3.27) is known as the **ratio asymptotics** of OPUC. As a matter of fact, there is a way to describe all possible limits for the ratio in (9.3.27). The result below is due to Khruschev (2002) and Barrios and López (1999).

Theorem 9.3.21 *Suppose*

$$\lim_{n\to\infty} \frac{\Phi_{n+1}^*(z,\mu)}{\Phi_n^*(z,\mu)} = G(z) \tag{9.3.28}$$

exists uniformly on compact subsets of $\mathbb{D}$. *Then either* $G \equiv 1$ *or*

$$G(z) = G_{a,\lambda}(z) = \frac{1 + \lambda z + \sqrt{(1 - \lambda z)^2 + 4a^2 \lambda z}}{2}$$

holds for some $\lambda \in \mathbb{T}$ *and* $a \in (0, 1]$. *Equation* (9.3.28) *holds with* $G = G_{a,\lambda}$ *if and only if* $\alpha_n(\mu)$ *obeys the López condition*

$$\lim_{n\to\infty} |\alpha_n| = a, \qquad \lim_{n\to\infty} \frac{\alpha_{n+1}}{\alpha_n} = \lambda. \tag{9.3.29}$$

In this case the essential support of μ *is a closed subinterval of* $[-\pi, \pi]$, *and* (9.3.28) *holds uniformly on compact subsets of* $\mathbb{C} \backslash e^{i\mathrm{supp}(\mu)}$.

The following extension of the above result, which can be viewed as the **relative ratio asymptotics**, is in Golinskii and Zlatoš (2007).

Theorem 9.3.22 *Let* μ *and* ν *be two nontrivial probability measures on* $\mathbb{T}$. *Let* $\{\alpha_n(\mu)\}$ *and* $\{\alpha_n(\nu)\}$, *respectively, be their Verblunsky coefficients, and let* $\Phi_n^*(\mu)$ *and* $\Phi_n^*(\nu)$, *respectively, be their reverse monic orthogonal polynomials. Then*

$$\frac{\Phi_{n+1}^*(z;\mu)}{\Phi_n^*(z;\mu)} - \frac{\Phi_{n+1}^*(z;\nu)}{\Phi_n^*(z;\nu)} \to 0 \tag{9.3.30}$$

uniformly on compact subsets of $\mathbb{D}$ *as* $n \to \infty$ *if and only if for any* $k \in \mathbb{N}$,

$$\lim_{n\to\infty} [\alpha_n(\mu)\bar{\alpha}_{n-k}(\mu) - \alpha_n(\nu)\bar{\alpha}_{n-k}(\nu)] = 0. \tag{9.3.31}$$

A closely related subject is the description of all possible limits in the Rakhmanov condition (9.3.22). A comprehensive study of this problem is in Khruschev (2002).

We conclude with a theorem of Bello and López (1998), which is analogous to Rakhmanov's theorem ($\mathcal{E} \subset \mathcal{N}$) but for any arc. Define

$$\theta_a = 2 \arcsin(a), \quad 0 < a < 1, \tag{9.3.32}$$

so $0 < \theta_a < \pi$. For $\lambda \in \mathbb{T}$ we let

$$\Gamma_{a,\lambda} = \{\zeta \in \mathbb{T} \colon |\arg(\lambda\zeta)| > \theta_a\}. \tag{9.3.33}$$

Theorem 9.3.23 *Let* $e^{i\mathrm{supp}(\mu)} = \Gamma_{a,\lambda}$ *and* $\mu' > 0$ *a.e. Then*

$$\lim_{n\to\infty} |\alpha_n(\mu)| = a, \qquad \lim_{n\to\infty} \bar{\alpha}_{n+1}(\mu)\alpha_n(\mu) = a\lambda^2. \tag{9.3.34}$$

An essential extension of this result due to Simon (2004b, Theorem 13.4.4) claims that supp can be relaxed to ess supp.

9.4 Zeros of OPUC

The structure of the zero sets for OPUC is another fascinating topic of the theory. Given a nontrivial probability measure μ, denote by $Z_n(\mu) = \{z_{jn}\}_{j=1}^n$ the zero set for the monic orthogonal polynomial $\Phi_n(\mu)$:

$$\Phi_n(z, \mu) = \prod_{j=1}^n (z - z_{jn}), \qquad |z_{nn}| \le |z_{n-1,n}| \le \cdots \le |z_{1,n}|.$$

As we have already seen, $Z_n(\mu) \subset \mathbb{D}$. Conversely, the following analogue of Wendroff's theorem was proved by Geronimus (1946).

Theorem 9.4.1 *Let π_n be any monic polynomial of degree n which has all its zeros inside $\mathbb{D}$. Then $\pi_n = \Phi_n(\mu)$ is a monic orthogonal polynomial for some $\mu \in \mathcal{P}$. Moreover, if μ and ν are any two such measures, we have*

(i) $\Phi_j(\mu) = \Phi_j(\nu)$, $j = 0, 1, \dots, n$;
(ii) $\alpha_j(\mu) = \alpha_j(\nu)$, $j = 0, 1, \dots, n-1$;
(iii) $\mu_j(\mu) = \mu_j(\nu)$, $j = 0, 1, \dots, n$.

It is clear that, unlike the case of orthogonal polynomials on the real line, the zeros need not be simple. The free case ($d\mu = dm = \frac{d\theta}{2\pi}$, the normalized Lebesgue measure on $[-\pi, \pi]$), where $\Phi_j(z, dm) = z^j$, illustrates this situation. Note also that the explicit measure in Theorem 9.4.1 can be easily constructed. Namely, the Bernstein–Szegő measure

$$d\sigma = \frac{C}{|\pi_n(\zeta)|^2}\, d\theta$$

is one, which satisfies $\alpha_j(\sigma) = 0$, $j \ge n$.

The fact that $Z_n(\mu) \subset \mathbb{D}$ reflects the following quite general situation (Fejér, 1922).

Theorem 9.4.2 (Fejér's theorem) *Let μ be a nontrivial probability measure on $\mathbb{C}$ so that*

$$\int |z|^k \, d\mu(z) < \infty, \qquad k = 0, 1, \dots, 2n-1.$$

Let Φ_n be the monic polynomial of degree n orthogonal to $\{1, \dots, z^{n-1}\}$ in $L^2(\mathbb{C}, \mu)$. Then all of the zeros of Φ_n lie in the convex hull of supp(μ). *Suppose further that* supp(μ) *is compact. Then no extreme point of the hull is a zero, and if* supp(μ) *does not lie on a straight line, all zeros lie in the interior of the convex hull.*

Fejér's theorem is optimal in the following sense. For the unit circle, $\Phi_1(z) = z - \bar{\alpha}_0 = z - \bar{\mu}_1$ has its zero at $\bar{\mu}_1$. But $\int_K \zeta \, d\mu$ runs through a dense set of the convex hull of K as μ runs through all probability measures on K.

If supp(μ) $\subset \mathbb{T}$, the interior of the convex hull is a subset of $\mathbb{D}$, so the zeros of Φ_n lie in $\mathbb{D}$. If supp(μ) is a proper subset of $\mathbb{T}$, then Fejér's theorem gives more information. For example, if $\zeta_0 \in \mathbb{T}$ and $d = \text{dist}(\zeta_0, \text{supp}(\mu)) > 0$, then a little geometry shows that the distance of zeros of Φ_n from ζ_0 is at least $d^2/2$.

Here is the result by Denisov–Simon (Simon, 2004a, Theorem 1.7.20) which provides some information about the zeros near isolated points of the support.

Theorem 9.4.3 *Let μ and Φ_n be as in the above theorem, and ζ_0 be an isolated point of* $\mathrm{supp}(\mu)$. *Let* $\Gamma = \mathrm{supp}(\mu)\backslash\{\zeta_0\}$ *and* $\mathrm{ch}(\Gamma)$ *its convex hull. Suppose*

$$\delta = \mathrm{dist}(\zeta_0, \mathrm{ch}(\Gamma)) > 0.$$

Then Φ_n has at most one zero in $\{|z - \zeta_0| < \delta/3\}$. In particular, if μ is supported on $\mathbb{T}$ and $d = \mathrm{dist}(\zeta_0, \Gamma) > 0$, *there is at most one zero in the circle of radius $d^2/6$ about ζ_0.*

If μ is an even measure with support $\{0\} \cup [1, 2] \cup [-2, -1]$, one can show that for n large enough and even, P_n has two zeros near 0. Thus, it is not enough that ζ_0 be isolated from Γ; it must be isolated from $\mathrm{ch}(\Gamma)$.

If $\mathrm{supp}(\mu) = \mathbb{T}$, the zeros of Φ_n may stay away from the support (take, for example, $d\mu = dm$). But when this set is a proper subset of the unit circle, it attracts zeros in the following sense (see Simon, 2004a, Theorems 8.1.11 and 8.1.12).

Theorem 9.4.4 *Suppose ζ_0 is an isolated point of* $\mathrm{supp}(\mu)$. *Then for any $\delta > 0$, there is N_δ so $\{|z - \zeta_0| < \delta\}$ has exactly one zero of Φ_n for $n > N_\delta$. If this zero is called z_n, there is an $a > 0$ so for all large enough n, $|z_n - \zeta_0| \le e^{-an}$.*

Theorem 9.4.5 *Let* $\mathrm{supp}(\mu) \ne \mathbb{T}$, *and ζ_0 be a nonisolated point of* $\mathrm{supp}(\mu)$. *Then for each $\delta > 0$,*

$$\sharp\{z\colon |z - \zeta_0| < \delta,\ \Phi_n(z) = 0\} \to \infty, \qquad n \to \infty.$$

The following question arises naturally: Is it possible that the bulk of zeros still stay away from the support in the latter case? A negative answer was given by Widom (1967).

Theorem 9.4.6 (Widom's zero theorem) *Let* $\mathrm{supp}\,\mu \ne \mathbb{T}$. *Then, for any compact set $K \subset \mathbb{D}$, there is a positive integer n_K, so that for each $j \in \mathbb{N}$,*

$$\sharp\{z\colon z \in K,\ \Phi_j(z) = 0\} \le n_K.$$

Here is another theorem on zeros of OPUC, which appeared in Alfaro and Vigil (1988).

Theorem 9.4.7 (Alfaro–Vigil theorem) *Let $\{z_n\}_{n=1}^{\infty}$ be a sequence of numbers in $\mathbb{D}$. Then there exists a unique nontrivial probability measure μ on $[-\pi, \pi]$ with $\Phi_n(z_n, \mu) = 0$.*

Alfaro and Vigil were answering the following question from P. Turán (Turán, 1980): Can the set $Z_\infty(\mu) = \cup_n Z_n(\mu)$ of all zeros of the Φ_n be dense in $\mathbb{D}$? The answer is clearly yes, and follows from this theorem. Such measures are called Turán measures. It is proved in Khruschev (2003) that there are absolutely continuous Turán measures with μ' a C^∞ function. This is especially interesting since if μ' is real analytic and nonvanishing then $\overline{Z_\infty(\mu)} \ne \mathbb{D}$ (see below).

It is known (Saff and Totik, 1992) that zeros of $\Phi_n(\mu)$ cluster to $\mathrm{supp}(\mu)$ as long as this set is a proper subset of the whole circle. The situation changes dramatically if $\mathrm{supp}(\mu) = \mathbb{T}$ (see, for example, $d\mu = dm$). By the Alfaro–Vigil theorem, zeros of Φ_n can cluster to all points of $\overline{\mathbb{D}}$. Denote by

$$Z_w(\mu) := \{z \in \overline{\mathbb{D}}:\ \liminf_{n\to\infty} \operatorname{dist}(z, Z_n) = 0\}$$

the set of limit points for the zeros of all Φ_n (weakly attracting points). Let $Z_w = \{Z_w(\mu)\}_\mu$ be the class of all such point sets. So $\overline{\mathbb{D}} \in Z_w$. It turns out that Z_w is rich enough. More precisely, each compact subset K of $\mathbb{D}$ belongs to Z_w, and the same is true if $K \supset \mathbb{T}$ (Simon and Totik, 2005, Theorem 4). On the other hand, $K = [1/2, 1]$ is not in Z_w.

Similarly, denote by

$$Z_s(\mu) := \{z \in \overline{\mathbb{D}}:\ \lim_{n\to\infty} \operatorname{dist}(z, Z_n) = 0\}$$

the point set of strongly attracting points, and Z_s the class of all such point sets. The structure of the latter is quite different from that of Z_w. For instance, it is proved in Alfaro et al. (2005) that if $0 \in Z_s(\mu)$ for some measure μ, then $Z_s(\mu)$ is at most a countable set which converges to the origin. So the disk $\{|z| \le 1/2\}$ is not in Z_s.

A significant generalization of the Alfaro–Vigil theorem is due to Simon and Totik (2005).

Theorem 9.4.8 *For an arbitrary sequence of points $\{z_k\}$ in $\mathbb{D}$ and an arbitrary sequence of positive integers $0 < m_1 < m_2 < \cdots$, there exists a measure μ on $[-\pi, \pi]$ such that*

$$\Phi_{m_k}(z_j, \mu) = 0, \quad j = m_{k-1} + 1, \ldots, m_k.$$

The following consequence of this result is surprising. Given a measure μ, consider the sequence $\{\nu_n(\mu)\}_{n\ge1}$ of **normalized counting measures for zeros** of Φ_n, that is,

$$\operatorname{supp} \nu_n = Z_n, \quad \nu_n\{z_{jn}\} = \frac{l(z_{jn})}{n} \tag{9.4.1}$$

with $l(z_{jn})$ equal to the multiplicity of the zero z_{jn}. Let $\mathcal{M}_+(\overline{\mathbb{D}})$ be a space of probability measures on $\overline{\mathbb{D}}$ endowed with the weak* topology. A measure μ is called **universal** if for each $\sigma \in \mathcal{M}_+(\overline{\mathbb{D}})$ there is a sequence of indices n_j such that $\nu_{n_j}(\mu)$ converges to σ as $j \to \infty$ in the weak* topology. The existence of universal measures is proved in Simon and Totik (2005, Corollary 3).

A remarkable theorem of Mhaskar and Saff (1990) provides some information about the limit points (in the space $\mathcal{M}_+(\overline{\mathbb{D}})$) of the sequence of counting measures of zeros associated with a measure μ in the case when Verblunsky coefficients tend to zero fast enough.

Theorem 9.4.9 (Mhaskar–Saff theorem) *Let*

$$A := \limsup_{n\to\infty} |\alpha_n(\mu)|^{\frac{1}{n}} = \lim_{j\to\infty} |\alpha_{n_j}(\mu)|^{\frac{1}{n_j}}. \tag{9.4.2}$$

Suppose that either $A < 1$, or $A = 1$ and $\sum_{j=0}^{n-1} |\alpha_j(\mu)| = o(n)$ as $n \to \infty$. Then $\{\nu_{n_j}(\mu)\}$ converges to the uniform measure on the circle of radius A.

A crucial feature of the Mhaskar–Saff theorem is its universality. Under its assumption the angular distribution is the same. To get certain quantitative bounds on the distance between zeros, Simon studied various more stringent conditions, and among them the so-called **Barrios–López–Saff** condition

$$\alpha_n(\mu) = Cb^n + O((b\Delta)^n), \quad C \in \mathbb{C}\setminus\{0\}, \; 0 < b, \Delta < 1. \tag{9.4.3}$$

The following result is proved in Simon (2006).

Theorem 9.4.10 *Under assumption* (9.4.3) *there is a bounded number J of "spurious" zeros of $\Phi_n(\mu)$ for all large n. Furthermore, for $j = 1, 2, \ldots, n-J$ let*

$$z_{jn} = |z_{jn}|e^{i\Theta_{jn}}, \quad 0 = \Theta_{0n} < \Theta_{1n} < \cdots < \Theta_{n-J,n} < 2\pi = \Theta_{n-J+1,n}$$

be the other zeros. Then the following limit relations hold:

$$\sup_{1\le j\le n-J} \big||z_{jn}| - b\big| = O\left(\frac{\log n}{n}\right), \quad n \to \infty; \tag{9.4.4}$$

$$\sup_{1\le j\le n-J} n\left|\Theta_{j+1,n} - \Theta_{jn} - \frac{2\pi}{n}\right| = o(1), \quad n \to \infty; \tag{9.4.5}$$

$$\frac{|z_{j+1,n}|}{|z_{jn}|} = 1 + O\left(\frac{1}{n\log n}\right), \quad n \to \infty. \tag{9.4.6}$$

Note that (9.4.5)–(9.4.6) imply $\lim_n n|z_{j+1,n} - z_{jn}| = 2\pi b$. Amazingly, the spurious zeros also follow the clock pattern!

Simon (2005a) treats the more general case

$$\alpha_n(\mu) = \sum_{k=1}^m C_k e^{in\Theta_k} b^n + O((b\Delta)^n).$$

The value A in (9.4.2) is closely related to some other characteristics in OPUC theory. Define the following "radii":

- $R(D^{-1})$ is the radius of convergence of the Taylor series for the inverse Szegő function D^{-1} about the origin, if $\mu_s = 0$ and the Szegő condition holds, and $R(D^{-1}) = 1$ otherwise;
- $R^* = \sup\{r\colon \sup_{n,|z|\le r} |\Phi^*(z,\mu)| < \infty\}$, if the Szegő condition holds, and $R^* = 1$ otherwise.

Let $N_n(r)$ be a number of zeros of $\Phi_n(\mu)$ in $\{r < |z| < 1\}$. Define the Nevai–Totik radius R_{NT} by

$$R_{\mathrm{NT}} = \inf\{r\colon N_n(r) = O(1), \; n \to \infty\}.$$

The next result is proved in Nevai and Totik (1989).

Theorem 9.4.11 (Nevai–Totik theorem) *For an arbitrary measure μ the following equalities hold:*

$$A = \limsup_{n\to\infty} |\alpha_n(\mu)|^{\frac{1}{n}} = R_{\mathrm{NT}} = \frac{1}{R(D^{-1})} = \frac{1}{R^*}. \tag{9.4.7}$$

If $A < 1$, then $\phi_n^* \to D^{-1}$ uniformly on compact subsets of $\{z\colon |z| < A^{-1}\}$.

9.5 CMV Matrices – Unitary Analogues of Jacobi Matrices

One of the key tools in the case of the real line, especially in perturbation theory, is the realization of a measure σ as the spectral measure of the Jacobi matrix, which comes in as a matrix of multiplication by x on $L^2_\sigma(\mathbb{R})$. Of course, in the OPUC case, μ is the spectral measure of multiplication by ζ on $L^2_\mu(\mathbb{T})$. That alone is not enough because $L^2_\mu(\mathbb{T})$ is μ-dependent, and we cannot connect different μ. What we need is a suitable matrix representation; in other words, we need to choose a convenient orthonormal basis. There is an "obvious" set to try, namely, $\{\phi_n(\mu)\}$, but the corresponding matrix, called the GGT (Geronimus–Gragg–Teplyaev) matrix in Simon (2004a), has two defects. First, a fundamental theorem by Szegő–Kolmogorov–Krein states that $\{\phi_n(\mu)\}$ is a basis (complete, orthonormal system) if and only if μ is outside the Szegő class, that is, $\log\mu' \notin L^1$, or equivalently, $\sum_{n=0}^{\infty}|\alpha_n(\mu)|^2 = \infty$, and if it is not, the matrix $\mathcal{G} = ((\zeta\phi_m, \phi_n))$ is not unitary. Second, even if it is, the matrix $\mathcal{G}$ is not of finite band width measured from the diagonal.

One of the most interesting developments in the theory of OPUC in recent years is the discovery by Cantero, Moral, and Velázquez (2003) of a matrix realization for multiplication by $\zeta = e^{i\theta}$ on $L^2_\mu(\mathbb{T})$ which is of finite band size (that is, $(\zeta\chi_m, \chi_n)_\mu = 0$ if $|m-n| > k$ for some k); in this case, $k = 2$ is to be compared with $k = 1$ for the Jacobi matrices, which correspond to the real line case. The **CMV basis** $\{\chi_n\}$ is obtained by orthonormalizing the sequence $1, \zeta, \zeta^{-1}, \zeta^2, \zeta^{-2}, \ldots$, and the matrix

$$\mathcal{C}(\mu) = (c_{n,m})_{m,n=0}^{\infty}, \quad c_{n,m} = (\zeta\chi_m, \chi_n)_\mu,$$

called the **CMV matrix**, is unitary and pentadiagonal. Remarkably, the χ can be expressed in terms of ϕ and ϕ^* by

$$\chi_{2n}(z) = z^{-n}\phi^*_{2n}(z), \quad \chi_{2n+1}(z) = z^{-n}\phi_{2n+1}(z), \quad n \in \mathbb{Z}_+,$$

and the matrix entries in terms of α and ρ:

$$C = \mathcal{L}\mathcal{M}, \tag{9.5.1}$$

where $\mathcal{L}$, $\mathcal{M}$ are 2×2 block diagonal matrices

$$\mathcal{L} = \Theta_0 \oplus \Theta_2 \oplus \Theta_4 \oplus \cdots, \quad \mathcal{M} = 1 \oplus \Theta_1 \oplus \Theta_3 \oplus \cdots, \tag{9.5.2}$$

with

$$\Theta_j = \begin{pmatrix} \overline{\alpha}_j & \rho_j \\ \rho_j & -\alpha_j \end{pmatrix}, \quad j \in \mathbb{Z}_+ \tag{9.5.3}$$

(the first block of $\mathcal{M}$ is 1×1). By C_0 we will denote the CMV matrix for the Lebesgue measure $\frac{1}{2\pi}\,d\theta$. For an exhaustive exposition of the theory of CMV matrices see Simon (2004a, Chapter 4) and Simon (2007a).

Expanding out the matrix product (9.5.1)–(9.5.3) is rather laborious and leads to a quite rigid structure

$$C(\mu) = \begin{pmatrix} * & * & + & & & & & \\ + & * & * & & & & & \\ & * & * & * & + & & & \\ & + & * & * & * & & & \\ & & & * & * & * & + & \cdots \\ & & & + & * & * & * & \cdots \\ \cdots & \cdots & \cdots & \cdots & \cdots & \cdots & \cdots & \cdots \end{pmatrix} \tag{9.5.4}$$

where + represents strictly positive entries, and $*$ generally nonzero ones. The entries marked + and called the exposed entries of the CMV matrix are precisely $(2,1)$ and those of the form $(2j-1,2j+1)$ and $(2j+2,2j)$ with $j \in \mathbb{N}$. Matrices of the form (9.5.4) are said to have CMV shape. Naturally, CMV matrices have CMV shape, and, what is more to the point, any unitary matrix (9.5.4) is actually a CMV matrix (9.5.1)–(9.5.3). Matrices (9.5.4) (of zigzag pentadiagonal form) appeared first in Watkins (1993), see also Bunse-Gerstner, Elsner (1991), who outlined the connection of such matrices with OPUC.

Yet, expanding out (9.5.1)–(9.5.3) can be carried out, and explicit formulas for the matrix entries c_{nm} in terms of the Verblunsky coefficients are available (cf. Golinskii, 2006). Let $2\lambda_m := 1-(-1)^m$, $m \in \mathbb{Z}_+$, and $\lambda_{-1} = 1$, so

$$\{\lambda_m\}_{m\ge 0} = \{0,1,0,1,\dots\},$$

$$\lambda_m + \lambda_{m+1} = 1, \quad \lambda_m\lambda_{m+1} = 0, \quad \lambda_m - \lambda_{m+1} = (-1)^{m+1}.$$

One has

$$\begin{gathered} c_{mm} = -\overline{\alpha}_m\alpha_{m-1}, \\ c_{m+2,m} = \rho_m\rho_{m+1}\lambda_m, \qquad c_{m,m+2} = \rho_m\rho_{m+1}\lambda_{m+1}, \end{gathered} \tag{9.5.5}$$

and

$$\begin{aligned} c_{m+1,m} &= \overline{\alpha}_{m+1}\rho_m\lambda_m - \alpha_{m-1}\rho_m\lambda_{m+1}, \\ c_{m,m+1} &= \overline{\alpha}_{m+1}\rho_m\lambda_{m+1} - \alpha_{m-1}\rho_m\lambda_m. \end{aligned} \tag{9.5.6}$$

Given an arbitrary sequence $\{\alpha_n\} \in \mathbb{D}^\infty$ one can construct a matrix $C = C(\alpha_n)$ by (9.5.1)–(9.5.3) (which generates a unitary operator in $\ell^2(\mathbb{N})$), and make sure that a distinguished unit vector $e_0 = (1,0,0,\dots)'$ is **cyclic**, that is, finite linear combinations of $\{C^n e_0\}_{n=-\infty}^{\infty}$ are dense in $\ell^2(\mathbb{N})$. So, C is unitarily equivalent to the multiplication by ζ on $L^2_\mu(\mathbb{T})$, μ being a spectral measure associated to C and e_0.

Theorem 9.5.1 *For an arbitrary sequence* $\{\alpha_n\} \in \mathbb{D}^\infty$ *a matrix* C *of* (9.5.4)–(9.5.6) *is the CMV matrix associated to the measure* μ, *that is,* C *takes the form* (9.5.1)–(9.5.3) *and* $\alpha_n = \alpha_n(\mu)$.

Clearly, it is just as natural to take the ordered set $1, \zeta^{-1}, \zeta, \zeta^{-2}, \zeta^{2}, \ldots$ in place of $1, \zeta, \zeta^{-1}, \zeta^{2}, \zeta^{-2}, \ldots$, and come to what is called the **alternate CMV basis** $\{x_n\}$ and the alternate CMV representation

$$\tilde{c_{ij}}(\mu) = (\zeta x_j, x_i)_\mu.$$

As it turns out, $\tilde{C}$ is just the transpose of C.

To state the analogue of Stone's self-adjoint cyclic model theorem, consider a cyclic unitary model, that is, a unitary operator U on a separable Hilbert space $\mathcal{H}$ with $\dim \mathcal{H} = \infty$, along with a cyclic unit vector v_0. Two cyclic unitary models $(\mathcal{H}, U, v_0)$ and $(\tilde{\mathcal{H}}, \tilde{U}, \tilde{v}_0)$ are called equivalent if there is a unitary W form $\mathcal{H}$ onto $\tilde{\mathcal{H}}$ so that

$$WUW^{-1} = \tilde{U}, \quad Wv_0 = \tilde{v}_0.$$

Theorem 9.5.2 *Each cyclic unitary model is equivalent to a unique CMV model* $(\ell^2(\mathbb{N}), C, e_0)$.

There is an important relation between CMV matrices and monic orthogonal polynomials akin to the well-known property of orthogonal polynomials on the real line:

$$\Phi_n(z) = \det(zI_n - C^{(n)}), \tag{9.5.7}$$

where $C^{(n)}$ is the principal $n \times n$ block of C. Note that $C^{(n)}$ is no more a unitary matrix. As a matter of fact, it is quite close to unitary in the sense that $C^{(n)}$ is a contraction with one-dimensional defect. So its eigenvalues (zeros of monic orthogonal polynomial Φ_n) are inside the unit disk.

There is another property of CMV matrices similar to the well-known property of Jacobi matrices. Given $\zeta_0 \in \mathbb{T}$, let

$$v = \{v_n\}_{n=0}^{\infty} = \{\overline{\chi_n(\zeta_0)}\}_{n=0}^{\infty}.$$

Then $Cv = \zeta_0 v$, which means $\sum_j c_{kj} v_j = \zeta_0 v_k$ for all k (because of the pentadiagonal structure of C this sum always makes sense). In general, $v \notin \ell^2$, but if it is, then $\zeta_0 = e^{i\theta_0}$ is an eigenvalue of C, or equivalently, θ_0 is a mass point of the measure μ (cf. (9.3.11)). Furthermore, we have the following theorem.

Theorem 9.5.3 *Let* $v \notin \ell^2$ *but* $\liminf |\phi_n(\zeta_0)|^{1/n} \le 1$. *Then* $\zeta_0 \in \sigma(C)$ *and it is a nonisolated point of the support of* μ.

There is an explicit formula for the resolvent of the CMV matrix C in the CMV basis. It has already proved useful in some applications of CMV matrices (see Golinskii, 2006). By the spectral theorem,

$$(C - zI)^{-1}_{mn} = \int_{-\pi}^{\pi} \frac{\chi_n(\zeta)\overline{\chi_m(\zeta)}}{\zeta - z}\, d\mu(\theta).$$

Let ϕ_n and ψ_n be orthonormal polynomials of the first and second kind, respectively, and F the Carathéodory function. Define

$$p_{2k}(z) = z^{-k}(F(z)\phi_{2k}(z) + \psi_{2k}(z)), \quad p_{2k-1}(z) = z^{-k}(F(z)\phi^*_{2k-1}(z) - \psi^*_{2k-1}(z)),$$
$$\pi_{2k}(z) = z^{-k}(F(z)\phi^*_{2k}(z) - \psi^*_{2k}(z)), \quad \pi_{2k-1}(z) = z^{-k+1}(F(z)\phi_{2k-1}(z) + \psi_{2k-1}(z)).$$

The following result is in Simon (2004a, Theorem 4.4.1).

Theorem 9.5.4 *For $z \in \mathbb{D}$,*

$$[(C - zI)^{-1}]_{mn} = \begin{cases} (2z)^{-1}\chi_n(z)p_m(z), & m > n, \\ (2z)^{-1}\pi_n(z)x_m(z), & n > m, \end{cases} \tag{9.5.8}$$

and

$$\begin{aligned} [(C - zI)^{-1}]_{2n-1,2n-1} &= (2z)^{-1}\chi_{2n-1}(z)p_{2n-1}(z), \\ [(C - zI)^{-1}]_{2n,2n} &= (2z)^{-1}\pi_{2n}(z)x_{2n}(z). \end{aligned} \tag{9.5.9}$$

9.6 Differential Equations

This section is based on Ismail and Witte (2001). It will be assumed that μ is absolutely continuous, that is, the orthogonality relation becomes

$$\int_{|\zeta|=1} \phi_m(\zeta)\overline{\phi_n(\zeta)}w(\zeta)\frac{d\zeta}{i\zeta} = \delta_{m,n}. \tag{9.6.1}$$

Following the notation in Section 2.8, we set

$$w(z) = e^{-v(z)}, \tag{9.6.2}$$

and assume that $w(z)$ is differentiable in a neighborhood of the unit circle, has moments of all integral orders, and the integrals

$$\int_{|\zeta|=1} \frac{v'(z) - v'(\zeta)}{z - \zeta}\zeta^n w(\zeta)\frac{d\zeta}{i\zeta}$$

exist for all integers n. Let

$$A_n(z) = n\frac{\kappa_{n-1}}{\kappa_n} + i\frac{\kappa_{n-1}}{\phi_n(0)}z \int_{|\zeta|=1} \frac{v'(z) - v'(\zeta)}{z - \zeta}\phi_n(\zeta)\overline{\phi^*_n(\zeta)}w(\zeta)\, d\zeta, \tag{9.6.3}$$

$$B_n(z) = -i \int_{|\zeta|=1} \frac{v'(z) - v'(\zeta)}{z - \zeta}\phi_n(\zeta)\left[\overline{\phi_n(\zeta)} - \frac{\kappa_n}{\phi_n(0)}\overline{\phi^*_n(\zeta)}\right] w(\zeta)\, d\zeta. \tag{9.6.4}$$

For future reference we note that $A_0 = B_0 = 0$ and

$$A_1(z) = \kappa_1 - \phi_1(z)v'(z) - \frac{\phi_1^2(z)}{\phi_1(0)}M_1(z), \tag{9.6.5}$$

$$B_1(z) = -v'(z) - \frac{\phi_1(z)}{\phi_1(0)}M_1(z), \tag{9.6.6}$$

where M_1 is defined by

$$M_1(z) = \int_{|\zeta|=1} \zeta \frac{v'(z) - v'(\zeta)}{z - \zeta} w(\zeta) \frac{d\zeta}{i\zeta}. \tag{9.6.7}$$

Theorem 9.6.1 *Under the above stated assumptions on w, the corresponding orthonormal polynomials satisfy the differential relation*

$$\phi_n'(z) = A_n(z)\phi_{n-1}(z) - B_n(z)\phi_n(z). \tag{9.6.8}$$

Define differential operators $L_{n,1}$ and $L_{n,2}$ by

$$L_{n,1} = \frac{d}{dz} + B_n(z), \tag{9.6.9}$$

and

$$L_{n,2} = -\frac{d}{dz} - B_{n-1}(z) + \frac{A_{n-1}(z)\kappa_{n-1}}{z\kappa_{n-2}} + \frac{A_{n-1}(z)\kappa_n\phi_{n-1}(0)}{\kappa_{n-2}\phi_n(0)}. \tag{9.6.10}$$

Then the operators $L_{n,1}$ and $L_{n,2}$ are annihilation and creation operators in the sense that they satisfy

$$\begin{aligned} L_{n,1}\phi_n(z) &= A_n(z)\phi_{n-1}(z), \\ L_{n,2}\phi_{n-1}(z) &= \frac{A_{n-1}(z)}{z}\frac{\phi_{n-1}(0)\kappa_{n-1}}{\phi_n(0)\kappa_{n-2}}\phi_n(z). \end{aligned} \tag{9.6.11}$$

This establishes the second-order differential equation

$$L_{n,2}\left(\frac{1}{A_n(z)}L_{n,1}\right)\phi_n(z) = \frac{A_{n-1}(z)}{z}\frac{\phi_{n-1}(0)\kappa_{n-1}}{\phi_n(0)\kappa_{n-2}}\phi_n(z). \tag{9.6.12}$$

Note that, unlike for polynomials orthogonal on the line, $L_{n,1}^*$ is not related to $L_{n,2}$.

When $v(z)$ is a meromorphic function in the unit disk then the following functional equation holds:

$$B_n + B_{n-1} - \frac{\kappa_{n-1}}{\kappa_{n-2}}\frac{A_{n-1}}{z} - \frac{\kappa_n}{\kappa_{n-2}}\frac{\phi_{n-1}(0)}{\phi_n(0)}A_{n-1} = \frac{1-n}{z} - v'(z). \tag{9.6.13}$$

Using (9.6.13) we simplify the expanded form of (9.6.12) to

$$\begin{aligned} &\phi_n'' - \left\{\frac{A_n'}{A_n} + v'(z) + \frac{n-1}{z}\right\}\phi_n' \\ &+ \left\{B_n' - \frac{B_nA_n'}{A_n} + B_nB_{n-1} - \frac{\kappa_{n-1}}{\kappa_{n-2}}\frac{A_{n-1}B_n}{z} - \frac{\kappa_n}{\kappa_{n-2}}\frac{\phi_{n-1}(0)}{\phi_n(0)}A_{n-1}B_n \right. \\ &\left. + \frac{\kappa_{n-1}}{\kappa_{n-2}}\frac{\phi_{n-1}(0)}{\phi_n(0)}\frac{A_{n-1}A_n}{z}\right\}\phi_n = 0. \end{aligned} \tag{9.6.14}$$

Recall that the zeros of the polynomial ϕ_n are denoted by $\{z_{jn}\}_{1\le j\le n}$ and are confined within the unit circle $|z| < 1$. One can construct a real function $|T(z_{1n}, \ldots, z_{nn})|$ from

$$T(z_{1n}, \ldots, z_{nn}) = \prod_{j=1}^{n} z_{jn}^{-n+1}\frac{e^{-v(z_{jn})}}{A_n(z_{jn})}\prod_{1\le j<k\le n}\left(z_{jn} - z_{kn}\right)^2, \tag{9.6.15}$$

such that the zeros are given by the stationary points of this function.

This function has the interpretation of being the total energy function for n mobile unit charges in the unit disk interacting with a one-body confining potential, $v(z) + \ln A_n(z)$, an attractive logarithmic potential with a charge $n-1$ at the origin, $(n-1)\ln z$, and repulsive logarithmic two-body potentials, $-\ln(z_i - z_j)$, between pairs of charges. However, all the stationary points are saddle points, a natural consequence of analyticity in the unit disk.

For more details and examples we refer the interested reader to Ismail (2005b, Chapter 8).

9.7 Examples of OPUC

In this section a number of examples are discussed, most of which are "exactly solvable" in the sense that there are explicit formulas for both moments and Verblunsky coefficients and, in most cases there are also explicit formulas for the actual orthogonal polynomials. A nice collection of examples is in Simon (2004a, Chapter 1.6); see also Ismail (2005b, Chapters 8 and 17).

Example 9.7.1 (Free case) Let $\mu = \frac{1}{2\pi}\,d\theta$; then for the moments, Verblunsky coefficients, and orthogonal polynomials we have, respectively,

$$\mu_n = \delta_{n0}, \quad \alpha_n \equiv 0, \quad \Phi_n = \phi_n = z^n, \quad n \in \mathbb{Z}_+.$$

In this case $\phi_n^* = 1$ for all n, so the Szegő function $D(z, dm) = 1$.

Example 9.7.2 (Bernstein–Szegő measures and polynomials) Let T be a positive trigonometric (Laurent) polynomial of degree n on $\mathbb{T}$. By the Fejér–Riesz theorem there is a unique algebraic polynomial p_n of degree n with a positive leading coefficient and all zeros inside $\mathbb{D}$ so that $T(\zeta) = |p_n(\zeta)|^2$. The measures $\mu = cT^{-1}(\zeta)\,d\theta$, $\zeta = e^{i\theta}$ constitute the class of **Bernstein–Szegő measures**, $c > 0$ is a normalizing constant,

$$c^{-1} = \frac{1}{2\pi}\int_{-\pi}^{\pi} \frac{d\theta}{|p_n(\zeta)|^2}.$$

The orthonormal polynomials and Verblunsky coefficients are

$$\phi_k(z,\mu) = c^{-1/2} z^{k-n} p_n(z), \quad \alpha_k(\mu) = 0, \quad k = n, n+1, \ldots.$$

Since $\phi_k^* = c^{-1/2} p_n^*$, $k \ge n$, we have $D(\mu) = \sqrt{c}\,(p_n^*)^{-1}$.

An important particular case is $p_1(z) = z - \bar{w}$, $w \in \mathbb{D}$. Now $c = 1 - |w|^2$,

$$\phi_k(z,\mu) = \frac{z^k - \bar{w} z^{k-1}}{\sqrt{1-|w|^2}}, \quad d\mu(\zeta) = \frac{1-|w|^2}{|1 - w\zeta|^2}\,dm, \quad D(z,\mu) = \frac{\sqrt{1-|w|^2}}{1 - wz}.$$

The Verblunsky coefficients are $\alpha_0 = w$, $\alpha_j = 0$ for $j \ge 1$. The moments $\mu_j = w^j$ for $j \ge 0$ and $\mu_j = \bar{w}^{|n|}$ for $n \le 0$.

The Bernstein–Szegő measures had already arisen in Szegő's work in the early 1920s (Szegő, 1920, 1921). Translated to the real line, they were studied by Bernstein about 10 years later.

Example 9.7.3 (Single nontrivial moment) This example goes back to Grenander and Szegő (1958, Section 5.3).

Let $\mu = |1-\zeta|^2 \frac{d\theta}{4\pi}$ and $\Phi_n(\mu)$ be monic orthogonal polynomials that satisfy

$$\int_{-\pi}^{\pi} \Phi_n(\zeta)\zeta^{-j}(2-\zeta-\zeta^{-1})\,d\theta = 0, \quad j = 0, 1, \ldots, n-1.$$

If

$$\Phi_n(z,\mu) = \sum_{k=0}^{n} f_{kn} z^k, \quad f_{nn} = 1,$$

we come to a simple boundary value problem for the second-order difference equation

$$2f_{kn} = f_{k-1,n} + f_{k+1,n}, \quad k = 0, 1, \ldots, n-1, \qquad f_{-1,n} = 0, \quad f_{nn} = 1,$$

so $f_{kn} = (k+1)(n+1)^{-1}$, and

$$\Phi_n(z,\mu) = \frac{1}{n+1}\sum_{k=0}^{n}(k+1)z^k, \quad \alpha_n(\mu) = -\frac{1}{n+2}, \quad n \in \mathbb{Z}_+.$$

By (9.2.4),

$$\|\Phi_n\|^2 = \prod_{k=0}^{n-1}(1-|\alpha_k|^2) = \frac{n+2}{2(n+1)}$$

so

$$\phi_n(z,\mu) = k_n \sum_{k=0}^{n}(k+1)z^k, \quad \phi_n^*(z,\mu) = k_n \sum_{k=0}^{n}(n-k+1)z^k, \quad k_n = \sqrt{\frac{2}{(n+1)(n+2)}},$$

and

$$D^{-1}(z,\mu) = \lim_{n\to\infty} \phi_n^*(z,\mu) = \sqrt{2}\sum_{k=0}^{\infty} z^k = \frac{\sqrt{2}}{1-z},$$

initially in the sense of Taylor coefficients, but then using the Szegő limit theorem, on all of $\mathbb{D}$. The Szegő function is $D(z,\mu) = (1-z)/\sqrt{2}$.

The general case $\mu = |1-r\zeta|^2 \frac{d\theta}{2\pi(1+r^2)}$, $0 < r \le 1$ can be handled in the same way (cf. Simon, 2004a, Example 1.6.4). For instance,

$$\alpha_n(\mu) = -\frac{r^{-1}-r}{r^{-n-2}-r^{n+2}}$$

so α_n decays exponentially,

$$\Phi_n(z,\mu) = \frac{1}{d_n^-}\sum_{k=0}^{n} d_k^- z^k, \quad d_k^- = \frac{r^{-k-1}-r^{k+1}}{r^{-1}-r},$$

the Szegő function is $D(z,\mu) = (1+r^2)^{-1/2}(1-rz)$.

Example 9.7.4 (Circular Jacobi polynomials) Let $\mu = w(\zeta)\,d\theta$ with

$$w(\zeta) = \frac{\Gamma^2(a+1)}{2\pi\Gamma(2a+1)}|1-\zeta|^{2a}, \quad a > -1$$

which for $a = 1$ is Example 9.7.3. Now the orthogonal polynomials are expressed in terms of the hypergeometric function

$$\phi_n(z,\mu) = \frac{(a)_n}{\sqrt{n!(2a+1)_n}}\,{}_2F_1(-n, a+1; -n+1-a; z), \quad (a)_n = a(a+1)\ldots(a+n-1),$$

and the Verblunsky coefficients are

$$\alpha_n(\mu) = -\frac{a}{n+a+1}, \qquad n \in \mathbb{Z}_+.$$

Example 9.7.5 (Rogers–Szegő polynomials) The example is from Szegő (1926) and the name comes from earlier consideration of Rogers (1894, 1895). Ismail (2005b) has a whole Chapter 17 on this example (see also Simon, 2004a, Example 1.6.5). This class of polynomials is parametrized by a number $q \in (0,1)$ (although the extension to $q \in \mathbb{D}$ is easy). The weight function is a "wrapped Gaussian." Let

$$q = e^{-a}, \quad a = \log\frac{1}{q} > 0.$$

The Gaussian measure on the real line of variance a is given by

$$d\nu_a(x) = (2\pi a)^{-1/2} e^{-x^2/2a}\,dx.$$

The wrapped Gaussian measure on $[-\pi,\pi]$ is defined by

$$\mu = \mu(q,\zeta) = v_q(\zeta)\,d\theta, \quad v_q(e^{i\theta}) = \frac{1}{\sqrt{2\pi a}}\sum_{j=-\infty}^{\infty} e^{-(\theta-2\pi j)^2/2a}. \tag{9.7.1}$$

It is a matter of direct calculation to find the moments $\mu_n = q^{n^2/2}$.

Identifying the orthogonal polynomials depends on the use of q-binomial coefficients defined by

$$\begin{bmatrix} n \\ j \end{bmatrix}_q = \frac{[n]_q}{[j]_q[n-j]_q}, \quad [n]_q = (1-q)(1-q^2)\cdots(1-q^n), \ [0]_q = 1.$$

The monic orthogonal with respect to the wrapped Gaussian measure (9.7.1) polynomials, known as the **Rogers–Szegő polynomials**, are

$$\Phi_n(z,\mu) = \sum_{j=0}^{n} (-1)^{n-j} \begin{bmatrix} n \\ j \end{bmatrix}_q q^{(n-j)/2} z^j, \tag{9.7.2}$$

so

$$\alpha_n(\mu) = (-1)^n q^{(n+1)/2}, \quad \|\Phi_n\|^2 = [n]_q.$$

The Szegő function is now

$$D(z,\mu) = \prod_{j=1}^{\infty}(1-q^j)^{1/2}\prod_{j=0}^{\infty}(1+zq^{j+1/2}).$$

An amazing property of zeros of the Rogers–Szegő polynomials (9.7.2) is proved in Mazel, Geronimo, and Hayes (1990): all of them have their zeros on the same circle $|z| = q^{1/2}$.

Example 9.7.6 (Geronimus measures and polynomials) This example is perhaps the most notable example of a measure outside the Szegő class. In this (and the next) examples it is more convenient to view a measure as one supported on the unit circle $\mathbb{T}$.

The Geronimus polynomials are those associated with constant Verblunsky coefficients $\alpha_n \equiv \alpha$, $\alpha \in \mathbb{D}\setminus\{0\}$. By Verblunsky's theorem (see Section 9.8) the corresponding measure μ_α, called the Geronimus measure, is uniquely determined. The measures and polynomials appeared in Geronimus (1977), and have been extensively studied over the past fifteen years (see Simon, 2004a, Example 1.6.12).

The Szegő recurrence relations (9.2.9) for orthonormal Geronimus polynomials and their reverse take the form

$$\begin{bmatrix}\phi_n(z)\\ \phi_n^*(z)\end{bmatrix} = T^n(z,\alpha)\begin{bmatrix}1\\1\end{bmatrix}, \quad T(z,\alpha) = \frac{1}{\sqrt{1-|\alpha|^2}}\begin{bmatrix} z & -\bar{\alpha}\\ -z\alpha & 1\end{bmatrix}. \tag{9.7.3}$$

It is not hard now to derive the expressions for Geronimus polynomials and their reverse. Denote by $r_{1,2}$ the eigenvalues of matrix T (cf. (9.7.3)), which are the roots of characteristic equation

$$r^2 - \frac{z+1}{\rho}r + z = 0, \quad \rho = \sqrt{1-|\alpha|^2},$$

so

$$r_{1,2}(z) = \frac{z+1\pm\sqrt{(z+1)^2-4\rho^2 z}}{2\rho} = \frac{z+1\pm\sqrt{(z-\zeta_\tau)(z-\zeta_\tau^{-1})}}{2\rho} \tag{9.7.4}$$

with $\zeta_\tau = e^{i\tau}$ and $\sin\frac{\tau}{2} = |\alpha|$, $0<\tau<\pi$, and the branch of the square root is taken so that $r_2(0) = 0$. It is clear that the spectrum of T depends only on $|\alpha|$. Define a circular arc Δ_τ closely related to T by

$$\Delta_\tau = \{\zeta = e^{it} : \tau \le t \le 2\pi - \tau\} \tag{9.7.5}$$

so

$$|r_2(z)| < 1 < |r_1(z)|, \quad z \in \mathbb{C}\setminus\Delta_\tau; \qquad |r_2(\zeta)| = |r_1(\zeta)| = 1, \quad \zeta\in\Delta_\tau,$$

and $r_1 = r_2$ only at the endpoints of Δ_τ. It follows from (9.7.3) that

$$\phi_n(z,\mu_\alpha) = \frac{z-\bar{\alpha}}{\rho}\frac{r_1^n - r_2^n}{r_1 - r_2} - z\frac{r_1^{n-1}-r_2^{n-1}}{r_1-r_2}, \tag{9.7.6}$$

$$\phi_n^*(z,\mu_\alpha) = \frac{1-\alpha z}{\rho}\frac{r_1^n - r_2^n}{r_1 - r_2} - z\frac{r_1^{n-1}-r_2^{n-1}}{r_1-r_2}. \tag{9.7.7}$$

There is another expression for Geronimus polynomials which holds on the arc Δ_τ. Indeed, for $e^{it} \in \Delta_\tau$,

$$r_{1,2}(e^{it}) = \frac{e^{i\frac{t}{2}}}{\rho}\left(\cos\frac{t}{2} \pm i\sqrt{\cos^2\frac{\tau}{2} - \cos^2\frac{t}{2}}\right) = e^{i\frac{t}{2}}(\cos\lambda \pm i\sin\lambda), \quad \cos\lambda = \frac{\cos\frac{t}{2}}{\cos\frac{\tau}{2}},$$

$0 \le \lambda \le \pi$, so one has

$$\phi_n(e^{it}, \mu_\alpha) = e^{in\frac{t}{2}}\left(\frac{e^{i\frac{t}{2}} - \bar\alpha e^{-i\frac{t}{2}}}{\rho}U_{n-1}(\cos\lambda) \quad U_{n-2}(\cos\lambda)\right),$$

where U_k are the Chebyshev polynomials of the second kind. In particular, there is a bound for Geronimus polynomials on Δ_τ,

$$|\phi_n(\zeta, \mu_\alpha)| \le C(\alpha)\min(n, v^{-1}(\zeta)), \quad v(z) = \sqrt{(z-\zeta_\tau)(z-\zeta_\tau^{-1})}, \quad n \in \mathbb{Z}_+,$$

and hence they are uniformly bounded inside the arc Δ_τ and

$$|\phi_n(e^{\pm i\tau})| = \left|\left(\frac{e^{i\frac{\tau}{2}} - \bar\alpha e^{-i\frac{\tau}{2}}}{\rho} \mp 1\right)n \pm 1\right|. \tag{9.7.8}$$

It is clear from the definition that the second kind measures and polynomials are also Geronimus measures and polynomials for the parameter $-\alpha$, so for $\psi_n(z, \mu_\alpha)$, $\psi_n^*(z, \mu_\alpha)$ the same formulas as (9.7.6) hold.

The Carathéodory function (9.2.18) can be now computed explicitly

$$F(z, \mu_\alpha) = 1 + \frac{z + 2\alpha z - 1 + \sqrt{(z-\zeta_\tau)(z-\zeta_\tau^{-1})}}{(1+\alpha)(\zeta_\beta - z)}, \quad \zeta_\beta = e^{i\beta} = \frac{1+\bar\alpha}{1+\alpha}. \tag{9.7.9}$$

Thus there is at most one mass point at $\zeta_\beta \notin \Delta_\tau$, and the actual value of this mass can be found from (9.2.23) and (9.7.9):

$$\mu_\alpha\{\zeta_\beta\} = \begin{cases} \frac{2}{|1+\alpha|^2}\left(\left|\alpha + \frac{1}{2}\right|^2 - \frac{1}{4}\right), & |\alpha + \frac{1}{2}| > \frac{1}{2}, \\ 0, & |\alpha + \frac{1}{2}| \le \frac{1}{2}. \end{cases} \tag{9.7.10}$$

As follows from (9.7.9), the measure μ_α is supported on Δ_τ along with a possible mass point at ζ_β, and $\mu_\alpha = \mu'_\alpha\, dm$ with

$$\mu'_\alpha(e^{it}) = \frac{1}{|1+\alpha|}\frac{\sqrt{\cos^2\frac{\tau}{2} - \cos^2\frac{t}{2}}}{\sin\frac{t-\beta}{2}}, \quad e^{it} \in \Delta_\tau.$$

Example 9.7.7 (Perturbed Geronimus measures) A measure μ is called a perturbed Geronimus measure if

$$\lim_{n\to\infty} \alpha_n(\mu) = \alpha, \quad 0 < |\alpha| < 1. \tag{9.7.11}$$

The following fundamental result about such measures is due to Geronimus (1941).

Theorem 9.7.8 *Let μ be a perturbed Geronimus measure. Then Δ_τ (cf. (9.7.5)) belongs to the support of μ, and the part of the support outside Δ_τ is at most a countable point set which can accumulate only to the endpoints of Δ_τ.*

Much more can be said about μ as long as some additional assumptions are imposed upon the rate of convergence in (9.7.11). The following results are in Golinskii, Nevai, and Van Assche (1995).

Theorem 9.7.9 *Let μ be a perturbed Geronimus measure.*

(i) *If $\{\alpha_n(\mu) - \alpha\} \in \ell^1$ then μ is absolutely continuous inside Δ_τ, and $1/\mu' \in L^\infty(\Delta)$ for any interior closed arc $\Delta \subset \Delta_\tau$.*
(ii) *If $\{\log n(\alpha_n(\mu) - \alpha)\} \in \ell^1$ then μ satisfies the Szegő condition for the arc*

$$\int_{\Delta_\tau} \frac{|\log \mu'(\zeta)|}{\sqrt{|\zeta - \zeta_\tau||\zeta - \zeta_\tau^{-1}|}}\, dm < \infty.$$

(iii) *If $\{n(\alpha_n(\mu) - \alpha)\} \in \ell^1$, then μ is absolutely continuous on the whole Δ_τ, and $\mu'(\zeta) \geq C|\zeta - \zeta_\tau||\zeta - \zeta_\tau^{-1}|$ a.e. on Δ_τ.*

The bounds for perturbed Geronimus polynomials are also available.

Theorem 9.7.10 *Let $\{\phi_n\}$ be perturbed Geronimus polynomials. If $\{\alpha_n(\mu) - \alpha\} \in \ell^1$, then*

$$\sup_n \max_{\zeta \in \Delta} |\phi_n(\zeta, \mu)| = C(\Delta) < \infty$$

for any interior closed arc $\Delta \subset \Delta_\tau$. If $\{n(\alpha_n(\mu) - \alpha)\} \in \ell^1$, then

$$\sup_n \max_{\zeta \in \Delta_\alpha} \frac{|\phi_n(\zeta, \mu)|}{n} < \infty.$$

Equation (9.7.8) shows that the latter bound is optimal.

The following result (Golinskii, 2000) provides a sufficient condition for the perturbed Geronimus measure to have finitely many mass points outside Δ_τ.

Theorem 9.7.11 *The portion of the support of μ outside Δ_τ is a finite set as long as $\{n(\alpha_n(\mu) - \alpha)\} \in \ell^1$.*

9.8 Modification of Measures

By Verblunsky's theorem each transformation in the class of nontrivial probability measures on $[-\pi, \pi]$ gives rise to a certain transformation in the space $\mathbb{D}^\infty$ of the Verblunsky coefficients and vice versa. We consider here the simplest such transformations when the explicit expressions are available. Again we will deal with measures on $\mathbb{T}$ rather than on $[-\pi, \pi]$.

Let S be a Borel transformation of $\mathbb{T}$ into itself. Such a transformation acts in the space of measures by $S\mu = \mu_S$, $\mu_S(E) = \mu(S^{-1}E)$. A key role is played by the change of variables formula

$$\int_{\mathbb{T}} h(\zeta)\, d\mu_S = \int_{\mathbb{T}} h(S(\zeta))\, d\mu.$$

9.8.1 Rotation of the Circle and Parameters

Let $\lambda \in \mathbb{T}$ and $S(\zeta) = \lambda\zeta$. It is clear that

$$\Phi_n(z, \mu_S) = \lambda^n \Phi_n(\bar{\lambda} z, \mu), \quad \alpha_n(\mu_S) = \lambda^{-n-1}\alpha_n(\mu)$$

for the monic orthogonal polynomials and Verblunsky coefficients, respectively. For the Carathéodory functions one has $F(z, \mu_S) = F(\bar{\lambda} z, \mu)$.

Conversely, the rotation of parameters leads to Aleksandrov measures $\{\mu_\lambda\}_{\lambda \in \mathbb{T}}$ with $\alpha_n(\mu_\lambda) = \lambda\alpha_n$. The second kind measure is included in the family with $\lambda = -1$. The rotation $\alpha_n \to \lambda\alpha_n$ can be viewed as a change of boundary conditions since (cf. (9.2.7))

$$\begin{bmatrix} \Phi_{n,\lambda}(z) \\ \bar{\lambda}\Phi^*_{n,\lambda}(z) \end{bmatrix} = T_{n+1}(z) \begin{bmatrix} 1 \\ \bar{\lambda} \end{bmatrix}, \tag{9.8.1}$$

where $\Phi_{n,\lambda}$ are monic orthogonal polynomials for μ_λ. Since the space of solutions of (9.2.6) is 2-dimensional, any solution can be written in terms of Φ and Ψ:

$$2\Phi_{n,\lambda}(z) = (1 + \bar{\lambda})\Phi_n(z) + (1 - \bar{\lambda})\Psi_n(z). \tag{9.8.2}$$

For the corresponding Carathéodory functions one has

$$F_\lambda(z) = \frac{(1 - \lambda) + (1 + \lambda)F(z)}{(1 + \lambda) + (1 - \lambda)F(z)}, \quad F_{-1}(z) = \frac{1}{F(z)}. \tag{9.8.3}$$

It is sometimes advisable to study spectral properties of the entire family of Aleksandrov measures. The following result is in Simon (2004a, Theorem 3.2.16).

Theorem 9.8.1 *Let the Lebesgue decomposition for Aleksandrov measures be*

$$d\mu_\lambda = w_\lambda(\zeta)\, dm + d\mu_{s,\lambda}.$$

Then

(i) μ_λ *have the same essential support, and* $\{\zeta : w_\lambda(\zeta) \neq 0\}$ *is a.e. independent of* λ.
(ii) *If* $\mathrm{supp}(\mu_1) \cap (\zeta_0, \zeta_1)$ *is a finite set, the same is true for* $\mathrm{supp}(\mu_\lambda) \cap (\zeta_0, \zeta_1)$ *for each* λ.
(iii) *The singular components* $\mu_{\lambda,s}$ *and* $\mu_{\lambda',s}$ *are mutually singular for* $\lambda \neq \lambda'$.

There is another important property of Aleksandrov measures, known as the "spectral averaging," which states that, roughly speaking, the average of μ_λ over λ is always the Lebesgue measure (Golinskii and Nevai, 2001). More precisely, for any Borel set $B \subset \mathbb{T}$,

$$\int_{\mathbb{T}} \mu_\zeta(B,)\, dm(\zeta) = m(B).$$

9.8.2 Sieved Measures and Polynomials

Let $N \geq 2$ be a positive integer, and $S(\zeta) = \zeta^N$. Now $\mu_S = \mu^{(N)}$ puts scaled copies of μ on each of the arcs $[\zeta_j, \zeta_{j+1}]$ with

$$\zeta_j = \exp(2\pi i j/N), \quad j = 0, 1, \ldots, N-1.$$

One can easily show that for Verblunsky coefficients,

$$\alpha_n(\mu^{(N)}) = \begin{cases} \alpha_r(\mu), & n = rN + N - 1, \\ 0, & \text{otherwise.} \end{cases}$$

For the monic orthogonal polynomials one has

$$\Phi_n(z, \mu^{(N)}) = z^k \Phi_r(z, \mu), \quad n = rN + k, \quad k = 0, 1, \ldots, N-1.$$

The Carathéodory function is $F(z, \mu^{(N)}) = F(z^N, \mu)$.

A process in this example is known as **sieving**, and these Φ_n are the sieved polynomials. They were systematically discussed in Badkov (1987) and Ismail and Li (1992b).

We complete with a particular example of the Al-Salam–Carlitz q-polynomials on the unit circle. Let

$$A_n(x) = \frac{U_n^{(-1)}(x; q)}{q^{\frac{n(n-1)}{4}} \sqrt{(q; q)_n}}$$

be orthonormal Al-Salam–Carlitz q-polynomials (see Section 6.7 of this volume). The orthogonality measure γ is concentrated on two sequences $\{\pm q^j\}$, which converge to zero, and is symmetric with respect to the origin: $\gamma\{q^j\} = \gamma\{-q^j\}$. The three-term recurrence relation is

$$xA_n(x) = a_{n+1}A_{n+1}(x) + a_n A_{n-1}(x), \quad a_n^2 = q^{n-1}(1 - q^n).$$

Going over first to the unit circle by the Szegő mapping theorem we end up with $\mu \in \mathcal{P}$ concentrating on a discrete point set $\{e^{\pm i\theta_j^{\pm}}\}$ with $\cos\theta_j^{\pm} = \pm q^j$, which has two limit points ± 1. The corresponding Verblunsky coefficients are

$$\alpha_{2k}(\mu) = 0, \quad \alpha_{2k+1} = 1 - 2q^{k+1}, \quad k \in \mathbb{Z}_+.$$

9.8.3 Inserting Point Mass

There is an interesting problem of comparing Verblunsky coefficients and orthogonal polynomials of two measures μ and ν. We consider here an obvious way of building ν from μ by adding a mass point (finitely many mass points). Such a transformation is known as the Jost–Kohn perturbation. Explicitly,

$$\nu = t\mu + (1-t)\sigma, \quad 0 < t < 1, \quad \sigma = \sum_{j=1}^{p} k_j \delta(\zeta_j) \tag{9.8.4}$$

is a finite linear combination of pure point masses adjusted so that σ is a probability measure. Jost–Kohn theory for OPUC appeared in Golinskii (1966), Geronimus (1961), Cachafeiro and Marcellán (1988, 1993), Marcellán and Maroni (1992), and Peherstorfer and Steinbauer (1999). In particular, the phenomenon discovered in Peherstorfer and Steinbauer (1999) says that it can happen that adding a point mass to a case with $\alpha_n(\mu) \to a$ can result in a ν obeying $\alpha_n(\nu) \to a' \neq a$.

The relation between the Carathéodory functions is simple:

$$F(z, \nu) = tF(z, \mu) + (1 - t) \sum_{j=1}^{p} k_j \frac{\zeta_j + z}{\zeta_j - z}.$$

For the case $p = 1$, $\nu = \nu(t, \zeta_1)$ the relation between orthogonal polynomials was obtained in Geronimus (1961, formula (3.30)),

$$\Phi_n(z, \nu) = \Phi_n(z, \mu) - \frac{s\Phi_n(\zeta_1, \mu)K_{n-1}(z, \zeta_1; \mu)}{1 + sK_{n-1}(\zeta_1, \zeta_1; \mu)}, \qquad s = \frac{t}{1 - t}, \tag{9.8.5}$$

where K_n is the Christoffel kernel (9.2.24). By using the complex conjugate of (9.2.28) we have for $z = 0$,

$$\alpha_n(\nu) - \alpha_n(\mu) = \frac{s\overline{\Phi_{n+1}(\zeta_1, \mu)\kappa_n}\phi_n(\zeta_1, \mu)}{1 + sK_n(\zeta_1, \zeta_1; \mu)}. \tag{9.8.6}$$

But

$$|\Phi_{n+1}(\zeta_1, \mu)| = \left|\zeta_1 \Phi_n(\zeta_1, \mu) - \bar{\alpha}_n(\mu)\Phi_n^*(\zeta_1, \mu)\right| \leq 2\,|\Phi_n(\zeta_1, \mu)|$$

so

$$|\alpha_n(\nu) - \alpha_n(\mu)| \leq \frac{2s|\phi_n(\zeta_1, \mu)|^2}{1 + sK_n(\zeta_1, \zeta_1; \mu)}. \tag{9.8.7}$$

Let us say that a class X of nontrivial probability measures on $\mathbb{T}$ is invariant with regard to addition of the mass points if $\mu \in X$ implies $\nu \in X$ for all $0 < t < 1$ and $\zeta_1 \in \mathbb{T}$. Clearly, both Szegő and Erdős classes are invariant (the addition of a mass point does not affect the absolutely continuous part of the measure). As a consequence of (9.8.7) and Theorem 9.3.18 one has a much more delicate result that the Nevai class is also invariant.

As far as the Rakhmanov class goes, the problem is still open. There is a partial result in this direction (Golinskii and Khrushchev, 2002) which claims that a proper subclass $\mathcal{R}_0 \subset \mathcal{R}$, which consists of measures $\mu \in \mathcal{R}$ with $\sup_n |\alpha_n(\mu)| < 1$, is invariant with regard to addition of the mass points.

9.8.4 Modification by a Rational Function

Let G be a rational function regular on $\mathbb{T}$ such that

$$\int_{-\pi}^{\pi} |G(\zeta)|^2 \, d\mu(\theta) = 1, \quad \zeta = e^{i\theta}.$$

Put $\nu = |G|^2\mu$, also known as the Christoffel–Bargmann perturbation.

We start with the unit circle analogue of the Christoffel formula.

Theorem 9.8.2 *Let $\{\phi_n\}$ be orthonormal with respect to μ, and G_{2m} be a polynomial of precise degree $2m$ such that*

$$\zeta^{-m} G_{2m}(\zeta) = |G_{2m}(\zeta)|, \quad |\zeta| = 1.$$

Let $\phi = \phi_{n+m}$. Define polynomials $\{\xi_n\}$ by

$$G_{2m}(z)\xi_n(z) \tag{9.8.8}$$
$$= \begin{vmatrix} \phi^*(z) & z\phi^*(z) & \cdots & z^{m-1}\phi^*(z) & \phi(z) & z\phi(z) & \cdots & z^m\phi(z) \\ \phi^*(z_1) & z_1\phi^*(z_1) & \cdots & z_1^{m-1}\phi^*(z_1) & \phi(z_1) & z_1\phi(z_1) & \cdots & z_1^m\phi(z_1) \\ \phi^*(z_2) & z_2\phi^*(z_2) & \cdots & z_2^{m-1}\phi^*(z_2) & \phi(z_2) & z_2\phi(z_2) & \cdots & z_2^m\phi(z_2) \\ \vdots & \vdots & & \vdots & \vdots & \vdots & & \vdots \\ \phi^*(z_{2m}) & z_{2m}\phi^*(z_{2m}) & \cdots & z_{2m}^{m-1}\phi^*(z_{2m}) & \phi(z_{2m}) & z_{2m}\phi(z_{2m}) & \cdots & z_{2m}^m\phi^*(z_{2m}) \end{vmatrix},$$

where $z_1, z_2, \ldots, z_{2m}$ are the zeros of G_{2m}.

For zeros of multiplicity r, $r > 1$, replace the corresponding rows in (9.8.8) by derivatives of order $0, 1, \ldots, r-1$ of the polynomials in the first row evaluated at that zero.

Then $\{\xi_n(z)\}$ are orthogonal with respect to $C\,|G_{2m}(\zeta)|\, d\mu(\theta)$.

A similar result holds when G goes in the denominator.

Theorem 9.8.3 *Let μ, $\{\phi_n\}$ be as in the above theorem. Let H_{2k} be a polynomial of precise degree $2k$ such that*

$$\zeta^{-k} H_{2k}(\zeta) = |H_{2k}(\zeta)| > 0, \qquad |\zeta| = 1,$$

and put $\phi = \phi_{n+k}$. Define a new system of polynomials $\{\eta_n\}$, $n = 2k, 2k+1, \ldots$ by

$$\eta_n(z) \tag{9.8.9}$$
$$= \begin{vmatrix} \phi^*(z) & z\phi^*(z) & \cdots & z^{k-1}\phi^*(z) & \phi(z) & z\phi(z) & \cdots & z^k\phi(z) \\ L_{w_1}(\phi^*) & L_{w_1}(z\phi^*) & \cdots & L_{w_1}(z^{k-1}\phi^*) & L_{w_1}(\phi) & L_{w_1}(z\phi) & \cdots & L_{w_1}(z^k\phi) \\ L_{w_2}(\phi^*) & L_{w_2}(z\phi^*) & \cdots & L_{w_2}(z^{k-1}\phi^*) & L_{w_2}(\phi) & L_{w_2}(z\phi) & \cdots & L_{w_2}(z^k\phi) \\ \vdots & \vdots & & \vdots & \vdots & \vdots & & \vdots \\ L_{w_{2k}}(\phi^*) & L_{w_{2k}}(z\phi^*) & \cdots & L_{w_{2k}}(z^{k-1}\phi^*) & L_{w_{2k}}(\phi) & L_{w_{2k}}(z\phi) & \cdots & L_{w_{2k}}(z^k\phi) \end{vmatrix},$$

where the zeros of H_{2k} are $w_1, w_2, \ldots, w_{2k}$, and we define

$$L_\beta(p) := \int_{-\pi}^{\pi} p(\zeta)\overline{\left(\frac{\zeta^k}{\zeta - \beta}\right)}\, d\mu(\theta), \quad \beta \notin \mathbb{T}.$$

For zeros of multiplicity h, $h > 1$, we replace the corresponding rows in the determinant (9.8.9) by

$$L_\beta^j(p) := \int_{-\pi}^{\pi} p(\zeta)\overline{\left(\frac{\zeta^k}{(\zeta - \beta)^j}\right)}\, d\mu(\theta), \quad j = 1, 2, \ldots, h$$

acting on the first row.

Under the above assumptions, for $n \geq 2k$, $\{\eta_n(z)\}$ *are orthogonal with respect to* $C\,|H_{2k}(\zeta)|^{-1}\, d\mu$.

A combination of Theorems 9.8.2 and 9.8.3 leads to the following result which covers the modification by a rational function.

Theorem 9.8.4 *Let* μ, $\{\phi_n(z)\}$, G_{2m}, H_{2k}, *and* $z_1, \ldots, z_{2m}$, $w_1, \ldots, w_{2k}$ *be as in Theorems* 9.8.2 *and* 9.8.3. *Let* ϕ *denote* ϕ_{n+m-k} *and* $s = m + k$. *For* $n \geq 2k$ *define* ψ_n *by*

$$G_{2m}(z)\psi_n(z) \tag{9.8.10}$$

$$= \begin{vmatrix} \phi^*(z) & z\phi^*(z) & \cdots & z^{s-1}\phi^*(z) & \phi(z) & z\phi(z) & \cdots & z^s\phi(z) \\ \phi^*(z_1) & z_1\phi^*(z_1) & \cdots & z_1^{s-1}\phi^*(z_1) & \phi(z_1) & z_1\phi(z_1) & \cdots & z_1^s\phi(z_1) \\ \phi^*(z_2) & z_2\phi^*(z_2) & \cdots & z_2^{s-1}\phi^*(z_2) & \phi(z_2) & z_2\phi(z_2) & \cdots & z_2^s\phi(z_2) \\ \vdots & \vdots & & \vdots & \vdots & \vdots & & \vdots \\ \phi^*(z_{2m}) & z_{2m}\phi^*(z_{2m}) & \cdots & z_{2m}^{s-1}\phi^*(z_{2m}) & \phi(z_{2m}) & z_{2m}\phi(z_{2m}) & \cdots & z_{2m}^s\phi(z_{2m}) \\ L_{w_1}(\phi^*) & L_{w_1}(z\phi^*) & \cdots & L_{w_1}(z^{s-1}\phi^*) & L_{w_1}(\phi) & L_{w_1}(z\phi) & \cdots & L_{w_1}(z^s\phi) \\ L_{w_2}(\phi^*) & L_{w_2}(z\phi^*) & \cdots & L_{z_2}(z^{s-1}\phi^*) & L_{w_2}(\phi) & L_{z_2}(z\phi) & \cdots & L_{w_2}(z^s) \\ \vdots & \vdots & & \vdots & \vdots & \vdots & & \vdots \\ L_{w_{2k}}(\phi^*) & L_{w_{2k}}(z\phi^*) & \cdots & L_{w_{2k}}(z^{s-1}\phi^*) & L_{w_{2k}}(\phi) & L_{w_{2k}}(z\phi) & \cdots & L_{w_{2k}}(z^s\phi) \end{vmatrix},$$

where we define

$$L_\beta(p) := \int_{-\pi}^{\pi} p(\zeta)\overline{\left(\frac{\zeta^s}{\zeta - \beta}\right)}\, d\mu(\theta), \quad \beta \notin \mathbb{T}.$$

For zeros of H_{2k} *of multiplicity* h, $h > 1$, *we replace the corresponding rows in the determinant* (9.8.10) *by*

$$L_\beta^j(p) := \int_{-\pi}^{\pi} p(\zeta)\overline{\left(\frac{\zeta^s}{(\zeta - \beta)^j}\right)}\, n d\mu(\theta), \quad j = 1, 2, \ldots, h$$

acting on the first row.

For zeros of G_{2m} *of multiplicity* h, $h > 1$, *we replace the corresponding row in the determinant* (9.8.10) *by the derivatives of order* $0, 1, 2, \ldots, h-1$ *of the polynomials in the first row, evaluated at that zero.* (*As usual,* $p_r^*(z) = z^r \bar{p}_r(z^{-1})$, *for* ψ_r *a polynomial of degree* r.)

Then $\{\psi_n\}$ *are orthogonal with respect to* $C|G_{2m}/H_{2k}|\, d\mu$ *on the unit circle.*

The results of this section are in Ismail and Ruedemann (1992), which contains explicit formulas for certain polynomials. For earlier partial results see Golinskii (1958), Mikaelyan (1978), and Godoy and Marcellán (1991).

9.8.5 Bessel Transformations and Schur Flows

Throughout the rest of the section we will view a nontrivial probability measure μ supported on $\mathbb{T}$. We define a family of measures which depends on parameter $t \ge 0$ by

$$\mu(\zeta,t) = C(t)e^{t(\zeta+\zeta^{-1})}\mu(\zeta,0), \quad C^{-1}(t) = \int_{\mathbb{T}} e^{t(\zeta+\zeta^{-1})}\, d\mu(\zeta,0) \tag{9.8.11}$$

is a normalizing factor. We refer to (9.8.11) as the **Bessel transformation** of the initial measure $\mu = \mu(\cdot,0)$. The main problem we deal with here is the dynamics of the corresponding orthogonal polynomials $\Phi_n(\cdot,t)$ and Verblunsky coefficients $\alpha_n(\mu(t)) = \alpha_n(t)$.

As far as the polynomials go, the following result is proved in Golinskii (2006).

Theorem 9.8.5 *The monic polynomials $\Phi_n(\cdot,t)$ orthogonal with respect to $\mu(t)$ (cf. (9.8.11)) satisfy the first-order differential equation*

$$\frac{d}{dt}\Phi_n(z,t) = \Phi_{n+1}(z,t) - (z+\overline{\alpha}_n(t)\alpha_{n-1}(t))\Phi_n(z,t) - (1-|\alpha_{n-1}(t)|^2)\Phi_{n-1}(z,t).$$

A comprehensive study of the asymptotic behavior of Verblunsky coefficients $\alpha_n(t)$ for each fixed n and $t \to \infty$ is accomplished in Simon (2007b). Moreover, in Simon (2007b) the asymptotics of the zeros $\{z_{j,n}(t)\}_{j=1}^n$ of Φ_n is examined, which yields the information about α_n via

$$\alpha_{n-1}(t) = (-1)^{n-1}\prod_{j=1}^{n}\overline{z_{j,n}(t)}. \tag{9.8.12}$$

The key tool is Theorem 9.1.2. As it turns out, the limit behavior of the α_n depends heavily on whether the point 1 belongs to the essential support of the initial measure $\mu(\zeta,0)$, that is, any punctured neighborhood of 1 has nonempty intersection with the support of μ, or not. The former case is rather simple, and here is the result.

Theorem 9.8.6 *Let* $1 \in \operatorname{supp}_{\mathrm{ess}}\mu$. *Then*

$$\lim_{t\to\infty} z_{j,n}(t) = 1 \ \textit{for all}\ n \in \mathbb{N},\ j = 1,2,\dots,n \quad \textit{implies} \quad \lim_{t\to\infty}\alpha_{n-1}(t) = (-1)^{n-1}.$$

The latter case is much more complicated, and a complete picture is available only for the case when μ is symmetric (and then so are all $\mu(t)$), and $\alpha_n(t)$ are real-valued functions. Now, there exists a unique open arc $\Gamma(\mu) = (\bar{\Theta},\Theta)$, $\Theta = \Theta(\mu)$ so that $\operatorname{Im}\Theta > 0$ and

(i) its endpoints $\bar{\Theta}$, Θ belong to the essential support of μ, and $1 \in \Gamma(\mu)$;
(ii) the portion of $\operatorname{supp}\mu$ on $\Gamma(\mu)$ is at most a countable set of mass points $\{\zeta_j\}_{j=1}^N$, $N \le \infty$, with no limit points inside Γ.

One can label ζ_j so that $\operatorname{Re}\zeta_1 \ge \operatorname{Re}\zeta_2 \ge \cdots$, and it is clear that this series of inequalities cannot have two equality signs in a row. Specifically, $\operatorname{Re}\zeta_n = \operatorname{Re}\zeta_{n+1}$ if and only if $\zeta_{n+1} = \bar{\zeta}_n$.

Theorem 9.8.7 *Suppose that* $1 \notin \operatorname{supp}_{\mathrm{ess}}\mu$, *and* $\Gamma(\mu)$ *has an infinity of mass points* ζ_j.

(i) *If* $1 \in \operatorname{supp}\mu$, *then* $1 = \operatorname{Re}\zeta_1 > \operatorname{Re}\zeta_2 = \operatorname{Re}\zeta_3 > \operatorname{Re}\zeta_4 = \operatorname{Re}\zeta_5 > \cdots$ *and*

$$\lim_{t\to\infty} \alpha_{2n}(t) = 1, \quad \lim_{t\to\infty} \alpha_{2n+1}(t) = -\operatorname{Re}\zeta_{2n+2}, \quad n \in \mathbb{Z}_+.$$

(ii) *If* $1 \notin \operatorname{supp}\mu$, *then* $1 > \operatorname{Re}\zeta_1 = \operatorname{Re}\zeta_2 > \operatorname{Re}\zeta_3 = \operatorname{Re}\zeta_4 > \cdots$ *and*

$$\lim_{t\to\infty} \alpha_{2n}(t) = \operatorname{Re}\zeta_{2n+1}, \quad \lim_{t\to\infty} \alpha_{2n+1}(t) = -1, \quad n \in \mathbb{Z}_+.$$

Theorem 9.8.8 *Suppose that* $1 \notin \operatorname{supp}_{\text{ess}}\mu$, *and* $\Gamma(\mu)$ *has* $N < \infty$ *mass points* ζ_j.

(i) *If* $N = 2m+1$, *then* $1 \in \operatorname{supp}\mu$,

$$1 = \operatorname{Re}\zeta_1 > \operatorname{Re}\zeta_2 = \operatorname{Re}\zeta_3 > \cdots > \operatorname{Re}\zeta_{N-1} = \operatorname{Re}\zeta_N > \operatorname{Re}\Theta(\mu)$$

and

$$\lim_{t\to\infty} \alpha_{2n}(t) = 1, \qquad n \in \mathbb{Z}_+,$$
$$\lim_{t\to\infty} \alpha_{2n+1}(t) = -\operatorname{Re}\zeta_{2n+2}, \quad n = 0, 1, \dots, m-1,$$
$$\lim_{t\to\infty} \alpha_{2n+1}(t) = -\operatorname{Re}\Theta(\mu), \quad n = m, m+1, \dots.$$

(ii) *If* $N = 2m$, *then* $1 \notin \operatorname{supp}\mu$,

$$1 > \operatorname{Re}\zeta_1 = \operatorname{Re}\zeta_2 > \operatorname{Re}\zeta_3 = \cdots > \operatorname{Re}\zeta_{N-1} = \operatorname{Re}\zeta_N > \operatorname{Re}\Theta(\mu)$$

and

$$\lim_{t\to\infty} \alpha_{2n+1}(t) = -1, \quad n \in \mathbb{Z}_+,$$
$$\lim_{t\to\infty} \alpha_{2n}(t) = -\operatorname{Re}\zeta_{2n+1}, \quad n = 0, 1, \dots, m-1,$$
$$\lim_{t\to\infty} \alpha_{2n}(t) = \operatorname{Re}\Theta(\mu), \quad n = m, m+1, \dots.$$

Some particular results for the general case are also obtained in Simon (2007b). For instance, an example of a measure μ with $\Theta(\mu) = i$ and no mass points in $\Gamma(\mu)$ is given for which $\alpha_0(t)$ has no limit as $t \to \infty$.

Note that the distinguished role of the point 1 is quite obvious: this is the only global maximum for the function $\operatorname{Re}\zeta$ on $\mathbb{T}$. If 1 is in the essential support, it attracts all zeros of all polynomials Φ_n. If 1 is an isolated mass point, it can attract only one zero by Theorem 9.4.4. The behavior of other zeros is in general rather chaotic.

One can think of the Bessel transformation (9.8.11) as the unit circle analogue of a Toda-type transformation from Theorem 2.5.3. Instead of Jacobi parameters and matrices the Verblunsky coefficients $\alpha_n(t)$ and CMV matrices $C(t)$ (9.5.1)–(9.5.3) appear on the central stage. So (9.8.11) plays the same role in the theory of discrete integrable systems as the Toda transformation. The result below is in Golinskii (2006).

Theorem 9.8.9 (Schur flows) *Let* $\mu(\cdot, t)$ *be a family of measures that depend on a real parameter* $t \geq 0$, *with Verblunsky coefficients* $\alpha_n(t)$. *The following three statements are equivalent:*

(i) *$\mu(\cdot, t)$ satisfy* (9.8.11);

(ii) *$\alpha_n(t)$ solve the system of differential-difference equations*

$$\frac{d}{dt}\alpha_n(t) = (1 - |\alpha_n(t)|^2)(\alpha_{n+1}(t) - \alpha_{n-1}(t)), \quad t > 0 \tag{9.8.13}$$

known as the Schur flow;

(iii) *the CMV matrices $C(t)$ satisfy the Lax equation*

$$\frac{d}{dt}C(t) = [A, C], \tag{9.8.14}$$

where $A(t)$ is an upper-triangular and tridiagonal matrix

$$A = \begin{pmatrix} \mathrm{Re}\,\overline{\alpha}_0 & \rho_0\overline{\Delta}_0 & \rho_0\rho_1 & & & \\ & -\mathrm{Re}\,\overline{\alpha}_1\alpha_0 & \rho_1\Delta_1 & \rho_1\rho_2 & & \\ & & -\mathrm{Re}\,\overline{\alpha}_2\alpha_1 & \rho_2\overline{\Delta}_2 & \rho_2\rho_3 & \\ & & & \ddots & \ddots & \ddots \end{pmatrix}, \tag{9.8.15}$$

where $\Delta_n = \alpha_{n+1}(t) - \alpha_{n-1}(t)$.

There is an equivalent form of (9.8.14):

$$\frac{d}{dt}C(t) = [B, C], \tag{9.8.16}$$

$$B = \frac{(C + C^*)_+ - (C + C^*)_-}{2} = \frac{1}{2}\begin{pmatrix} 0 & \rho_0\overline{\Delta}_0 & \rho_0\rho_1 & & & \\ -\rho_0\Delta_0 & 0 & \rho_1\Delta_1 & \rho_1\rho_2 & & \\ -\rho_0\rho_1 & -\rho_1\Delta_1 & 0 & \rho_2\overline{\Delta}_2 & \rho_2\rho_3 & \\ \vdots & \vdots & \vdots & \ddots & \ddots & \ddots \end{pmatrix} \tag{9.8.17}$$

$$= A - \frac{C + C^*}{2} = -B^*,$$

where we use the standard notation $X_\pm$ for the upper (lower) projection of a matrix X. The latter form of the Lax equation is closer to its counterpart in the Toda lattices setting.

Thereby, the solution of the initial–boundary value problem for the Schur flow (9.8.13) with arbitrary initial data

$$|\alpha_n(0)| < 1, \quad n \in \mathbb{Z}_+, \quad \alpha_{-1} \equiv -1 \tag{9.8.18}$$

amounts to a combination of the direct and inverse spectral problems for the unit circle (from Verblunsky coefficients to orthogonality measures and backwards) with (9.8.11) in between.

The Schur flow (9.8.13) appeared in Ablowitz and Ladik (1976a,b) under the name "discrete modified KdV equation," as a spatial discretization of the modified Korteweg–de Vries equation

$$\partial_t f = 6f^2\partial_x f - \partial_x^3 f.$$

The name "Schur flow" is suggested in Faybusovich and Gekhtman (1999), where the authors consider finite real Schur flows and suggest two more Lax equations based upon the Hessenberg matrix representation of the multiplication operator (see also Ammar and Gragg, 1994). In Mukaihira and Nakamura (2000, 2002) the Bessel modification of measures appeared, and a part of the results from Theorem 9.8.9 is proved. In Nenciu (2005) (see also Killip and Nenciu, 2005) the authors deal with the Poisson structure and Lax pairs for the Ablowitz–Ladik systems closely related to the Schur flows. The latter can also be viewed as the zero-curvature equation for the Szegő matrices (cf. Geronimo, Gesztesy, and Holden, 2005)

$$\frac{d}{dt}T_n(z,t) + T_n(z,t)W_n(z,t) - W_{n+1}(z,t)T_n(z,t) = 0,$$

$$W_n(z,t) := \begin{pmatrix} z+1-\alpha_{n-1}\overline{\alpha}_n & -\overline{\alpha}_n - \overline{\alpha}_{n-1}z^{-1} \\ -\alpha_{n-1}z - \alpha_n & 1-\overline{\alpha}_{n-1}\alpha_n + z^{-1} \end{pmatrix}.$$

It might be worth pointing out that some properties of Verblunsky coefficients for the Bessel transformed measures (such as the rate of decay) are inherited from those of the initial data (see Golinskii, 2006).

Theorem 9.8.10 *Let $\alpha_n(t)$ solve the Schur flow equations* (9.8.13). *Then*

(i) $\{\alpha_n(0)\} \in \ell^p$ *implies* $\{\alpha_n(t)\} \in \ell^p$ *for all* $t > 0$, $p = 1, 2$;
(ii) $|\alpha_n(0)| \le Ke^{-an}$ *implies* $|\alpha_n(t)| \le K(t)e^{-an}$ *for all* $t > 0$, $\alpha > 0$.

Because of the boundary condition $\alpha_{-1} \equiv -1$ the initial–boundary value problem (9.8.13)/(9.8.18) with zero initial conditions

$$\alpha_0(0) = \alpha_1(0) = \cdots = 0$$

has a nontrivial solution. We are now dealing with the Bessel transformation of the Lebesgue measure

$$\mu(\zeta,t) = C(t)e^{t(\zeta+\zeta^{-1})}\,dm,$$

called the **modified Bessel measures** on the unit circle. Denote by $\beta_n(t)$ the Verblunsky coefficients of $\mu(\cdot,t)$, which are clearly real. The corresponding system of orthogonal polynomials has arisen from studies of the length of longest increasing subsequences of random words (Baik, Deift, and Johansson, 1999) and matrix models (Periwal and Shevitz, 1990).

Note first that the normalizing constant $C(t)$ can be easily computed

$$C^{-1}(t) = \int_{\mathbb{T}} e^{t(\zeta+\zeta^{-1})}\,dm = \frac{1}{2\pi}\int_0^{2\pi} e^{2t\cos x}\,dx = \sum_{n=0}^{\infty}\frac{t^{2n}}{(n!)^2} = I_0(2t),$$

where I_k is the modified Bessel function of order k. Similarly, for the moments of the measure we have

$$\mu_p(t) = \int_{\mathbb{T}} \zeta^{-p}\,d\mu(\zeta,t) = \frac{I_p(2t)}{I_0(2t)}, \quad p \in \mathbb{Z}_+, \quad \mu_{-p} = \mu_p. \tag{9.8.19}$$

The explicit expression for Verblunsky coefficients as a ratio of two determinants follows from (9.1.7) with $z = 0$ and (9.8.19),

$$\beta_n(t) = (-1)^n \frac{\det(I_{k-j-1}(2t))_{0\le k,j\le n}}{\det(I_{k-j}(2t))_{0\le k,j\le n}}, \quad n \in \mathbb{Z}_+. \tag{9.8.20}$$

There is an important feature of the modified Bessel measures proved in Periwal and Shevitz (1990).

Theorem 9.8.11 (Periwal–Shevitz) *The Verblunsky coefficients $\beta_n(t)$ for the modified Bessel measures satisfy a form of the discrete Painlevé II equation*

$$-(n+1)\frac{\beta_n(t)}{t(1-\beta_n^2(t))} = \beta_{n+1}(t) + \beta_{n-1}(t), \quad n \in \mathbb{Z}_+, \tag{9.8.21}$$

with $\beta_{-1} = -1, \beta_0 = I_1(2t)/I_0(2t)$.

There are also differential relations satisfied by modified Bessel polynomials, their leading coefficients, and Verblunsky coefficients, specific for this particular case. For instance (see Ismail, 2005b, Lemma 8.3.6),

$$\frac{2}{\kappa_n(t)}\frac{d}{dt}\kappa_n(t) = \frac{I_1(t)}{I_0(t)} + \alpha_n(t)\alpha_{n-1}(t),$$
$$\frac{d}{dt}\alpha_n(t) = \frac{I_1(t)}{I_0(t)} + \alpha_{n+1}(t) - (1 - |\alpha_n(t)|^2)\alpha_{n-1}(t).$$

Concerning the long-time behavior of Verblunsky coefficients, the following result is proved in Simon (2007b).

Theorem 9.8.12 *Let $\mu(\cdot, t)$ be the Bessel transformation* (9.8.11). *Suppose that $\mu(\zeta, 0) = w(\zeta)dm$, w is a positive and continuous function on $\mathbb{T}$. Then for the Verblunsky coefficients one has*

$$(-1)^n\alpha_n(t) = 1 - \frac{n+1}{4t} + O\left(\frac{1}{t}\right), \quad t \to \infty. \tag{9.8.22}$$

In particular, (9.8.22) holds for $\beta_n(t)$. It might be a challenging problem to find an asymptotic series expansion for β_n from (9.8.20) and the expansion for the modified Bessel function

$$I_k(t) \simeq \frac{e^t}{\sqrt{2\pi t}} \sum_{j=0}^{\infty} (-1)^j \frac{(4k^2 - 1^2)\cdots(4k^2 - (2j-1)^2)}{j!(8t)^j}, \quad t \to \infty.$$

10

Zeros of Orthogonal Polynomials

Andrea Laforgia and Martin E. Muldoon

10.1 Introduction

We begin by outlining some methods for getting information on zeros of orthogonal polynomials. Some of the main ones depend on the orthogonality measure, the recurrence relation and, if available, the differential equation for the polynomials.

In Section 10.2 we give results obtained by these methods for general classes of orthogonal polynomials. Sections 10.3 to 10.8 give specific applications to, and special results for, Jacobi, ultraspherical, Legendre, Laguerre, Hermite, and other polynomials.

Other chapters contain material on zeros. Zeros of Bessel polynomials are discussed in Section 3.13.

In some classical cases, it is convenient to approximate zeros of orthogonal polynomials by zeros of Bessel functions or of Airy functions. These are discussed in Section 1.2. For $\nu > -1$, we use $j_{\nu k}$ or $j_{\nu,k}$ for the kth positive zero of J_ν. We use i_k for the kth positive zero of the Airy function A (1.2.35) in Szegő's notation (Szegő, [1939] 1975, pp. 18–19). We use a_k for the kth positive zero of $\mathrm{Ai}(-x)$ (Olver et al., 2010, ch. 9). We have $\mathrm{Ai}(x) = 3^{1/3}\pi^{-1}A(-3^{1/3}x)$ and, for the zeros, $a_k = 3^{-1/3}i_k$.

10.2 General Results on Zeros

10.2.1 Using the Orthogonality Measure

Many properties of zeros of orthogonal polynomials can be obtained from a knowledge of the orthogonality measure. Some of these are outlined in Sections 2.1, 2.2 and, occasionally, in other parts of Chapter 2.

As was pointed out in Theorems 2.1.3 and 2.1.4, the zeros of p_n are simple and occur within any closed interval containing the support of the measure of orthogonality. Some typical additional questions relate to the spacing of the zeros and to their behavior (monotonicity, convexity, etc.) with respect to a parameter.

Concerning the distance between consecutive zeros we have the following theorem (Szegő, [1939] 1975, Thm. 6.11.1):

Theorem 10.2.1 *Let $w(x) \geq \mu > 0$ be a weight function on the finite interval $[a, b]$. Let*

$$x_\nu = \frac{1}{2}(a+b) + \frac{1}{2}(b-a)\cos\theta_\nu, \quad 0 < \theta_\nu < \pi, \ \nu = 1, 2, \ldots, n \tag{10.2.1}$$

be the zeros, in decreasing order, of the corresponding p_n. Then, with $\theta_0 = 0$, $\theta_{n+1} = \pi$, we have

$$\theta_{\nu+1} - \theta_\nu < K\frac{\log n}{n}, \quad \nu = 0, 1, 2, \ldots, n, \tag{10.2.2}$$

where the constant K depends only on μ, a, and b.

See Szegő ([1939] 1975, §6.11) for various related results.

The following result on variation of the zeros with a parameter is an extension of a useful theorem of A. Markov (Szegő, [1939] 1975, Theorem 6.12.1) which is stated as Freud (1971, Problem 15, Chapter III) and proved in Ismail (2005b, pp. 204–205).

Theorem 10.2.2 *Let $\{p_n(x;\tau)\}$ be orthogonal polynomials with respect to*

$$d\alpha(x;\tau) = \rho(x;\tau)\, d\alpha(x) \tag{10.2.3}$$

on an interval $I = (a, b)$ and assume that, for each $x \in I$, $\rho(x;\tau)$ is positive and has a continuous first derivative with respect to τ, $\tau \in T = (\tau_1, \tau_2)$. Furthermore, assume that for each $j = 0, 1, \ldots, 2n-1$, the integral

$$\int_a^b x^j \rho_\tau(x;\tau)\, d\alpha(x)$$

converges uniformly for τ in every compact subinterval of T. Let $\partial\{\ln\rho(x;\tau)\}/\partial\tau$ be an increasing (decreasing) function of x, $x \in I$. Then the zeros of $p_n(x;\tau)$ are increasing (decreasing) functions of τ, $\tau \subset T$.

The reference Lubinsky (1994) includes a survey of properties of zeros of orthogonal polynomials associated with the weight function $\exp(-|x|^\lambda)$.

In the sections that follow, we give a number of specific results where properties of zeros of orthogonal polynomials are deduced from properties of the corresponding measure.

10.2.2 Recurrence Relation and Tridiagonal Matrices

Let the sequence of monic polynomials $\{P_n(x)\}$ satisfy (2.1.14) and (2.1.15). The determinant representation (2.1.22) shows that the zeros of s_n are the eigenvalues of the tridiagonal matrix

$$H_n = \left[h_{ij}\right] = \begin{bmatrix} \alpha_0 & 1 & 0 & 0 & \cdots \\ \beta_1 & \alpha_1 & 1 & 0 & \cdots \\ 0 & \beta_2 & \alpha_2 & 1 & \cdots \\ \cdots & \cdots & \cdots & \cdots & \cdots \\ \cdots & \cdots & 0 & \beta_{n-1} & \alpha_{n-1} \end{bmatrix}. \tag{10.2.4}$$

This is a useful characterization of the zeros including applications to their numerical approximation (Gil, Segura, and Temme, 2007, §7.2).

10.2.3 The Hellmann–Feynman Theorem

The **Hellmann–Feynman theorem** provides a formula for the derivative of an eigenvalue with respect to a parameter. Let S_ν be an inner product space with an inner product $\langle .,.\rangle_\nu$. The inner product depends on a parameter ν which is assumed to vary continuously in an open interval (a,b). We assume that there is a fixed set (independent of ν) which is dense in S_ν for all $\nu \in (a,b)$. The following version of the Hellmann–Feynman theorem was proved in Ismail and Zhang (1988, §2). See also Ismail (2005b, Theorem 7.3.1).

Theorem 10.2.3 *Let H_ν be a symmetric operator defined on S_ν and let ψ_ν be an eigenfunction of H_ν corresponding to an eigenvalue λ_ν. Furthermore, assume that*

$$\lim_{\mu\to\nu} \left\langle \psi_\mu, \psi_\nu \right\rangle_\nu = \langle \psi_\nu, \psi_\nu \rangle_\nu, \tag{10.2.5}$$

and that

$$\lim_{\mu\to\nu} \left\langle \frac{H_\mu - H_\nu}{\mu - \nu} \psi_\mu, \psi_\nu \right\rangle_\nu \text{ exists.} \tag{10.2.6}$$

Define the action of $\partial H_\nu / \partial \nu$ on the eigenspaces by

$$\left\langle \frac{\partial H_\nu}{\partial \nu} \psi_\nu, \psi_\nu \right\rangle_\nu := \lim_{\mu\to\nu} \left\langle \frac{H_\mu - H_\nu}{\mu - \nu} \psi_\mu, \psi_\nu \right\rangle_\nu . \tag{10.2.7}$$

Then $d\lambda_\nu / d\nu$ exists for $\nu \in I$ and

$$\frac{d\lambda_\nu}{d\nu} = \frac{\left\langle \frac{\partial H_\nu}{\partial \nu} \psi_\nu, \psi_\nu \right\rangle_\nu}{\langle \psi_\nu, \psi_\nu \rangle_\nu}. \tag{10.2.8}$$

As an immediate consequence of Theorem 10.2.3, we have the following corollary (Ismail, 2005b, Corollary 7.3.2):

Corollary 10.2.4 *If $\partial H_\nu / \partial \nu$ is positive (negative) definite then the eigenvalues of H_ν increase (decrease) with ν.*

Theorem 10.2.3 has the following application to the case where H_ν is a tridiagonal matrix operator.

Theorem 10.2.5 (Ismail, 1987, Thm. 4) *Consider the class of polynomials $\{h_n(x)\}$ generated by*

$$\begin{aligned} h_0(x) = 1, \quad h_1(x) = x a_1(\tau), \\ x a_n(\tau) h_n(x) = h_{n+1}(x) + h_{n-1}(x), \quad n = 1, 2, \ldots, N-1, \end{aligned} \tag{10.2.9}$$

where $\{a_n(\tau)\}$ is a sequence of positive numbers for each τ in an interval T. The polynomials $h_n(x)$ of odd (even) degrees are odd (even) functions. The positive zeros of $h_n(x)$ are increas-

ing (decreasing) differentiable functions of τ, $\tau \in T$, if $a_n(\tau)$ is a decreasing (increasing) differentiable function of τ, $\tau \in T$, $0 \leq n < N$. Moreover, if λ is a positive zero of h_N then

$$\frac{1}{\lambda}\frac{d\lambda}{d\tau} = -\frac{\sum_{n=0}^{N-1} a_n'(\tau)h_n^2(\lambda)}{\sum_{n=0}^{N-1} a_n(\tau)h_n^2(\lambda)}. \tag{10.2.10}$$

10.2.4 Chain Sequences

The concept of a **chain sequence** is another useful tool in the study of the location of zeros. It is defined as follows:

Definition 10.2.6 A (finite or infinite) sequence $\{c_n\colon n = 1,2,\ldots,N\}$, $N \leq \infty$ is called a chain sequence if there exists another sequence $\{g_n\colon n = 0,1,2,\ldots,N\}$ such that

$$c_n = g_n(1-g_{n-1}), \quad n > 0, \text{ with } 0 \leq g_0 < 1 \text{ and } 0 < g_n < 1,\ n = 1,2\ldots,N.$$

The idea of a chain sequence is due to Wall and Wetzel (1944). We follow the definition of Chihara (1962, 1978). Some examples of (infinite) chain sequences are given by $c_n = 1/4$, $n = 1,2,\ldots$ and $c_1 = 1/2$, $c_n = 1/4$, $n = 2,3,\ldots$. It is notable that if $\{c_n\}$ is a chain sequence, then so is $\{d_n\}$, where $0 < d_n \leq c_n$ for all n (Ismail, 2005b, Theorem 7.2.2). A positive constant sequence $\{c\}_1^{N-1}$ is a chain sequence if and only if $0 < c \leq 1/[4\cos^2(\pi/(N+1))]$. For an infinite sequence, this becomes $c \leq 1/4$ (Ismail, 2005b, Theorem 7.2.6).

Let A_N be a symmetric tridiagonal matrix

$$A_N = \left[h_{ij}\right] = \begin{bmatrix} b_0 & a_1 & 0 & 0 & \cdots \\ a_1 & b_1 & a_2 & 0 & \cdots \\ 0 & a_2 & b_2 & a_3 & \cdots \\ \cdots & \cdots & \cdots & \cdots & \cdots \\ \cdots & \cdots & 0 & a_{N-1} & b_{N-1} \end{bmatrix}. \tag{10.2.11}$$

Then A_n is positive definite if and only if (i) $b_j > 0$ for $0 \leq j < N$ and (ii) $\{a_j^2/(b_jb_{j-1})\colon 1 \leq j < N\}$ is a chain sequence (Ismail, 2005b, Theorem 7.2.1).

The following result follows from Chihara (1978, Ch. 4); see Ismail (2005b, Theorem 7.2.3).

Theorem 10.2.7 *Let a sequence of polynomials be given by*

$$p_{-1} = 0, \quad p_0 = 1, \quad a_np_n = (x-b_{n-1})\,p_{n-1} - a_{n-1}p_{n-2}, \qquad n = 1,2,\ldots,N,$$

where $b_j > 0$, $a_j \neq 0$, $0 \leq j < N$. Then the zeros of p_N belong to (a,b) if and only if

(i) *$b_j \in (a,b)$, for $0 \leq j < N$ and*
(ii) *the sequence*

$$\frac{a_j^2}{\left(b_j - x\right)\left(b_{j-1} - x\right)}, \qquad 0 < j < N$$

is a chain sequence at $x = a$ and $x = b$.

The following theorem gives upper and lower bounds for zeros of certain orthogonal polynomials.

Theorem 10.2.8 *Let $\{p_n(x)\colon 1 \le n < N\}$ be a sequence of real polynomials satisfying $p_0(x) = 1$, $p_1(x) = x - \alpha_0$, and*

$$xp_n(x) = p_{n+1}(x) + \alpha_n p_n(x) + \beta_n p_{n-1}(x), \quad n > 0, \tag{10.2.12}$$

with $\beta_n > 0$, $0 < n < N$, and let $\{c_n\}$ be a chain sequence. Then the zeros of p_N lie in the interval between $A := \min\{y_n\colon 0 < n < N\}$ and $B := \max\{x_n\colon 0 < n < N\}$ where x_n, y_n are the roots of the equation

$$(x - \alpha_n)(x - \alpha_{n-1})c_n = \beta_n, \tag{10.2.13}$$

that is,

$$x_n, y_n = \frac{1}{2}(\alpha_n + \alpha_{n-1}) \pm \frac{1}{2}\sqrt{(\alpha_n - \alpha_{n-1})^2 + 4\beta_n/c_n}\,. \tag{10.2.14}$$

Theorem 10.2.8 is due to Ismail and Li (1992a).

10.2.5 Electrostatic Interpretations and Related Matters

According to Theorem 2.9.3, under certain conditions, the zeros of a polynomial satisfying a second-order linear differential equation may be identified as equilibrium positions of n unit charges under the influence of an external logarithmic potential related to the coefficients in the differential equation. This characterization may be used to obtain bounds and other properties for the zeros. See Forrester and Rogers (1986); Ismail (2000a,b); Valent and Van Assche (1995); Van Assche (1993); Van Deun (2007), and references. This is closely related to the subject of functional equations satisfied by zeros of some orthogonal polynomials, discussed following Theorem 2.1.4.

The following theorem generalizes a result of Stieltjes for the Jacobi polynomials.

Theorem 10.2.9 (Muldoon, 1993) *Let $y(x,\alpha)$ be a polynomial solution of*

$$y'' + P(x,\alpha)y' + Q(x,\alpha)y = 0, \tag{10.2.15}$$

where P and Q are meromorphic in x for each α. We suppose in addition that a polynomial solution $y(x,\alpha)$ has simple zeros $x_1(\alpha), \ldots, x_n(\alpha)$, none of which coincides with a singularity of P or Q. We suppose that $P(x,\alpha)$ is a decreasing function of x for each α and a decreasing (increasing) function of α for each x. Then each zero of $x_k(\alpha)$ decreases (increases) as α increases.

The usual monotonicity properties of zeros of Jacobi, ultraspherical, and Laguerre polynomials (Szegő, [1939] 1975, §6.21) with respect to parameters follow from this theorem.

10.2.6 The Sturm Comparison Theorem

The **Sturm comparison theorem** (Szegő, [1939] 1975, §1.82) is one of the main tools used in connection with zeros of orthogonal polynomials satisfying a second-order differential equation. The main idea is that if y, Y satisfy differential equations $y'' + fy = 0$, $Y'' + FY = 0$ where f and F are continuous and $F > f$ on an interval (a, b), then $(y'Y - yY')' = (F - f)Yy$ so

$$[y'Y - yY']_x^z = \int_x^z (F - f)yY\,dx. \tag{10.2.16}$$

Thus $y'Y - yY'$ is increasing in any subinterval of (a, b) where Y and y have the same sign. Furthermore, $Y'' + FY = 0$ is said to be a **Sturm majorant** of $y'' + fy = 0$.

(i) **In particular, Y has a zero between every pair of consecutive zeros of y and if y, Y have a common zero at a, then the next zero of Y to the right of a will occur before the next zero of y.**

This can be used as a tool to generate upper and lower bounds for zeros of y in terms of zeros of Y, and vice versa.

(ii) **Similarly, if $y'' + L^2y = 0$, then the spacing between successive zeros of Y is $\leq \pi/L$.**

A useful result, following from the above, is the "Sturm convexity theorem" (Szegő, [1939] 1975, Theorem 1.82.2):

(iii) **If ϕ is continuous and decreasing (increasing) in an interval I, the consecutive zeros x_1, x_2, x_3 of a nontrivial solution of $y'' + \phi(x)y = 0$ in I satisfy $x_2 - x_1 < (>)\, x_3 - x_2$. The result remains true even under the less restrictive condition $\phi(\xi) > (<)\, \phi(x_2) > (<)\, \phi(\eta)$ for $x_1 < \xi < x_2 < \eta < x_3$.**

The first part of (iii) is a consequence of a more general result formulated by E. Makai (1952):

(iv) **Let x_1, x_2, x_3 be successive zeros of y. If ϕ is decreasing, and if we rotate the arch of the graph of y between x_1 and x_2 through $180°$ about x_2, the resulting arch will be contained entirely within the arch joining x_2 to x_3.**

These and similar results are given in a simple form in Laforgia and Muldoon (1986).

A consequence of the above discussion is that if $f < 0$ in (a, b), then a nontrivial solution of $y'' + f(x)y = 0$ cannot have more than one zero in (a, b). This is achieved by a comparison with $y'' = 0$.

10.2.7 Other Results Using Differential Equations

Volkmer (2008) considers properties of zeros of orthogonal polynomials that converge to eigenvalues of certain differential equations including the Mathieu, Whittaker–Hill, and Lamé equations.

10.2.8 Some Methods Originating with Laguerre

The following result due to Laguerre is proved in Pólya and Szegő (1976, Problem 118 in Part 5, Chapter 2) and Szegő ([1939] 1975, §6.2).

Theorem 10.2.10 *Let f be a polynomial of exact degree n and x_0 one of its simple zeros. Then any circle through the points*

$$x_0 \quad \text{and} \quad x_0' = x_0 - 2(n-1)f'(x_0)/f''(x_0)$$

contains some zeros of f in both domains bounded by it, unless all the zeros lie on the circumference of this circle. The same is true if $f''(x_0) = 0$ and a straight line replaces the circle.

This theorem can be used to get bounds for the zeros in certain cases.

Another simple method, due to Laguerre (Szegő, [1939] 1975, p. 120) is based on the result that if f is a polynomial with n real and distinct zeros, and x_0 is one such zero, then

$$3(n-2)\left\{f''(x_0)^2\right\} - 4(n-1)f'(x_0)f'''(x_0) \geq 0. \tag{10.2.17}$$

In certain special cases this enables us to assert that the zeros of f satisfy a certain inequality (Szegő, [1939] 1975, p. 119). Various extensions have been found. For example, it is known (Foster and Krasikov, 2002; Jensen, 1913) that

$$\sum_{j=0}^{2m}(-1)^{m+j}\binom{2m}{j} f^{(j)} f^{(2m-j)} \geq 0, \quad m = 1, 2, \ldots \tag{10.2.18}$$

at the zeros of f, where f belongs to a class of functions that includes the polynomials with exclusively real zeros. This has been extended recently (Nikolov and Uluchev, 2004) to

$$\sum_{j=0}^{2m}(-1)^{m+j}\binom{2m}{j}\frac{(n-j)!(n-2m+j)!}{(n-m)!(n-2m)!} f^{(j)} f^{(2m-j)} \geq 0, \quad m = 1, 2, \ldots. \tag{10.2.19}$$

For applications, see, for example, Dimitrov and Nikolov (2010) and Krasikov (2003, 2007).

10.2.9 Asymptotic Formulas and Expansions

There is a method due to Tricomi for deriving an asymptotic expansion of the zeros of a function from the asymptotic expansion of the function; see Tricomi (1947) or Tricomi (1954, p. 151). It has been applied successfully by Gatteschi (1949a,b, 1950, 1972) to zeros of some of the classical polynomials.

The method of Liouville–Steklov uses the differential equation satisfied by a sequence of orthogonal polynomials to find an asymptotic expansion for the polynomials in terms of Bessel functions (Szegő, [1939] 1975, §8.61). These so-called Hilb-type formulas can then be used to find asymptotic formulas and expansions for the zeros. L. Gatteschi and others (Baratella and Gatteschi, 1988; Gatteschi, 1952, 1967/1968) have provided several asymptotic formulas in this way.

10.2.10 Zeros of General Polynomials

In many special cases the weight functions depend on parameters and the corresponding polynomials are orthogonal only for certain values of these parameters. This raises the question of the location and behavior of the (generally complex) zeros in these cases. In some of the following sections, we describe some of what is known in this area for the Jacobi, ultraspherical, and Laguerre polynomials.

10.3 Jacobi Polynomials

For $\alpha,\beta > -1$, the zeros of the Jacobi polynomial $P_n^{(\alpha,\beta)}$ lie in $(-1,1)$ (Theorem 2.1.4). We denote them by $x_k = x_{nk}(\alpha,\beta)$, $k = 1,\ldots,n$ in decreasing order, so

$$1 > x_{n1}(\alpha,\beta) > x_{n2}(\alpha,\beta) > \cdots > x_{nn}(\alpha,\beta) > -1, \quad \alpha,\beta > -1.$$

For $\alpha,\beta > -1$, the θ-zeros of $P_n^{(\alpha,\beta)}(\cos\theta)$ lie in the interval $(0,\pi)$. We denote these zeros in increasing order by $\theta_k = \theta_{nk}(\alpha,\beta)$, $k = 1,\ldots,n$. Clearly, $x_k = \cos\theta_k$.

As shown in Szegő ([1939] 1975, §6.2), the largest zero has the lower bound

$$x_{n1}(\alpha,\beta) > \frac{\beta - \alpha + 2n - 2}{\beta + \alpha + 2n}. \tag{10.3.1}$$

This follows from Theorem 10.2.10.

Theorem 10.2.2 may be used (Szegő, [1939] 1975, p. 121) to give one of the proofs (another, due to Stieltjes, is related to the electrostatic interpretation in Section 10.2.5) that for $\alpha,\beta > -1$, the zeros of a Jacobi polynomial $P_n^{(\alpha,\beta)}$ increase as β increases and decrease as α increases. This is because for Jacobi polynomials, $\rho(x;\alpha,\beta) = (1-x)^\alpha(1+x)^\beta$ and $\alpha(x) = x$, and hence

$$\frac{\partial \ln \rho(x;\alpha,\beta)}{\partial\beta} = \ln(1+x),$$

which increases with x, $x > -1$. The monotonicity in α follows similarly.

Let α, β be fixed numbers > -1. Dimitrov and Rodrigues (2002) point out that it is a consequence of Markov's theorem (Theorem 10.2.2) that the zeros of $P_n^{(a,b)}$ are larger or smaller than those of $P_n^{(\alpha,\beta)}$ according to whether the vector $(\alpha,\beta) - (a,b)$ is in the second or fourth quadrant. They show that the inequalities $a > \alpha$, $b < \beta$ are necessary in order that all the zeros of all the polynomials $P_n^{(\alpha,\beta)}$ precede the corresponding zeros of $P_n^{(a,b)}$. They use the Routh–Hurwitz criterion (Gantmacher, 1959, Ch. 15; Marden, 1966, Ch. 9) among other things.

Dimitrov and Rafaeli (2007) investigate the question of how fast the functions $1 - x_{nk}(\alpha,\beta)$ decrease as β increases. Using Sturmian methods (among other things), they prove that, with

$$f_n(\alpha,\beta) = 2n^2 + 2n(\alpha+\beta+1) + (\alpha+1)(\beta+1),$$

the products

$$t_{nk}(\alpha,\beta) := f_n(\alpha,\beta)(1 - x_{nk}(\alpha,\beta))$$

are increasing functions of β and that, for any fixed $\alpha > -1$, $f_n(\alpha,\beta)$ is the asymptotically extremal (with respect to n) function of β that forces the products $t_{nk}(\alpha,\beta)$ to increase.

An elementary consequence of the monotonicity of the zeros with respect to α and β is the set of inequalities (Szegő, [1939] 1975, Theorem 6.21.2)

$$\frac{k-\frac{1}{2}}{n+\frac{1}{2}}\pi < \theta_{n,k} < \frac{k}{n+\frac{1}{2}}\pi, \quad -\frac{1}{2} < \alpha,\beta < \frac{1}{2}. \tag{10.3.2}$$

This is obtained by using a comparison with the special (Chebyshev polynomial) cases $\alpha = -\frac{1}{2}, \beta = \frac{1}{2}$ and $\alpha = \frac{1}{2}, \beta = -\frac{1}{2}$.

We have

$$\lim_{\alpha\to -1^+} x_{n1}(\alpha,\beta) = 1, \quad \lim_{\beta\to 1^-} x_{nn}(\alpha,\beta) = -1. \tag{10.3.3}$$

The first relation in (10.3.3) follows from the inequalities

$$\frac{2(\alpha+1)}{n(n+\alpha+\beta+1)} < 1 - x_{n1} < \frac{2(\alpha+1)(\alpha+2)}{n(n+\alpha+\beta+1)+\alpha(\alpha+\beta+3)+\beta+2}, \tag{10.3.4}$$

which follow from the explicit representation of $P_n^{(\alpha,\beta)}(1-x)$, using a method outlined in Ismail and Muldoon (1995, Lemma 3.2).

The Sturm comparison theorem yields (Szegő, [1939] 1975, Theorem 6.3.1), for $-\frac{1}{2} \le \alpha$, $\beta \le \frac{1}{2}$,

$$\theta_k - \theta_{k-1} \le \frac{1}{n+(\alpha+\beta+1)/2}\pi, \quad k = 1,2,\ldots,n+1, \tag{10.3.5}$$

with equality only for $\alpha^2 = \beta^2 = \frac{1}{4}$. Here the notation is extended slightly:

$$\theta_0 = \begin{cases} 0, & \alpha > -\frac{1}{2}, \\ -\theta_1, & \alpha = -\frac{1}{2}, \end{cases} \quad \text{and} \quad \theta_{n+1} = \begin{cases} \pi, & \beta > -\frac{1}{2}, \\ 2\pi - \theta_n, & \beta = -\frac{1}{2}. \end{cases} \tag{10.3.6}$$

Hence (Szegő, [1939] 1975, Theorem 6.3.2), for $-\frac{1}{2} < \alpha,\beta < \frac{1}{2}$, we have

$$\frac{k+(\alpha+\beta-1)/2}{n+(\alpha+\beta+1)/2}\pi < \theta_k < \frac{k}{n+(\alpha+\beta+1)/2}\pi, \quad k = 1,2,\ldots,n. \tag{10.3.7}$$

The result (10.3.5) is obtained by writing

$$N(\alpha,\beta) = n + \frac{\alpha+\beta+1}{2} \tag{10.3.8}$$

and using Section 10.2.6(ii) to compare the normalized Jacobi equation (3.1.19)

$$\frac{d^2u}{d\theta^2} + \left[N^2(\alpha,\beta) + \frac{1/4-\alpha^2}{4\sin^2\theta/2} + \frac{1/4-\beta^2}{4\cos^2\theta/2}\right]u = 0, \tag{10.3.9}$$

satisfied by

$$u = \left(\sin\frac{\theta}{2}\right)^{\alpha+1/2}\left(\cos\frac{\theta}{2}\right)^{\beta+1/2} P_n^{(\alpha,\beta)}(\cos\theta),$$

with the trigonometric equation

$$\frac{d^2u}{d\theta^2} + N^2(\alpha,\beta)u = 0. \tag{10.3.10}$$

We have the asymptotic formula (Szegő, [1939] 1975, Theorem 8.1.2)

$$\lim_{n\to\infty} n\theta_{n,k}(\alpha,\beta) = j_{\alpha,k}, \tag{10.3.11}$$

and Sturmian methods show that

$$\theta_k(\alpha,\beta) < \frac{2j_{\alpha k}}{2n+\alpha+\beta+1}, \quad k = 1,2,\ldots,n, \ -\frac{1}{2} < \alpha,\beta < \frac{1}{2}, \tag{10.3.12}$$

with the notation (1.2.27) for the zeros of Bessel functions.

Better, though more complicated, results can be obtained if we use more complicated comparison equations. The Sturm comparison theorem suggests a comparison between equation (10.3.9) and the equations

$$\frac{d^2v}{d\theta^2} + \left[N^2(\alpha,\beta) + \frac{1-\alpha^2-3\beta^2}{12} + \frac{1/4-\alpha^2}{\theta^2}\right]v = 0 \tag{10.3.13}$$

and

$$\frac{d^2w}{d\theta^2} + \left[N^2(\alpha,\beta) + \frac{1}{4} - \frac{\alpha^2+\beta^2}{2} - \frac{1-4\alpha^2}{\pi^2} + \frac{1/4-\alpha^2}{\theta^2}\right]w = 0, \tag{10.3.14}$$

satisfied by

$$v = \theta^{1/2}J_\alpha\left(\left[N^2(\alpha,\beta) + \frac{1-\alpha^2-3\beta^2}{12}\right]^{1/2}\theta\right)$$

and

$$w = \theta^{1/2}J_\alpha\left(\left[N^2(\alpha,\beta) + \frac{1}{4} - \frac{\alpha^2+\beta^2}{2} - \frac{1-4\alpha^2}{\pi^2}\right]^{1/2}\theta\right),$$

respectively. Here N is given by (10.3.8). Thus, one obtains the bounds (Gautschi and Giordano, 2008)

$$\begin{aligned} j_{\alpha,k}&\left[N^2(\alpha,\beta) + \frac{1}{4} - \frac{\alpha^2+\beta^2}{2} - \frac{1-4\alpha^2}{\pi^2}\right]^{-1/2} \\ &< \theta_{nk}^{(\alpha,\beta)} < j_{\alpha,k}\left[N^2(\alpha,\beta) + \frac{1-\alpha^2-3\beta^2}{12}\right]^{-1/2}, \end{aligned} \tag{10.3.15}$$

for $k = 1,2,\ldots,\lfloor n/2\rfloor$ with $-1/2 \le \alpha,\beta \le 1/2$, $\alpha^2+\beta^2 < 1/2$.

A similar analysis shows that when $\alpha,\beta \ge 1/2$, $\alpha^2+\beta^2 > 1/2$, the bounds in (10.3.15) are reversed.

Gatteschi (1987) makes ingenious use of the Sturm comparison theorem to provide very sharp upper and lower bounds for the zeros of the Jacobi polynomial $P_n^{(\alpha,\beta)}(\cos\theta)$, in the case $-\frac{1}{2} \le \alpha,\beta \le \frac{1}{2}$. He shows that an asymptotic formula, involving zeros of Bessel functions, due to Frenzen and Wong (1985, 1988), in fact provides a lower bound for these zeros (and also an upper bound, using $P_n^{(\alpha,\beta)}(x) = (-1)^n P_n^{(\beta,\alpha)}(-x)$). Specifically, Gatteschi shows that

$$z'' + F(\theta)z = 0 \tag{10.3.16}$$

is a Sturm majorant for (10.3.9), where

$$F(\theta) = \frac{1}{2}\frac{f'''}{f'} - \frac{3}{4}\left(\frac{f''}{f'}\right)^2 + \left(\frac{1}{4} - \alpha^2\right)\left(\frac{f'}{f}\right)^2 + f'^2$$

and

$$f(\theta) = N\theta + \frac{1}{4N}\left[\left(\frac{1}{4}\alpha^2\right)\left(\frac{2}{\theta} - \cot\frac{\theta}{2}\right) + \left(\frac{1}{4} - \beta^2\right)\tan\frac{\theta}{2}\right].$$

This leads to the following lower bound for the zeros (Gatteschi, 1987, Theorem 2.1):

$$\theta_{n,k} \geq t - \frac{1}{4N^2}\left[\left(\frac{1}{4} - \alpha^2\right)\left(\frac{2}{t} - \cot\frac{t}{2}\right) + \left(\frac{1}{4} - \beta^2\right)\tan\frac{t}{2}\right], \tag{10.3.17}$$

valid for $|\alpha|, |\beta| \leq \frac{1}{2}$, $r = 1, 2, \ldots, n$, with equality for $\alpha^2 = \beta^2 = \frac{1}{4}$, where $N = n+(\alpha+\beta+1)/2$, $t = j_{\alpha,k}/N$, and $k = 1, 2, \ldots, n$.

Using the method of Tricomi (Section 10.2.9), Gatteschi and Pittaluga (Gatteschi and Pittaluga, 1985; Gautschi and Giordano, 2008, p. 14) provided the following asymptotic result: For $|\alpha|, |\beta| \leq 1/2$, and for the zeros contained in any compact subinterval of $(-1, 1)$, we have

$$\begin{aligned}\theta_{n,r}(\alpha,\beta) &= \vartheta_{n,r}(\alpha,\beta) + \frac{1}{(2n+\alpha+\beta+1)^2} \\ &\times \left[\left(\frac{1}{4} - \alpha^2\right)\cot\left(\frac{1}{2}\vartheta_{n,r}(\alpha,\beta)\right) - \left(\frac{1}{4} - \beta^2\right)\tan\left(\frac{1}{2}\vartheta_{n,r}(\alpha,\beta)\right)\right] + O\left(n^{-4}\right),\end{aligned} \tag{10.3.18}$$

where $\vartheta_{n,r}(\alpha,\beta) = (2r + \alpha - 1/2)\pi/(2n + \alpha + \beta + 1)$.

Using a Hilb-type asymptotic formula, Gatteschi (1985, Theorem 3.2) showed that for $|\alpha|, |\beta| \leq 1/2$,

$$\theta_{n,r}(\alpha,\beta) = \frac{j_{\alpha,r}}{\nu}\left[1 - \frac{4 - \alpha^2 - 15\beta^2}{720\nu^4}\left(\frac{1}{2}j_{\alpha,r}^2 + \alpha^2 - 1\right)\right] + j_{\alpha,r}^5 O\left(n^{-7}\right), \tag{10.3.19}$$

for $r = 1, 2, \ldots, \lfloor \gamma n \rfloor$, with γ fixed in $(0, 1)$ and

$$\nu = \left(N^2 + \frac{1 - \alpha^2 - 3\beta^2}{12}\right)^{\frac{1}{2}}. \tag{10.3.20}$$

The expository article of Gautschi and Giordano (2008) summarizes much of Gatteschi's work in this and related areas.

Several results like (10.3.5) are known on the spacing of zeros of Jacobi polynomials. Here we present some of these.

Theorem 10.3.1 (Ahmed, Laforgia, and Muldoon, 1982) *Let $\theta_{nk} = \theta_{nk}^{(\alpha,\beta)}$, $k = 1, 2, \ldots, n$ be the zeros of $P_n^{(\alpha,\beta)}(\cos\theta)$ in increasing order on $(0, \pi)$ where $\alpha > -1, \beta > -1$, and for each n let k_n be the largest value of k for which $\theta_{nk} < \pi/2$. Then the following results hold for each fixed n:*

(i) *If $\alpha^2 \le \beta^2$, $\alpha^2 \le \frac{1}{4}$, then*

$$\theta_{n+1,k+1} - \theta_{n+1,k} < \theta_{n,k+1} - \theta_{nk}, \quad k = 1, 2, \ldots, k_n - 1.$$

(ii) *If $\alpha^2 \le 1/4, \beta^2 \le 1/4, \alpha^2 + \beta^2 \ne 1/2$, and $\gamma = (\alpha + \beta + 1)/2$, then*

$$(n + \gamma)\theta_{nk} < (n + \gamma + 1)\theta_{n+1,k}, \quad k = 1, 2, \ldots, n.$$

(iii) *If $\alpha^2 \le \beta^2 \le 1/4, \alpha^2 + \beta^2 \ne 1/2$, and $\gamma = (\alpha + \beta + 1)/2$, then*

$$(n + \gamma + 1)(\theta_{n+1,k+1} - \theta_{n+1,k}) > (n + \gamma)(\theta_{n,k+1} - \theta_{nk}), \quad k = 1, 2, \ldots, k_n - 1.$$

Gautschi (2009) advances conjectures on the possible monotonicity of $(n + \gamma)\,\theta_{nk}$ for other values of α and β.

Deaño, Gil, and Segura (2004) use the Sturm comparison theorem in a different way to get the following results on the spacing of zeros of Jacobi polynomials. We add the hypothesis $\alpha^2 + \beta^2 \ne \frac{1}{2}$ to several parts of Abramov (1989, Theorem 4) in order to get strict inequality in the results.

Theorem 10.3.2 *Let n, α, and β satisfy*

$$n > 0, \quad n + \alpha + \beta > 0, \quad n + \alpha > 0, \quad n + \beta > 0$$

and suppose that $L = 2n + \alpha + \beta + 1$.

(i) *If $|\alpha| = |\beta| = \frac{1}{2}$, then*

$$\Delta\theta_k = \frac{2\pi}{L}.$$

(ii) *If $|\alpha| \le \frac{1}{2}$, $|\beta| \le \frac{1}{2}$, and $\alpha^2 + \beta^2 \ne \frac{1}{2}$ then*

$$\Delta\theta_k < \frac{2\pi}{\sqrt{L^2 + \left(\sqrt{\frac{1}{4} - \alpha^2} + \sqrt{\frac{1}{4} - \beta^2}\right)^2}}.$$

(iii) *If $|\alpha| \ge \frac{1}{2}$, $|\beta| \ge \frac{1}{2}$, and $\alpha^2 + \beta^2 \ne \frac{1}{2}$ then*

$$\Delta\theta_k > \frac{2\pi}{\sqrt{L^2 - \left(\sqrt{\alpha^2 - \frac{1}{4}} + \sqrt{\beta^2 - \frac{1}{4}}\right)^2}}.$$

(iv) *If $|\alpha| \ge \frac{1}{2}$, $|\beta| \le \frac{1}{2}$, and $\alpha^2 + \beta^2 \ne \frac{1}{2}$ then*

$$\Delta^2\theta_k < 0.$$

(v) *If $|\alpha| \le \frac{1}{2}$, $|\beta| \ge \frac{1}{2}$, and $\alpha^2 + \beta^2 \ne \frac{1}{2}$ then*

$$\Delta^2\theta_k > 0.$$

Theorem 10.3.3 *Let n, α, and β satisfy*

$$n > 0, \quad n + \alpha + \beta > 0, \quad n + \alpha > 0, \quad n + \beta > 0, \ -1 \leq \beta \leq 1.$$

Then the zeros of the Jacobi polynomials satisfy

$$\Delta^2 \ln(1 - x_k) < 0, \quad k = 1, \dots, n-2$$

or

$$(1 - x_k)^2 > (1 - x_{k-1})(1 - x_{k+1}), \quad k = 2, \dots, n-1.$$

This result was proved in Deaño, Gil, and Segura (2004, Corollary 6(1)) generalizing the Legendre case (Grosjean, 1987).

Using an extension of the Laguerre method of Section 10.2.8, Dimitrov and Nikolov (2010) show for $\alpha, \beta > -1$ that all the zeros of the Jacobi polynomial lie between the roots of a particular quadratic function; specifically

$$\frac{B - 4(n-1)\sqrt{\Delta}}{A} \leq x_{n,k}(\alpha, \beta) \leq \frac{B + 4(n-1)\sqrt{\Delta}}{A}, \tag{10.3.21}$$

where

$$B = (\beta - \alpha)((\alpha + \beta + 6)n + 2(\alpha + \beta)),$$

$$A = (2n + \alpha + \beta)(n(2n + \alpha + \beta) + 2(\alpha + \beta + 2)),$$

$$\Delta = n^2(n + \alpha + \beta + 1)^2 + (\alpha + 1)(\beta + 1)\left(n^2 + (\alpha + \beta + 4)n + 2(\alpha + \beta)\right).$$

For the zeros $x_k = x_{nk}(\alpha, \beta)$, $k = 1, \dots, n$ of the Jacobi polynomials, we have (Stieltjes, 1887; Szegő, [1939] 1975, §6.7)

$$\sum_{k \neq j} \frac{1}{x_j - x_k} + \frac{1}{2} \frac{\alpha + 1}{x_j - 1} + \frac{1}{2} \frac{\beta + 1}{x_j + 1} = 0, \quad j = 1, \dots, n, \tag{10.3.22}$$

and these equations characterize the zeros in question. That is, n numbers x_k, $k = 1, 2, \dots, n$ satisfying equations (10.3.22) must be the zeros of $P_n^{(\alpha,\beta)}$.

Many results of the kind just given are known. See for instance Ahmed et al. (1979) and Calogero (2001, Appendix C).

Although the Jacobi polynomials are orthogonal with respect to $(1 - x)^\alpha (1 + x)^\beta$ on $[-1, 1]$ only for $\alpha, \beta > -1$, it is of interest to examine the locations of their zeros when one or both of these conditions fails. As pointed out in Section 10.3.7, the zeros decrease (increase) as α (β) decreases with the largest (smallest) zero being 1 (-1) when $\alpha = -1$ ($\beta = -1$). When either α or β is < -1 the zeros are generally complex, the number of zeros in each of the real intervals $(-\infty, -1)$, $(-1, 1)$, and $(1, \infty)$ is given by results of Hilbert and Klein (Szegő, [1939] 1975, Theorem 6.7.2).

10.4 Ultraspherical Polynomials

Since $\left|C_n^\lambda(x)\right|$ is an even function of x (Szegő, [1939] 1975, (4.7.4)), we find that for $\lambda > -\frac{1}{2}$, the zeros of the ultraspherical polynomial

$$C_n^\lambda(x) = P_n^{(\lambda-1/2,\lambda-1/2)}(x)$$

are located symmetrically on $(-1, 1)$. According to Szegő ([1939] 1975, (4.1.5)), the ultraspherical polynomials may be expressed in terms of Jacobi polynomials with $\beta = \pm\frac{1}{2}$. Then we can use some of the results of the previous section to show (Szegő, [1939] 1975, §6.21) that the positive (negative) zeros of C_n^λ decrease (increase) as λ increases (decreases). In particular, the largest zero increases to 1 as $\lambda \to -\frac{1}{2}^+$.

Throughout this section we use the notation $x_{nk}(\lambda)$, $k = 1, 2, \ldots, n$ for the kth nonnegative zero in *decreasing* order of C_n^λ. We use $\theta_{nk}(\lambda)$ for the kth positive zero in increasing order of $C_n^\lambda(\cos\theta)$, $k = 1, 2, \ldots, n$.

The methods of Section 10.2.8 show that we have (Szegő, [1939] 1975, (6.2.13))

$$x_{n1}(\lambda) > \left(\frac{n-1}{n+2\lambda}\right)^{\frac{1}{2}}, \quad \lambda > -\frac{1}{2}. \tag{10.4.1}$$

The monotonicity in λ gives (Szegő, [1939] 1975, Theorem 6.3.2)

$$(k - 1/2)\frac{\pi}{n} \le \theta_{nk}(\lambda) \le k\frac{\pi}{n+1}, \quad 0 \le \lambda \le 1,\ k = 1, 2, \ldots, \lfloor n/2 \rfloor, \tag{10.4.2}$$

with equality only if $\lambda = 0, 1$.

We have (Szegő, [1939] 1975, (6.3.8))

$$\theta_{n,k}(\lambda) > \frac{k - (1-\lambda)/2}{n+\lambda}\pi, \quad 0 < \lambda < 1,\ k = 1, 2, \ldots, \lfloor n/2 \rfloor. \tag{10.4.3}$$

We get equality for $\lambda = 0, 1$.

Szegő ([1939] 1975, §6.6) provides a comparison between various simple bounds for the zeros of C_n^λ in the case $0 < \lambda < 1$.

From the Sturm convexity theorem (Szegő, [1939] 1975, Theorem 1.82.2) applied to equation (3.5.18), the sequence $\theta_0, \theta_1, \ldots, \theta_{\lfloor n/2 \rfloor + 1}$ is convex. Similarly, from (3.5.17), the sequence $x_0, x_1, \ldots, x_{\lfloor n/2 \rfloor + 1}$ is convex.

If $0 < \lambda < 1$, then (Ahmed, Laforgia, and Muldoon, 1982, Theorem 3.3(ii)) for each k $(= 1, \ldots, n)$,

$$(n + \lambda)\theta_{nk} \text{ increases with } n. \tag{10.4.4}$$

We have (Szegő, [1939] 1975, Theorem 6.3.4)

$$\theta_{nk}(\lambda) < \frac{j_{\alpha k}}{n+\lambda}, \quad 0 < \lambda < 1,\ k = 1, 2, \ldots, n, \tag{10.4.5}$$

with the notation (1.2.27) for the zeros of Bessel functions.

For the *positive* zeros of the ultraspherical polynomials, we have Gatteschi's result (Gatteschi, 1987, Corollary 3.2)

$$\phi_{nk}(\lambda) \le \theta_{nk}(\lambda) \le \phi_{nk}(\lambda) + \frac{\lambda - \lambda^2}{2(n+\lambda)^2} \cot \phi_{nk}(\lambda), \tag{10.4.6}$$

where $\phi_{nk}(\lambda) = (k + (\lambda - 1)/2)\pi/(n + \lambda)$. Gatteschi (1987) makes comparisons with known bounds and gives numerical examples to illustrate the sharpness of his inequalities. Using the Tricomi method (Section 10.2.9), Gatteschi (Gatteschi, 1972; Gautschi and Giordano, 2008, (2.1.1)) shows that for $0 < \lambda < 1$ and $r = 1, 2, \ldots, \lfloor n/2 \rfloor$, we have

$$\theta_{n.r}(\lambda) = \vartheta_{n.r}(\lambda) + \frac{\lambda(1-\lambda)}{2(n+\lambda)(n+\lambda+1)} \cot \vartheta_{n.r}(\lambda) + \rho, \tag{10.4.7}$$

where $\vartheta_{n.r}(\lambda) = (r - (1 - \lambda)/2)\pi/(n + \lambda)$, and

$$|\rho| < \frac{\lambda(1-\lambda)}{(n+\lambda+1)(2r+\lambda-1)^2}, \qquad 0 < \lambda < 1. \tag{10.4.8}$$

For the zeros x_{nk}, $k = 1, 2 \ldots, \lfloor n/2 \rfloor$, we have (Gautschi and Giordano, 2008, (3))

$$x_{n.r}(\lambda) = \xi_{n.r}(\lambda) \left[1 - \frac{\lambda(1-\lambda)}{2(n+\lambda)(n+\lambda+1)} \right] + \varepsilon, \tag{10.4.9}$$

where $\xi_{n.r}(\lambda) = \cos \vartheta_{n.r}(\lambda)$, and

$$|\varepsilon| < \frac{1.55\lambda(1-\lambda)}{(n+\lambda)(n+\lambda+1)(2r+\lambda-1)}, \qquad 0 < \lambda < 1. \tag{10.4.10}$$

In Area et al. (2004) there are sharp upper bounds for the zeros of the ultraspherical polynomials C_n^λ, $\lambda > -1/2$. The proof is based on a result by Obrechkoff and Descartes' rule of signs among other things.

Dimitrov (2003) shows that, for any $n \in \mathbb{N}$, the product $(\lambda + 1)^{3/2} x_{n1}(\lambda)$ is a convex function of λ if $\lambda \ge 0$. The result is applied to obtain some inequalities for $x_{n1}(\lambda)$.

Using Sturmian methods, Elbert and Laforgia (1990) obtained some inequalities valid without major restrictions on the parameter λ:

Theorem 10.4.1 *For each of the following choices of A and B:*

(i) $A = n^2 + 2\lambda n$, $B = (n + \lambda)^2$, $\lambda > 0$,
(ii) $A = n^2 + 2\lambda n + \lambda$, $B = (n + \lambda)^2 + \lambda$, $\lambda > 0$,
(iii) $A = n^2 + 2\lambda n + \lambda$, $B = (n + \lambda)^2$, $-1/3 \le \lambda < 0$, *or* $\lambda \ge 1$,

we have, for $k = 1, 2, \ldots, \lfloor n/2 \rfloor$,

$$x_{nk}(\lambda) \le \sqrt{\frac{A}{B}} \sin \vartheta_{nk}(\lambda) = \sqrt{\frac{A}{B}} \sin I^{-1} \left(\frac{(n+1-2k)\pi}{2\sqrt{B}} \right), \tag{10.4.11}$$

with

$$I(\vartheta) = \vartheta - \sqrt{\frac{B-A}{B}} \arctan \left(\sqrt{\frac{B-A}{B}} \tan \vartheta \right).$$

Corollary 10.4.2 *Let A and B be chosen as in Theorem* 10.4.1. *Then*

$$x_{nk}(\lambda) \leq \sqrt{\frac{A}{B}} \sin\left(\frac{(n+1-2k)\pi}{2(\sqrt{B}-\sqrt{A/B})}\right), \quad k = 1,2\ldots,\lfloor n/2 \rfloor. \tag{10.4.12}$$

By Theorem 10.4.1(i) and inequality (10.4.12) we get, for example, for the largest zero

$$x_{n1}(\lambda) < \frac{\sqrt{n^2+2\lambda n}}{n+\lambda}, \quad \lambda \geq 0.$$

Corollary 10.4.3 *Let A, B, and λ satisfy the conditions of Theorem* 10.4.1. *Then for $x_{nk}(\lambda)$ we have the approximation formula*

$$x_{nk}(\lambda) \doteq \sqrt{\frac{A}{B}}\left[\tau + \frac{2m^2-1}{6}\tau^3 + \frac{16m^4-4m^2+1}{120}\tau^5\right],$$

where

$$\tau = (n+1-2k)\sqrt{B}\,\pi/(2A), \quad m^2 = 1 - A/B.$$

For the zeros $\theta_{nk}(\lambda)$ of $C_n^\lambda(\cos\theta)$, (10.3.15) gives

$$\begin{aligned} j_{\lambda-1/2,k}\left[(n+\lambda)^2 + \left(1-\frac{4}{\pi^2}\right)\lambda(1-\lambda)\right]^{-1/2} &< \theta_{nk}(\lambda) \\ &< j_{\lambda-1/2,k}\left[(n+\lambda)^2 + \frac{1}{3}\lambda(1-\lambda)\right]^{-1/2}, \end{aligned} \tag{10.4.13}$$

for $0 < \lambda < 1$, $k = 1, \ldots, \lfloor n/2 \rfloor$.

For a positive zero $x = x(\lambda)$ of C_n^λ, we have (Elbert and Muldoon, 1994, p. 357) the formulas

$$\frac{dx}{d\lambda} = -\left(1-x^2\right)^{-\lambda-1/2}\left[C_n^{\lambda\prime}(x)\right]^{-2}\int_0^x \frac{2n+1-2(n+\lambda)s^2}{(1-s^2)^{-\lambda+3/2}}\left[C_n^\lambda(s)\right]^2 ds, \quad \lambda > -\frac{1}{2} \tag{10.4.14}$$

and

$$\frac{dx}{d\lambda} = \left(1-x^2\right)^{-\lambda-1/2}\left[C_n^{\lambda\prime}(x)\right]^{-2}\int_x^1 \frac{2n+1-2(n+\lambda)s^2}{(1-s^2)^{-\lambda+3/2}}\left[C_n^\lambda(s)\right]^2 ds, \quad \lambda > \frac{1}{2}. \tag{10.4.15}$$

As shown in Elbert and Muldoon (1994), the first of these formulas recovers the result (Szegő, [1939] 1975, Theorem 6.21.1) that all the positive zeros of C_n^λ are decreasing functions of λ, $\lambda > -\frac{1}{2}$; see Elbert and Muldoon (1994, p. 357) for details.

On the other hand, one can use Sturm methods (Laforgia, 1981) to show that $\lambda x_{n,k}(\lambda)$ increases as λ increases, $0 < \lambda < 1$. Elbert and Siafarikas (1999) used (10.4.14) to prove the stronger result that for $n \geq 3$ and $1 \leq k \leq \lfloor k/2 \rfloor$, the function $[\lambda + (2n^2+1)/(4n+2)]^{1/2} x_{n,k}(\lambda)$ increases as λ increases for $\lambda > -1/2$. In particular, $\sqrt{\lambda + 1/2}\, x_{n,k}(\lambda)$ increases as λ increases, $\lambda > -1/2$. Dimitrov and Rodrigues (2002) show that these results are quite sharp.

Using scaling and the Sturm theorem, Elbert and Laforgia (1986) proved the following two results.

Theorem 10.4.4 *Let i, j, n, m be natural numbers such that $m \geq n \geq 2$, $1 \leq i \leq [n/2]$, $1 \leq j \leq [n/2]$. Then, for $-1/2 < \lambda < 3/2$,*

$$\operatorname{sign} l \cdot \begin{vmatrix} x_{ni}(\lambda) & x_{n,i+l}(\lambda) \\ x_{mj}(\lambda) & x_{m,j+l}(\lambda) \end{vmatrix} > 0$$

provided $x_{mj}(\lambda) > x_{ni}(\lambda)$ and the integer $l \neq 0$ satisfies the relations

$$-\min(i-1, j-1) < l < \min(n-i, m-j).$$

Theorem 10.4.5 *Suppose that $-1/2 < \lambda < \lambda'$, $0 \leq \lambda' \leq 1/2$, and $n \geq 4$. Then*

$$\operatorname{sign}(i-k) \cdot \begin{vmatrix} \theta_{ni}(\lambda) & \theta_{ni}(\lambda') \\ \theta_{nk}(\lambda) & \theta_{nk}(\lambda') \end{vmatrix} > 0$$

provided $1 \leq k \leq [n/2]$, $i \neq k$.

If, in Theorem 10.4.4, we put $i = j$, $m > n$, we know that $x_{ni}(\lambda) < x_{mi}(\lambda)$ and we get

$$\operatorname{sign}(k-i) \cdot \begin{vmatrix} x_{ni}(\lambda) & x_{nk}(\lambda) \\ x_{mi}(\lambda) & x_{mk}(\lambda) \end{vmatrix} > 0$$

provided $1 \leq i \leq \lfloor n/2 \rfloor$ and $-1/2 < \lambda \leq 3/2$.

Similarly, the quotient $x_{n,i+l}(\lambda)/x_{n,j+l}(\lambda)$ increases when the integer l increases from 0 to $n+1-i$, provided $1 \leq j < i \leq \lfloor n/2 \rfloor$ and $-1/2 < \lambda \leq 3/2$.

Durand (1975, p. 367) showed that, if θ_k is the kth zero on $(0, \pi)$ of $C_n^\lambda(\cos\theta)$, or indeed of any nontrivial solution of (3.5.18) (with ν replaced by λ), then for $\lambda > 1$, $\Delta\theta_k > 0$, $\Delta^2\theta_k < 0$, $\Delta^3\theta_k > 0$. The results on third differences go beyond what would be expected from using the Sturm comparison theorem.

Using results of the kind described in Section 10.2.8, Dimitrov and Nikolov (2010) showed

$$\frac{b-(n-2)\sqrt{\delta}}{a} \leq x_{nk}^2(\lambda) \leq \frac{b+(n-2)\sqrt{\delta}}{a},$$

where

$$a = 2(n+\lambda-1)\left(n^2 + n(\lambda-1) + 4(\lambda+1)\right),$$

$$b = n^3 + 2(\lambda-1)n^2 - (3\lambda-5)n + 4(\lambda-1),$$

and

$$\delta = n^2(n+2\lambda)^2 + (2\lambda+1)\left(n^2 + 2(\lambda+3)n + 8(\lambda-1)\right).$$

Although it is only for $\lambda > -\frac{1}{2}$ that the ultraspherical polynomials are orthogonal with respect to $(1-x^2)^{\lambda-1/2}$ on $[-1, 1]$, it is of interest to examine their zeros for $\lambda \leq -\frac{1}{2}$. As pointed out in Section 10.4, the zeros increase in absolute value as λ $(> -\frac{1}{2})$ decreases, the extreme zeros being at ± 1 for $\lambda = -\frac{1}{2}$. Driver and Duren (2000) showed that when $\lambda < 1-n$, all of the zeros lie on the imaginary axis. In Driver and Duren (2001), they describe the trajectories of the zeros in the complex plane as λ decreases from $-\frac{1}{2}$ to $1-n$. This description helps to explain the formulas of Hilbert and Klein (see Szegő, [1939] 1975, p. 144 for the more general Jacobi case) for the number of zeros of the ultraspherical polynomial C_n^λ lying in each of the real intervals $(-\infty, -1), (-1, 1)$, and $(1, \infty)$.

10.5 Legendre Polynomials

The Legendre polynomials are defined in Section 3.6. They correspond to the case $\alpha = \beta = 0$ of the Jacobi polynomials and the case $\lambda = \frac{1}{2}$ of the ultraspherical polynomials. We use $x_{n1}, \ldots, x_{nn}$ for their zeros in *decreasing* order. Here, we emphasize special results known for the Legendre case only.

The largest zero of the Legendre polynomial satisfies (Szegő, [1939] 1975, (6.2.17))

$$x_{n1} \leq (n-1)\left(\frac{n+2}{n\,(n^2+2)}\right)^{\frac{1}{2}} = 1 - \frac{5}{2n^2} + \cdots. \tag{10.5.1}$$

We have (see Gautschi and Giordano, 2008, (19))

$$x_{n,r} = 1 - \frac{j_{0,r}^2}{2\left(n+\frac{1}{2}\right)^2} + \frac{j_{0,r}^2 + j_{0,r}^4}{24\left(n+\frac{1}{2}\right)^4} + O\left(n^{-6}\right), \quad r = 1, 2 \ldots, \lfloor n/2 \rfloor. \tag{10.5.2}$$

The associated Legendre functions are defined by (Abramowitz and Stegun, 1965, Ch. 8; Olver et al., 2010, 14.3.1)

$$P_\lambda^\mu(z) = \frac{1}{\Gamma(1-\mu)}\left[\frac{1+z}{1-z}\right]^{\mu/2} {}_2F_1\left(-\lambda, \lambda+1; 1-\mu; \frac{1-z}{2}\right). \tag{10.5.3}$$

When m and n are nonnegative integers, we have

$$P_n^m(x) = (-1)^m \left(1 - x^2\right)^{m/2} \frac{d^m}{dx^m} P_n(x). \tag{10.5.4}$$

Baginski (1991) considers the problem of ordering the elements of the set $\{\nu_j^m(z_0)\}$ of ν-zeros of the Legendre functions $P_\nu^m(z_0)$ for $m = 0, 1, \ldots$ and $z_0 \in (-1, 1)$. By Sturm methods, he shows that

$$\nu_j^m(z_0) < \nu_j^{m+1}(z_0) < \nu_{j+1}^m(z_0), \quad -1 < z_0 < 1. \tag{10.5.5}$$

Some additional results proved in Baginski (1991) are

$$\nu_2^3(z_0) < \nu_1^6(z_0), \qquad -1 < z_0 < 1, \tag{10.5.6}$$

$$\nu_2^0(z_0) < \nu_1^3(z_0), \quad \nu_2^1(z_0) < \nu_1^4(z_0), \quad \nu_3^0(z_0) < \nu_1^5(z_0), \quad -1 < z_0 < 1, \tag{10.5.7}$$

$$\nu_j^{m+1}(z_0) < \nu_{j+1}^m(z_0) < \nu_j^{m+2}(z_0) < \nu_{j+1}^{m+1}(z_0), \qquad -1 < z_0 < 0, \tag{10.5.8}$$

$$\nu_j^{m+2}(0) = \nu_{j+1}^m(0) = m + 2j - 1, \tag{10.5.9}$$

$$\nu_j^{m+1}(z_0) < \nu_j^{m+2}(z_0) < \nu_{j+1}^m(z_0) < \nu_{j+1}^{m+1}(z_0), \qquad 0 < z_0 < 1. \tag{10.5.10}$$

10.6 Laguerre Polynomials

For $\alpha > -1$ the zeros of the Laguerre polynomial $L_n^{(\alpha)}$ lie in the orthogonality interval $(0, \infty)$. We denote them by $x_{nk}(\alpha)$, $k = 1, \ldots, n$, in *increasing* order. By the methods of Section 10.2.8, we have (Szegő, [1939] 1975, (6.2.14))

$$x_{nn}(\alpha) > 2n + \alpha - 1, \quad \alpha > -1. \tag{10.6.1}$$

It is shown in Szegő ([1939] 1975, Theorem 6.21.4) that if $\alpha \in [-\frac{1}{2}, \frac{1}{2}]$, then the zeros of $L_n^{(\alpha)}$, in increasing order, have the bounds

$$\xi_k^2 \le x_\nu \le \eta_k^2, \tag{10.6.2}$$

where ξ_k and η_k denote the kth positive zeros of the Hermite polynomials H_{2n} and H_{2n+1}, respectively.

Each zero is an increasing function of α, $-1 < \alpha < \infty$ (Szegő, [1939] 1975, §6.21 (4)). On the other hand (Szegő, [1939] 1975, p. 128–129), for fixed α and k, $(n + (\alpha + 1)/2)x_{nk}(\alpha)$ decreases to its limiting value $j_{\nu k}^2/4$ as n increases. Thus

$$\left(n + \frac{\alpha+1}{2}\right)x_{nk} > \left(n + 1 + \frac{\alpha+1}{2}\right)x_{n+1,k}, \quad \alpha > -1,\ k = 1, \ldots, n.$$

This inequality can be complemented by a result of Elbert and Laforgia (1987):

$$\left(n + 1 + \frac{\alpha+1}{2}\right)x_{n+1,k} - \frac{1}{4}x_{n+1,k}^2 > \left(n + \frac{\alpha+1}{2}\right)x_{nk} - \frac{1}{4}x_{nk}^2.$$

Lorch (1977, (4.2)) used a Sturm method to show that for $-1 < \alpha < 1$,

$$\Delta_n\Delta_k x_{nk} = x_{n+1,k+1} - x_{n+1,k} - x_{n,k+1} + x_{n,k} < 0, \quad k = 1, \ldots, n-1.$$

Also, it is known (Ahmed, Laforgia, and Muldoon, 1982, Theorem 5.1) that, for $-1 < \alpha \le 1$, and for each $k = 1, 2, \ldots, n-1$, the quantity

$$(n + (\alpha + 1)/2)(x_{n,k+1}(\alpha) - x_{nk}(\alpha))$$

decreases to its limiting value $(j_{\alpha,k+1}^2 - j_{\alpha,k}^2)/4$ as n increases.

Gatteschi (Gatteschi, 1988a; Gautschi and Giordano, 2008, §3.2.2) uses Sturm methods to get inequalities for zeros of Laguerre polynomials. Here are some of the results:

If $-1 < \alpha \le 1$, then

$$x_{nk}(\alpha) < \nu\cos^2\left(\frac{1}{2}x_{nk}^{(\alpha)}\right), \quad r = 1, 2, \ldots, n, \tag{10.6.3}$$

where $x_{nk}^{(\alpha)}$ is the root of the equation

$$x - \sin x = \pi - \frac{4j_{\alpha k}}{\nu} \tag{10.6.4}$$

and $\nu = 4n + \alpha + 2$. He also shows that

$$x_{nk}(\alpha) > \nu\cos^2\left(\frac{1}{2}x_{nk}^{*(\alpha)}\right) \quad \text{if } -\frac{1}{2} \le \alpha \le \frac{1}{2}, \tag{10.6.5}$$

and

$$x_{nk}(\alpha) < \nu \cos^2\left(\frac{1}{2}x_{nk}^{*(\alpha)}\right) \quad \text{if } -1 < \alpha \le -\frac{2}{3} \text{ or } \alpha \ge \frac{2}{3}, \tag{10.6.6}$$

where $x_{nk}^{*(\alpha)}$ is the root of the equation

$$x - \sin x = \frac{8}{3\sqrt{3}\,\nu} i_{n+1-k}^{3/2} \tag{10.6.7}$$

and a_k is the kth zero in ascending order of $\mathrm{Ai}(-x)$.

By Sturmian methods, we have (Szegő, [1939] 1975, Theorem 6.31.1)

$$x_{nk}(\alpha) > \frac{j_{\alpha k}^2}{2(2n+\alpha+1)}, \quad k = 1, 2, \ldots, n. \tag{10.6.8}$$

For the largest zero, we have (Szegő, [1939] 1975, Theorem 6.31.2) the inequality

$$x_{nn}(\alpha) < 2n + \alpha + 1 + \{(2n+\alpha+1)^2 + 1/4 - \alpha^2\}^{1/2} \cong 4n. \tag{10.6.9}$$

For all the zeros we have (Szegő, [1939] 1975, Theorem 6.31.3) the inequality

$$x_{nk}(\alpha) < \{k + (\alpha+1)/2\}\frac{2k + \alpha + 1 + \{(2n+\alpha+1)^2 + 1/4 - \alpha^2\}^{1/2}}{n + (\alpha+1)/2}. \tag{10.6.10}$$

If $|\alpha| \ge 1/4$, $\alpha > -1$, we have (Szegő, [1939] 1975, Theorem 6.32)

$$x_{nk}^{1/2} < (4n + 2\alpha + 2)^{1/2} - 6^{-1/3}(4n + 2\alpha + 2)^{-1/6} i_k, \quad k = 1, 2, \ldots, \tag{10.6.11}$$

where i_k is the kth positive zero of the Airy function $A(x)$.

Dimitrov (2003) showed that $x_{n1}(\alpha)/(\alpha+1)$ is a convex function of α for $\alpha > -1$.

Elbert and Laforgia (1987) proved the determinantal inequality

$$\begin{vmatrix} x_{nk}(\alpha) & x_{n,k+l}(\alpha) \\ x_{n+m,k}(\alpha) & x_{n+m,k+l}(\alpha) \end{vmatrix} < 0, \tag{10.6.12}$$

for $\alpha > -1$, $1 \le k < k + l \le n$, $m = 1, 2, \ldots$, and $l = 1, 2, \ldots, n - k$. In particular, with $l = m = 1$, this gives

$$\Delta_n \Delta_k \log x_{nk}(\alpha) < 0, \quad \alpha > -1,\ k = 1, \ldots, n-1.$$

The following result (Ismail and Li, 1992a) is a consequence of Theorem 10.2.8:

Theorem 10.6.1 *We have*

$$x_{nn}(\alpha) < 2N + \alpha - 2 + \sqrt{1 + a(N-1)(N+\alpha-1)} \tag{10.6.13}$$

for $\alpha > -1$, *and*

$$x_{n1}(\alpha) > 2N + \alpha - 2 - \sqrt{1 + 4(N-1)(N+\alpha-1)} \tag{10.6.14}$$

for $\alpha \ge 1$, *where*

$$a = 4\cos^2(\pi/(N+1)). \tag{10.6.15}$$

Segura (2003, Lemma 8) gives further bounds for zeros of Laguerre polynomials. See also Krasikov (2003, 2007).

The expressions

$$\frac{2n^2 + n(\alpha - 1) + 2(\alpha + 1) \pm 2(n-1)\sqrt{n^2 + (n+2)(\alpha + 1)}}{n+2}$$

provide upper and lower bounds for the zeros of Laguerre polynomials $\alpha > -1$ (Dimitrov and Nikolov, 2010, Theorem 1).

Dimitrov and Rafaeli (2009) prove monotonicity results for functions involving the zeros of $L_n^{(a)}(x)$, with respect to α. This leads to the inequalities

$$x_{nk}(\alpha) \le 2n + \alpha - 1 + \sqrt{2n + \alpha - 1}\, h_{nk}, \quad n \ge 2,\ k = 1, \ldots, n,\ \alpha \ge 1/(n-1), \tag{10.6.16}$$

$$x_{n1}(\alpha) \le 2n + \alpha - 1 + \sqrt{2n + \alpha - 1}\, h_{n1}, \quad n = 1, 2, \ldots,\ \alpha > -1, \tag{10.6.17}$$

involving zeros of Hermite polynomials.

Quite sharp inequalities can be obtained for the smallest zero when $\alpha > -1$ is close to -1. In this case, the zeros are all positive and by the method outlined in Ismail and Muldoon (1995, §3) we have, for $n \ge 2$,

$$S_m^{-1/m} < x_{n1}(\alpha) < S_m / S_{m+1}, \quad m = 1, 2, \ldots, \tag{10.6.18}$$

where $S_m = \sum_{k=1}^{n} x_{nk}(\alpha)^{-m}$. These upper and lower bounds give successively improving upper and lower bounds for x_1. For example, for $\alpha > -1$, $n \ge 2$, we get, for the smallest zero $x_{n1}(\alpha)$ (Gupta and Muldoon, 2007),

$$\frac{1}{n} < \frac{x_{n1}(\alpha)}{\alpha + 1} < \frac{(\alpha + 2)}{(\alpha + 1 + n)}, \tag{10.6.19}$$

$$\left[\frac{\alpha + 2}{n(n + \alpha + 1)}\right]^{1/2} < \frac{x_{n1}(\alpha)}{\alpha + 1} < \frac{(\alpha + 3)}{(\alpha + 1 + 2n)}, \tag{10.6.20}$$

where the upper bound recovers that in Szegő ([1939] 1975, (6.31.12)), and

$$\begin{aligned} \left[\frac{(\alpha + 2)(\alpha + 3)}{n(n + \alpha + 1)(2n + \alpha + 1)}\right]^{1/3} &< \frac{x_{n1}(\alpha)}{\alpha + 1} \\ &< \frac{(\alpha + 2)(\alpha + 4)(\alpha + 2n + 1)}{\alpha^3 + 4\alpha^2 + 5\alpha + 2 + 5n\alpha^2 + 16n\alpha + 11n + 5n^2\alpha + 11n^2}. \end{aligned} \tag{10.6.21}$$

Further such bounds may be found but they become successively more complicated. From the higher estimates we can produce a series expansion valid for $-1 < \alpha < 0$. The first five terms, obtained with the help of MAPLE, are (Gupta and Muldoon, 2007)

$$
\begin{aligned}
x_1(\alpha) = {} & \frac{\alpha+1}{n} + \frac{n-1}{2}\left(\frac{\alpha+1}{n}\right)^2 - \frac{n^2+3n-4}{12}\left(\frac{\alpha+1}{n}\right)^3 \\
& + \frac{7n^3+6n^2+23n-36}{144}\left(\frac{\alpha+1}{n}\right)^4 \\
& - \frac{293n^4+210n^3+235n^2+990n-1728}{8640}\left(\frac{\alpha+1}{n}\right)^5 + \cdots .
\end{aligned} \tag{10.6.22}
$$

Gatteschi (Gatteschi, 1988b; Gautschi and Giordano, 2008, §4.1) gave some uniform asymptotic formulas for the zeros of Laguerre polynomials. With $x_{n,r}^{(\alpha)}$ denoting the root of (10.6.4) and $\tau_{n,r}^{(\alpha)} = \cos^2(x_{n,r}^{(\alpha)}/2)$, he showed

$$
\begin{aligned}
x_{n,r}(\alpha) = \nu\tau_{n,r}^{(\alpha)} - \frac{1}{2\nu}\Bigg[& \frac{(1-4\alpha^2)\nu}{2j_{\alpha,r}}\left(\frac{\tau_{n,r}^{(\alpha)}}{1-\tau_{n,r}^{(\alpha)}}\right)^{1/2} + \frac{4\alpha^2-1}{2} \\
& + \frac{\tau_{n,r}^{(\alpha)}}{1-\tau_{n,r}^{(\alpha)}} + \frac{5}{6}\left(\frac{\tau_{n,r}^{(\alpha)}}{1-\tau_{n,r}^{(\alpha)}}\right)^2\Bigg] + O(\nu)^{-3},
\end{aligned} \tag{10.6.23}
$$

where $\nu = 4n + 2\alpha + 2$ and the O-term is uniformly bounded for all $r = 1, 2, \dots, \lfloor qn \rfloor$, with q $(0 < q < 1)$ fixed.

For the range $r = \lfloor pn \rfloor, \lfloor pn \rfloor + 1, \dots, n$, with $0 < p < 1$ and with $x_{n,r}^{*(\alpha)}$ denoting the root of (10.6.7) and $\tau_{n,r}^{*(\alpha)} = \cos^2(x_{n,r}^{*(\alpha)}/2)$, we have

$$
\begin{aligned}
x_{n,r}(\alpha) = \nu\tau_{n,r}^{*(\alpha)} + \frac{1}{\nu}\Bigg[& \frac{5\nu}{24a_{n+1-r}^{3/2}}\left(\frac{\tau_{n,r}^{*(\alpha)}}{1-\tau_{n,r}^{*(\alpha)}}\right)^{1/2} + \frac{1}{4} - \alpha^2 \\
& - \frac{1}{2}\frac{\tau_{n,r}^{*(\alpha)}}{1-\tau_{n,r}^{*(\alpha)}} - \frac{5}{12}\left(\frac{\tau_{n,r}^{(\alpha)}}{1-\tau_{n,r}^{(\alpha)}}\right)^2\Bigg] + O(\nu)^{-3}.
\end{aligned} \tag{10.6.24}
$$

Gautschi and Giordano (2008, p. 24) point out that when r is fixed and $\nu \to \infty$, estimate (10.6.23) can be sharpened to

$$
x_{n,r}(\alpha) = \frac{j_{\alpha,r}^2}{\nu}\left[1 + \frac{j_{\alpha,r}^2 + 2(\alpha^2-1)}{3\nu^2}\right] + O\left(\nu^{-5}\right), \quad r \text{ fixed}. \tag{10.6.25}
$$

Similarly, for $r = n + 1 - s$ and s fixed, (10.6.24) can be sharpened to

$$
x_{n,n+1-s}(\alpha) = \nu - 2^{2/3}a_s\nu^{1/3} + \frac{1}{5}2^{4/3}a_s^2\nu^{-1/3} + O\left(\nu^{-1}\right), \tag{10.6.26}
$$

where $a_s = 3^{-1/3}i_s$ and i_s is the sth positive zero of the Airy function $A(x)$. Equations (10.6.25) and (10.6.26) are in the work of Tricomi. In Gatteschi (2002), there is the more exact result

$$
\begin{aligned}
x_{n,n+1-s}(\alpha) = {} & \nu - 2^{2/3}a_s\nu^{1/3} + \frac{1}{5}2^{4/3}a_s^2\nu^{-1/3} \\
& + \left(\frac{11}{35} - \alpha^2 + \frac{12}{175}a_s^3\right)\nu^{-1} + \left(\frac{92}{7875}a_s^4 - \frac{16}{1575}a_s\right)2^{2/3}\nu^{-5/3} \\
& + \left(\frac{15152}{3031875}a_s^5 - \frac{1088}{121275}a_s^2\right)2^{1/3}\nu^{-7/3} + O(\nu^{-3}).
\end{aligned} \tag{10.6.27}
$$

The Laguerre polynomials are orthogonal on $[0, \infty)$ only for $\alpha > -1$. As pointed out in Szegő ([1939] 1975, §6.21, (4)), it follows from Markov's theorem, Theorem 10.2.2, that the zeros decrease as α decreases. The smallest zero becomes 0 when $\alpha = -1$. When $\alpha < -1$ the zeros are generally complex. Note that 0 is a zero of $L_n^{(\alpha)}$ when and only when $\alpha = -1, -2, \ldots, -n$ (Szegő, [1939] 1975, §6.73). Its multiplicity is $|\alpha|$. For all other values of α the zeros are simple. The number of positive zeros is n if $\alpha > -1$; it is $n + [\alpha] + 1$ if $-n < \alpha < -1$ and it is 0 if $\alpha < -n$ and the number of negative zeros is 0 or 1 (Szegő, [1939] 1975, Theorem 6.73). By an argument of the kind given in Szegő ([1939] 1975, §6.72), it can be seen that as α decreases through each $-k$ $(k = 1, 2, \ldots, n)$, k zeros approach the origin at angles $2m\pi/k$, $m = 0, \ldots, k-1$, collide (if $k \geq 2$), and emerge at angles $\pi - 2m\pi/k$, $m = 0, \ldots, k-1$. For $n \geq 2$, although the reciprocal of each zero becomes infinite, as $\alpha \to -k$, the sum of their reciprocals approaches a finite negative value: Let $x_1(\alpha), \ldots, x_k(\alpha)$ be the k $(2 \leq k \leq n)$ zeros of $L_n^{(\alpha)}(x)$ in a neighborhood of $x = 0$ for $\alpha \sim -k$. Then (DeFazio, Gupta, and Muldoon, 2007)

$$\lim_{\alpha \to -k} \sum_{m=1}^{k} \frac{1}{x_m(\alpha)} = \frac{k(k-2n-1)}{k^2-1}. \tag{10.6.28}$$

The number of zeros in each of the real intervals $(-\infty, -1), (-1, 1)$, and $(1, \infty)$ is given by results of Hilbert and Klein (Szegő, [1939] 1975, Theorem 6.7.2).

10.7 Hermite Polynomials and Functions

The zeros of the Hermite polynomial H_n are located symmetrically in the interval $(-\infty, \infty)$. We will use the notation x_k or x_{nk} for the kth positive zero, in increasing order, of H_n. We will also use the notation $h_{n1}, h_{n2}, \ldots, h_{nn}$ for the zeros in *decreasing* order. Although $x_k = h_{\lfloor n/2 \rfloor - k+1}$, $k = 1, 2, \ldots, \lfloor n/2 \rfloor$, it will be convenient to retain both notations.

Since $w(x) = \exp(-x^2) H_n(x)$ satisfies

$$w'' + \left(2n + 1 - x^2\right) w = 0, \tag{10.7.1}$$

all the points of inflection of w are either zeros of H_n or equal to $\pm(2n+1)^{1/2}$, we see that the largest zero is $< (2n+1)^{1/2}$. However, better upper bounds are given in (10.7.4).

It follows from Section 10.2.6(iii) that the sequence $x_{n0}, x_{n1}, x_{n2}, \ldots$ is convex where $x_{n0} = 0$ if n is odd and $x_{n0} = -x_{n1}$ if n is even.

Simple Sturmian comparisons give (Szegő, [1939] 1975, (6.31.19))

$$\left.\begin{array}{l} \dfrac{k-\frac{1}{2}}{(2n+1)^{\frac{1}{2}}}\pi \\ \dfrac{k}{(2n+1)^{\frac{1}{2}}}\pi \end{array}\right\} < x_{nk} < \left\{\begin{array}{l} \dfrac{4k+1}{(2n+1)^{\frac{1}{2}}}\pi, \\ \dfrac{4k+3}{(2n+1)^{\frac{1}{2}}}\pi, \quad \nu = 1, 2, \ldots, [n/2]. \end{array}\right. \tag{10.7.2}$$

For the least positive zero x_{n1}, we have $x_{21} = 2^{-1/2}$, $x_{31} = (3/2)^{1/2}$, and

$$x_{n1} < \left\{\begin{array}{l} \left(\frac{5/2}{2n+1}\right)^{\frac{1}{2}}, \\ \left(\frac{21/2}{2n+1}\right)^{\frac{1}{2}}, \quad n \geq 2. \end{array}\right. \tag{10.7.3}$$

These inequalities and further refinements are in Szegő ([1939] 1975, §6.31).

Since the Hermite polynomials are a special case of the Laguerre polynomials, via the relations (3.8.10), (3.8.11), we can get various relations from the results of Section 10.6. We have (Szegő, [1939] 1975, (6.2.14), (6.2.18)), for the largest zero,

$$\left(\frac{n-1}{2}\right)^{1/2} < h_{1n} \leq \frac{n-1}{(n/2+1)^{1/2}}. \tag{10.7.4}$$

Gatteschi (see Gatteschi, 1988b and Gautschi and Giordano, 2008, p. 21) gives the following upper and lower bounds for zeros of Hermite polynomials:

$$h_{n,\lfloor (n+1)/2\rfloor + r} < \sqrt{2n+1} \times \begin{cases} \cos\left[\frac{1}{2}x\left(\frac{2n-4r+3}{2n+1}\pi\right)\right], & n \text{ even}, \\ \cos\left[\frac{1}{2}x\left(\frac{2n-4r+1}{2n+1}\pi\right)\right], & n \text{ odd}, \end{cases} \tag{10.7.5}$$

where $x = x(y)$ is the inverse function of $y = \sin x - x$, and

$$h_{n,\lfloor (n+1)/2\rfloor + r} > \sqrt{2n+1}\left[\frac{1}{2}x\left(\frac{8}{3(2n+1)}a^{3/2}_{\lfloor n/2\rfloor+1-r}\right)\right], \quad r = 1, 2, \ldots, \lfloor n/2\rfloor. \tag{10.7.6}$$

It is shown by Dimitrov and Nikolov (2010) that, for $k = 1, 2, \ldots, \lfloor n/2\rfloor$, the quantities x_k^2 have the upper and lower bounds

$$\frac{n^2 - \frac{3}{2}n + 2 \pm (n-2)\sqrt{n^2+n+4}}{n+4}.$$

When n is odd, this can be improved to

$$\frac{n^2 - \frac{5}{2}n + \frac{15}{2} \pm (n-3)\sqrt{n^2+n+10}}{n+3}.$$

In Szegő ([1939] 1975, Theorem 6.32) we find upper bounds for the largest (and other) zeros of Hermite polynomials in terms of zeros of Airy functions.

In Area et al. (2004) sharp bounds for the zeros of the Hermite polynomials are obtained. For example, it is shown that

$$h_{nk} \leq \sqrt{2n-2}\cos\frac{(k-1)\pi}{n-1}, \quad k = 1, 2, \ldots, \lfloor n/2\rfloor. \tag{10.7.7}$$

Using (10.6.12), we obtain the determinantal inequality

$$\begin{vmatrix} x_{nk} & x_{n,k+l} \\ x_{n+2m,k} & x_{n+2m,k+l} \end{vmatrix} < 0,$$

for $k = 1, 2, \ldots, \lfloor n/2\rfloor$, $m = 1, 2, \ldots,$ and $l = 1, 2, \ldots, n-k$.

The Hermite functions $H_\lambda(t)$ and $G_\lambda(t)$ are defined in Section 3.8. It is shown in Elbert and Muldoon (1999, Theorem 3.1) that, for $n < \lambda \leq n+1$, $n = 0, 1, \ldots, H_\lambda$ has $n+1$ real zeros and it has no real zeros when $\lambda \le 0$. Furthermore, if $h = h(\lambda, k)$ is the kth largest real zero of $H_\lambda(t)$, then (Elbert and Muldoon, 1999, Theorem 7.1)

$$\frac{dh}{d\lambda} = \frac{\sqrt{\pi}}{2} \int_0^\infty e^{-(2\lambda+1)\tau} \phi\left(h\sqrt{\tanh \tau}\right) \frac{d\tau}{\sqrt{\sinh \tau \cosh \tau}}, \tag{10.7.8}$$

where $\phi(x) = e^{x^2} \operatorname{erfc}(x)$, and erfc is the complementary error function,

$$\operatorname{erfc}(x) = 1 - \operatorname{erf}(x) = \frac{2}{\sqrt{\pi}} \int_x^\infty e^{-t^2}\, dt.$$

From this it follows that each $h(\lambda, k)$ is an increasing function of λ on its interval of definition and (Elbert and Muldoon, 1999, Corollary 7.2)

$$(-1)^{r+1} \frac{d^r}{d\lambda^r} h(\lambda, k) > 0, \quad \lambda > k - 1,\ r = 1, 2, \ldots. \tag{10.7.9}$$

In addition,

$$\operatorname{erf}[h(\lambda, 1)] \le \lambda - 1, \quad 0 < \lambda < \infty, \tag{10.7.10}$$

with equality only for $\lambda = 1$, and for $\lambda \to 0^+$. This was shown in Elbert and Muldoon (1999, Theorem 8.1), following a partially numerical proof in Hayman and Ortiz (1975/76).

We may, as in Muldoon (2008), define the zeros $h(\lambda, \kappa)$ for continuous rank κ by

$$\int_{h(\lambda,\kappa)}^\infty \frac{du}{p_\lambda(u)} = \kappa\pi, \tag{10.7.11}$$

where

$$p_\lambda(x) = \frac{2^{-\lambda-1}\sqrt{\pi}}{\Gamma(\lambda + 1)} e^{-x^2} \left[H_\lambda^2(x) + G_\lambda^2(x)\right]. \tag{10.7.12}$$

Then we have the following result:

Theorem 10.7.1

$$(-1)^r \frac{d^r}{d\kappa^r} h(\lambda, \kappa) > 0, \quad 0 < \kappa < \frac{\lambda + 1}{2}, \ r = 0, 1, 2, \ldots. \tag{10.7.13}$$

Theorem 10.7.1 is a continuous analogue of Durand's result (Durand, 1975, pp. 371–372) that for the positive zeros of $H_n(x)$, with fixed n,

$$(-1)^r \Delta_{(k)}^r h_{nk} > 0, \quad k = 1, 2, \ldots, \lfloor n/2 \rfloor. \tag{10.7.14}$$

(It should be noted that Durand lists the zeros in increasing order so the $(-1)^r$ does not appear in his result.)

Elbert and Muldoon (2008) find a convergent series expansion of the zeros of Hermite functions in the form

$$h(\lambda) = \Lambda + \Lambda^{-1/3} \sum_{k=1}^\infty a_k \Lambda^{-4(k-1)/3},$$

for large λ, where $\Lambda = \sqrt{2\lambda+1}$ and a_k is a polynomial in the positive zeros of the Airy function. The first few terms are given by

$$h(\lambda) = \Lambda + a_1\Lambda^{-1/3} - \frac{1}{10}a_1^2\Lambda^{-5/3} + \left[\frac{9}{280} + \frac{11}{350}a_1^3\right]\Lambda^{-3} + \cdots, \tag{10.7.15}$$

where $\Lambda = \sqrt{2\lambda+1}$, $a_1 = -6^{-1/3}i_1$, and i_1 is the first positive zero of the Airy function $A(x)$.

For results for Hermite polynomials, similar to (10.3.22) for Jacobi polynomials, see Calogero (2001, Appendix C).

10.8 Other Orthogonal Polynomials

We conclude this chapter with some miscellaneous applications of the methods of Section 10.2 to the zeros of other orthogonal polynomials.

We can use Theorem 10.2.7 in the case of the polynomials of Section 4.1 to get the following theorem:

Theorem 10.8.1 *The zeros of birth and death process polynomials belong to* $(0,\infty)$ *while the zeros of random walk polynomials belong to* $(-1,1)$.

For the associated Laguerre polynomials of Section 4.3, we can use Corollary 10.2.4 of the Hellmann–Feynman theorem to get the following result:

Theorem 10.8.2 *The zeros of the associated Laguerre polynomials* $L_n^{(\alpha)}(x;c)$ *increase with* α *for* $\alpha \geq 0$ *and* $c > -1$.

The special case $c = 0$ recovers the result that the zeros of Laguerre polynomials increase with α, $\alpha \geq 0$. However, it is known (Szegő, [1939] 1975, §6.21,(4)) that this increase holds for $\alpha > -1$.

We also have, using Theorem 10.2.8 on chain sequences, the following theorem:

Theorem 10.8.3 *Let* $L^{(c)}(N,\alpha)$ *and* $I^{(c)}(N,\alpha)$ *be the largest and smallest zeros for an associated Laguerre polynomial* $L_n^{(\alpha)}(x;c)$. *Then*

$$L^{(c)}(N,\alpha) < 2N + 2c + \alpha - 2 + \sqrt{1 + a(N+c-1)(N+c+\alpha-1)}, \tag{10.8.1}$$

$$I^{(c)}(N,\alpha) > 2N + 2c + \alpha - 2 - \sqrt{1 + 4(N+c-1)(N+c+\alpha-1)}, \tag{10.8.2}$$

where a is as in (10.6.15).

The associated Laguerre polynomials do not satisfy a second-order differential equation, so the Sturmian techniques of Section 10.2.6 are not applicable.

For the Meixner polynomials of Section 6.1, we have the next result, using Theorem 10.2.8:

Theorem 10.8.4 *Let* $m_{N,1}$ *be the largest zero of* $M_N(x\sqrt{c}/(1-c);\beta,c)$. *Then, with a defined by* (10.6.15), *we have*

$$m_{N,1} \leq \sqrt{c}\beta + \left(N - \frac{1}{2}\right)\frac{1+c}{\sqrt{c}} + \frac{1}{2\sqrt{c}}\sqrt{(1+c)^2 + 4acN(N+\beta-1)}. \tag{10.8.3}$$

The bound (10.8.3) *is sharp in the sense that*

$$m_{N,1} = \frac{\left(1+\sqrt{c}\right)^2}{\sqrt{c}} N(1+o(N)), \quad \text{as } N \to \infty. \tag{10.8.4}$$

Ismail and Li (1992a) also contains bounds on the largest and smallest zeros of Meixner–Pollaczek polynomials of Section 5.1.

As another example, consider the Meixner polynomials. The corresponding Jacobi matrix $A_N = (a_{j,k})$ is

$$\begin{aligned} a_{j,k} = & \frac{\sqrt{c(j+1)(j+\beta)}}{1-c} \delta_{j,k-1} + \frac{j + c(j+\beta)}{1-c} \delta_{j,k} \\ & + \frac{\sqrt{cj(j+\beta-1)}}{1-c} \delta_{j,k+1}. \end{aligned} \tag{10.8.5}$$

It follows from Corollary 10.2.4 that the zeros of the Meixner polynomial $M_n(x;\beta,c)$ increase with β when $\beta > 1$. To study the dependence of the zeros on the parameter c, it is convenient to use the renormalization

$$p_n(x;\beta,c) := (-1)^n c^{n/2} \sqrt{\frac{(\beta)_n}{n!}} M_n\left(\frac{x\sqrt{c}}{1-c};\beta,c\right).$$

In view of (6.1.14), the zeros $x_{n,k}(\beta,c)$ of $p_n(x;\beta,c)$ converge to the corresponding zeros $l_{n,1}(\beta-1)$ of $L_n^{(\beta-1)}(x)$, as $c \to 1^-$.

Theorem 10.8.5 *The quantities $x_{n,j}(\beta,c)$ increase with c on the interval $(n-1)/(\beta+n-1) < c < 1$ and converge to $l_{n,j}(\beta-1)$ as $c \to 1^-$.*

The rate of convergence is also estimated in Ismail and Muldoon (1991).

It follows from Theorem 10.2.2 that the zeros of the Hahn polynomial $Q_n(x;\alpha,\beta,N)$ increase (decrease) as β (α) increases (Ismail, 2005b, Theorem 7.1.2).

The Lommel polynomials of Section 6.4 and the q-Lommel polynomials are among those to which Theorem 10.2.5 may be applied to get monotonicity properties of their zeros with respect to ν.

Acknowledgment. We are grateful to Dharma P. Gupta and Kathy Driver for some corrections and comments.

11
The Moment Problem

Christian Berg and Jacob S. Christiansen

11.1 Hamburger Moment Problems

The **moment problem** is the characterization of those real sequences that can appear as moment sequences together with the problem of recovering a positive measure from its moments. So given a sequence of real numbers $(m_n)_n$ we wish to find out if there exists a positive measure μ such that

$$m_n = \int_{\mathbb{R}} x^n \, d\mu(x), \quad n \geq 0, \tag{11.1.1}$$

and in the affirmative case to find all such measures. This is the **Hamburger moment problem** if there is no restriction imposed on the support of μ. One usually normalizes the problem by requiring $m_0 = 1$. Hamburger proved that a sequence of real numbers $(m_n)_n$ is a Hamburger moment sequence if and only if the sequence is **positive definite** in the sense that

$$\sum_{k,l=0}^{n} m_{k+l} c_k c_l \geq 0 \quad \text{for all } c_0, c_1, \ldots, c_n \in \mathbb{R} \text{ and for all } n \geq 0, \tag{11.1.2}$$

or equivalently that all the Hankel matrices H_n, $n \geq 0$,

$$H_n = \begin{pmatrix} m_0 & m_1 & \cdots & m_n \\ m_1 & m_2 & \cdots & m_{n+1} \\ \vdots & \vdots & & \vdots \\ m_n & m_{n+1} & \cdots & m_{2n} \end{pmatrix}, \tag{11.1.3}$$

are positive semidefinite.

A Hamburger moment sequence $(m_n)_n$ is called **degenerate** if (11.1.1) has a solution μ with finite support. If the support has $k \geq 0$ points then $\det(H_n) = 0$ for $n \geq k$ and $\det(H_n) > 0$ for $n = 0, 1, \ldots, k-1$. If $(m_n)_n$ is nondegenerate then $\det(H_n) > 0$ for all $n \geq 0$ and any solution of (11.1.1) has infinite support.

In the **Stieltjes moment problem** the support of μ is restricted to being a subset of $[0, \infty)$. Stieltjes (1894) proved that $(m_n)_n$ is a Stieltjes moment sequence if and only if $(m_n)_n$ as well as $(m_{n+1})_n$ are positive definite sequences. It is interesting to note that Stieltjes solved the moment problem named after him in 1894 while Hamburger's work appeared in 1920.

When a solution μ to a Hamburger problem (Stieltjes problem) is unique, the moment problem is called **determinate** in the sense of Hamburger (in the sense of Stieltjes), otherwise it is called **indeterminate** in the sense of Hamburger (Stieltjes). The word Hamburger or Stieltjes is omitted if no ambiguity is possible. There exist Stieltjes moment sequences which are indeterminate in the sense of Hamburger but determinate in the sense of Stieltjes. For an example of this depending on two parameters, with complete description of the various possibilities in terms of the parameters, see Berg and Valent (1994).

A Hamburger moment sequence satisfies $m_n = O(r^n)$, $n \to \infty$ for some $r > 0$ if and only if the measure μ is uniquely determined and supported within the interval $[-r, r]$.

In the **Hausdorff moment problem** the support of μ is required to be a subset of $[0, 1]$. A necessary and sufficient condition is that it is a bounded Stieltjes moment sequence. Hausdorff found necessary and sufficient conditions in terms of iterated differences

$$(-1)^k(\Delta^k m)_n = \sum_{j=0}^{k}(-1)^j\binom{k}{j}m_{n+j} \geq 0, \quad k, n \geq 0. \tag{11.1.4}$$

These conditions are usually expressed by saying that the sequence $(m_n)_n$ is **completely monotonic**. Here $(\Delta m)_n = m_{n+1} - m_n$ and $(\Delta^k m)_n = \Delta^{k-1}(\Delta m)_n$.

Our principal references on the moment problem are Akhiezer (1965), Shohat and Tamarkin (1950), Simon (2005b), and Stone (1932) and in an abstract setting Berg, Christensen, and Ressel (1984). Results in the following without references can be found in these monographs.

Let $(m_n)_n$ denote a nondegenerate Hamburger moment sequence. An important tool is the corresponding sequence $(p_n)_n$ of orthonormal polynomials.

Theorem 11.1.1 (Riesz, 1923) *Assume that μ is the unique solution to the Hamburger moment sequence $(m_n)_n$ and $\mu(\mathbb{R}) = 1$. Then the following hold:*

(i) *$(p_n)_n$ is an orthonormal basis for the Hilbert space $L^2(\mathbb{R}, \mu)$, hence for any $f \in L^2(\mathbb{R}, \mu)$ the orthogonal expansion*

$$f \sim \sum_{k=0}^{\infty} c_k p_k, \quad c_k = \int_{-\infty}^{\infty} f(x)p_k(x)\,d\mu(x)$$

converges to f in $L^2(\mathbb{R}, \mu)$, that is,

$$\lim_{n\to\infty} \int_{-\infty}^{\infty} \left| f(x) - \sum_{k=0}^{n} c_k p_k(x) \right|^2 d\mu(x) = 0 \tag{11.1.5}$$

and **Parseval's formula** *holds*

$$\int_{-\infty}^{\infty} |f(x)|^2\, d\mu(x) = \sum_{k=0}^{\infty} |c_k|^2. \tag{11.1.6}$$

(ii) *μ has an atom at $x = u$ if and only if*

$$S := \sum_{n=0}^{\infty} p_n^2(u) < \infty. \tag{11.1.7}$$

Furthermore, if μ has an atom at $x = u$, then

$$\mu(\{u\}) = 1/S. \tag{11.1.8}$$

The following theorem gives necessary and sufficient conditions for the determinacy/indeterminacy of a Hamburger moment problem. (See also the necessary and sufficient condition at the end of Section 2.3.)

Theorem 11.1.2 *The Hamburger moment problem is indeterminate if and only if the series*

$$\sum_{n=0}^{\infty} |p_n(z)|^2 \tag{11.1.9}$$

converges for all z, uniformly on compact subsets of $\mathbb{C}$. For determinacy it is sufficient that it diverges for one $z \notin \mathbb{R}$ and then it diverges for all nonreal z and it converges only at those at most countably many points of the real axis where μ has atoms.

The theorem above may be difficult to apply unless the asymptotics of p_n is known. When applicable, as in Askey and Ismail (1984), the result is powerful. The following two criteria are very useful.

Theorem 11.1.3 (Carleman) *Let $(m_n)_n$ be a Hamburger moment sequence and let $(a_n)_n$ be given by* (2.1.17).

If

$$\sum_{n=1}^{\infty} \frac{1}{a_n} = \infty \tag{11.1.10}$$

or if

$$\sum_{n=0}^{\infty} \frac{1}{\sqrt[2n]{m_{2n}}} = \infty, \tag{11.1.11}$$

then the moment problem is determinate.

Remark 11.1.4 By an inequality of Carleman,

$$\sum_{n=1}^{\infty} \frac{1}{\sqrt[2n]{m_{2n}}} \le e \sum_{n=1}^{\infty} \frac{1}{a_n}$$

so condition (11.1.10) is better than (11.1.11).

The reader is warned that the conditions in Carleman's criterion are sufficient but by no means necessary. See, however, the comment after Theorem 11.1.22. Similarly, the following result contains a sufficient but not necessary condition for indeterminacy.

Theorem 11.1.5 (Krein) *Assume that the probability measure μ on $\mathbb{R}$ has a density $d\mu = \omega(x)\,dx$ with $\omega(x) > 0$ for $x \in \mathbb{R}$ satisfying*

$$\int_{-\infty}^{\infty} \frac{\log \omega(x)}{1+x^2}\,dx > -\infty. \tag{11.1.12}$$

Then μ is indeterminate.

Proofs of Krein's theorem are in Berg (1995) and Simon (2005b). Krein's theorem can be extended: in inequality (11.1.12) it is enough to integrate over $\mathbb{R} \setminus (-K, K)$ for K sufficiently large to ensure the indeterminacy; see Pedersen (1998) for proof and further refinements.

It follows from Theorems 11.1.3 and 11.1.5 that the **Freud weight functions**

$$\omega_\alpha(x) = C_\alpha \exp\left(-|x|^\alpha\right), \quad x \in \mathbb{R} \tag{11.1.13}$$

(C_α is a normalization factor) are determinate if and only if $\alpha \geq 1$.

The determinate case is important for the following result.

Theorem 11.1.6 (Method of moments) *Let $\{\mu_n\}$ and μ be probability measures on $\mathbb{R}$ with moments of any order and assume that μ is determinate.*

If

$$\lim_{n\to\infty} \int x^k\,d\mu_n(x) = \int x^k\,d\mu(x) \quad \text{for all } k \geq 0, \tag{11.1.14}$$

then μ_n converges weakly to μ, that is,

$$\lim_{n\to\infty} \int f(x)\,d\mu_n(x) = \int f(x)\,d\mu(x) \tag{11.1.15}$$

for all continuous and bounded functions $f\colon \mathbb{R} \to \mathbb{C}$.

As an application of the method of moments one gets the following theorem:

Theorem 11.1.7 (Markov) *Let $(m_n)_n$ be a determinate Hamburger moment sequence. The Stieltjes transform of the unique solution μ to* (11.1.1) *satisfies*

$$\lim_{n\to\infty} \frac{P_n^*(z)}{P_n(z)} = \int_{\mathbb{R}} \frac{d\mu(x)}{z-x}, \quad z \in \mathbb{C} \setminus \Lambda, \tag{11.1.16}$$

where

$$\Lambda = \cap_{N=1}^{\infty} \overline{M_N}, \quad M_N = \cup_{n=N}^{\infty} \{x_{n,1}, \ldots, x_{n,n}\}, \tag{11.1.17}$$

and $x_{n,j}$ are the zeros of the nth monic orthogonal polynomial P_n. The convergence in (11.1.16) *is uniform on compact subsets of $\mathbb{C} \setminus \Lambda$.*

This result is an extension of Theorem 2.4.3. For historical comments see Berg (1994).

Indeterminacy of a nondegenerate Hamburger moment problem can be characterized in terms of the behavior of the smallest eigenvalue Λ_n of the Hankel matrix H_n given in (11.1.3). Since H_n is positive definite, the smallest eigenvalue is positive and given as

$$\Lambda_n = \min\left\{\sum_{k,l=0}^{n} m_{k+l}c_k c_l \;\middle|\; c_0, c_1, \ldots, c_n \in \mathbb{R},\ \sum_{k=0}^{n} c_k^2 = 1\right\}. \tag{11.1.18}$$

Since H_n is a submatrix of H_{n+1} we have $\Lambda_n \ge \Lambda_{n+1}$, so $\Lambda_\infty := \lim_{n\to\infty} \Lambda_n$ exists and is ≥ 0.

Theorem 11.1.8 (Berg, Chen, and Ismail, 2002) *The Hamburger problem is determinate if and only if* $\Lambda_\infty = 0$.

In the indeterminate case, characterized by $\Lambda_\infty > 0$, *we have* $\Lambda_\infty \ge \rho_0^{-1}$, *where*

$$\rho_0 = \frac{1}{2\pi}\int_0^{2\pi} \sum_{n=0}^{\infty} |p_n(e^{it})|^2\, dt. \tag{11.1.19}$$

Many results about the behavior of the sequence $(\Lambda_n)_n$ of smallest eigenvalues of H_n can be found in Berg and Szwarc (2011).

11.1.1 Indeterminate Moment Problems

Consider the polynomials

$$A_{n+1}(z) = z\sum_{k=0}^{n} \frac{P_k^*(0)P_k^*(z)}{\zeta_k} = \frac{P_{n+1}^*(z)P_n^*(0) - P_{n+1}^*(0)P_n^*(z)}{\zeta_n}, \tag{11.1.20}$$

$$B_{n+1}(z) = -1 + z\sum_{k=0}^{n} \frac{P_k^*(0)P_k(z)}{\zeta_k} = \frac{P_{n+1}(z)P_n^*(0) - P_{n+1}^*(0)P_n(z)}{\zeta_n}, \tag{11.1.21}$$

$$C_{n+1}(z) = 1 + z\sum_{k=0}^{n} \frac{P_k(0)P_k^*(z)}{\zeta_k} = \frac{P_{n+1}^*(z)P_n(0) - P_{n+1}(0)P_n^*(z)}{\zeta_n}, \tag{11.1.22}$$

$$D_{n+1}(z) = z\sum_{k=0}^{n} \frac{P_k(0)P_k(z)}{\zeta_k} = \frac{P_{n+1}(z)P_n(0) - P_{n+1}(0)P_n(z)}{\zeta_n}. \tag{11.1.23}$$

The above and the Casorati determinant (Wronskian) evaluation imply

$$A_n(z)D_n(z) - B_n(z)C_n(z) = 1. \tag{11.1.24}$$

Note that the formulas above simplify if the monic polynomials P_n and P_n^* are replaced by the orthonormal polynomials $p_n(x) = P_n(x)/\sqrt{\zeta_n}$ and the corresponding numerator polynomials $p_n^*(x) = P_n^*(x)/\sqrt{\zeta_n}$.

Theorem 11.1.9 (Nevanlinna) *For an indeterminate Hamburger moment problem the polynomials* $A_n(z), B_n(z), C_n(z), D_n(z)$ *converge locally uniformly in* $\mathbb{C}$ *to real entire functions* $A(z)$, $B(z)$, $C(z)$, $D(z)$, *respectively. Each of these functions has infinitely many real and simple zeros and no other zeros. The zeros of A and C (respectively B and D) interlace.*

The **Nevanlinna matrix** is

$$\begin{pmatrix} A(z) & C(z) \\ B(z) & D(z) \end{pmatrix} \tag{11.1.25}$$

and its determinant is 1.

The point $z = 0$ plays a special role in the above formulas and therefore it is convenient to introduce entire functions of two complex variables

$$\mathcal{A}(u, v) = (v - u) \sum_{k=0}^{\infty} \frac{P_k^*(u) P_k^*(v)}{\zeta_k}, \tag{11.1.26}$$

$$\mathcal{B}(u, v) = -1 + (v - u) \sum_{k=0}^{\infty} \frac{P_k^*(u) P_k(v)}{\zeta_k}, \tag{11.1.27}$$

$$C(u, v) = 1 + (v - u) \sum_{k=0}^{\infty} \frac{P_k(u) P_k^*(v)}{\zeta_k}, \tag{11.1.28}$$

$$\mathcal{D}(u, v) = (v - u) \sum_{k=0}^{\infty} \frac{P_k(u) P_k(v)}{\zeta_k}, \tag{11.1.29}$$

where the series converge uniformly on compact subsets of $\mathbb{C} \times \mathbb{C}$. Clearly,

$$\mathcal{A}(u, v) = -\mathcal{A}(v, u), \quad \mathcal{B}(u, v) = -C(v, u), \quad \mathcal{D}(u, v) = -\mathcal{D}(v, u), \tag{11.1.30}$$

and the functions satisfy the fundamental identity

$$\mathcal{A}(u, v)\mathcal{D}(u, v) - \mathcal{B}(u, v)C(u, v) = 1; \tag{11.1.31}$$

see Buchwalter and Cassier (1984a). If the first variable is put equal to zero we get the functions

$$A(z) = \mathcal{A}(0, z), \quad B(z) = \mathcal{B}(0, z), \quad C(z) = C(0, z), \quad D(z) = \mathcal{D}(0, z) \tag{11.1.32}$$

from the Nevanlinna matrix. On the other hand, the functions $\mathcal{A}$, $\mathcal{B}$, C, $\mathcal{D}$ can be expressed in a simple way using the functions in the Nevanlinna matrix (see Buchwalter and Cassier, 1984a):

$$\mathcal{A}(u, v) = A(v)C(u) - A(u)C(v), \tag{11.1.33}$$

$$\mathcal{B}(u, v) = B(v)C(u) - A(u)D(v), \tag{11.1.34}$$

$$\mathcal{D}(u, v) = B(v)D(u) - B(u)D(v). \tag{11.1.35}$$

A **Pick function** is a holomorphic function $\varphi\colon \mathbb{H} \to \mathbb{C}$ satisfying $\operatorname{Im} \varphi(z) \geq 0$ for $z \in \mathbb{H}$, where $\mathbb{H}$ denotes the open upper half-plane. Pick functions are extended to the open lower half-plane by reflection, that is, $\varphi(\overline{z}) = \overline{\varphi(z)}$. Pick functions also occur under the names **Nevanlinna functions** or **Herglotz functions** and are treated in Donoghue (1974). They have the integral representation

$$\varphi(z) = \alpha z + \beta + \int_{-\infty}^{\infty} \frac{tz + 1}{t - z} d\tau(t), \quad z \in \mathbb{C} \setminus \mathbb{R}, \tag{11.1.36}$$

where $\alpha \geq 0, \beta \in \mathbb{R}$, and τ is a positive measure on $\mathbb{R}$ with finite total mass. Also α, β, and τ are uniquely determined by φ.

The set of Pick functions is denoted $\mathcal{P}$. A Pick function is either a real constant ($\alpha = 0$, $\tau = 0$) and is called degenerate, or it maps the open upper half-plane into itself and is called nondegenerate. The composition of two nondegenerate Pick functions is again a nondegenerate Pick function. The space of holomorphic functions on a domain G carries the topology of locally uniform convergence on G. With this topology $\mathcal{P}$ becomes a locally compact space and we let $\mathcal{P} \cup \{\infty\}$ denote the one-point compactification by adding a function which is ∞ at all points of the cut plane. Details are in Berg and Christensen (1981). This compact space is the parameter space for Nevanlinna's parametrization of the indeterminate moment problem from 1922:

Theorem 11.1.10 (Nevanlinna parametrization) *The formula*

$$\int_{\mathbb{R}} \frac{d\mu_\varphi(x)}{x-z} = -\frac{A(z)\varphi(z) - C(z)}{B(z)\varphi(z) - D(z)}, \quad z \in \mathbb{C} \setminus \mathbb{R} \tag{11.1.37}$$

establishes a one-to-one correspondence between solutions μ of the moment problem and functions φ in $\mathcal{P} \cup \{\infty\}$. The correspondence is also a homeomorphism when the set of solutions carries the weak topology.

A proof can be found in Akhiezer (1965), Buchwalter and Cassier (1984a), and Simon (2005b). We recall that the weak topology on the set V of solutions to an indeterminate moment problem is the coarsest topology such that the map $\mu \to \int_{\mathbb{R}} f\, d\mu$ of V to $\mathbb{C}$ is continuous for all continuous and bounded functions $f: \mathbb{R} \to \mathbb{C}$.

A solution of the moment problem is called **N-extremal** (N for Nevanlinna) if φ is a real constant t, respectively $\varphi = \infty$. In this case the right-hand side of (11.1.37) is the meromorphic function

$$-\frac{A(z)t - C(z)}{B(z)t - D(z)}, \quad \text{respectively} \quad -\frac{A(z)}{B(z)}. \tag{11.1.38}$$

For $t \in \mathbb{R}$, the zeros of $B(z)t - D(z)$ are real and simple and

$$\Lambda_t = \{z \in \mathbb{C} \mid B(z)t - D(z) = 0\}, \quad \Lambda_\infty = \{z \in \mathbb{C} \mid B(z) = 0\} \tag{11.1.39}$$

are countable discrete sets satisfying

$$\Lambda_t \cap \Lambda_s = \emptyset \text{ for } t \neq s, \quad \bigcup_{t \in \mathbb{R} \cup \{\infty\}} \Lambda_t = \mathbb{R}. \tag{11.1.40}$$

From (11.1.37) it now follows that μ_t is a discrete measure concentrated on Λ_t, that is,

$$\mu_t = \sum_{\lambda \in \Lambda_t} c_\lambda \delta_\lambda, \tag{11.1.41}$$

and since $B(\lambda)t = D(\lambda)$, we find

$$c_\lambda = \frac{1}{B'(\lambda)D(\lambda) - D'(\lambda)B(\lambda)} = \left(\sum_{n=0}^{\infty} p_n^2(\lambda)\right)^{-1}. \tag{11.1.42}$$

Every $x_0 \in \mathbb{R}$ is an atom for precisely one N-extremal solution μ_t, where t is determined from $B(x_0)t - D(x_0) = 0$ (with $t = \infty$ if $B(x_0) = 0$).

A Hamburger moment problem is indeterminate if and only if the matrix operator T defined by the action of the Jacobi matrix on ℓ^2 has deficiency indices $(1,1)$. The spectral measures of the self-adjoint extensions of this operator in ℓ^2 are in one-to-one correspondence with the N-extremal solutions of the indeterminate moment problem, while the other solutions of the moment problem come from the spectral measures of self-adjoint extensions which go out of the space. The details are in Akhiezer (1965, Chapter 4). An up-to-date account is in Simon (2005b).

Theorem 11.1.11 *The set V of solutions to an indeterminate Hamburger moment problem is a compact convex set in the weak topology. For $\mu \in V$ the following holds:*

(1) (Riesz, 1923) *The polynomials are dense in $L^2(\mathbb{R}, \mu)$, that is, the corresponding orthonormal polynomials form a complete system in $L^2(\mathbb{R}, \mu)$ if and only if μ is an N-extremal solution.*

(2) (Naimark, 1947) *The polynomials are dense in $L^1(\mathbb{R}, \mu)$ if and only if μ is an extreme point of V.*

Recall that a point in a convex set is called extreme if it is not an interior point of a segment belonging to the convex set. Thus, the extreme points of a convex polygon or polyhedron are the corners. The set V is however of infinite dimension. The reader is warned that the N-extremal solutions are called **extremal** in Shohat and Tamarkin (1950, p. 60). They are extreme points in V because denseness in L^2 is a stronger property than denseness in L^1, but there are many extreme points in V which are not N-extremal solutions, as discussed below.

For a solution μ to an indeterminate moment problem the polynomials are never dense in $L^p(\mathbb{R}, \mu)$ when $p > 2$; cf. Berg and Christensen (1981).

Determinacy can be characterized by a polynomial denseness property equivalent to the assertion that the operator T has deficiency indices $(0, 0)$:

Theorem 11.1.12 (Riesz, 1923) *A positive measure μ with moments of any order is determinate if and only if the polynomials are dense in $L^2(\mathbb{R}, (1 + x^2)\, d\mu(x))$.*

A solution μ to an indeterminate Hamburger moment problem is called ***n*-canonical**, $n \geq 0$, if the closure of the polynomials in $L^2(\mathbb{R}, \mu)$ has codimension n. By the theorem of M. Riesz the N-extremal solutions are the same as the 0-canonical. The n-canonical solutions are always discrete measures and via the Nevanlinna parametrization they correspond exactly to the Pick functions which are rational functions of degree n, the degree of a rational function $\varphi = p/q$ being $\deg(\varphi) = \max(\deg(p), \deg(q))$, when p, q are polynomials without common zeros. See Buchwalter and Cassier (1984b) for details.

A measure μ with moments of any order can be changed by considering the sequence of measures

$$d\mu_n = (1 + x^2)^n \, d\mu(x), \quad n \in \mathbb{Z}. \tag{11.1.43}$$

If μ is assumed to be N-extremal, then μ_n is n-canonical for $n = 0, 1, \ldots$, while μ_n is determinate by Theorem 11.1.12 for $n = -1$ and hence for $n < -1$. We can characterize the measures μ_n for $n < 0$ by an index of determinacy introduced below.

An N-extremal measure can also be changed by adding or removing a finite number of atoms (= point masses). Stieltjes knew already that if the atom at zero is removed from the N-extremal solution μ_0, then the new measure becomes determinate. Stieltjes formulated this in Stieltjes (1894, Sect. 65) for the case of Stieltjes moment problems.

In order to clarify what happens if more than one atom is removed from an N-extremal measure, Berg and Durán (1995, 1996) introduced an **index of determinacy** $\operatorname{ind}_a(\mu)$ of a determinate measure μ with respect to a point $a \in \mathbb{C}$:

$$\operatorname{ind}_a(\mu) = \sup\{k \in \mathbb{N} \colon |x - a|^{2k}\, d\mu(x) \text{ is determinate}\}. \tag{11.1.44}$$

Here $\mathbb{N} = \{0, 1, \ldots\}$, so the values of $\operatorname{ind}_a(\mu)$ are in $\mathbb{N} \cup \{\infty\}$.

In particular, for $a = i$ we have

$$\operatorname{ind}_i(\mu) = \sup\{k \in \mathbb{N} \colon (1 + x^2)^k\, d\mu(x) \text{ is determinate}\}. \tag{11.1.45}$$

Theorem 11.1.13 *Let $k \in \mathbb{N}$ and assume that μ is derived from an N-extremal measure by removing $k + 1$ of its atoms. Then μ is determinate and*

$$\operatorname{ind}_a(\mu) = \begin{cases} k & \text{for } a \notin \operatorname{supp}(\mu), \\ k + 1 & \text{for } a \in \operatorname{supp}(\mu). \end{cases} \tag{11.1.46}$$

The index of determinacy of a determinate measure μ is either infinite at every point in $\mathbb{C}$, and we then define $\operatorname{ind}(\mu) = \infty$, *or it is finite at some point $a \in \mathbb{C}$, and then it has the form* (11.1.46). *Moreover, μ is derived from an N-extremal measure by removing $k + 1$ atoms. In this case we define* $\operatorname{ind}(\mu) = k$.

The determinate measures of index 0 are the measures derived from an N-extremal measure by removing 1 atom.

If μ is a determinate measure of $\operatorname{ind}(\mu) = k < \infty$, then the measures μ_n are determinate for $n \le k$ and indeterminate for $n > k$.

Adding atoms to N-extremal measures leads to n-canonical measures.

Theorem 11.1.14 (Buchwalter and Cassier, 1984b) *Assume that μ is derived from an N-extremal measure by adding n atoms of arbitrary size at points disjoint from the existing atoms. Then μ is n-canonical. Conversely, if n atoms are removed from an n-canonical measure we obtain an N-extremal measure.*

Theorem 11.1.15 *Let μ be a measure with moments of any order and let μ_n be defined by* (11.1.43). *The following conditions are equivalent for $n \in \mathbb{N}$ fixed:*

(i) *μ is n-canonical;*
(ii) *μ_{-n} is N-extremal;*
(iii) *μ_{-n-1} is determinate with* $\operatorname{ind}(\mu_{-n-1}) = 0$.

Measures of finite index of determinacy have played an important role in the solution of the **Challifour problem**: there exist measures in $\mathbb{R}^k$, $k \geq 2$ which are uniquely determined by their moments but for which the polynomials in k variables are not dense in $L^2(\mathbb{R}^k, \mu)$. See Berg and Thill (1991) for details.

The compact convex set V of solutions to an indeterminate Hamburger moment problem has many interesting and surprising properties. While the set of N-extremal solutions is a closed curve in V and hence a compact set, the set $e(V)$ of extreme points of V is dense in V because it contains the set of solutions which are n-canonical for some n. This comes from the fact that any Pick function can be approximated by rational Pick functions. In particular, the set of discrete solutions is dense in V. Also the set of solutions $\mu \in V$ which are absolutely continuous with a C^∞-density with respect to Lebesgue measure is dense in V. The set of solutions $\mu \in V$ which are continuous singular with respect to Lebesgue measure is also dense in V. For details see Berg and Christensen (1981).

Theorem 11.1.16 (Gabardo, 1992) *If the absolutely continuous part of $\mu \in V$ has density ω with respect to Lebesgue measure, then the* **entropy integral**

$$\frac{1}{\pi}\int_{\mathbb{R}} \frac{\log \omega(x)}{x^2+1}\,dx$$

attains its maximum on V when

$$d\mu(x) = \frac{1}{\pi}\frac{1}{B^2(x)+D^2(x)}\,dx.$$

In general, the functions A and C are harder to find than the functions B and D, so it is desirable to find ways of determining measures from (11.1.37) without knowledge of A and C.

Let $F(z)$ denote either side of (11.1.37). The **Perron–Stieltjes inversion formula** shows that

$$\lim_{y\to 0^+} \frac{1}{\pi}\,\mathrm{Im}\, F(x+iy)\,dx = d\mu_\varphi \tag{11.1.47}$$

weakly. If the integrand to the left of (11.1.47) converges to a continuous function $h(x)$ for x in an interval I and the convergence is uniform on compact subsets of I, we can conclude that $d\mu_\varphi = h(x)\,dx$ on the interval I.

This leads to the following theorem, where only B, D enter.

Theorem 11.1.17 (Berg and Christensen, 1981; Ismail and Masson, 1994) *Let φ be a Pick function such that $\varphi(x+i0) = \lim_{y\to 0^+}\varphi(x+iy)$ exists uniformly for x in compact subsets of the interval I. If $\mathrm{Im}\,\varphi(x+i0) > 0$ for all $x \in I$, then as measures on I,*

$$d\mu_\varphi = \frac{\mathrm{Im}\,\varphi(x+i0)}{\pi\,|B(x)\varphi(x+i0) - D(x)|^2}\,dx. \tag{11.1.48}$$

Corollary 11.1.18 (Berg and Christensen, 1981) *Let $\beta \in \mathbb{R}$, $\gamma > 0$. The solution to the indeterminate moment problem corresponding to the Pick function $\varphi(z) = \beta + i\gamma$, $z \in \mathbb{H}$ is given by*

$$d\mu_\varphi = \frac{\gamma/\pi}{\gamma^2 B^2(x) + (\beta B(x) - D(x))^2}\, dx. \tag{11.1.49}$$

The maximizing solution of Theorem 11.1.16 corresponds to the Pick function $\varphi(z) = i$ for $z \in \mathbb{H}$.

In case the measure μ_φ from (11.1.37) has an atom at $u \in \mathbb{R}$, the mass depends only on B, D as the following theorem shows.

Theorem 11.1.19 (Ismail and Masson, 1994) *Let $F(z)$ denote either side of* (11.1.37). *If F has a simple pole at $z = u$ and $B(u) \neq 0$, then*

$$\mu_\varphi(\{u\}) = -\operatorname{Res}(F(z), u) = -\operatorname{Res}\left(\frac{1}{B(z)(B(z)\varphi(z) - D(z))}, u\right). \tag{11.1.50}$$

Theorem 11.1.20 (Akhiezer, 1965) *Assume that an N-extremal measure μ has an atom at $x = u$. Then*

$$\mu(\{u\}) = \left(\sum_{n=0}^{\infty} p_n^2(u)\right)^{-1} \tag{11.1.51}$$

and any other solution ν of the moment problem with $\nu \neq \mu$ satisfies $\nu(\{u\}) < \mu(\{u\})$.

Theorem 11.1.21 (Berg and Pedersen, 1994) *The entire functions A, B, C, D have the same order, type, and Phragmén–Lindelöf indicator.*

The common order, type, and Phragmén–Lindelöf indicator of the functions A, B, C, D is called the order, type, and Phragmén–Lindelöf indicator of the moment problem. The order ρ satisfies $0 \leq \rho \leq 1$ and if $\rho = 1$ then the type is $\sigma = 0$ by a theorem of M. Riesz. Concrete examples of moment problems where the order of A, B, C, D is $1/4$ (resp. $1/3$) are in Berg and Valent (1994) (resp. Gilewicz et al., 2006). Many indeterminate moment problems involving basic hypergeometric functions have order 0; see Section 11.3.2. In this case it is of interest to study a refined scale of growth called logarithmic order and type; see Berg and Pedersen (2007), where it is proved that the functions A, B, C, D have the same logarithmic order and type, called the logarithmic order and type of the moment problem. There exist indeterminate moment problems with prescribed order and type or prescribed logarithmic order and type within the obvious limitations; see Pedersen (2009).

The order of an indeterminate moment problem can be calculated directly from the coefficients in the three-term recurrence relation as the following theorem shows.

Theorem 11.1.22 (Berg and Szwarc, 2014) *Let $(p_n)_n$ denote the orthonormal polynomials of an indeterminate Hamburger moment problem with the three-term recurrence relation*

$$xp_n(x) = a_{n+1}p_{n+1}(x) + b_n p_n(x) + a_n p_{n-1}(x).$$

Then the order ρ of the moment problem is equal to the exponent of convergence

$$\mathcal{E}(a_n) = \inf\left\{\alpha > 0 \,\middle|\, \sum_{n=0}^{\infty} a_n^{-\alpha} < \infty\right\} \tag{11.1.52}$$

*of the sequence $(a_n)_n$, provided that the recurrence coefficients $\{a_n, b_n\}$ satisfy a **finiteness** condition*

$$\sum_{n=0}^{\infty} \frac{1+|b_n|}{\sqrt{b_n b_{n+1}}} < \infty \tag{11.1.53}$$

*and a **regularity** condition: $(a_n)_n$ is either eventually log-convex or eventually log-concave, that is,*

$$a_n^2 \le a_{n-1}a_{n+1} \quad \text{for all sufficiently large } n \tag{11.1.54}$$

or

$$a_n^2 \ge a_{n-1}a_{n+1} \quad \text{for all sufficiently large } n. \tag{11.1.55}$$

In the symmetric case, where $b_n = 0$ for all n, the first condition is equivalent to $\sum 1/a_n < \infty$, provided that the regularity condition for $(a_n)_n$ is fulfilled, so in this case Carleman's condition (11.1.10) is necessary and sufficient for determinacy.

For moment problems of order zero the logarithmic order is equal to the exponent of convergence of the sequence $\{\log a_n\}$ under the same two conditions on the recurrence coefficients.

A Hamburger moment sequence $(m_n)_n$ is called **symmetric** if all odd moments vanish. In terms of the corresponding orthogonal polynomials $(P_n)_n$ the symmetry means that $P_{2n}(x)$ (resp. $P_{2n+1}(x)$) contains only even (resp. odd) powers of x. In the determinate case, symmetry means that the unique solution is a symmetric measure. In the indeterminate case there will be symmetric as well as nonsymmetric solutions.

For a symmetric indeterminate Hamburger moment sequence $(m_n)_n$, the functions A, D from (11.1.32) are odd while B, C are even. For details see (11.2.13). The N-extremal solutions μ_0, μ_∞ are symmetric while the other N-extremal solutions are nonsymmetric. The solution μ_φ corresponding to the Pick function φ is symmetric if and only if $\varphi(-z) = -\varphi(z)$. In the representation (11.1.36) this corresponds to $\beta = 0$, τ symmetric.

11.2 Stieltjes Moment Problems

If $(m_n)_n$ is a symmetric Hamburger moment sequence

$$m_n = \int_{-\infty}^{\infty} x^n \, d\mu(x), \tag{11.2.1}$$

where μ is symmetric, then $s_n = m_{2n}$, $n \ge 0$ is a Stieltjes moment sequence

$$s_n = \int_0^{\infty} t^n \, d\nu(t) \tag{11.2.2}$$

with respect to the image measure ν of μ under the function x^2 mapping $\mathbb{R}$ onto $[0,\infty)$, that is, for Borel sets $B \subseteq [0,\infty)$,

$$\nu(B) = \mu\left(\{x \in \mathbb{R} \mid x^2 \in B\}\right). \tag{11.2.3}$$

Conversely, every Stieltjes moment sequence $(s_n)_n$ is associated to a symmetric Hamburger moment sequence $(m_n)_n$ defined by $m_{2n} = s_n$, $m_{2n+1} = 0$.

Theorem 11.2.1 *The symmetric Hamburger moment sequence $(m_n)_n$ is determinate if and only if $(s_n)_n$ is determinate in the sense of Stieltjes. In the case of indeterminacy, $\mu \mapsto \nu$ defined in* (11.2.3) *is a bijection of the set of symmetric solutions to the Hamburger problem* (11.2.1) *onto the set of solutions to the Stieltjes problem* (11.2.2).

If $(s_n)_n$ is a Stieltjes moment sequence which is indeterminate in the sense of Hamburger, there also exist solutions to the moment equations which are not supported in the half-line.

We shall now relate the orthonormal polynomials $(r_n)_n$ of the symmetric measure μ and the orthonormal polynomials $(p_n)_n$ of the measure ν on $[0,\infty)$. For this we also need the orthonormal polynomials $(q_n)_n$ with respect to the measure $x\,d\nu(x)$ (notice that $m_2 = \int x\,d\nu(x)$ need not be 1), so we have

$$\int_{\mathbb{R}} r_m(x)r_n(x)\,d\mu(x) = \int_0^\infty p_m(x)p_n(x)\,d\nu(x) = \int_0^\infty q_m(x)q_n(x)x\,d\nu(x) = \delta_{m,n}.$$

The relation is

$$r_{2n}(x) = p_n(x^2), \quad r_{2n+1}(x) = xq_n(x^2). \tag{11.2.4}$$

Assume that the three-term recurrence relation for the symmetric polynomials is

$$xr_n(x) = \sigma_{n+1}r_{n+1} + \sigma_n r_{n-1}. \tag{11.2.5}$$

Expressing this for even and odd n and using (11.2.4), we get the following relations with the coefficients in the three-term recurrence relation (2.1.17) for $(p_n)_n$:

$$a_n = \sigma_{2n}\sigma_{2n-1}, \quad n > 0, \qquad b_n = \sigma_{2n+1}^2 + \sigma_{2n}^2, \quad n \ge 1, \quad b_0 = \sigma_1^2. \tag{11.2.6}$$

Using the above we can formulate analogues of Carleman's and Krein's criteria in the Stieltjes case.

Theorem 11.2.2 (Carleman) *Let $\{s_n\}$ be a Stieltjes moment sequence and let $(a_n)_n$ be given by* (2.1.17) *for the corresponding orthonormal polynomials $(p_n)_n$.*

If

$$\sum_{n=1}^{\infty} \frac{1}{\sqrt{a_n}} = \infty \tag{11.2.7}$$

or if

$$\sum_{n=0}^{\infty} \frac{1}{\sqrt[2n]{s_n}} = \infty \tag{11.2.8}$$

then the moment problem is determinate in the sense of Stieltjes.

Theorem 11.2.3 (Krein) *Assume that the probability measure ν on $[0, \infty)$ has a density $d\nu = \omega(x)\,dx$ with $\omega(x) > 0$ for $x > 0$ satisfying*

$$\int_0^\infty \frac{\log \omega(x)}{\sqrt{x}(1+x)}\,dx > -\infty. \tag{11.2.9}$$

Then ν is indeterminate in the sense of Stieltjes.

It follows from Theorems 11.2.2 and 11.2.3 that the measure on the half-line with density

$$\omega_\gamma(x) = C_\gamma \exp(-x^\gamma), \quad x > 0 \tag{11.2.10}$$

(C_γ is a normalization factor) is determinate in the sense of Stieltjes if and only if $\gamma \geq 1/2$. Stieltjes (1894) considered this example with $\gamma = 1/4$.

From an operator point of view a Stieltjes moment sequence $(s_n)_n$ defines a nonnegative Jacobi matrix (2.3.9) since

$$\langle T_0 p, p\rangle = \langle xp(x), p(x)\rangle = \sum_{k,l=0}^{n} s_{k+l+1} c_k c_l \geq 0$$

for any polynomial $p(x) = \sum_{k=0}^{n} c_k x^k$ with real coefficients.

We seek self-adjoint extensions of T_0 which are positive operators. In the literature there is a description of two positive self-adjoint extensions: the Friedrichs and the Krein extensions T_{F} and T_{K}. They coincide if and only if the problem is determinate in the sense of Stieltjes. In the indeterminate case the spectral measures of T_{F} and T_{K} lead to two N-extremal solutions ν_{F} and ν_{K} supported on $[0, \infty)$. They will be called the **Friedrichs** and **Krein** measures.

For a Stieltjes problem the limit

$$\alpha = \lim_{n\to\infty} \frac{P_n(0)}{P_n^*(0)} \tag{11.2.11}$$

always exists and $\alpha \leq 0$. If $\alpha = 0$ the problem is determinate in the sense of Stieltjes and $T_{\mathrm{F}} = T_{\mathrm{K}}$, $\nu_{\mathrm{F}} = \nu_{\mathrm{K}}$. The problem can nevertheless be indeterminate in the sense of Hamburger, and $\nu_{\mathrm{F}} = \nu_{\mathrm{K}}$ is then the N-extremal solution corresponding to the Pick function $\varphi = 0$ in the Nevanlinna parametrization (11.1.37). All the solutions corresponding to $\varphi \neq 0$ have mass on the negative half-axis. If $\alpha < 0$ the problem is indeterminate in the sense of Stieltjes. For details see Berg and Valent (1994), Pedersen (1995, 1997), and Simon (2005b).

To distinguish the Nevanlinna parametrization of the Stieltjes problem with moments (11.2.2) from the Nevanlinna parametrization of the corresponding symmetric Hamburger moment problem with moments (11.2.1), we use (11.1.25) as the Nevanlinna matrix for the Stieltjes case and let

$$\begin{pmatrix} A_{\mathrm{s}}(z) & C_{\mathrm{s}}(z) \\ B_{\mathrm{s}}(z) & D_{\mathrm{s}}(z) \end{pmatrix} \tag{11.2.12}$$

denote the Nevanlinna matrix of the latter case (s for symmetric). The relations between the two Nevanlinna matrices are

$$\begin{aligned} A_s(z) &= zA(z^2) - \frac{z}{\alpha}C(z^2), \qquad B_s(z) = B(z^2) - \frac{1}{\alpha}D(z^2), \\ C_s(z) &= C(z^2), \qquad zD_s(z) = D(z^2); \end{aligned} \tag{11.2.13}$$

cf. Chihara (1982). Note that A_s, D_s are odd and B_s, C_s even functions.

If ρ, τ, h (resp. ρ_s, τ_s, h_s) denote the order, type, and Phragmén–Lindelöf indicator of the Stieltjes problem (resp. symmetric Hamburger problem), then because of (11.2.13) we see that

$$\rho = \rho_s/2, \quad \tau = \tau_s, \quad h(\theta) = h_s(\theta/2).$$

In particular, the order of an indeterminate Stieltjes moment problem is always $\leq 1/2$.

We call the Hamburger solutions to the Stieltjes case ν_f parametrized by $f \in \mathcal{P} \cup \{\infty\}$ and the solutions to the symmetric Hamburger problem μ_φ parametrized by $\varphi \in \mathcal{P} \cup \{\infty\}$; hence

$$\int_{\mathbb{R}} \frac{d\nu_f(x)}{x - z} = -\frac{A(z)f(z) - C(z)}{B(z)f(z) - D(z)}, \quad z \in \mathbb{C} \setminus \mathbb{R} \tag{11.2.14}$$

and

$$\int_{\mathbb{R}} \frac{d\mu_\varphi(x)}{x - z} = -\frac{A_s(z)\varphi(z) - C_s(z)}{B_s(z)\varphi(z) - D_s(z)}, \quad z \in \mathbb{C} \setminus \mathbb{R}. \tag{11.2.15}$$

The N-extremal solutions ν_t ($t \in \mathbb{R} \cup \{\infty\}$) to the Stieltjes problem are supported by $[0, \infty)$ if and only if $\alpha \leq t \leq 0$. For all the other values of t the measure ν_t has one negative mass-point. Furthermore, $\nu_F = \nu_\alpha$ and $\nu_K = \nu_0$. This observation is a special case of the following result.

Theorem 11.2.4 (Pedersen, 1997) *A solution ν_f, $f \in \mathcal{P}$ to the Stieltjes problem is supported on $[0, \infty)$ if and only if one of the following equivalent conditions holds:*

(i) *f has a holomorphic extension to $\mathbb{C} \setminus [0, \infty)$ satisfying $\alpha \leq f(x) \leq 0$ for $x < 0$;*
(ii) *f has the form*

$$f(z) = b + \int_0^\infty \frac{d\tau(t)}{t - z}, \quad z \in \mathbb{C} \setminus [0, \infty),$$

where τ is a positive measure on $[0, \infty)$ such that

$$\gamma = \int_0^\infty \frac{d\tau(t)}{t} < \infty, \quad \alpha \leq b \leq -\gamma.$$

A symmetric solution μ_φ to (11.2.15) determines via (11.2.3) a solution ν_f to (11.2.14) supported on $[0, \infty)$. The relation between f and φ is given by

$$\varphi(z) = \frac{\alpha f(z^2)}{z\,(\alpha - f\,(z^2))}. \tag{11.2.16}$$

The N-extremal symmetric solutions μ_0 and μ_∞ correspond to the solutions with $f = 0$ and $f = \alpha$, which are precisely the Krein and the Friedrichs measures. The N-extremal solutions to (11.2.2) which are associated with a constant Pick function f satisfying $\alpha < f < 0$ correspond to the 1-canonical solutions μ_φ to (11.2.1) with $\varphi(z) = -c/z$, where $c = f\alpha/(f-\alpha)$.

In the case of an indeterminate Stieltjes moment problem it is of interest to give a formula for all the solutions in analogy with the Nevanlinna parametrization, that is, to single out those solutions to the corresponding Hamburger problem which are supported on $[0,\infty)$. To formulate this we need the **class $\mathcal{S}$ of Stieltjes transforms**, that is, functions of the form

$$\sigma(w) = a + \int_0^\infty \frac{d\zeta(u)}{u+w}, \qquad w \in \mathbb{C} \setminus (-\infty, 0], \tag{11.2.17}$$

where $a \geq 0$ and ζ is a positive measure on $[0,\infty)$ such that

$$\int_0^\infty \frac{d\zeta(u)}{1+u} < \infty. \tag{11.2.18}$$

Defining

$$\begin{aligned} P(w) &= A(-w) - \frac{1}{\alpha}C(-w), & R(w) &= C(-w), \\ Q(w) &= -\left(B(-w) - \frac{1}{\alpha}D(-w)\right), & S(w) &= -D(-w), \end{aligned} \tag{11.2.19}$$

we have a theorem:

Theorem 11.2.5 (Krein parametrization) *The formula*

$$\int_0^\infty \frac{d\nu_\sigma(t)}{w+t} = \frac{P(w)+\sigma(w)R(w)}{Q(w)+\sigma(w)S(w)}, \qquad w \in \mathbb{C} \setminus (-\infty, 0] \tag{11.2.20}$$

establishes a one-to-one correspondence between solutions ν to the Stieltjes moment problem and functions $\sigma \in \mathcal{S} \cup \{\infty\}$. The correspondence is also a homeomorphism when the set of solutions carries the weak topology and $\mathcal{S}$ the topology of locally uniform convergence on the cut plane $\mathbb{C} \setminus (-\infty, 0]$.

This result is given in Krein and Nudel'man (1977, p. 199) and in Berg (1995). The solutions ν_σ, where $\sigma \geq 0$ or $\sigma = \infty$, are the N-extremal solutions supported by $[0,\infty)$. In particular, ν_0 and ν_∞ are the Friedrichs and the Krein measures. The parameters σ in (11.2.20) and f in (11.2.14) are related by the equation

$$f(z) = \frac{\alpha}{1-\alpha\sigma(-z)}, \qquad z \in \mathbb{C} \setminus [0,\infty). \tag{11.2.21}$$

11.3 Examples of Indeterminate Moment Problems

In this section we present a collection of examples of indeterminate moment problems.

As follows from Carleman's criterion (11.1.10) (and (11.2.7) in the Stieltjes case), the sequence $(a_n)_n$ of recurrence coefficients for the orthonormal polynomials must tend to infinity sufficiently fast in order for the moment problem to be indeterminate.

The present section is divided into two parts. First we consider examples where the coefficients a_n grow **polynomially** as $n \to \infty$. These examples originate from birth and death processes with quartic (or cubic) rates and hence start out as Stieltjes problems. But as we will see, it pays to consider the associated symmetric Hamburger problems as well.

Second, we consider examples where the coefficients a_n grow **exponentially** as $n \to \infty$, more precisely, where $a_n = f_n q^{-n}$ for some sequence $(f_n)_n$ bounded away from 0 and ∞, and some fixed $q \in (0, 1)$. These examples are all associated with polynomials that fit into the Askey scheme of basic hypergeometric orthogonal polynomials.

11.3.1 Polynomially Growing Recurrence Coefficients

The link between orthogonal polynomials and birth and death processes goes back at least to Karlin and McGregor (1957a,b). If $(\lambda_n)_{n\ge0}$ and $(\mu_n)_{n\ge0}$ denote the birth and death rates, the polynomials in question are generated by

$$(\lambda_n + \mu_n - x) F_n(x) = \mu_{n+1} F_{n+1}(x) + \lambda_{n-1} F_{n-1}(x), \quad n \ge 0, \tag{11.3.1}$$

with initial conditions $F_{-1}(x) = 0$ and $F_0(x) = 1$. They are orthogonal on $[0, \infty)$ since $\lambda_n > 0$ for $n \ge 0$ and $\mu_0 \ge 0$, $\mu_n > 0$ for $n \ge 1$. The monic orthogonal polynomials are given by

$$P_n(x) = (-1)^n \mu_1 \cdots \mu_n F_n(x)$$

and they satisfy (2.1.14) with $\alpha_n = \lambda_n + \mu_n$ and $\beta_n = \lambda_{n-1}\mu_n$.

We also mention that the quantity α introduced in (11.2.11) can be expressed as

$$-\frac{1}{\alpha} = \sum_{n=1}^{\infty} \frac{1}{\mu_n \pi_n},$$

where

$$\pi_0 = 1, \quad \pi_n = \frac{\lambda_0 \cdots \lambda_{n-1}}{\mu_1 \cdots \mu_n}, \quad n \ge 1.$$

The Quartic Cases

The **Valent–Berg polynomials**: Valent (1994, 1995, 1996a,b) considered the quartic rates

$$\begin{cases} \lambda_n = (4n+4c+1)(4n+4c+2)^2(4n+4c+3), \\ \mu_n = (4n+4c-1)(4n+4c)^2(4n+4c+1)\,(1-\delta_{n,0}). \end{cases} \tag{11.3.2}$$

If we restrict to $c > -1/4$, there are two values of c for which $\mu_0 = 0$ even without the factor $1 - \delta_{n,0}$, namely $c = 0$ and $c = 1/4$. In the case $c = 0$, two of the N-extremal solutions were found in Valent (1994, Thms. 6, 7). (We will consider the case $c = 1/4$ at the very end of the section.) Letting δ_x denote the unit mass at the point x, they can be written as

$$\nu_\infty = \frac{\pi}{K_0^2}\delta_{x_0} + \frac{4\pi}{K_0^2}\sum_{n=1}^{\infty}\frac{2n\pi}{\sinh(2n\pi)}\delta_{x_n}, \quad x_n = \left(\frac{2n\pi}{K_0}\right)^4 \tag{11.3.3}$$

and

$$\nu_0 = \frac{4\pi}{K_0^2}\sum_{n=0}^{\infty}\frac{(2n+1)\pi}{\sinh((2n+1)\pi)}\delta_{x_n}, \quad x_n = \left(\frac{(2n+1)\pi}{K_0}\right)^4, \tag{11.3.4}$$

where

$$\frac{K_0}{\sqrt{2}} = \int_0^1 \frac{du}{\sqrt{1-u^4}} \quad \text{or equivalently,} \quad 4K_0 = \frac{\Gamma(1/4)^2}{\sqrt{\pi}}. \tag{11.3.5}$$

Berg and Valent (1994, Prop. 3.3.2) computed the Nevanlinna matrix and showed that the order, type, and indicator function for the entire functions in question are given by

$$\rho = \frac{1}{4}, \quad \sigma = \frac{K_0}{\sqrt{2}}, \quad h(\theta) = \frac{K_0}{2}\left(|\cos(\theta/4)| + |\sin(\theta/4)|\right). \tag{11.3.6}$$

They used this result to derive the two N-extremal measures mentioned above and also gave examples of solutions that are not N-extremal.

The Nevanlinna matrix for general $c > 0$ was computed by Valent (1996a) based on the generating functions in Valent (1996a, Thms. 9, 10). The expressions involve the Jacobi elliptic functions and the trigonometric functions of order 4 defined by

$$\delta_l(z) = \sum_{n=0}^{\infty}(-1)^n\frac{z^{4n+l}}{(4n+l)!}, \quad l = 0, 1, 2, 3.$$

Theorem 11.3.1 (Valent, 1996a, Thm. 11) *The Nevanlinna matrix for the Valent–Berg polynomials is given by*

$$\begin{cases} A(z) - \dfrac{C(z)}{\alpha} = \dfrac{\Gamma\left(c+\frac{1}{2}\right)}{(4c+1)c\Gamma(c)} \dfrac{\Delta_2\left(c+\frac{1}{2}; \frac{z^{1/4}K_0}{\sqrt{2}}\right)}{\sqrt{\pi z}}, \\[2ex] C(z) = \dfrac{4c\Gamma(c)}{\Gamma\left(c+\frac{1}{2}\right)} \dfrac{\Delta_0\left(c+\frac{1}{2}; \frac{z^{1/4}K_0}{\sqrt{2}}\right)}{\sqrt{\pi}}, \\[2ex] B(z) - \dfrac{D(z)}{\alpha} = -\dfrac{4\Gamma\left(c+\frac{1}{2}\right)}{\Gamma(c)} \dfrac{\Delta_0\left(c; \frac{z^{1/4}K_0}{\sqrt{2}}\right)}{\sqrt{\pi}}, \\[2ex] D(z) = \dfrac{(4c)^2(4c+1)\Gamma(c)}{\Gamma\left(c+\frac{1}{2}\right)} \dfrac{\Delta_2\left(c; \frac{z^{1/4}K_0}{\sqrt{2}}\right)}{\sqrt{\pi/z}}, \end{cases}$$

with

$$-\frac{\lambda_0}{\alpha} = {}_4F_3\left(\left.\begin{matrix}1, c+1, c+1, c+3/4\\ c+3/2, c+3/2, c+7/4\end{matrix}\right| 1\right)$$

and

$$\Delta_l(c;z) = \int_0^{K_0} (\mathrm{cn}\, v)^{4c-1}\delta_l\left(\frac{zv}{K_0}\right)\frac{dv}{\sqrt{2}}, \qquad l = 0, 1, 2, 3.$$

It turns out that the order, type, and indicator function are independent of c and coincide with (11.3.6). The result of Berg and Valent (1994, Prop. 3.3.2) can be obtained in the limit $c \downarrow 0$ (see also (11.3.9) below).

The **Chen–Ismail polynomials**: When $c = 0$, Chen and Ismail (1998) considered the associated symmetric Hamburger problem. The monic orthogonal polynomials satisfy the three-term recurrence relation

$$xP_n(x) = P_{n+1}(x) + \beta_n P_{n-1}(x), \quad n \geq 1, \tag{11.3.7}$$

with

$$\beta_n = (2n-1)(2n)^2(2n+1). \tag{11.3.8}$$

Based on the results of Berg and Valent (1994) they proved that the Nevanlinna matrix is given by

$$\begin{cases} A(z) = \Delta_2\left(K_0\sqrt{z/2}\right), & B(z) = -\delta_0\left(K_0\sqrt{z/2}\right), \\ C(z) = \dfrac{4}{\pi}\Delta_0\left(K_0\sqrt{z/2}\right), & D(z) = \dfrac{4}{\pi}\delta_2\left(K_0\sqrt{z/2}\right), \end{cases} \tag{11.3.9}$$

where K_0 is defined in (11.3.5) and

$$\Delta_l(z) = \frac{K_0}{\sqrt{2}}\int_0^1 \delta_l(uz)\,\mathrm{cn}\,(K_0 u)\,du, \quad l = 0, 1, 2, 3.$$

Besides the two symmetric N-extremal solutions corresponding to (11.3.3) and (11.3.4), Chen and Ismail (1998, Thm. 5.1) also found the weight function

$$w(x) = \frac{1/2}{\cos\left(K_0\sqrt{x}\right) + \cosh\left(K_0\sqrt{x}\right)}, \quad x \in \mathbb{R}. \tag{11.3.10}$$

This symmetric solution fits into the following family $(-1 < c < 1)$ of solutions:

$$\frac{\sqrt{1-c^2}\,/2}{\cos\left(K_0\sqrt{x}\right) + \cosh\left(K_0\sqrt{x}\right) + c\sin\left(K_0\sqrt{x}\right)\sinh\left(K_0\sqrt{x}\right)}, \quad x \in \mathbb{R}. \tag{11.3.11}$$

Note that (11.3.11) is symmetric if $c = 0$ and nonsymmetric otherwise. Since the weight functions are normalized by $\int_{\mathbb{R}} w(x)\,dx = 1$ this establishes the **Chen–Ismail integrals**

$$\int_{\mathbb{R}} \frac{1/2}{\cos\left(K_0\sqrt{x}\right) + \cosh\left(K_0\sqrt{x}\right)}\,dx = 1, \tag{11.3.12}$$

$$\int_{\mathbb{R}} \frac{\sqrt{1-c^2}\,/2}{\cos\left(K_0\sqrt{x}\right) + \cosh\left(K_0\sqrt{x}\right) + c\sin\left(K_0\sqrt{x}\right)\sinh\left(K_0\sqrt{x}\right)}\,dx = 1. \tag{11.3.13}$$

Recently, direct proofs of the above integral evaluations were given in Berndt (2016) and Kuznetsov (2017). We note the work Kuznetsov (2016) where a general construction for moment evaluations is given.

The **Ismail–Valent polynomials**: Ismail and Valent (1998) considered a one-parameter extension ($a \geq 0$) of the polynomials in (11.3.7), namely

$$xQ_n(x) = Q_{n+1}(x) + 2a(2n+1)^2Q_n(x) + \beta_n Q_{n-1}(x), \quad n \geq 0, \tag{11.3.14}$$

with β_n as in (11.3.8). The associated moment problem remains indeterminate when $a < 1$, but is only symmetric if $a = 0$. It is still possible to give simple expressions for D and B (see Ismail and Valent, 1998, Thm. 4.4), whereas C and A get more complicated. With

$$2K = K\left(\frac{1+a}{2}\right) \quad \text{and} \quad 2K' = K\left(\frac{1-a}{2}\right), \tag{11.3.15}$$

where $K(\cdot)$ is the first complete elliptic integral,

$$K(k^2) = \int_0^1 \frac{dt}{\sqrt{(1-t^2)(1-k^2t^2)}}, \quad k^2 < 1, \tag{11.3.16}$$

we have

$$\begin{cases} D(z) = \dfrac{4}{\pi}\sin\left(K\sqrt{z}\right)\sinh\left(K'\sqrt{z}\right), & (11.3.17)\\[2ex] B(z) + \dfrac{1}{4}\log\left(\dfrac{1+a}{1-a}\right)D(z) = -\cos\left(K\sqrt{z}\right)\cosh\left(K'\sqrt{z}\right). & (11.3.18)\end{cases}$$

As a consequence, Ismail and Valent (1998, Thm. 4.5) could write down two N-extremal solutions generalizing (11.3.3) and (11.3.4); see Theorem 11.3.2 below. Moreover, they proved that the polynomials in (11.3.14) are orthogonal with respect to a weight function similar to the one in (11.3.10), namely

$$w(x;a) = \frac{1/2}{\cos\left(2K\sqrt{x}\right) + \cosh\left(2K'\sqrt{x}\right)}, \quad x \in \mathbb{R}. \tag{11.3.19}$$

The above solution fits into the family ($\xi > 0$) of solutions

$$\frac{\xi/4}{\cos^2\left(K\sqrt{x}\right)\cosh^2\left(K'\sqrt{x}\right) + \xi^2\sin^2\left(K\sqrt{x}\right)\sinh^2\left(K'\sqrt{x}\right)}, \quad x \in \mathbb{R}, \tag{11.3.20}$$

as the special case $\xi = 1$; see Ismail and Valent (1998, (4.30)). This proves the **Ismail–Valent integral evaluations**

$$\begin{gathered}\int_{\mathbb{R}} \frac{1/2}{\cos\left(2K\sqrt{x}\right) + \cosh\left(2K'\sqrt{x}\right)}\, dx = 1, \\ \int_{\mathbb{R}} \frac{\xi/4}{\cos^2\left(K\sqrt{x}\right)\cosh^2\left(K'\sqrt{x}\right) + \xi^2 \sin^2\left(K\sqrt{x}\right)\sinh^2\left(K'\sqrt{x}\right)}\, dx = 1.\end{gathered} \tag{11.3.21}$$

The results in Ismail and Valent (1998) depend on the generating function

$$\sum_{n=0}^{\infty} (-1)^n \frac{Q_n(x)}{(2n)!} t^n = \frac{\cos\left(g(t)\sqrt{x}\right)}{\sqrt{1 - 2at + t^2}}, \quad |t| < 1, \tag{11.3.22}$$

where g is the elliptic integral

$$g(t) = \frac{1}{2} \int_0^t \frac{du}{\sqrt{u(1 - 2au + u^2)}}.$$

Another generating function is

$$\sum_{n=0}^{\infty} (-1)^n \frac{Q_n(x)}{(2n+1)!} t^n = \frac{\sin\left(g(t)\sqrt{x}\right)}{\sqrt{xt}}, \quad |t| < 1. \tag{11.3.23}$$

Ismail, Valent, and Yoon (2001, Thm. 2.2.1) characterized the orthogonal polynomials with similar generating functions. Besides the polynomials defined by (11.3.14), the polynomials generated by

$$\begin{aligned} xR_n(x) &= R_{n+1}(x) + 2a(2n+2)^2 R_n(x) \\ &\quad + 2n(2n+1)^2(2n+2)R_{n-1}(x), \quad n \ge 1, \end{aligned} \tag{11.3.24}$$

also lead to an indeterminate moment problem when $a < 1$. We shall refer to these polynomials as the **IVY polynomials**. Based on the generating function

$$\sum_{n=0}^{\infty} (-1)^n \frac{R_n(x)}{(2n+1)!} t^n = \frac{\sin\left(g(t)\sqrt{x}\right)}{\sqrt{xt(1 - 2at + t^2)}}, \quad |t| < 1, \tag{11.3.25}$$

Ismail, Valent, and Yoon showed that the entire functions D and B from the Nevanlinna matrix can be written as

$$D(z) = \frac{8K'}{\pi} \frac{\sin\left(K\sqrt{z}\right)}{\sqrt{z}} \cosh\left(K'\sqrt{z}\right) - \frac{8K}{\pi} \frac{\sinh\left(K'\sqrt{z}\right)}{\sqrt{z}} \cos\left(K\sqrt{z}\right), \tag{11.3.26}$$

$$B(z) = \frac{U(a)}{\pi} \frac{\sinh\left(K'\sqrt{z}\right)}{\sqrt{z}} \cos\left(K\sqrt{z}\right) - \frac{V(a)}{\pi} \frac{\sin\left(K\sqrt{z}\right)}{\sqrt{z}} \cosh\left(K'\sqrt{z}\right), \tag{11.3.27}$$

where K and K' are defined in (11.3.15), and

$$U(a) + iV(a) = 2\int_0^{e^{i\phi}} \frac{\sqrt{u}\,du}{\sqrt{1-2au+u^2}}, \qquad a = \cos\phi.$$

With (11.3.17)–(11.3.18) and (11.3.26)–(11.3.27) in mind we get the following general result about N-extremal solutions; see Ismail and Valent (1998, Thm. 4.5) and Ismail, Valent, and Yoon (2001, Thm. 4.4.5).

Theorem 11.3.2 *Set*

$$-\frac{4}{\sigma} = \log\left(\frac{1+a}{1-a}\right) \quad \textit{and} \quad \sigma_+ = -\frac{8K}{U(a)}, \quad \sigma_- = -\frac{8K'}{V(a)}.$$

Moreover, for $n \geq 0$ set

$$x_n = \frac{n^2\pi^2}{K^2}, \quad x_n' = -\frac{n^2\pi^2}{(K')^2}$$

and

$$y_n = \frac{(n+1/2)^2\pi^2}{K^2}, \quad y_n' = -\frac{(n+1/2)^2\pi^2}{(K')^2}.$$

Then two of the N-extremal solutions to the moment problem associated with the polynomials in (11.3.14) *are given by*

$$\mu_0 = \frac{\pi}{4KK'}\delta_0 + \frac{\pi^2}{K^2}\sum_{n=1}^{\infty}\frac{n\delta_{x_n}}{\sinh(2n\pi K'/K)} + \frac{\pi^2}{(K')^2}\sum_{n=1}^{\infty}\frac{n\delta_{x_n'}}{\sinh(2n\pi K/K')}, \tag{11.3.28}$$

$$\mu_\sigma = \frac{\pi^2}{K^2}\sum_{n=0}^{\infty}\frac{(n+1/2)\delta_{y_n}}{\sinh((2n+1)\pi K'/K)} + \frac{\pi^2}{(K')^2}\sum_{n=0}^{\infty}\frac{(n+1/2)\delta_{y_n'}}{\sinh((2n+1)\pi K/K')}, \tag{11.3.29}$$

and two of the N-extremal solutions to the moment problem associated with the polynomials in (11.3.24) *are given by*

$$\mu_{\sigma_+} = \frac{\pi^5}{L}\left(\frac{2}{K^4}\sum_{n=1}^{\infty}\frac{n^3\delta_{x_n}}{\sinh(2n\pi K'/K)} + \frac{1}{4(K')^4}\sum_{n=0}^{\infty}\frac{(2n+1)^3\delta_{y_n'}}{\sinh((2n+1)\pi K/K')}\right), \tag{11.3.30}$$

$$\mu_{\sigma_-} = \frac{\pi^5}{L}\left(\frac{2}{(K')^4}\sum_{n=1}^{\infty}\frac{n^3\delta_{x_n'}}{\sinh(2n\pi K/K')} + \frac{1}{4K^4}\sum_{n=0}^{\infty}\frac{(2n+1)^3\delta_{y_n}}{\sinh((2n+1)\pi K'/K)}\right), \tag{11.3.31}$$

where $L = 4(KV(a) - K'U(a))$.

Note that when $a = 0$, we have $K = K'$ $(= K_0/2)$, $U(a) = V(a)$ and $x_n' = -x_n$, $y_n' = -y_n$. So in this case, (11.3.28) and (11.3.29) are the two symmetric N-extremal solutions whereas $\sigma_+ = -\sigma_-$ so that $\frac{1}{2}(\mu_{\sigma_+} + \mu_{\sigma_-})$ is symmetric.

Ismail, Valent, and Yoon also gave examples of weight functions similar to the ones in (11.3.19); see Ismail, Valent, and Yoon (2001, p. 270–271) and (11.3.33) below.

As pointed out by Christiansen (2005), the special case $a = 0$ in (11.3.24) corresponds to the symmetric Hamburger problem associated with $c = 1/4$ in (11.3.2). For completeness, and to underline the fact that all δ_l and Δ_l play a role, we note that the Nevanlinna matrix in this case is given by

$$\begin{cases} A(z) = \dfrac{\Delta_3\left(K_0\sqrt{z/2}\right)}{K_0\sqrt{z/2}}, & B(z) = -\dfrac{\delta_1\left(K_0\sqrt{z/2}\right)}{K_0\sqrt{z/2}}, \\ C(z) = \dfrac{4K_0}{\pi}\dfrac{\Delta_1\left(K_0\sqrt{z/2}\right)}{\sqrt{z/2}}, & D(z) = \dfrac{4K_0}{\pi}\dfrac{\delta_3\left(K_0\sqrt{z/2}\right)}{\sqrt{z/2}}; \end{cases} \tag{11.3.32}$$

see Christiansen (2005, Thm. 1). Finally, we mention that a special case of the weight functions in Ismail, Valent, and Yoon (2001, p. 270–271) reduces to

$$v(x) = \frac{x/4}{\cosh\left(K_0\sqrt{x}\right) - \cos\left(K_0\sqrt{x}\right)}, \quad x \in \mathbb{R} \tag{11.3.33}$$

when $a = 0$. One should compare (11.3.32) with (11.3.9) and (11.3.33) with (11.3.10).

Table of Formulas

Quartic rates	Nevanlinna matrix	Solutions
$c > 0$	Theorem 11.3.1	
$c = 0$		(11.3.3)–(11.3.4)
$c = 0$, symmetric	(11.3.9)	(11.3.10)–(11.3.11)
$c = 0$, extension	(11.3.17)–(11.3.18)	(11.3.19)–(11.3.20), (11.3.28)–(11.3.29)
$c = 1/4$, symmetric	(11.3.32)	(11.3.33)
$c = 1/4$, extension	(11.3.26)–(11.3.27)	(11.3.30)–(11.3.31)

Orthogonality Relations

Three-term recurrence	Orthogonality ($\mu(\mathbb{R}) = 1$)
(11.3.14)	$\int Q_n Q_m \, d\mu = (2n+1)((2n)!)^2 \delta_{n,m}$
(11.3.24)	$\int R_n R_m \, d\mu = (n+1)((2n+1)!)^2 \delta_{n,m}$

The Cubic Cases

For cubic rates the results are not as simple and explicit as for quartic rates. In particular, none of the N-extremal solutions are explicitly known.

Recent works of Gilewicz et al. (2006) and Gilewicz, Leopold, and Valent (2005) studied the rates

$$\begin{cases} \lambda_n = (3n+3c+1)^2(3n+3c+2), \\ \mu_n = (3n+3c-1)(3n+3c)^2\,(1-\delta_{n,0}) \end{cases} \tag{11.3.34}$$

and

$$\begin{cases} \lambda_n = (3n+3c+1)(3n+3c+2)^2, \\ \mu_n = (3n+3c)^2(3n+3c+1)\,(1-\delta_{n,0})\,, \end{cases} \tag{11.3.35}$$

assuming that $c > 0$. They computed the Nevanlinna matrices in Gilewicz et al. (2006, Props. 8, 10) based on certain generating functions. The expressions involve integrals of the trigonometric functions of order 3,

$$\sigma_l(z) = \sum_{n=0}^{\infty}(-1)^n \frac{z^{3n+l}}{(3n+l)!}, \qquad l = 0, 1, 2.$$

The order, type, and Phragmén–Lindelöf indicator function for the entire functions in question are also determined in Gilewicz et al. (2006):

$$\rho = \frac{1}{3}, \quad \sigma = \theta_0 := \int_0^1 \frac{du}{(1-u^3)^{2/3}}, \quad \text{and} \quad h(\theta) = \theta_0 \cos\left(\frac{\theta-\pi}{3}\right),$$

respectively.

When $c = 0, 1/3$ in (11.3.34) and $c = 0$ in (11.3.35), the expressions for B and D simplify to read (see Gilewicz, Leopold, and Valent, 2005, (10)–(12))

$$c = 0 \text{ in } (11.3.34): \quad B(z) - \frac{D(z)}{\alpha} = -\sigma_0\left(z^{1/3}\theta_0\right), \quad D(z) = \frac{3\sqrt{3}}{2\pi} z^{2/3}\sigma_1\left(z^{1/3}\theta_0\right),$$

$$c = 0 \text{ in } (11.3.35): \quad B(z) - \frac{D(z)}{\alpha} = -\sigma_0\left(z^{1/3}\theta_0\right), \quad D(z) = \frac{3\sqrt{3}}{2\pi} z^{1/3}\sigma_2\left(z^{1/3}\theta_0\right),$$

$$c = \tfrac{1}{3} \text{ in } (11.3.34): \quad B(z) - \frac{D(z)}{\alpha} = -\frac{\sigma_1\left(z^{1/3}\theta_0\right)}{z^{1/3}}, \quad D(z) = \frac{3z^{1/3}\sigma_2\left(z^{1/3}\theta_0\right)}{B(2/3, 2/3)}.$$

The General Cases and Valent's Conjecture

The quartic and cubic rates are a special case of the polynomial rates

$$\begin{cases} \lambda_n = (pn+e_1)(pn+e_2)\dots(pn+e_p), \\ \mu_n = (pn+d_1)(pn+d_2)\dots(pn+d_p), \end{cases} \tag{11.3.36}$$

where $0 < e_1 \le e_2 \le \cdots \le e_p$, $-p < d_1 \le d_2 \le \cdots \le d_p$ with the condition $d_1 \cdots d_p \ge 0$ so that $\lambda_n, \mu_{n+1} > 0$, $\mu_0 \ge 0$. Define

$$E = e_1 + \cdots + e_p, \quad D = d_1 + \cdots + d_p. \tag{11.3.37}$$

Valent (1999) formulated the following conjecture:

Conjecture 11.3.3 (Valent) *The Stieltjes moment problem with the rates* (11.3.36) *satisfying the condition*

$$1 < \frac{E-D}{p} < p-1 \tag{11.3.38}$$

(hence $p \geq 3$) is indeterminate in the sense of Stieltjes, and the order, type, and Phragmén-Lindelöf indicator are

$$\rho = 1/p, \quad \tau = \int_0^1 \frac{du}{(1-u^p)^{2/p}}, \quad h(\theta) = \tau \cos((\theta - \pi)/p), \ \theta \in [0, 2\pi]. \tag{11.3.39}$$

Valent proved the indeterminacy and based the conjecture on the calculations done in the cubic and quartic cases. The conjecture that the order is $1/p$ has been proved in Romanov (2017) and Berg and Szwarc (2017), the conjecture about type has been proved by Bochkov (2019), but the conjecture about the Phragmén–Lindeloöf indicator is still open.

11.3.2 Exponentially Growing Recurrence Coefficients

In this section we will exclusively be dealing with polynomials that appear in the Askey scheme of basic hypergeometric orthogonal polynomials; see Koekoek and Swarttouw (1998) or Koekoek et al. (2010). Unless otherwise stated, q will always denote a fixed number in the interval $(0, 1)$.

Just below the Askey–Wilson polynomials, as the special case $d = 0$, one finds the **continuous dual q-Hahn polynomials**, denoted $p_n(x \mid q) \equiv p_n(x; a, b, c \mid q)$. Replacing q by $1/q$, we get the **continuous dual q^{-1}-Hahn polynomials** on the *real line*, generated by the three-term recurrence relation

$$\begin{aligned} 2xp_n(x \mid 1/q) &= p_{n+1}(x \mid 1/q) \\ &+ q^{-2n}\left[(a+b+c)q^n - abc(1+q-q^{n+1})\right] p_n(x \mid 1/q) \\ &+ q^{-4n+3}(1-q^n)(ab-q^{n-1})(ac-q^{n-1})(bc-q^{n-1})p_{n-1}(x \mid 1/q). \end{aligned} \tag{11.3.40}$$

As usual, the initial conditions are $p_{-1}(x \mid 1/q) = 0$ and $p_0(x \mid 1/q) = 1$. They are orthogonal if (but *not* only if) $a + b + c \in \mathbb{R}$ and $ab, ac, bc > 1$. We also introduce $P_n(x \mid 1/q) \equiv (-i)^n p_n(ix; ia, ib, ic \mid 1/q)$, the continuous dual q^{-1}-Hahn polynomials on the *imaginary axis*. They satisfy the three-term recurrence relation

$$\begin{aligned} 2xP_n(x \mid 1/q) &= P_{n+1}(x \mid 1/q) \\ &+ q^{-2n}\left[(a+b+c)q^n + abc(1+q-q^{n+1})\right] P_n(x \mid 1/q) \\ &+ q^{-4n+3}(1-q^n)(ab+q^{n-1})(ac+q^{n-1})(bc+q^{n-1})P_{n-1}(x \mid 1/q), \end{aligned} \tag{11.3.41}$$

and are orthogonal if (but again *not* only if) $a + b + c \in \mathbb{R}$ and $ab, ac, bc \geq 0$. As can be read off using a criterion of Chihara (see Chihara, 1989, §3), the associated Hamburger moment problems are indeterminate whenever $abc \neq 0$.

The Al-Salam–Chihara Tableau

The special case $c = 0$ leads to the **Al-Salam–Chihara polynomials**, denoted $Q_n(x \mid q) \equiv Q_n(x; a, b \mid q)$. They satisfy the three-term recurrence relation

$$\begin{aligned} 2xQ_n^{(\mp)}(x \mid q) &= Q_{n+1}^{(\mp)}(x \mid q) + q^{-n}(a+b)Q_n^{(\mp)}(x \mid q) \\ &\quad + q^{-2n+1}\left(1-q^n\right)\left(ab \mp q^{n-1}\right)Q_{n-1}^{(\mp)}(x \mid q), \end{aligned} \tag{11.3.42}$$

where the $\mp$ is the only thing that differentiates the Al-Salam–Chihara polynomials on the real line $(-)$ from the ones on the imaginary axis $(+)$. We also have the generating function

$$\sum_{n=0}^{\infty} \frac{q^{\binom{n}{2}}}{(q;q)_n} Q_n^{(\mp)}(x \mid q)t^n = \prod_{n=0}^{\infty} \frac{1 + 2xtq^n \mp t^2q^{2n}}{(1-atq^n)(1-btq^n)}, \qquad |t| < \min(1/|a|, 1/|b|). \tag{11.3.43}$$

The conditions

$$a + b \in \mathbb{R} \quad \text{and} \quad \begin{cases} ab > 1 & (-), \\ ab \geq 0 & (+) \end{cases} \tag{11.3.44}$$

are necessary and sufficient for orthogonality. Askey and Ismail (see Askey and Ismail, 1984, §3) proved that the associated moment problem is indeterminate if and only if, in addition to (11.3.44), $\overline{a} = b$ or $q < a/b < 1/q$. Of special interest are the two cases

$$\begin{cases} q < a/b < 1/q & (-), \\ \overline{a} = b & (+), \end{cases}$$

to be considered below.

The polynomials

$$R_n(x) = \frac{(-1)^n q^{\binom{n+1}{2}}}{a^n(q;q)_n} Q_n^{(-)}\left(\frac{x + a + 1/a}{2}\right)$$

are birth–death polynomials for $a > 0$ and $b > 1/a$ as they satisfy the recurrence relation (11.3.1) with

$$\lambda_n = bq^{-n} - 1/a \quad \text{and} \quad \mu_n = a\left(q^{-n} - 1\right).$$

The associated Stieltjes moment problem is indeterminate if and only if $a > 0$, $ab > 1$, and $q < a/b < 1$. Based on generating functions one finds that the entire functions from the Nevanlinna parametrization are given by

$$\begin{cases} A(z) - \dfrac{C(z)}{\alpha} = \dfrac{1}{a-b}\, {}_2\phi_1\left(\begin{matrix} aqe^y, aqe^{-y} \\ aq/b \end{matrix} \,\middle|\, q; 1/ab\right), \\ C(z) = {}_2\phi_1\left(\begin{matrix} e^y/a, e^{-y}/a \\ bq/a \end{matrix} \,\middle|\, q; q\right), \\ B(z) - \dfrac{D(z)}{\alpha} = -\dfrac{(e^y/b, e^{-y}/b; q)_\infty}{(1/ab, a/b; q)_\infty}, \\ D(z) = -a\dfrac{(e^y/a, e^{-y}/a; q)_\infty}{(bq/a, q; q)_\infty}, \end{cases} \tag{11.3.45}$$

where $z = \sinh y$ and

$$-\frac{1}{\alpha} = \frac{1}{b} \sum_{n=0}^{\infty} \frac{(q;q)_n}{(1/ab;q)_{n+1}} (a/b)^n.$$

Writing the zeros of D and $B - D/\alpha$ as

$$x_n(a) = aq^{-n} + q^n/a - (a + 1/a), \quad n \geq 0,$$

respectively

$$x_n(b) = bq^{-n} + q^n/b - (a + 1/a), \quad n \geq 0,$$

we are led to the two N-extremal solutions

$$\nu_\infty = \frac{(bq/a;q)_\infty}{(1/a^2;q)_\infty} \sum_{n=0}^{\infty} \frac{\left(1/a^2, 1/ab; q\right)_n}{(bq/a, q;q)_n} (b/a)^n \left(1 - q^{2n}/a^2\right) q^{n^2} \delta_{x_n(a)} \tag{11.3.46}$$

and

$$\nu_0 = \frac{(aq/b;q)_\infty}{(1/b^2;q)_\infty} \sum_{n=0}^{\infty} \frac{\left(1/b^2, 1/ab; q\right)_n}{(aq/b, q;q)_n} (a/b)^n \left(1 - q^{2n}/b^2\right) q^{n^2} \delta_{x_n(b)}. \tag{11.3.47}$$

While the Nevanlinna matrix (11.3.45) does not seem to be in the literature, the measures in (11.3.46) and (11.3.47) are contained in Koornwinder (2004) and Atakishiyev and Klimyk (2004).

When $\overline{a} = b$, the polynomials $Q_n^{(+)}(x \mid q)$ reduce to the q^{-1}**-Meixner–Pollaczek polynomials**. It is convenient to write $a = \sqrt{\beta}\, e^{i\theta}$ for $\beta \geq 0$ and $\theta \in [0, \pi/2]$. Moreover, we define (with $x = \sinh y$)

$$\frac{1}{2} R(x) e^{i\zeta(x)} = \frac{\left(e^{y-i\theta}/\sqrt{\beta}, -e^{-y-i\theta}/\sqrt{\beta}; q\right)_\infty}{(e^{-2i\theta}, q;q)_\infty}$$

and

$$\frac{1}{2} S(x) e^{i\eta(x)} = {}_2\phi_1\left(\left. \begin{matrix} e^{y-i\theta}/\sqrt{\beta}, -e^{-y-i\theta}/\sqrt{\beta} \\ qe^{-2i\theta} \end{matrix} \right| q;q \right),$$

where $R(x), S(x) > 0$ and $\zeta(x), \eta(x) \in \mathbb{R}$. In the case $\beta > 0$ and $0 < \theta \leq \pi/2$, Chihara and Ismail (1993) proved that the Nevanlinna matrix takes the form (for $x \in \mathbb{R}$)

$$\begin{cases} A(x) = \dfrac{(q;q)_\infty}{2(-1/\beta;q)_\infty} \dfrac{1}{\sqrt{\beta}\, \sin\theta} S(0) S(x) \sin\left(\eta(x) - \eta(0)\right), \\ B(x) = \dfrac{-(q;q)_\infty}{2(-1/\beta;q)_\infty} S(0) R(x) \cos\left(\zeta(x) - \eta(0)\right), \\ C(x) = \dfrac{(q;q)_\infty}{2(-1/\beta;q)_\infty} R(0) S(x) \cos\left(\eta(x) - \zeta(0)\right), \\ D(x) = \dfrac{(q;q)_\infty}{2(-1/\beta;q)_\infty} \sqrt{\beta}\, \sin\theta\, R(0) R(x) \sin\left(\zeta(x) - \zeta(0)\right). \end{cases} \tag{11.3.48}$$

The special case $a = -b$, corresponding to the **symmetric Al-Salam–Chihara polynomials** $Q_n(x;\beta)$ was studied by Christiansen and Ismail (2006). The expressions in (11.3.48) reduce to

$$\begin{cases} A(x) = \dfrac{2\left(q^2;q^2\right)_\infty}{\sqrt{\beta}\,(-1/\beta;q^2)_\infty} S(x)\sin\eta(x), \\ B(x) = -\dfrac{\left(q^2;q^2\right)_\infty}{(-1/\beta;q^2)_\infty} R(x)\cos\zeta(x), \\ C(x) = \dfrac{\left(q;q^2\right)_\infty}{(-q/\beta;q^2)_\infty} S(x)\cos\eta(x), \\ D(x) = \dfrac{\sqrt{\beta}\left(q;q^2\right)_\infty}{2(-q/\beta;q^2)_\infty} R(x)\sin\zeta(x), \end{cases} \tag{11.3.49}$$

where $R(x), S(x) > 0$ and $\zeta(x), \eta(x) \in \mathbb{R}$ are given by

$$R(x)e^{i\zeta(x)} = \frac{\left(-ie^y/\sqrt{\beta}, ie^{-y}/\sqrt{\beta}; q\right)_\infty}{(q^2;q^2)_\infty}$$

and

$$S(x)e^{i\eta(x)} = {}_2\phi_1\left(\begin{matrix} -ie^y/\sqrt{\beta}, ie^{-y}/\sqrt{\beta} \\ -q \end{matrix}\,\middle|\, q;q\right).$$

With reference to Corollary 11.1.18, we conclude that the weight function

$$w^{(\beta)}(x) = \frac{2}{\pi\sqrt{\beta}} \frac{\left(q^2;q^2\right)_\infty}{(q;q^2)_\infty} \frac{(-1/\beta;q)_\infty}{(-e^{2y}/\beta, -e^{-2y}/\beta;q^2)_\infty}, \qquad x = \sinh y \in \mathbb{R} \tag{11.3.50}$$

is a solution to the moment problem. The symmetric Al-Salam–Chihara polynomials satisfy the q-Sturm–Liouville equation

$$\mathcal{D}_q\left(w^{(\beta/q)}(x)\mathcal{D}_q Q_n(x;\beta)\right) + \frac{4q(1-q^n)}{(\beta+1)(1-q)^2} w^{(\beta)}(x)Q_n(x;\beta) = 0, \tag{11.3.51}$$

where $\mathcal{D}_q$ is the analogue of the Askey–Wilson divided difference operator on $\mathbb{R}$. This operator was introduced by Ismail (1993) and is defined as in (1.4.27), but with $x = \sinh y$ and $\breve{f}(e^y) = f(x)$. To be precise,

$$\left(\mathcal{D}_q f\right)(x) = \frac{\breve{f}\left(q^{1/2}e^y\right) - \breve{f}\left(q^{-1/2}e^y\right)}{\breve{e}\left(q^{1/2}e^y\right) - \breve{e}\left(q^{-1/2}e^y\right)} = \frac{\breve{f}\left(q^{1/2}e^y\right) - \breve{f}\left(q^{-1/2}e^y\right)}{(q^{1/2}-q^{-1/2})\cosh y}, \tag{11.3.52}$$

where $e(x) = x$. One can obtain (11.3.51) by combining the lowering operator

$$\mathcal{D}_q Q_{n+1}(x;\beta) = \frac{2q^{n/2}}{1-q} Q_n(x;\beta/q)$$

and the raising operator

$$\frac{1}{w^{(\beta q)}(x)} \mathcal{D}_q\left(w^{(\beta)}(x)Q_{n-1}(x;\beta)\right) = \frac{2\sqrt{q}\left(q^{n/2}-q^{-n/2}\right)}{(\beta+1/q)(1-q)} Q_n(x;\beta q),$$

and it serves as an alternative tool for deriving orthogonality relations; see Christiansen and Ismail (2006) for details. Note that (11.3.51) can also be written

$$\left(2x^2+1+\beta\right)\mathcal{D}_q^2 Q_n(x;\beta) - \frac{4x\sqrt{q}}{1-q}\mathcal{A}_q\mathcal{D}_q Q_n(x;\beta) + \frac{4\sqrt{q}\,(1-q^n)}{(1-q)^2} Q_n(x;\beta) = 0, \quad (11.3.53)$$

where $\mathcal{A}_q$ is the average operator defined by

$$\left(\mathcal{A}_q f\right)(x) = \frac{1}{2}\left(\breve{f}\left(q^{1/2}e^y\right) + \breve{f}\left(q^{-1/2}e^y\right)\right). \quad (11.3.54)$$

In addition to $w^{(\beta)}$, we have the following family $(q < \xi \le 1)$ of discrete solutions:

$$\lambda_\xi^{(\beta)} = \frac{\left(-\xi^2\beta q^2, -\beta q^2/\xi^2; q^2\right)_\infty}{(-\xi^2, -q/\xi^2, -\beta q, q; q)_\infty} \sum_{n=-\infty}^{\infty} \frac{\left(-\xi^2/\beta; q^2\right)_n}{(-\xi^2\beta q^2; q^2)_n} \xi^{2n}\beta^n \left(1+\xi^2 q^{2n}\right) q^{n^2} \delta_{x_n(\xi)}, \quad (11.3.55)$$

where

$$x_n(\xi) = \frac{1}{2}\left(\frac{1}{\xi q^n} - \xi q^n\right), \quad n \in \mathbb{Z}.$$

In Christiansen and Koelink (2008) it is proved that these solutions are N extremal only in the limit $\beta \to 0$, where the symmetric Al-Salam–Chihara polynomials reduce to the **continuous q^{-1}-Hermite polynomials**; see Ismail and Masson (1994) for a detailed study of these polynomials and the associated moment problem.

The q^{-1}-Hermite polynomials, denoted $h_n(x \mid q)$, satisfy the three-term recurrence relation

$$2xh_n(x \mid q) = h_{n+1}(x \mid q) + q^{-n}(1-q^n)\,h_{n-1}(x \mid q) \quad (11.3.56)$$

and have the generating function

$$\sum_{n=0}^{\infty} \frac{q^{\binom{n}{2}}}{(q;q)_n} h_n(x \mid q) t^n = (-te^y, te^{-y}; q)_\infty, \quad t \in \mathbb{C}, \quad (11.3.57)$$

as well as the Poisson kernel

$$\sum_{n=0}^{\infty} \frac{q^{\binom{n}{2}}}{(q;q)_n} h_n(x \mid q) h_n(x' \mid q) t^n = \frac{\left(-te^{y+y'}, -te^{-y-y'}, te^{y-y'}, te^{-y+y'}; q\right)_\infty}{(t^2/q;q)_\infty}, \quad |t| < \sqrt{q}, \quad (11.3.58)$$

where $x = \sinh y$ and $x' = \sinh y'$.

The Nevanlinna matrix for the q^{-1}-Hermite moment problem has the form

$$\begin{cases} A(x) = \dfrac{4qx}{1-q}\,{}_2\phi_1\left(\begin{matrix} q^2e^{2y}, q^2e^{-2y} \\ q^3 \end{matrix}\,\middle|\, q^2; q\right), \\ B(x) = -\dfrac{\left(qe^{2y}, qe^{-2y}; q^2\right)_\infty}{(q;q^2)_\infty^2}, \\ C(x) = {}_2\phi_1\left(\begin{matrix} e^{2y}, e^{-2y} \\ q \end{matrix}\,\middle|\, q^2; q^2\right), \\ D(x) = x\dfrac{\left(q^2e^{2y}, q^2e^{-2y}; q^2\right)_\infty}{(q;q)_\infty}, \end{cases} \tag{11.3.59}$$

and as pointed out by Ismail and Masson (1994), B and D are essentially theta functions,

$$B(x) = -\frac{\vartheta_4(iy)}{(q;q)_\infty\,(q;q^2)_\infty} \quad \text{and} \quad D(x) = \frac{\vartheta_1(iy)}{2iq^{1/4}(q;q)_\infty\,(q^2;q^2)_\infty}.$$

It is remarkable that *all* of the N-extremal solutions can be found explicitly; see Ismail and Masson (1994, §6) or Christiansen and Koelink (2008, §4). They are given by

$$\mu_{t(\xi)} = \frac{1}{(-\xi^2, -q/\xi^2, q;q)_\infty} \sum_{n=-\infty}^{\infty} \xi^{4n} q^{\binom{2n}{2}} \left(1 + \xi^2 q^{2n}\right) \delta_{x_n(\xi)}, \tag{11.3.60}$$

where

$$t(\xi) = \frac{\xi - 1/\xi}{2}\,\frac{\left(q^2\xi^2, q^2/\xi^2, q; q^2\right)_\infty}{\left(q\xi^2, q/\xi^2, q^2; q^2\right)_\infty} \quad \text{for } q < \xi \le 1.$$

Moreover, several absolutely continuous solutions are explicitly known. For $s = re^{i\theta}$ with $q < r \le 1$ and $\theta \in (0, \pi/2]$, the weight functions

$$v_s(x) = \frac{s}{\pi i}\,\frac{(-s\bar{s}, -q/s\bar{s}, \bar{s}/s, qs/\bar{s}, q;q)_\infty}{|\,(se^y, -se^{-y}, -qe^y/s, qe^{-y}/s; q)_\infty\,|^2}, \quad x = \sinh y \in \mathbb{R} \tag{11.3.61}$$

are solutions to the moment problem; see Ismail and Masson (1994, §7). Also, the weight functions $(0 < t \le 1)$

$$w_t(x) = \frac{\sin \pi t}{\pi}\,\frac{(q;q)_\infty}{(q^t, q^{1-t}; q)_\infty}\,\frac{2q^{t(1-t)}e^{y(2t-1)}}{(-q^{1-t}e^{2y}, -q^te^{-2y}; q)_\infty}, \quad x = \sinh y \in \mathbb{R} \tag{11.3.62}$$

have the right moments (see Christiansen and Ismail, 2006, §5), as well as the very simple one,

$$w(x) = \frac{2q^{1/8}}{\sqrt{2\pi \log q^{-1}}} \exp\left\{\frac{2}{\log q}\left(\log\left(x + \sqrt{x^2+1}\,\right)\right)^2\right\}, \quad x \in \mathbb{R} \tag{11.3.63}$$

found in Atakishiyev, Frank, and Wolf (1994). The limiting case $t = 1$ in (11.3.62) is due to Askey (1989b) and has the form

$$w_1(x) = \frac{1}{\log q^{-1}(q;q)_\infty}\,\frac{1/\cosh y}{(-qe^{2y}, -qe^{-2y}; q)_\infty}, \quad x = \sinh y \in \mathbb{R}. \tag{11.3.64}$$

The polynomial $h_n(x \mid q)$ solves the equation

$$\sqrt{q}\left(1+2x^2\right)\mathcal{D}_q^2 f(x) - \frac{4xq}{1-q}\mathcal{A}_q\mathcal{D}_q f(x) = \lambda f(x) \tag{11.3.65}$$

with $\lambda = \lambda_n$,

$$\lambda_n = -\frac{4q\,(1-q^n)}{(1-q)^2}.$$

Note that (11.3.65) can be written in the symmetric q-Sturm–Liouville form

$$\mathcal{D}_q\left(v(x)\mathcal{D}_q h_n(x \mid q)\right) + \frac{4q\,(1-q^n)}{(1-q)^2}v(x)h_n(x \mid q) = 0, \quad n \geq 0, \tag{11.3.66}$$

where v is any of the weight functions in (11.3.61)–(11.3.64). We stress that the indeterminacy of the moment problem is manifested in the fact that the equations

$$\frac{1}{v(x)}(\mathcal{D}_q v)(x) = -\frac{4xq}{1-q}, \qquad \frac{1}{v(x)}(\mathcal{A}_q v)(x) = \sqrt{q}\left(2x^2+1\right)$$

hold for any of the weight functions in (11.3.61)–(11.3.64).

The q-Meixner Tableau

Another special case of the continuous dual q^{-1}-Hahn polynomials is the q-**Meixner polynomials**, $M_n(x;q) \equiv M_n(x;b,c;q)$, generated by

$$\begin{aligned} q^{2n+1}(1-x)M_n(x;q) &= c(1-bq^{n+1})M_{n+1}(x;q) + q\,(1-q^n)\,(c+q^n)\,M_{n-1}(x;q) \\ &\quad - \left[c(1-bq^{n+1}) + q\,(1-q^n)\,(c+q^n)\right]M_n(x;q). \end{aligned} \tag{11.3.67}$$

When $b < 1/q$ and $c > 0$, the polynomials $M_n(x+1;q)$ are birth–death polynomials corresponding to the rates

$$\lambda_n = cq^{-n}(q^{-n-1} - b) \quad \text{and} \quad \mu_n = (1+cq^{-n})\,(q^{-n}-1)\,.$$

The associated Stieltjes problem is always indeterminate and the entire functions from the Krein parametrization, computed in Christiansen (2004), are given by

$$\begin{cases} P(z) = -\dfrac{(-q/c;q)_\infty}{(bq;q)_\infty}\displaystyle\sum_{n=1}^{\infty}\frac{(-1)^n q^{\binom{n+1}{2}}}{(c+q^n)\,(q;q)_n}\sum_{k=0}^{n-1}\frac{(-\frac{bcq^{1-n}}{1-z};q)_k}{(-cq^{1-n};q)_k}(1-z)^k, \\ Q(z) = {}_1\phi_1\left(\begin{matrix}1-z\\ bq\end{matrix}\,\middle|\, q;-q/c\right), \\ R(z) = 1 - \dfrac{z}{(q;q)_\infty}\displaystyle\sum_{n=1}^{\infty}\frac{(-1)^n q^{\binom{n+1}{2}+n}}{(c+q^n)\,(q;q)_n}\sum_{k=0}^{n-1}\frac{(-\frac{bcq^{1-n}}{1-z};q)_k}{(-cq^{1-n};q)_k}(1-z)^k, \\ S(z) = \dfrac{z\left(bq^2;q\right)_\infty}{(-q/c,q;q)_\infty}\,{}_1\phi_1\left(\begin{matrix}q(1-z)\\ bq^2\end{matrix}\,\middle|\, q;-q/c\right). \end{cases} \tag{11.3.68}$$

For $b \geq 0$, the q-Meixner polynomials are orthogonal with respect to the discrete probability measure

$$\lambda_{b,c} = \frac{(-bcq;q)_\infty}{(-c;q)_\infty} \sum_{n=0}^{\infty} \frac{(bq;q)_n}{(-bcq,q;q)_n} c^n q^{\binom{n}{2}} \delta_{q^{-n}}, \tag{11.3.69}$$

cf. Koekoek and Swarttouw (1998). The special case $b = 0$ gives the q-**Charlier polynomials**. When $b = 0$ and $c = 1$, the expressions for Q and S in (11.3.68) simplify to

$$Q(z) = \frac{\left(q(1-z);q^2\right)_\infty}{(q;q^2)_\infty}, \quad S(z) = \frac{\left(1-z;q^2\right)_\infty}{(q;q)_\infty}, \tag{11.3.70}$$

and one arrives at the two N-extremal solutions

$$\nu_\infty = (q;q^2)_\infty \sum_{n=0}^{\infty} \frac{q^{\binom{2n}{2}}}{(q;q)_{2n}} \delta_{q^{-2n}-1} \tag{11.3.71}$$

and

$$\nu_0 = (q;q^2)_\infty \sum_{n=0}^{\infty} \frac{q^{\binom{2n+1}{2}}}{(q;q)_{2n+1}} \delta_{q^{-(2n+1)}-1}. \tag{11.3.72}$$

In general, it seems very hard to explicitly find the zeros of Q and S from (11.3.68). Using results of Bergweiler and Hayman (2003), one can show that there exists a constant $A > 0$ such that

$$1 - x_n \sim Aq^{-2n} \quad \text{as } n \to \infty,$$

where $0 \geq x_1 > x_2 > \cdots$ are the zeros of Q (or S).

If we set $b = q^\alpha$, replace x by $cq^\alpha x$, and let $c \to \infty$, we obtain the q-**Laguerre polynomials** $L_n^{(\alpha)}(x;q)$, studied by Moak (1981); see also Ismail and Rahman (1998) and Christiansen (2003a). They are orthogonal on $[0,\infty)$ for $\alpha > -1$ and the associated Stieltjes problem is always indeterminate. In fact, Askey (1980) pointed out that the weight function

$$v^{(\alpha)}(x) = -\frac{\sin \pi\alpha}{\pi} \frac{(q;q)_\infty}{(q^{-\alpha},q)_\infty} \frac{x^\alpha}{(-x;q)_\infty}, \quad x > 0 \tag{11.3.73}$$

has the same moments as each $(q < s \leq 1)$ of the discrete measures,

$$\kappa_s^{(\alpha)} = \frac{\left(-q/s, q^{\alpha+1};q\right)_\infty}{(-sq^{\alpha+1}, -1/sq^\alpha, q;q)_\infty} \sum_{n=-\infty}^{\infty} (-s;q)_n q^{n(\alpha+1)} \delta_{sq^n}. \tag{11.3.74}$$

Moak (1981) then proved directly that the q-Laguerre polynomials are orthogonal with respect to $\nu^{(\alpha)}$. Actually, it fits into the family ($0 < s \le 1$) of solutions

$$v_s^{(\alpha)}(x) = q^{s(\alpha+1-s)} \frac{\sin \pi s}{\pi} \frac{\left(q, q^{\alpha+1}; q\right)_\infty}{(q^s, q^{1-s}; q)_\infty} \frac{(-q/x; q)_\infty x^{s-1}}{(-q^{\alpha+1-s}x, -q^{-\alpha+s}/x; q)_\infty}, \quad x > 0; \tag{11.3.75}$$

see Berg (1998) and Christiansen (2003a). The entire functions Q and S from the Krein parametrization are given by

$$\begin{cases} Q(z) = \displaystyle\sum_{n=0}^{\infty} \frac{q^{n(n+\alpha)}}{(q^{\alpha+1}, q; q)_n} z^n, & (11.3.76) \\ S(z) = \displaystyle\frac{z\left(q^{\alpha+2}; q\right)_\infty}{(q; q)_\infty} \sum_{n=0}^{\infty} \frac{q^{n(n+\alpha+1)}}{(q^{\alpha+2}, q; q)_n} z^n, & (11.3.77) \end{cases}$$

and as pointed out in Moak (1981), they are closely related to the second q-Bessel function $J_\nu^{(2)}$ defined in (1.4.47). The zeros of Q (and S) are very well separated, that is, $x_{n+1}/x_n > q^{-2}$, where $0 \ge x_1 > x_2 > \cdots$ denote the zeros of Q (or S). For more precise results on the behavior of x_n as $n \to \infty$, the reader is referred to Hayman (2005).

The **Stieltjes–Wigert polynomials**, denoted $S_n(x; q)$, are a limiting case of both the q-Laguerre and the q-Charlier polynomials; see Koekoek and Swarttouw (1998) or Koekoek et al. (2010) for details. They are orthogonal with respect to the log-normal density

$$w(x) = \frac{q^{1/8}}{\sqrt{2\pi \log q^{-1}}} \frac{1}{\sqrt{x}} \exp\left\{\frac{1}{2} \frac{(\log x)^2}{\log q}\right\}, \quad x > 0. \tag{11.3.78}$$

In his famous memoir *Recherches sur les fractions continues* from 1894–95 (see, for example, Stieltjes, 1993), Stieltjes pointed out that for $\lambda \in [-1, 1]$, the densities

$$w_s(x) = w(x)\left(1 + \lambda \sin\left(2\pi \tfrac{\log x}{\log q}\right)\right), \quad x > 0 \tag{11.3.79}$$

all have the same moments, namely $q^{-n(n+1)/2}$. As a consequence, the associated moment problem is indeterminate. It turns out that the functional equation

$$xf(x) = f(xq), \quad x > 0, \tag{11.3.80}$$

which is also the q-Pearson equation, plays an important role. As proved in Christiansen (2003b), any probability measure on $(0, \infty)$ of the form $d\mu = f(x)\,dx$, with f satisfying (11.3.80), has the moments $q^{-n(n+1)/2}$. Hence it follows from the Askey–Roy q-beta integral (see Askey and Roy, 1986) that the family ($0 < c \le 1$) of weight functions

$$v_c(x) = \frac{\sin \pi c}{\pi} \frac{(q; q)_\infty}{(q^c, q^{1-c}; q)_\infty} \frac{q^{c(1-c)} x^{c-1}}{(-q^{1-c}x, -q^c/x; q)_\infty}, \quad x > 0 \tag{11.3.81}$$

are solutions to the Stieltjes–Wigert moment problem; cf. Berg (1998). Moreover, it follows that new solutions can be obtained by multiplying $w(x)$ in (11.3.78) with q-periodic functions, that is, functions having the same value at x and xq for all $x > 0$. This explains how Stieltjes was led to (11.3.79).

We mention in passing that there are absolutely continuous solutions that do not satisfy (11.3.80). The following example is due to Gómez and López-García (2007). Fix $\gamma \in \mathbb{R}\setminus 2\pi\mathbb{Q}$ and consider the function

$$h(x) = \cos(\gamma x)\,(1 + k\cos(2\pi x))\,, \quad x \in \mathbb{R},$$

where

$$k = \frac{-\int_{\mathbb{R}} \exp\left(\frac{x^2 \log q}{2}\right)\cos(\gamma x)\,dx}{\int_{\mathbb{R}} \exp\left(\frac{x^2 \log q}{2}\right)\cos(\gamma x)\cos(2\pi x)\,dx} < 0.$$

As h is not periodic, the function $h(\frac{\log x}{\log q})$ is not q-periodic. Yet, the density

$$w(x)\left(1 + \lambda h\left(\tfrac{\log x}{\log q}\right)\right), \quad x > 0 \tag{11.3.82}$$

is a solution to the Stieltjes–Wigert moment problem whenever $|\lambda| \le 1/(1-k)$.

One can generalize (11.3.80) to an equation for general probability measures. Letting $\tau_a(\mu)$ denote the image measure of μ under $\tau_a \colon x \mapsto ax$, it is not hard to prove that any probability measure on $[0, \infty)$ satisfying the equation

$$\tau_{q^{-1}}(\mu) = qx\,d\mu(x) \tag{11.3.83}$$

has the moments $q^{-n(n+1)/2}$. Following Chihara (1970) (see also Leipnik, 1981), we therefore have the family $(q < c \le 1)$ of discrete solutions,

$$\lambda_c = \frac{1}{(-cq, -1/c, q; q)_\infty} \sum_{n=-\infty}^{\infty} c^n q^{\binom{n+1}{2}} \delta_{cq^n}. \tag{11.3.84}$$

None of the measures in (11.3.84) are N-extremal (or canonical) as the mass points accumulate at 0. In fact, Christiansen and Koelink (2006) proved that the closure of the polynomials has codimension $+\infty$ in $L^2(\mathbb{R}, \lambda_c)$ by finding an explicit orthogonal basis (consisting of q-Bessel functions) for the orthogonal complement. See Ciccoli, Koelink, and Koornwinder (1999) and Christiansen and Koelink (2008) for similar results on the q-Laguerre polynomials and the symmetric Al-Salam–Chihara polynomials.

The entire functions Q and S from the Krein parametrization can both be written in terms of the Ramanujan entire function

$$\Phi(z) = \sum_{n=0}^{\infty} \frac{q^{n^2} z^n}{(q;q)_n}. \tag{11.3.85}$$

We simply have

$$Q(z) = \Phi(z), \quad S(z) = z\Phi(zq)/(q;q)_\infty. \tag{11.3.86}$$

Wigert (1923) showed that the limit of $S_n(x;q)$ as $n \to \infty$ is closely related to Φ and, as pointed out in Christiansen (2003b), particularly the zeros of Φ play an all-important role in the study of the N-extremal and canonical solutions. The reader is referred to Andrews (2005), Ismail (2005a), and Ismail and Zhang (2007) for recent (as well as old) results about Φ and its zeros.

We can push equation (11.3.83) one step further and consider the transformation T given by

$$T : \mu \mapsto \tau_q\left(qx\, d\mu(x)\right).$$

Whenever μ is a solution to the Stieltjes–Wigert moment problem, so is $T(\mu)$. While all the above solutions ((11.3.78), (11.3.79), (11.3.81), (11.3.84)) are fixed points of T, none of the canonical solutions are. The transformation T is described completely at the level of Pick functions in Christiansen (2003b). It is interesting that one can explicitly find the Pick function corresponding to the canonical solution $T^{(m)}(\mu_0)$, $m \geq 0$, where μ_0 is the N-extremal solution with zero in its support. In the limit $m \to \infty$,

$$T^{(2m+1)}(\mu_0) \to \kappa_{-1} = \frac{\left(q;q^2\right)_\infty}{\left(q^2;q^2\right)_\infty} \sum_{n=-\infty}^{\infty} q^{\binom{2n+2}{2}} \delta_{q^{2n+1}} \tag{11.3.87}$$

and

$$T^{(2m+2)}(\mu_0) \to \kappa_1 = \frac{\left(q;q^2\right)_\infty}{\left(q^2;q^2\right)_\infty} \sum_{n=-\infty}^{\infty} q^{\binom{2n+1}{2}} \delta_{q^{2n}}. \tag{11.3.88}$$

The solutions in (11.3.87)–(11.3.88) were constructed by Berg (1995) to illustrate that infinitely many solutions to the Stieltjes–Wigert moment problem are supported on the geometric progression $\{q^n \mid n \in \mathbb{Z}\}$. Just take convex combinations of κ_{-1} and κ_1.

The **discrete q-Hermite polynomials**, denoted by $\tilde{h}_n(x;q)$, can be obtained from the q-Laguerre polynomials in the same way as one can obtain the Hermite polynomials from the Laguerre polynomials. The relationship is given by

$$\tilde{h}_{2n}(x;q) = (-1)^n \frac{\left(q^2;q^2\right)_n}{q^{n(2n-1)}} L_n^{(-1/2)}\left(x^2;q^2\right)$$

and

$$\tilde{h}_{2n+1}(x;q) = (-1)^n \frac{\left(q^2;q^2\right)_n}{q^{n(2n-1)}} x L_n^{(1/2)}\left(x^2;q^2\right).$$

Appealing to (11.2.13), we see that the entire functions B and D from the Nevanlinna matrix are closely related to the basic trigonometric functions Cos_q and Sin_q. To be precise, we have

$$B(z) = -\,\mathrm{Cos}_q(z), \quad D(z) = \frac{\left(q;q^2\right)_\infty}{\left(q^2;q^2\right)_\infty}\,\mathrm{Sin}_q(z). \tag{11.3.89}$$

Using the identity

$$\mathrm{Cos}_q^2(z) + \mathrm{Sin}_q^2(z) = \left(-z^2;q^2\right)_\infty,$$

Ismail and Rahman (1998) arrived at the weight function (cf. Corollary 11.1.18)

$$w(x) = \frac{1}{\pi}\frac{\left(q^2;q^2\right)_\infty}{(q;q^2)_\infty}\frac{1}{(-x^2;q^2)_\infty}, \quad x \in \mathbb{R}. \tag{11.3.90}$$

More solutions to the discrete q-Hermite moment problem can be obtained by symmetrizing solutions to the q-Laguerre moment problem when $\alpha = -1/2$ (and replacing q by q^2). See also Groenevelt (2015) for examples of nonsymmetric solutions.

Another special case of the q-Meixner polynomials, obtained by setting $b = -a/c$ and letting $c \to 0$, are the **Al-Salam–Carlitz polynomials**, denoted $V_n^{(a)}(x;q)$. The polynomials

$$F_n(x) := \frac{(-1)^n q^{\binom{n+1}{2}}}{(q;q)_n} V_n^{(a)}(x+1;q)$$

are birth–death polynomials for $a > 0$ as they satisfy the recursion (11.3.1) with

$$\lambda_n = aq^{-n} \quad \text{and} \quad \mu_n = q^{-n} - 1.$$

The associated Stieltjes problem is indeterminate if and only if $1 < a < 1/q$. Moreover, the associated Hamburger problem remains indeterminate for $q < a \le 1$, but is determinate otherwise. See Chihara (1982) and Berg and Valent (1995). Berg and Valent (1994) computed the Nevanlinna matrix which is given by

$$\begin{cases} A(z) - \dfrac{C(z)}{\alpha} = \dfrac{(q;q)_\infty}{a-1}\,{}_2\phi_1\left(\left.\begin{matrix}\frac{1+z}{a}, 0 \\ q/a\end{matrix}\,\right|\, q;q\right), \\ C(z) = {}_2\phi_1\left(\left.\begin{matrix}1+z, 0 \\ aq\end{matrix}\,\right|\, q;q\right), \\ B(z) - \dfrac{D(z)}{\alpha} = -\dfrac{\left(\frac{1+z}{a};q\right)_\infty}{(1/a;q)_\infty}, \\ D(z) = -\dfrac{(1+z;q)_\infty}{(aq,q;q)_\infty}, \end{cases} \tag{11.3.91}$$

where

$$-\frac{1}{\alpha} = \frac{(q;q)_\infty}{a-1} {}_2\phi_1\left(\left.\begin{matrix} 1/a, 0 \\ q/a \end{matrix}\right| q; q\right).$$

The expressions for A and B in (11.3.91) have to be interpreted as certain limits when $a = 1$. For $q < a < 1/q$, the two discrete measures

$$\nu_\infty = (aq;q)_\infty \sum_{n=0}^{\infty} \frac{a^n q^{n^2}}{(aq, q;q)_n} \delta_{q^{-n}-1} \tag{11.3.92}$$

and

$$\nu_0 = (q/a;q)_\infty \sum_{n=0}^{\infty} \frac{a^{-n} q^{n^2}}{(q/a, q;q)_n} \delta_{aq^{-n}-1} \tag{11.3.93}$$

are N-extremal solutions. When $a > 1$, then ν_∞ is the Krein measure and ν_0 is the Friedrich measure discussed after Theorem 112.3. When $a = 1$ the two measures coincide and represent the unique solution on $[0, \infty)$. Furthermore, when $a < 1$ the measure in (11.3.92) is still the unique solution on $[0, \infty)$, whereas the measure in (11.3.93) has exactly one negative mass point, namely $a - 1$. The solution in (11.3.92) was discovered by Al-Salam and Carlitz (1965) and pointed out to be N-extremal by Chihara (1968).

Table of Formulas

Orthogonal polynomials	Nevanlinna matrix	Solutions
Al-Salam–Chihara (on $\mathbb{R}$)	(11.3.45)	(11.3.46)–(11.3.47)
q^{-1}-Meixner–Pollaczek	(11.3.48)	
Symmetric Al-Salam–Chihara	(11.3.49)	(11.3.50)–(11.3.55)
Continuous q^{-1}-Hermite	(11.3.59)	(11.3.60)–(11.3.64)
	Krein matrix	
q-Meixner	(11.3.68)	(11.3.69)
q-Charlier (special case)	(11.3.70)	(11.3.71)–(11.3.72)
q-Laguerre	(11.3.76)–(11.3.77)	(11.3.73)–(11.3.75)
Stieltjes–Wigert	(11.3.85)–(11.3.86)	(11.3.78)–(11.3.79), (11.3.81)–(11.3.82), (11.3.84), (11.3.87)–(11.3.88)
	Nevanlinna matrix	
Discrete q-Hermite	(11.3.89)	(11.3.90)
Al-Salam–Carlitz	(11.3.91)	(11.3.92)–(11.3.93)

Orthogonality Relations

Orthogonal polynomials	Orthogonality ($\mu(\mathbb{R}) = 1$)
Al-Salam–Chihara (on $\mathbb{R}$)	$\int_0^\infty R_n R_m \, d\mu = \frac{(bq)^n (1/ab;q)_n}{a^n (q;q)_n} \delta_{n,m}$
Symmetric Al-Salam–Chihara	$\int_{\mathbb{R}} Q_n^{(+)} Q_m^{(+)} \, d\mu = \frac{c^{2n}(-1/c^2, q; q)_n}{q^{n^2}} \delta_{n,m}$
Continuous q^{-1}-Hermite	$\int_{\mathbb{R}} h_n h_m \, d\mu = q^{-\binom{n+1}{2}} (q;q)_n \delta_{n,m}$
q-Meixner	$\int_1^\infty M_n M_m \, d\mu = \frac{(-q/c, q; q)_n}{q^n (bq;q)_n} \delta_{n,m}$
q-Laguerre	$\int_0^\infty L_n^{(\alpha)} L_m^{(\alpha)} \, d\mu = \frac{(q^{\alpha+1};q)_n}{q^n (q;q)_n} \delta_{n,m}$
Stieltjes–Wigert	$\int_0^\infty S_n S_m \, d\mu = \frac{1}{q^n (q;q)_n} \delta_{n,m}$
Discrete q-Hermite	$\int_{\mathbb{R}} \tilde{h}_n \tilde{h}_m \, d\mu = q^{-n^2} (q;q)_n \delta_{n,m}$
Al-Salam–Carlitz	$\int_{\mathbb{R}} V_n^{(a)} V_m^{(a)} \, d\mu = a^n q^{-n^2} (q;q)_n \delta_{n,m}$

12

Matrix-Valued Orthogonal Polynomials and Differential Equations

Antonio J. Durán and F. Alberto Grünbaum

12.1 Matrix Polynomials and Matrix Orthogonality

12.1.1 Basic Definitions and Properties

A (square) **matrix polynomial** P of size N and degree n is a square matrix of size $N \times N$ whose entries are polynomials in $t \in \mathbb{R}$ (with complex coefficients) of degree less than or equal to n (with at least one entry of degree n):

$$P(t) = \begin{pmatrix} p_{11}(t) & \cdots & p_{1N}(t) \\ \vdots & & \vdots \\ p_{N1}(t) & \cdots & p_{NN}(t) \end{pmatrix},$$

or equivalently, a polynomial of the form

$$P(t) = A_n t^n + A_{n-1} t^{n-1} + \cdots + A_0,$$

where $A_0, \ldots, A_n$ are matrices of size $N \times N$ and complex entries, $A_n \neq 0$. We denote the linear space of (square) matrix polynomials by $\mathbb{C}^{N\times N}[t]$.

A matrix of measures

$$W = \begin{pmatrix} w_{11} & \cdots & w_{1N} \\ \vdots & & \vdots \\ w_{N1} & \cdots & w_{NN} \end{pmatrix}$$

on the real line is a square matrix of size $N \times N$ whose entries $w_{i,j}$, $i, j = 1, \ldots, N$ are Borel complex measures on the real line.

We say that the matrix of measures W is positive definite if for any Borel set Ω the matrix $W(\Omega)$ is positive semidefinite, and positive definite at least for one Borelian.

The **support of a matrix of measures** W is by definition

$$\operatorname{supp}(W) = \{x \in \mathbb{R} \colon W((x - \epsilon, x + \epsilon)) \neq 0 \text{ for any } \epsilon > 0\}.$$

We say that a matrix of measures W has finite moments of any order if the integral $\int_{\mathbb{R}} t^n \, dW(t)$ (called the nth moment of W) exists and is finite for any $n \in \mathbb{N}$.

A Hermitian sesquilinear form in the linear space $\mathbb{C}^{N\times N}[t]$ of matrix polynomials can be associated to a positive definite matrix of measures with finite moments of any order:

$$\langle P, Q\rangle = \int_{\mathbb{R}} P(t)\, dW(t) Q^*(t) \in \mathbb{C}^{N\times N}. \tag{12.1.1}$$

Observe that the definition of the Hermitian sesquilinear form above gives **priority** to the multiplication on the left with respect to that on the right. Indeed, we have $\langle AP, Q\rangle = A\langle P, Q\rangle$, $\langle P, AQ\rangle = \langle P, Q\rangle A^*$ but, in general, $\langle PA, Q\rangle \neq A\langle P, Q\rangle$, $\langle P, QA\rangle \neq \langle P, Q\rangle A^*$.

We say that a sequence of matrix polynomials $(P_n)_n$ is orthogonal with respect to W if each polynomial P_n has degree n with nonsingular leading coefficient and they are orthogonal with respect to $\langle \cdot, \cdot\rangle$, that is, $\langle P_n, P_k\rangle = \Gamma_n \delta_{n,k}$, with Γ_n nonsingular. If $\Gamma_n = I$ (I stands for the $N\times N$ identity matrix), we say that the sequence $(P_n)_n$ is orthonormal with respect to W. The theory of matrix-valued orthogonal polynomials was started by M. G. Krein in 1949 (Krein, 1949a, b; see also Berezanskii, 1968; Atkinson, 1964; or Damanik, Pushnitski, and Simon, 2008).

Since each orthogonal polynomial P_n has degree n with nonsingular leading coefficient, any matrix polynomial of degree less than or equal to n can be expressed as a linear combination of P_k, $0 \leq k \leq n$, with matrix coefficients (multiplying on the left or on the right). This property, together with the orthogonality, defines the sequence of orthogonal polynomials uniquely from W up to multiplication on the left by a sequence of nonsingular matrices (multiplication by unitary matrices for the orthonormal polynomials).

We are interested in positive definite matrices of measures W having a sequence of orthogonal polynomials; this is equivalent to the nondegenerateness of the Hermitian sesquilinear form associated to W:

Definition 12.1.1 We say that a matrix of measures W is a **weight matrix** if

(1) W is positive definite;
(2) W has finite moments of any order;
(3) W is nondegenerate, that is, $\langle P, P\rangle = 0$ only when $P = 0$.

All of the examples of matrices of measures considered in this contribution will have a (smooth) density with respect to the Lebesgue measure: $W = W(t)\,dt$. An easy-to-check sufficient condition to guarantee the nondegenerateness of $W = W(t)\,dt$ is that the density $W(t)$ is positive definite in a set with positive Lebesgue measure contained in supp W.

We say that W reduces to scalar weights if there exists a nonsingular matrix T independent of t for which

$$W(t) = T D(t) T^*,$$

with D a diagonal weight matrix. Notice that the orthonormal matrix polynomials with respect to W are then

$$P_n(t) = \begin{pmatrix} p_{n,1}(t) & \cdots & 0 \\ \vdots & & \vdots \\ 0 & \cdots & p_{n,N}(t) \end{pmatrix} T^{-1}, \quad n \geq 0,$$

where $(p_{n,i})_n$ are the orthonormal polynomials with respect to the positive measure D_{ii} (the ith entry on the diagonal of D), $i = 1, \ldots, N$. Orthogonal matrix polynomials with respect to a weight matrix that reduces to scalar weights are essentially N scalar families of orthogonal polynomials. This is the case for many examples of orthogonal matrix polynomials which can be found in the literature.

A convenient way of checking whether one is dealing with a weight matrix $W = W(t)\,dt$ that reduces to scalar weights is the following (Durán and Grünbaum, 2004):

Lemma 12.1.2 (Reducibility to scalar case) *Assume that the matrix of functions $W(t)$ satisfies $W(a) = I$, for some real number a. Then W reduces to scalar weights if and only if $W(t)W(s) = W(s)W(t)$ for all t, s in the support of W.*

The condition $W(a) = I$, for some real number a, is not an important restriction. Indeed, as far as one has that $W(a)$ is nonsingular for some a, we can take $\tilde{W} = (W(a))^{-1/2}W(W(a))^{-1/2}$, so that $\tilde{W}(a) = I$, and it is clear that W reduces to scalar weights if and only if $\tilde{W}$ does. None of the examples of weight matrices considered in the second part of this contribution reduces to scalar weights.

12.1.2 The Three-Term Recurrence Relation and Other Related Algebraic Formulas

As in the scalar case, the orthonormality of a sequence of polynomials $(P_n)_n$ with respect to a weight matrix W is equivalent to an algebraic formula known as the three-term recurrence relation. This equivalence is, in fact, a consequence of the symmetry of the operator of multiplication by t with respect to the Hermitian sesquilinear form defined in the linear space of matrix polynomials by a weight matrix supported on the real line (see (12.1.1)).

Moreover, if the sequence $(P_n)_n$ is orthonormal with respect to W, then it satisfies a **matrix three-term recurrence formula**

$$tP_n(t) = A_{n+1}P_{n+1}(t) + B_nP_n(t) + A_n^*P_{n-1}(t), \tag{12.1.2}$$

where we take $P_{-1} = 0$. In this formula the coefficients B_n have to be Hermitian, and the coefficients A_n have to be nonsingular.

The three-term recurrence relation (12.1.2) for a sequence of matrix polynomials $(P_n)_n$ is equivalent to their orthonormality with respect to a weight matrix. In the scalar case, this result is known as Favard's theorem (Favard, 1935) but it is equivalent, both in the scalar and in the matrix case, to a spectral theorem.

We have already pointed out that a sequence $(P_n)_n$ of orthonormal polynomials with respect to a weight matrix W is defined up to multiplication on the left by a sequence of unitary matrices, that is, if $(U_n)_n$ is a sequence of unitary matrices, then the sequence of matrix polynomials $(\mathcal{P}_n)_n$ given by $\mathcal{P}_n(t) = U_nP_n(t)$ is also orthonormal with respect to the weight W. These polynomials satisfy the following three-term recurrence relation:

$$t\mathcal{P}_n(t) = \left(U_nA_{n+1}U_{n+1}^*\right)\mathcal{P}_{n+1}(t) + (U_nB_nU_n^*)\mathcal{P}_n(t) + (U_{n-1}A_nU_n^*)\mathcal{P}_{n-1}(t).$$

The different factorizations of a matrix allow us to choose the nonsingular matrices A_n in the three-term recurrence formula so as to have additional properties. For instance, the QR-factorization allows us to choose each A_n to be lower or upper triangular with positive entries on its diagonal. This, in particular, shows that any weight matrix W has a sequence of orthonormal polynomials having upper-triangular leading coefficient with positive entries on its diagonal. On the other hand, the polar decomposition allows us to choose each A_n to be positive definite.

For a given weight matrix W, the semi-infinite Hermitian matrix

$$J = \begin{pmatrix} B_0 & A_1 & 0 & 0 & \cdots \\ A_1^* & B_1 & A_2 & 0 & \cdots \\ 0 & A_2^* & B_2 & A_3 & \\ \vdots & & \ddots & \ddots & \ddots \end{pmatrix}$$

constructed with the matrices A_n and B_n, which appear in the three-term recurrence relation of a sequence of orthonormal matrix polynomials $(P_n)_n$ with respect to W, is called a Jacobi matrix associated to W. Observe that a weight matrix always has several associated Jacobi matrices but all of them are unitarily equivalent.

The Jacobi matrix associated to a sequence of orthogonal polynomials $(P_n)_n$ plays an important role in the theory. For instance, we will see in the next section that the zeros of the polynomial P_n are the eigenvalues of the Jacobi matrix truncated to size $nN \times nN$.

The polynomials of the second kind, denoted by $(Q_n)_n$ (we assume $P_0 = I$), are defined by

$$Q_n(t) = \int \frac{P_n(t) - P_n(x)}{t - x}\, dW(x), \quad n \in \mathbb{N}. \tag{12.1.3}$$

These polynomials play a fundamental role in the theory. They also satisfy the same three-term recurrence relation satisfied by $(P_n)_n$,

$$tQ_n(t) = A_{n+1}Q_{n+1}(t) + B_nQ_n(t) + A_n^*Q_{n-1}(t),$$

but with different initial conditions: $Q_0(t) = 0$, $Q_1(t) = A_1^{-1}$.

By using the three-term recurrence formula it is straightforward to prove some algebraic relations which are very useful (Durán, 1996), such as

$$P_{n-1}^*(t)A_nP_n(w) - P_n^*(t)A_n^*P_{n-1}(w) = (w - t)\sum_{k=0}^{n-1} P_k^*(t)P_k(w), \quad t, w \in \mathbb{R}, \tag{12.1.4}$$

which is known as the Christoffel–Darboux formula, with the particular cases

$$P_{n-1}^*(t)A_nP_n(t) - P_n^*(t)A_n^*P_{n-1}(t) = 0 \tag{12.1.5}$$

and

$$P_{n-1}^*(t)A_nP_n'(t) - P_n^*(t)A_n^*P_{n-1}'(t) = \sum_{k=0}^{n-1} P_k^*(t)P_k(t). \tag{12.1.6}$$

Another such relation is

$$P^*_{n-1}(t)A_nQ_n(t) - P^*_n(t)A^*_nQ_{n-1}(t) = I, \tag{12.1.7}$$

and the Liouville–Ostrogradsky formula

$$Q_n(t)P^*_{n-1}(t) - P_n(t)Q^*_{n-1}(t) = A_n^{-1}. \tag{12.1.8}$$

12.1.3 Zeros and Gaussian Quadrature Formulas

Definition 12.1.3 We say that a complex number a is a zero of the matrix polynomial P if it is a zero of the scalar polynomial $\det P$. The multiplicity of a as a zero of P is the multiplicity of a as a zero of $\det P$.

We admit a small abuse of language in the previous definition. Another possibility would have been to call a an eigenvalue of P (as in Gohberg, Lancaster, and Rodman, 1982, the standard reference for matrix polynomials), but there is here an abuse of language as well, and the first one is more natural in the context of matrix orthogonality. According to this definition, a is a zero of P if the matrix $P(a)$ is singular, or equivalently, if 0 is an eigenvalue of the matrix $P(a)$.

For a matrix polynomial $P(t) = A_nt^n + A_{n-1}t^{n-1} + \cdots + A_0$ we have

$$\det P(t) = \det(A_n)\, t^{nN} + \cdots ;$$

consequently, if A_n is nonsingular, the polynomial P has nN zeros, taking into account their multiplicities.

The description of a as a zero of P requires the full description of the singular matrix $P(a)$. In particular, we should say something about the eigenvectors of $P(a)$ with respect to the eigenvalue 0. In general, for a matrix polynomial of size N we can state only that

$$N - \operatorname{rank}(P(a)) = \dim\left\{v \in \mathbb{C}^N : P(a)v = 0\right\} \le \text{multiplicity of } a.$$

A zero a of an orthogonal matrix polynomial P always satisfies $N - \operatorname{rank}(P(a)) =$ multiplicity of a, and then the description of the eigenvectors of $P(a)$ associated to 0 is simpler (it is not necessary to use the concept of Jordan chain for P corresponding to a; see Gohberg, Lancaster, and Rodman, 1982, p. 23–24).

Results for the zeros of orthogonal matrix polynomials (obtained by using different techniques) can be found in (the list is not exhaustive) Atkinson (1964), Basu and Bose (1983), Durán (1996), Durán and López-Rodríguez (1996), Durán and Polo (2002), Sinap and Van Assche (1994), and Zhani (1984).

Given a weight matrix, we choose a sequence $(P_n)_n$ of orthonormal matrix polynomials with respect to W having lower triangular leading coefficients. This implies that P_n satisfies the three-term recurrence relation

$$tP_n(t) = A_{n+1}P_{n+1}(t) + B_nP_n(t) + A_n^*P_{n-1}(t),$$

where A_n are lower triangular (and nonsingular) and B_n are Hermitian.

We use the notation J for the Jacobi matrix

$$J = \begin{pmatrix} B_0 & A_1 & 0 & 0 & \cdots \\ A_1^* & B_1 & A_2 & 0 & \cdots \\ 0 & A_2^* & B_2 & A_3 & \\ \vdots & & \ddots & \ddots & \ddots \end{pmatrix},$$

which in this case is a $(2N+1)$-diagonal Hermitian matrix. This Jacobi matrix plays a fundamental role in the study of the zeros of the matrix polynomials $(P_n)_n$. In fact, the main properties are a consequence of the following (Durán and López-Rodríguez, 1996; Sinap, 1995).

Theorem 12.1.4 *For $n \in \mathbb{N}$, the zeros of the matrix polynomial P_n are the same as those of the polynomial* $\det(tI_{nN} - J_{nN})$ *(with the same multiplicity), where I_{nN} is the identity matrix of size nN and J_{nN} is the truncated Jacobi matrix of size nN. The multiplicity of a is just equal to $N - \operatorname{rank}(P_n(a))$.*

The main properties of the zeros of $(P_n)_n$ are contained in the following theorem (Durán, 1996):

Theorem 12.1.5 *If a is a zero of P_n of multiplicity p, then we have the following properties:*

(1) *$a \in \mathbb{R}$, its multiplicity p is less than or equal to N,*

$$\operatorname{rank}(P_n(a)) = N - p, \text{ and } \dim(R(a, P_n)) = \dim(L(a, P_n)) = p$$

(where $R(a, P_n)$ and $L(a, P_n)$ are, respectively, the spaces of right and left eigenvectors of $P_n(a)$ associated to the eigenvalue 0).

(2) *If a is a common zero of P_n and P_{n+1}, then $P_n(a)$ and $P_{n+1}(a)$ do not have any common right eigenvector associated to 0.*

(3) *The matrix $A_n^* P_{n-1}(a)$ defines an isomorphism from $R(a, P_n)$ into $L(a, P_n)$. Its inverse mapping is the isomorphism defined by the matrix $Q_n^*(a)$ (recall that Q_n is the nth polynomial of the second kind associated to $(P_n)_n$ (see (12.1.3)) and that $\operatorname{degr}(Q_n) = n - 1$).*

(4) *For $l = 0, \ldots, p-2$, $(\operatorname{Adj}(P_n(t)))^{(l)}(a) = 0$ and $(\operatorname{Adj}(P_n(t)))^{(p-1)}(a) \neq 0$. Moreover, the matrix $(\operatorname{Adj}(P_n(t)))^{(p-1)}(a)$ defines a linear mapping from $\mathbb{C}^N$ onto $R(a, P)$ which is an isomorphism from $L(a, P)$ into $R(a, P)$. Its inverse is given by the matrix $\frac{p}{(\det P(t))^{(p)}(a)} P'(a)$.*

(5) *We have*

$$(\operatorname{Adj}(P_n(t)))^{(p-1)}(a) P_n(a) = P_n(a) (\operatorname{Adj}(P_n(t)))^{(p-1)}(a) = 0,$$

$$\operatorname{rank}(\operatorname{Adj}(P_n(t)))^{(p-1)}(a) = p,$$

and the p linearly independent columns of $(\operatorname{Adj}(P_n(t)))^{(p-1)}(a)$ form a basis of the linear space of right eigenvectors of $P_n(a)$ associated to 0.

(6) *If we write* $x_{n,k}, y_{n,k}$, $k = 1, \dots, nN$ *for the zeros of* P_n *and* Q_{n+1} *ordered in increasing size and taking into account their multiplicities then*

$$x_{n+1,k} \leq x_{n,k} \leq x_{n+1,k+N}, \tag{12.1.9}$$

$$x_{n+1,k} \leq y_{n,k} \leq x_{n+1,k+N}. \tag{12.1.10}$$

(7) *The zeros of* P_n *are in the convex hull of the support of* W.

Quadrature formulas are one of the most interesting applications of orthogonal polynomials to approximate integrals (this goes back to Gauss). A quadrature formula with degree of precision n for a matrix weight W consists of numbers (real or complex) x_k, $k = 1, \dots, m$, called the nodes, and matrices G_k, $k = 1, \dots, m$, called the quadrature coefficients, such that

$$\int P(t)\, dW(t) = \sum_{k=1}^{m} P(x_k)\, G_k,$$

for any matrix polynomial P of degree less than or equal to n (let us note that this is equivalent to saying that $\int P(t)\, dW(t) Q^{\star}(t) = \sum_{k=1}^{m} P(x_k) G_k Q^{\star}(x_k)$, for any matrix polynomials P, Q satisfying $\text{degr}(P) + \text{degr}(Q) \leq n$).

It is proved in Durán and Polo (2002) that the degree of precision n for a quadrature formula with $\sum_{k=1}^{m} \text{rank}\,(G_k) = nN + h$, $0 \leq h \leq N - 1$ is not larger than $2n - 1$. For $h = 0$ a quadrature formula of the highest degree of precision $2n - 1$ does not hold for any matrix polynomial P of degree $2n$. Here we concentrate on Gaussian quadrature formulas for the case when $\sum_{k=1}^{m} \text{rank}\,(G_k) = nN$, that is, quadrature formulas with degree of precision $2n - 1$. For the general case of $\sum_{k=1}^{m} \text{rank}\,(G_k) = nN + h$, $0 \leq h \leq N - 1$, see Durán and Polo (2002).

Gaussian quadrature formulas with $\sum_{k=1}^{m} \text{rank}\,(G_k) = nN$ have been found by several authors: all these formulas have nodes at the zeros of P_n (the nth orthonormal matrix polynomial with respect to W) and these authors have used different approaches to find them. Indeed, Basu and Bose (1983) used the Gram–Schmidt orthogonalization procedure applied to the eigenvectors of P_n associated to the zeros, Sinap and Van Assche (1994) used matrix polynomial interpolation, Durán and López-Rodríguez (1996) used the matrix expression for $\det(P_n)$ (Theorem 12.1.4 above), Durán (1996) used the structural properties of $(P_n)_n$, and Dette and Studden (2003) used geometrical properties of the set of solutions of a truncated moment problem. A closed expression for the quadrature coefficients in these formulas is given only in Durán (1996), where quadrature formulas with degree of precision $2n - 2$ are also given (we here follow Durán, 1996).

Theorem 12.1.6 *Let* W *and* n *be a weight matrix and a nonnegative integer. We write* $x_{n,k}$, $k = 1, \dots, m$ *for the different zeros of an nth orthonormal polynomial* P_n *with respect to* W *ordered in increasing size (hence* $m \leq nN$*) and* $\Gamma_{n,k}$ *for the matrices*

$$\Gamma_{n,k} = \frac{l_k}{(\det(P_n(t)))^{(l_k)}(x_{n,k})} (\text{Adj}\,(P_n(t)))^{(l_k - 1)}(x_{n,k})\, Q_n(x_{n,k}), \quad k = 1, \dots, m, \tag{12.1.11}$$

where l_k *is the multiplicity of* $x_{n,k}$.

(1) *For any polynomial P with* $\deg(P) \leq 2n-1$ *the following formula holds:*

$$\int P(t)\,dW(t) = \sum_{k=1}^{m} P(x_{n,k})\Gamma_{n,k}.$$

(2) *The matrices* $\Gamma_{n,k}$ *are positive semidefinite matrices of rank* l_k, $k = 1,\dots,m$, *so that* $\sum_{k=1}^{m} \operatorname{rank} \Gamma_{n,k} = nN$.

Theorem 12.1.6 is sharp in the sense that any quadrature formula with quadrature coefficients G_k, $k = 1,\dots,m$ satisfying $\sum_{k=1}^{m} \operatorname{rank} G_k = nN$ and degree of precision $2n-1$ (the highest possible) necessarily has nodes at the zeros of P_n and the quadrature coefficients are then given by (12.1.11) (Durán and Defez, 2002).

The Gaussian quadrature formula appearing in Theorem 12.1.6 is a powerful tool to prove some asymptotic results for orthogonal matrix polynomials. Using it one can prove

(1) Favard's theorem (see Durán and López-Rodríguez, 1996; see also Durán, 1993, 1995 for another approach);
(2) Markov's theorem giving the limit behavior of the ratio $P_n^{-1}Q_n$ between the nth orthonormal polynomial and the corresponding nth polynomial of the second kind (see Durán, 1996);
(3) ratio asymptotics, that is, the limit behavior of the ratio $P_{n-1}P_n^{-1}$ between two consecutive orthonormal polynomials (see Durán, 1999; Durán and Daneri-Vias, 2001);
(4) zero behavior, that is, the asymptotic behavior of the counting measures

$$\frac{1}{nN}\sum_{k=1}^{m} l_k \delta_{x_{k,n}},$$

where $x_{k,n}$, l_k, $k = 1,\dots,m$ are, respectively, the zeros of P_n and their corresponding multiplicities (see Durán, López-Rodríguez, and Saff, 1999).

Favard's theorem has been proved for any family of matrix polynomials satisfying a three-term recurrence relation, while Markov's theorem has been proved for determinate weight matrices (that is, weight matrices W which are uniquely determined by the moments $\int t^n\,dW(t)$, $n \geq 0$). Ratio and zero asymptotics have been proved for weight matrices having orthonormal polynomials with convergent recurrence coefficients. In the scalar case, Rakhmanov's theorem guarantees this assumption for positive measure supported on a bounded interval $[a,b]$ and with a positive Radon–Nikodym derivative with respect to the Lebesgue measure a.e. in $[a,b]$ (see Rakhmanov, 1977; Rakhmanov, 1982b; and also Máté and Nevai, 1982). Rakhmanov's theorem has an analogue for weight matrices which can be used to guarantee that a weight matrix has a sequence of orthonormal matrix polynomials with convergent recurrence coefficients: if a weight matrix W supported on a bounded interval $[a,b]$ has a Radon–Nikodym derivative W' with respect to the Lebesgue measure satisfying $\det(W') > 0$ a.e. in $[a,b]$, then W has a sequence of orthonormal polynomials with convergent recurrence coefficients (see Van Assche, 2007).

12.2 Matrix-Valued Orthogonal Polynomials Satisfying Second-Order Differential Equations

12.2.1 Symmetric Second-Order Differential Operators

In the rest of this contribution we will consider orthogonal matrix polynomials $(P_n)_n$ satisfying right-hand-side second-order differential equations of the form

$$P_n''(t)F_2(t) + P_n'(t)F_1(t) + P_n(t)F_0(t) = \Gamma_n P_n(t), \quad n \geq 0, \tag{12.2.1}$$

where the differential coefficients F_2, F_1, and F_0 are matrix polynomials (which do not depend on n) of degrees not bigger than 2, 1, and 0, and Γ_n are matrices. In the matrix orthogonality situation, these families are most likely going to play a role similar to that of the classical families of Hermite, Laguerre, or Jacobi in the scalar case. The equation (12.2.1) for the polynomial P_n is equivalent to saying that P_n is a left eigenfunction of the right-hand-side second-order differential operator

$$D = \partial^2 F_2(t) + \partial^1 F_1(t) + \partial^0 F_0(t), \quad \partial = \frac{d}{dt} \tag{12.2.2}$$

if we allow for matrix-valued eigenvalues.

These operators could be made to act on our functions either on the left or on the right. One finds a discussion of these two actions in Durán (1997). The conclusion there is that if one wants to have matrix weights W that do not reduce to scalar weights and that have matrix polynomials as their eigenfunctions, one should settle for right-hand-side differential operators and left-hand-side eigenfunctions.

The key concept to study orthogonal matrix polynomials $(P_n)_n$ satisfying second-order differential equations of the form (12.2.1) is that of the symmetry of a differential operator with respect to a weight matrix:

Definition 12.2.1 We say that the second-order differential operator D (12.2.2) is symmetric with respect to the weight matrix W if

$$\int D(P)WQ^* \, dt = \int PW(D(Q))^* \, dt,$$

for all matrix polynomials P, Q.

One then has that D is symmetric with respect to W if and only if the orthonormal matrix polynomials $(P_n)_n$ with respect to W satisfy (12.2.1), where Γ_n, $n \geq 0$ are Hermitian matrices (see Durán, 1997).

Given a sequence of orthogonal polynomials $(P_n)_n$ one can consider (see Grünbaum and Tirao, 2007) the algebra $\mathcal{D}(W)$ of all right-hand-side differential operators that have the polynomials $(P_n)_n$ as their left eigenfunctions.

Note that if $D(P_n) = \Gamma_n P_n$ for some eigenvalue matrix $\Gamma_n \in \mathbb{C}^{N\times N}$, then Γ_n is determined by the differential operator D. Thus we have

$$\mathcal{D}(W) = \left\{D\colon D(P_n) = \Gamma_n(D)P_n, \ \Gamma_n(D) \in \mathbb{C}^{N\times N} \text{ for all } n \geq 0\right\}. \tag{12.2.3}$$

The definition of the algebra $\mathcal{D}(W)$ depends only on the weight matrix W and not on the sequence $(P_n)_n$.

The statement in Durán (1997) can be rephrased as saying that if D is symmetric then D belongs to the algebra introduced above. In Grünbaum and Tirao (2007) one proves that any element in $\mathcal{D}(W)$ can be decomposed in the form $D = D_1 + iD_2$ with D_1, D_2 symmetric.

The main point of the second part of this contribution is to consider those W such that $\mathcal{D}(W)$ is a nontrivial algebra. This is a matrix version of what is usually called the Bochner classification problem (although this classification problem was considered before, see Routh, 1884), and was raised in Durán (1997). It is important to observe that in the matrix case the structure of the algebra $\mathcal{D}(W)$ can be much richer than in the scalar case. In the classical examples of the scalar case the algebra is always commutative and generated by the second-order differential operator associated to each family. The situation in the matrix case is much richer. There are examples where $\mathcal{D}(W)$ contains operators of order 1, examples where $\mathcal{D}(W)$ contains several linearly independent operators of order 2 (modulo the identity matrix), examples where, for instance, one can conjecture a complete set of generators formed by operators of orders 2 and 3 (and the identity matrix), examples where the algebra $\mathcal{D}(W)$ is commutative and examples where it is not (see Castro and Grünbaum, 2005; Castro and Grünbaum, 2006; Durán and de la Iglesia, 2008b). Algebras of this kind are associated with certain algebraic–geometric objects such as a curve and a bundle on it. For one example (see (12.2.36) below) the detailed structure of $\mathcal{D}(W)$ has been established in Tirao (2011).

There is a dual notion to that described in the previous paragraph. For a fixed differential operator D of the form (12.2.2), we define the set of weight matrices

$$\Upsilon(D) = \{W\colon D \text{ is symmetric with respect to } W\}. \tag{12.2.4}$$

One straightforwardly has that if $\Upsilon(D) \neq \emptyset$ then it is a convex cone: if $W_1, W_2 \in \Upsilon(D)$ and $\gamma, \zeta \geq 0$ (one of them not zero), then $\gamma W_1 + \zeta W_2 \in \Upsilon(D)$.

The weight matrices W going along with a symmetric second-order differential operator D provide examples where $\Upsilon(D) \neq \emptyset$. In these examples $\Upsilon(D)$ contains, at least, a half-line: γW, $\gamma > 0$. For the second-order differential operators D associated to the classical families of the scalar case, $\Upsilon(D)$ always reduces to a half-line. In the matrix case the situation is, again, much richer because there are examples where this convex cone $\Upsilon(D)$ has dimension bigger than 1 (see Durán and de la Iglesia, 2008a). We will give examples of a fixed second-order differential operator D such as (12.2.2) for which there exist two weight matrices W_1 and W_2, $W_1 \neq \alpha W_2$ for any $\alpha > 0$, such that D is symmetric with respect to any of the weight matrices $\gamma W_1 + \zeta W_2$, $\gamma, \zeta \geq 0$. That means, in particular, that the corresponding monic matrix polynomials $(P_{n,\zeta/\gamma})_n$ orthogonal with respect to $\gamma W_1 + \zeta W_2$ (they depend only on W_1, W_2 and the ratio ζ/γ) are eigenfunctions of D,

$$D\left(P_{n,\zeta/\gamma}\right) = \Gamma_n P_{n,\zeta/\gamma}, \quad n = 0, 1, \ldots, \gamma > 0, \zeta \geq 0,$$

where D and Γ_n do not depend on γ, ζ.

In contrast to the scalar case we are very far from achieving a complete classification of all the weight matrices having symmetric second-order differential operators. Starting in the next section, we give a first approach to the classification problem. We obtain a set of linear differential equations which implies the symmetry of a second-order differential operator, we give several methods to solve this set of differential equations in a variety of cases, and illustrate them by displaying some of the examples produced.

A second approach leading to a general characterization result is given below but this cannot be considered a solution of the problem. This general result reads exactly as the one in Duistermaat and Grünbaum (1986), where the resulting nonlinear equations are solved in detail. Nothing comparable to this is available in the matrix case.

This result requires that we start with $W(t)$ and consider its unique sequence of monic orthogonal polynomials $(P_n)_n$, together with the three-term recursion relation

$$tP_n(t) = P_{n+1}(t) + B_nP_n(t) + A_nP_{n-1}(t), \quad n \geq 0,$$

where we put $P_{-1} = 0$.

It is now convenient to introduce the block tridiagonal matrix L,

$$L = \begin{pmatrix} B_0 & I & & \\ A_1 & B_1 & I & \\ & \ddots & \ddots & \ddots \end{pmatrix},$$

where all the matrices A_i, B_i are in $\mathbb{C}^{N\times N}$ and I denotes the $N \times N$ identity matrix.

The recursion relation now takes the form

$$LP = tP,$$

where P stands for the vector

$$P(t) = \begin{pmatrix} P_0(t) \\ P_1(t) \\ P_2(t) \\ \vdots \end{pmatrix}.$$

One can now show the following result:

Theorem 12.2.2 *Let $W(t)$ be a weight matrix on the real line, $(P_n)_n$ the corresponding sequence of monic orthogonal polynomials, and L the block tridiagonal matrix that gives $LP = tP$. If $D \in \mathcal{D}(W)$ and Λ is a block diagonal matrix as above with $\Lambda_n = \Lambda_n(D)$, we have*

$$(\mathrm{ad}\, L)^{m+1}(\Lambda) = 0 \tag{12.2.5}$$

for some m. Conversely, if L is a block tridiagonal matrix and Λ is a block diagonal matrix satisfying this condition for some $m \geq 0$, then there is a unique differential operator D in $\mathcal{D}(W)$ such that $\Lambda_n = \Lambda_n(D)$ for all $n \geq 0$. Moreover, the order of D is equal to the minimum m satisfying (12.2.5).

A third approach to obtaining examples that would be part of a complete solution of the matrix version of the Bochner problem comes from exploiting the connection between matrix-valued spherical functions and matrix-valued orthogonal polynomials, see Grünbaum, Pacharoni, and Tirao (2001, 2002, 2004, 2005) and Pacharoni and Tirao (2007a).

A fourth approach using moment sequences of weight matrices is considered in Durán (1997), Durán and de la Iglesia (2008a,b), and Durán and Grünbaum (2004).

12.2.2 The Differential Equations for the Weight Matrix

The basic idea for the study of the symmetry of a second-order differential operator with respect to a weight matrix W is to convert this condition of symmetry into a set of differential equations relating the weight matrix and the differential coefficients of the differential operator (see Durán and Grünbaum, 2004; Grünbaum, Pacharoni, and Tirao, 2001):

Theorem 12.2.3 *Let W be a weight matrix with support on an interval $[a, b]$ of the real line (bounded or unbounded). Assume that W is twice differentiable in the interior (a, b) of its support, and the boundary conditions that*

$$\lim_{t\to a^+,b^-} t^n F_2(t)W(t) = 0 \quad \textit{and} \quad \lim_{t\to a^+,b^-} t^n \left[(F_2(t)W(t))' - (F_1(t)W(t))\right] = 0, \tag{12.2.6}$$

for any $n \geq 0$. If the weight matrix W satisfies

$$F_2 W = W F_2^*, \tag{12.2.7}$$

as well as

$$2(F_2 W)' = W F_1^* + F_1 W \tag{12.2.8}$$

and

$$(F_2 W)'' - (F_1 W)' + F_0 W = W F_0^*, \tag{12.2.9}$$

then the second-order differential operator D (see (12.2.2)*) is symmetric with respect to W.*

We want to stress the importance of the boundary conditions (12.2.6). Consider the following nice weight matrix:

$$W(t) = e^{-t}\begin{pmatrix} 1 + |a|^2\log^2(t) & a\log(t) \\ \bar{a}\log(t) & 1 \end{pmatrix}, \quad t > 0,\ a \in \mathbb{C}.$$

W satisfies the differential equations (12.2.7), (12.2.8), and (12.2.9) for $F_2 = tI$, $F_1 = -tI + 2A + I$, and $F_0 = -A$, where $A = \left(\begin{smallmatrix} 0 & a \\ 0 & 0 \end{smallmatrix}\right)$. But unfortunately the corresponding second-order differential operator is not symmetric for W because W does not satisfy the second of the boundary conditions (12.2.6).

An extension of Theorem 12.2.3 for higher-order differential operators can be found in Durán and de la Iglesia (2008b).

12.2.3 A Method for Solving the Differential Equations for the Weight Matrix When F_2 Is Scalar

In this section we describe a method to find solutions of equations (12.2.7), (12.2.8), and (12.2.9) above when $F_2 = f_2 I$, where f_2 is a scalar polynomial of degree at most 2. The method is very fruitful: in this section we will show how the method can be used to find many families of weight matrices having symmetric second-order differential operators. As a first conclusion, we can affirm that the subtlety of the noncommutative algebra of matrices can be exploited to yield many different such families, almost dwarfing by comparison the scalar situation (where only the classical examples of Hermite, Laguerre, and Jacobi have such kinds of differential operators).

The method is the following (see Durán and Grünbaum, 2004):

Theorem 12.2.4 *Let Ω be an open set of the real line, and fix $a \in \Omega$. Let ρ, f_2, F_1, and F_0 be two (real) scalar functions, and two matrix functions, respectively. We assume $\rho(t), f_2(t) \neq 0$, $t \in \Omega$, and that all the functions are twice differentiable in Ω. Define $F_2(t) = f_2(t)I$, the scalar function c as*

$$c(t) = \frac{(\rho(t) f_2(t))'}{\rho(t)}, \tag{12.2.10}$$

and the matrix function F as

$$F_1(t) = 2 f_2(t) F(t) + c(t) I. \tag{12.2.11}$$

Write T for the solution of the differential equation

$$T'(t) = F(t)T(t), \quad T(a) = I,$$

and define the matrix function

$$W(t) = \frac{\rho(t)}{\rho(a)} T(t) W(a) T^*(t).$$

Then this matrix function satisfies the differential equation (12.2.8)*. Define the matrix function χ by means of*

$$\chi(t) = T^{-1}(t)\left(f_2(t)F'(t) + f_2(t)F^2(t) + c(t)F(t) - F_0\right)T(t). \tag{12.2.12}$$

Then the matrix function $\chi(t)W(a)$ is Hermitian for $t \in \Omega$ if and only if W satisfies the differential equation (12.2.9) *in Ω.*

Although we are interested in the case when W is a weight matrix and F_2, F_1, and F_0 are matrix polynomials of degrees not larger than 2, 1, and 0, respectively (otherwise one cannot apply Theorem 12.2.3), the method given above to solve (12.2.8) and (12.2.9) does not require these restrictions on W, F_2, F_1, and F_0.

We now illustrate how to use this method to find examples of weight matrices having symmetric second-order differential operators. First we introduce two auxiliary matrices that we will need to describe the examples. The first one is the nilpotent matrix

$$\mathscr{N} = \begin{pmatrix} 0 & v_1 & 0 & \cdots & 0 \\ 0 & 0 & v_2 & \cdots & 0 \\ \vdots & \vdots & \vdots & & \vdots \\ 0 & 0 & 0 & \cdots & v_{N-1} \\ 0 & 0 & 0 & \cdots & 0 \end{pmatrix}, \tag{12.2.13}$$

where $v_1, \ldots, v_{N-1}$ are complex parameters. The second one is the diagonal matrix

$$\mathscr{J} = \begin{pmatrix} N-1 & 0 & \cdots & 0 & 0 \\ 0 & N-2 & \cdots & 0 & 0 \\ \vdots & \vdots & & \vdots & \vdots \\ 0 & 0 & \cdots & 1 & 0 \\ 0 & 0 & \cdots & 0 & 0 \end{pmatrix}. \tag{12.2.14}$$

Taking into account the scalar case, it seems rather natural to choose the classical weights of Hermite, Laguerre, and Jacobi as the first term in the couple (ρ, f_2) as the most promising ones for applying our method. Formula (12.2.11) gives that the differential coefficient F of T $(T' = FT)$ is equal to

$$F(t) = \frac{1}{2f_2(t)}(F_1(t) - c(t)I).$$

Taking into account that F_1 is a matrix polynomial of degree not bigger than 1, we get that when $\rho(t) = e^{-t^2}$ and $f_2(t) = 1$, the matrix function F is a polynomial of degree 1, which we write as $2Bt + A$. For $\rho(t) = t^\alpha e^{-t}$ and $f_2(t) = t$, the function F is of the form $A + B/t$. For $\rho(t) = (1-t)^\alpha(1+t)^\beta$ and $f_2(t) = 1 - t^2$, the function F is given by $-A/(1-t) + B/(1+t)$. For $\rho(t) = e^{-t^2}$ and $\rho(t) = (1-t)^\alpha(1+t)^\beta$ we take $a = 0$ and for $\rho(t) = t^\alpha e^{-t}$ we take $a = 1$. We can summarize this as follows:

$$\begin{cases} (1)\ \rho(t) = e^{-t^2},\ F_2(t) = I\ \ (t \in \mathbb{R}), \text{ and then} \\ \qquad T'(t) = (2Bt + A)T(t), \quad F_1(t) = 2(2B - I)t + 2A; \\ (2)\ \rho(t) = t^\alpha e^{-t},\ F_2 = tI\ \ (t > 0,\ \alpha > -1), \text{ and then} \\ \qquad T'(t) = \left(A + \frac{B}{t}\right)T(t), \quad F_1(t) = (2A - I)t + 2B + (\alpha + 1)I; \\ (3)\ \rho(t) = (1-t)^\alpha(1+t)^\beta,\ F_2 = (1-t^2)I\ \ (t \in (-1,1),\ \alpha, \beta > -1), \text{ and then} \\ \qquad T'(t) = \left(\frac{-A}{1-t} + \frac{B}{1-t}\right)T(t), \\ \qquad F_1(t) = -(2A + 2B + (\alpha + \beta + 2)I)t - 2A + 2B - (\alpha - \beta)I. \end{cases} \tag{12.2.15}$$

To simplify, we assume $W(a) = I$. According to Theorem 12.2.4, the symmetry of the second-order differential operator (12.2.2) with respect to $W = \rho TT^*$ is then equivalent to the Hermiticity of the matrix function:

$$\chi(t) = T^{-1}(t)\left(F_2(t)F'(t) + f_2(t)F^2(t) + c(t)F(t) - F_0\right)T(t). \tag{12.2.16}$$

Write

$$X(t) = f_2(t)F'(t) + f_2(t)F^2(t) + c(t)F(t) - F_0$$

so that $\chi(t) = T^{-1}(t)X(t)T(t)$ (see (12.2.16)). The expression of the matrix function X for each of the classical weights of Hermite, Laguerre, and Jacobi is then as follows:

(1) For $\rho(t) = e^{-t^2}$,

$$X(t) = 2B + A^2 - F_0 + (2AB + 2BA - 2A)t + \left(4B^2 - 4B\right)t^2. \tag{12.2.17}$$

(2) For $\rho(t) = t^{\alpha}e^{-t}$,

$$X(t) = \left(\alpha B + B^2\right)\frac{1}{t} + AB + BA + (\alpha + 1)A - F_0 - B + \left(A^2 - A\right)t. \tag{12.2.18}$$

(3) For $\rho(t) = (1 - t)^{\alpha}(1 + t)^{\beta}$,

$$\begin{aligned} X(t) = {} & \frac{2t}{1-t}\left(\alpha A + A^2\right) - \frac{2t}{1+t}\left(\beta B + B^2\right) \\ & + A^2 + B^2 - AB - BA - (\alpha - \beta + 1)B - (\beta - \alpha + 1)A - F_0. \end{aligned} \tag{12.2.19}$$

To illustrate the method we single out the case when $\rho = e^{-t^2}$ and $f_2 = 1$, that is, $W(t) = e^{-t^2}T(t)T^*(t)$, $F_2(t) = I$, and thus

$$T'(t) = (2Bt + A)T(t).$$

Before moving to more general cases, we restrict ourselves to the rather modest case when B vanishes (this example is taken from Durán and Grünbaum, 2004). Assuming that $B = 0$, we have to check whether there is a convenient choice of F_0 so that the matrix function (12.2.16) is Hermitian; in our particular case, this matrix function is equal to (see (12.2.17))

$$\chi(t) = A^2 - 2tA - e^{-At}F_0e^{At}, \tag{12.2.20}$$

which can be written as

$$\begin{aligned} \chi(t) &= A^2 - 2At - e^{-t\,\mathrm{ad}_A}(F_0) \\ &= \left(A^2 - F_0\right) - t\left(2A - \mathrm{ad}_A F_0\right) - \sum_{n\geq 2}\frac{(-1)^n t^n}{n!}\,\mathrm{ad}_A^n F_0, \end{aligned}$$

where we use the standard notation $\mathrm{ad}_X^0 Y = Y$, $\mathrm{ad}_X Y = [X, Y] = XY - YX$, $\mathrm{ad}_X^{n+1} Y = [X, \mathrm{ad}_X^n Y]$, $n \geq 2$.

The existence of a matrix F_0 such that (12.2.20) is Hermitian is equivalent to the existence of a matrix F_0 such that the matrices

$$A^2 - F_0, \quad 2A - [A, F_0], \quad \mathrm{ad}_A^n F_0, \quad n \geq 2 \quad \text{are Hermitian.} \tag{12.2.21}$$

We now exhibit a solution for these equations: we define

$$A = \mathscr{N}, \tag{12.2.22}$$

$$F_0 = A^2 - 2\mathscr{J}, \tag{12.2.23}$$

where the nilpotent matrix $\mathscr{N}$ and the diagonal matrix $\mathscr{J}$ are defined by (12.2.13) and (12.2.14), respectively. The conditions (12.2.21) follow easily since $\mathrm{ad}_A F_0 = 2A$.

Theorems 12.2.4 and 12.2.3 show now that the second-order differential operator

$$D = \partial^2 I - \partial(2tI - 2A) + F_0$$

is symmetric with respect to the weight matrix $W(t) = e^{-t^2} e^{At} e^{A^* t}$.

If A and B do not commute, then one can only give in general a formal series expansion for the solution T of each one of the first-order differential equations (12.2.15). Such expressions in terms of time ordering are familiar in many fields, including quantum field theory, but make it rather laborious (if not impossible) to check whether the function χ in (12.2.16) is Hermitian.

We show now a method that does not require integrating these equations to check whether the function χ is Hermitian or not. The starting point is the following trivial observation. Assume that A, B, and F_0 are upper triangular. This implies that the solution T for each one of the differential equations (12.2.15), together with the function X in (12.2.17), (12.2.18), or (12.2.19), is also upper triangular. As a consequence the function $\chi(t) = T^{-1}(t)X(t)T(t)$ is upper triangular as well. Since the function χ has to be Hermitian if we want the weight matrix W to satisfy the differential equation (12.2.9), we can conclude that actually χ has to be diagonal. Moreover, $\chi(t) = \mathrm{diag}(X(t))$.

Under additional assumptions on the diagonal of the matrices A and B, we have that $f_2\chi$ is then a polynomial of degree at most 1, and so $(f_2\chi)'' = 0$. Using now the first-order differential equation for T (see (12.2.15)), we can manage to get from $(f_2\chi)'' = 0$ a set of finitely many equations for A, B, and F_0 which are equivalent to the Hermiticity of χ (for more details see Durán, 2009a).

We show here how to proceed when $\rho(t) = e^{-t^2}$ and $f_2(t) = 1$. In addition to the assumption that the matrices A, B, and F_0 are upper triangular, we assume here that B is nilpotent. Under those assumptions, the description of our method above says that if the equation (12.2.9) is satisfied then the matrix function $\chi(t) = T^{-1}(t)X(t)T(t)$ has to be diagonal, and conversely. According to (12.2.17) we then get that

$$\chi(t) = \left(A^2_{[1]} - F_{0,[1]}\right) - 2A_{[1]}t, \tag{12.2.24}$$

where $M_{[1]}$ denotes the diagonal matrix with diagonal entries equal to those of M.

We then have that $\chi'' = 0$. Using the differential equation for T (see (12.2.15)) to compute χ'', we can reduce formula (12.2.24) to the following set of equations for the matrices A, B, and F_0:

$$F_0 - 2B - A^2 + A_{[1]}^2 - F_{0,[1]} = 0, \tag{12.2.25}$$

$$2AB + 2BA - 2A - \mathrm{ad}_A\left(A_{[1]}^2 - F_{0,[1]}\right) + 2A_{[1]} = 0, \tag{12.2.26}$$

$$2\,\mathrm{ad}_A(A_{[1]}) - \mathrm{ad}_B\left(A_{[1]}^2 - F_{0,[1]}\right) + 4B^2 - 4B - \frac{1}{2}\mathrm{ad}_A^2(A_{[1]}^2 - F_{0,[1]}) = 0, \tag{12.2.27}$$

$$\mathrm{ad}_B(A_{[1]}) + \frac{1}{2}\mathrm{ad}_A^2(A_{[1]}) = 0, \tag{12.2.28}$$

$$\mathrm{ad}_B(\mathrm{ad}_A(A_{[1]})) = 0. \tag{12.2.29}$$

In other words, the second-order differential equation (12.2.9) for the weight matrix $W = e^{-t^2}TT^*$, $T'(t) = (2Bt + A)T(t)$ is equivalent to the matrix equations (12.2.25), (12.2.26), (12.2.27), (12.2.28), and (12.2.29) for the upper-triangular matrices A, B, and F_0, B nilpotent.

Of course, the assumption that B is nilpotent is not essential in our procedure: it simply leads to a simplification of the equations for A, B, and F_0. Using the same approach, if we do not assume that B is nilpotent, we get a set of seven (more) complicated equations (one more derivative is necessary since χ is then a polynomial of degree 2) which are equivalent to the Hermiticity of χ.

In the next theorem we display a solution of equations (12.2.25), (12.2.26), (12.2.27), (12.2.28), and (12.2.29) assuming that the entries of the main diagonal of A are in arithmetic progression (see Durán, 2009a).

Theorem 12.2.5 *Let $\mathcal{N}$ and $\mathcal{J}$ be the nilpotent matrix and the diagonal matrix defined by* (12.2.13) *and* (12.2.14)*, respectively. Define the matrices A and B by*

$$A = \nu_0\mathcal{J} + \sum_{k=2}^{N}\frac{(k-1)^{k-2}}{(k-1)!}\nu_0^{k-2}\mathcal{N}^{k-1}, \tag{12.2.30}$$

$$B = \frac{1}{2}\left[A, \nu_0\mathcal{J}\right], \tag{12.2.31}$$

where ν_0 is a real number. Then the solution T of the differential equation $T'(t) = (2Bt + A)T(t)$, $T(0) = I$ is the matrix

$$T(t) = e^{(A-\nu_0 J)t}e^{\nu_0 tJ},$$

and the weight matrix $W(t) = e^{-t^2}T(t)T^*(t)$ has a symmetric second-order differential operator like (12.2.2) *with coefficients*

$$F_2 = I, \quad F_1 = 2(2B - I)t + 2A, \quad F_0 = 2B + A^2 - \nu_0^2\mathcal{J}^2 - 2\mathcal{J}.$$

For $\nu_0 = 0$, this is the example involving (12.2.22) and (12.2.23).

Just a couple of comments: First, notice that the function $e^{(A-J_{\nu_0})t}e^{J_{\nu_0}t}$ is not equal to e^{At} because $A - J_{\nu_0}$ and J_{ν_0} do not commute. Second, $e^{(A-J_{\nu_0})t}$ is a matrix polynomial of degree at most $N-1$ because $A - J_{\nu_0}$ is nilpotent.

The classical weights of Laguerre and Jacobi lead to a different set of matrix equations which, as it happens with equations (12.2.25), (12.2.26), (12.2.27), (12.2.28), and (12.2.29), are rather difficult to solve. In the next two theorems, we include a couple of the examples that have been found by solving the corresponding matrix equations for Laguerre and Jacobi weights (see Durán, 2009a).

Theorem 12.2.6 *Let $\mathscr{N}$ and $\mathscr{J}$ be the nilpotent matrix and the diagonal matrix defined by* (12.2.13) *and* (12.2.14), *respectively. Define the matrices A and B by*

$$A = -\left(1 - \frac{1}{v_0}\right)\mathscr{N} - \sum_{k=2}^{N-1}\left(1 - \frac{1}{kv_0}\right)\frac{1}{v_0^{k-1}}\binom{2kv_0 - 2}{k-1}\mathscr{N}^k, \tag{12.2.32}$$

$$B = v_0\mathscr{J} + \mathscr{N} + \sum_{k=2}^{N-1}\frac{1}{v_0^{k-1}}\binom{2kv_0 - 2}{k-1}\mathscr{N}^k, \tag{12.2.33}$$

where $v_0 > (-1-\alpha)/2(N-1)$ (this condition is needed to avoid integration problems at $t = 0$). Then the solution T of the differential equation $T'(t) = (A + B/t)T(t)$, $T(1) = I$, $t > 0$ is the matrix

$$T(t) = e^{(A+B-v_0J)(t-1)}t^{v_0J},$$

and the weight matrix $W(t) = t^{\alpha}e^{-t}T(t)T^*(t)$ has a symmetric second-order differential operator like (12.2.2) *with coefficients $F_2 = tI$ and*

$$F_1 = 2B + (\alpha + 1)I + (2A - I)t,$$
$$F_0 = (\alpha - 1)B - (\alpha - 1)v_0\mathscr{J} + B^2 - v_0^2\mathscr{J}^2 + AB + BA + \alpha A + A^2 - \mathscr{J}.$$

(Notice that the assumption $v_0 > (-1 - \alpha)/2(N - 1)$ guarantees that the weight matrix W satisfies the boundary conditions (12.2.6)).

Theorem 12.2.7 *Let $\mathscr{N}$ and $\mathscr{J}$ be the nilpotent matrix and the diagonal matrix defined by* (12.2.13) *and* (12.2.14), *respectively. Define the matrices A and B by*

$$A = \frac{1}{2}\mathscr{J} + \mathscr{N}, \tag{12.2.34}$$

$$B = \frac{1}{2}\mathscr{J} - \mathscr{N}. \tag{12.2.35}$$

Then the solution T of the differential equation $T'(t) = (-A/(1-t) + B/(1+t))T(t)$, $T(0) = I$, $t \in (-1, 1)$ is the matrix

$$T(t) = (1 + t)^{2B}\left(\frac{1 - t}{1 + t}\right)^{\frac{1}{2}\mathscr{J}},$$

and the weight matrix $W(t) = (1-t)^{\alpha}(1+t)^{\beta}T(t)T^*(t)$, $\alpha, \beta > -1$ has a symmetric second-order differential operator like (12.2.2) *with coefficients $F_2 = \left(1 - t^2\right)I$ and*

$$F_1 = -(1 + t)(2A + (\alpha + 1)I) + (1 - t)(2B + (\beta + 1)I),$$
$$F_0 = -(\alpha + \beta + 1)\mathscr{J} - \mathscr{J}^2 - 2(\alpha - \beta)\mathscr{N} + 4\mathscr{N}^2.$$

More examples with $F_2 = f_2 I$ can be found in (the list is not exhaustive) Durán (2009a), Durán and de la Iglesia (2008b), Durán and Grünbaum (2004, 2005a), Grünbaum (2003, 2011), Grünbaum, Pacharoni, and Tirao (2001, 2002, 2004, 2005), and Pacharoni and Tirao (2007a).

12.2.4 A Catalog of Examples of Size 2×2

To illustrate the richness of examples of weight matrices having symmetric second-order differential operators, we give in this section a catalog of examples of weight matrices for size 2×2 and the corresponding symmetric differential operators (the catalog is not exhaustive). Each operator generates a second-order differential equation as (12.2.1) satisfied by any sequence of orthogonal polynomials with respect to the weight matrix (the matrix eigenvalues for the monic family are then given by $\binom{n}{2}F_2'' + nF_1' + F_0$).

We start with three examples associated with the Hermite weight e^{-t^2} and with $F_2 = I$ (in these examples the parameter a is a complex number, b is a real number, and $t \in \mathbb{R}$).

(1) See Durán and Grünbaum (2004, 2005c):

$$W_{1,h}(t) = e^{-t^2}\begin{pmatrix} 1 + |a|^2t^2 & at \\ \bar{a}t & 1 \end{pmatrix}, \tag{12.2.36}$$

$$\partial^2 + \partial\begin{pmatrix} -2t & 2a \\ 0 & -2t \end{pmatrix} + \begin{pmatrix} -2 & 0 \\ 0 & 0 \end{pmatrix}, \quad \partial = \frac{d}{dt}. \tag{12.2.37}$$

(2) See Durán and Grünbaum (2004, 2005c):

$$W_{2,h}(t) = e^{-t^2}\begin{pmatrix} 1 + |a|^2t^4 & at^2 \\ \bar{a}t^2 & 1 \end{pmatrix}, \tag{12.2.38}$$

$$\partial^2 + \partial\begin{pmatrix} -2t & 4at \\ 0 & -2t \end{pmatrix} + \begin{pmatrix} -4 & 2a \\ 0 & 0 \end{pmatrix}.$$

(3) See Durán and Grünbaum (2007):

$$W_{3,h}(t) = e^{-t^2}\begin{pmatrix} e^{2bt} + |a|^2t^2 & at \\ \bar{a}t & 1 \end{pmatrix},$$

$$\partial^2 + \partial\begin{pmatrix} 2b - 2t & 2a - 2abt \\ 0 & -2t \end{pmatrix} + \begin{pmatrix} -2 & 0 \\ 0 & 0 \end{pmatrix}.$$

We now continue with three more examples associated with the Laguerre weight $t^\alpha e^{-t}$, $\alpha > -1$, and with $F_2 = tI$ (in these examples the parameter a is a complex number and the parameters b and c are positive numbers and $t \in (0, +\infty)$).

(4) See Durán and Grünbaum (2004) and Durán and López-Rodríguez (2007):

$$W_{1,l}(t) = t^\alpha e^{-t}\begin{pmatrix} t^2 + |a|^2(t-1)^2 & a(t-1) \\ \bar{a}(t-1) & 1 \end{pmatrix},$$

$$\partial^2 t + \partial\begin{pmatrix} \alpha+3-t & 2a \\ 0 & \alpha+1-t \end{pmatrix} + \begin{pmatrix} -1 & a\alpha \\ 0 & 0 \end{pmatrix}.$$

(5) See Cantero, Moral, and Velázquez (2007) and Durán and de la Iglesia (2008b):

$$W_{2,l}(t) = t^\alpha e^{-t}\begin{pmatrix} t(1+|a|^2 t) & at \\ \bar{a}t & 1 \end{pmatrix}, \tag{12.2.39}$$

$$\partial^2 t + \partial\begin{pmatrix} \alpha+2-t & at \\ 0 & \alpha+1-t \end{pmatrix} + \begin{pmatrix} -1 & (1+\alpha)a \\ 0 & 0 \end{pmatrix}. \tag{12.2.40}$$

(6) See Durán and Grünbaum (2007):

$$W_{3,l}(t) = t^\alpha e^{-t}\begin{pmatrix} t^b + \frac{4|a|^2 t^c (t-1)^2}{(c-b)^2} & -\frac{2at^c(t-1)}{c-b} \\ -\frac{2\bar{a}t^c(t-1)}{c-b} & t^c \end{pmatrix}, \tag{12.2.41}$$

$$\partial^2 t + \partial\begin{pmatrix} -t+b+\alpha+1 & -\frac{2a(c-b+2)}{c-b}t + 2a \\ 0 & -t+c+\alpha+1 \end{pmatrix} + \begin{pmatrix} -1 & -\frac{2a(\alpha+c)}{c-b} \\ 0 & 0 \end{pmatrix}.$$

We close with two examples associated with the Jacobi weight $(1-t)^\alpha(1+t)^\beta$, $\alpha, \beta > -1$ and with $F_2 = (1-t^2)I$ (in these examples the parameter a is a complex number and $t \in (-1,1)$).

(7) See Durán and Grünbaum (2004) and Durán and López-Rodríguez (2007):

$$W_{1,j}(t) = (1-t)^\alpha(1+t)^\beta\begin{pmatrix} (1+t)^2 + |a|^2 t^2 & at \\ \bar{a}t & 1 \end{pmatrix},$$

$$\partial^2(1-t^2) + \partial\begin{pmatrix} -(\alpha+\beta+4)t + 2 - \alpha + \beta & 2a(1-t) \\ 0 & -(\alpha+\beta+2)t - \alpha + \beta \end{pmatrix}$$

$$+\begin{pmatrix} -(\alpha+\beta+2) & -(\alpha-\beta)a \\ 0 & 0 \end{pmatrix}.$$

(8) See Durán (2009a) ($\alpha, \alpha+\beta \neq 0$):

$$W_{2,j}(t) = (1-t)^\alpha(1+t)^\beta\begin{pmatrix} (1-t)^{\frac{2\alpha}{\alpha+\beta}}(1+t)^{\frac{2\beta}{\alpha+\beta}} + \frac{|a|^2(\alpha+\beta)^2 t^2}{\alpha^2} & -\frac{a(\alpha+\beta)t}{\alpha} \\ -\frac{\bar{a}(\alpha+\beta)t}{\alpha} & 1 \end{pmatrix},$$

$$\partial^2\left(1-t^2\right) + \partial\begin{pmatrix} -(\alpha+\beta+4)t - \frac{(\alpha+\beta+2)(\alpha-\beta)}{\alpha+\beta} & -\frac{2a(\alpha(1+t)+\beta(1-t))}{\alpha} \\ 0 & -(\alpha+\beta+2)t - \alpha + \beta \end{pmatrix}$$

$$+\begin{pmatrix} -(\alpha+\beta+2) & \left(\alpha^2-\beta^2\right)a/\alpha \\ 0 & 0 \end{pmatrix}.$$

As we mentioned in Section 12.2.1, the structure of the algebras of differential operators associated with a weight matrix can be much richer than in the scalar case. This is the case of some of the examples displayed in this section. To illustrate this situation we consider here the weight matrices $W_{1,h}$ (12.2.36) and $W_{3,l}$ (12.2.39).

It was conjectured in Castro and Grünbaum (2005) ($a \in \mathbb{R}$) that the algebra $\mathcal{D}(W_{1,h})$ is generated by the identity I, the operator (12.2.37), and the following symmetric second-order differential operator:

$$\partial^2 \begin{pmatrix} -at & a^2t^2-1 \\ -1 & at \end{pmatrix} + \partial \begin{pmatrix} -2a & 2\left(a^2+2\right)t \\ 0 & 0 \end{pmatrix} + \begin{pmatrix} 0 & 2\frac{(a^2+2)}{a^2} \\ \frac{4}{a^2} & 0 \end{pmatrix}. \tag{12.2.42}$$

The conjecture was proved in Tirao (2011).

The algebra $\mathcal{D}(W_{2,l})$ (see (12.2.39) above) contains differential operators of odd order (something impossible in the scalar case). For instance, the following third-order differential operator is symmetric with respect to $W_{2,l}$:

$$\begin{aligned} &\partial^3 \begin{pmatrix} -|a|^2t^2 & at^2\left(1+|a|^2t\right) \\ -\bar{a}t & |a|^2t^2 \end{pmatrix} \\ &+ \partial^2 \begin{pmatrix} -t(2+|a|^2(\alpha+5)) & at(2\alpha+4+t(1+|a|^2(\alpha+5))) \\ -\bar{a}(\alpha+2) & t(2+|a|^2(\alpha+2)) \end{pmatrix} \\ &+ \partial \begin{pmatrix} t-2(\alpha+2)\left(1+|a|^2\right) & \frac{|a|^2(\alpha+1)(\alpha+2)+t(1+2|a|^2(1+|a|^2(\alpha+2)))}{\bar{a}} \\ -\frac{1}{a} & 2\alpha+2-t \end{pmatrix} \\ &+ \begin{pmatrix} 1+\alpha & -\frac{1}{\bar{a}}(1+\alpha)(|a|^2\alpha-1) \\ \frac{1}{a} & -(1+\alpha) \end{pmatrix}. \end{aligned} \tag{12.2.43}$$

It was conjectured in Durán and de la Iglesia (2008b) that, for $\alpha \neq 1+\frac{1}{|a|^2}, -2+\frac{1}{|a|^2}$, the algebra $\mathcal{D}(W_{2,l})$ is generated by the identity I, the operator (12.2.40), and the third-order differential operator (12.2.43). For the two exceptional values of α it seems that another second-order differential operator is needed to generate the whole algebra.

We also mentioned in Section 12.2.1 that the convex hull of weight matrices $\Upsilon(D)$ associated to a second-order differential operator can have a richer structure than the half-line to which it reduces in the classical examples of the scalar case. This is the case for some of the second-order differential operators given in this section. To illustrate this situation we consider two symmetric operators with respect to the weight matrix $W_{1,h}$ (12.2.36) above.

Fix $a \in \mathbb{C} \setminus \{0\}$ and let us consider first the second-order differential operator (12.2.37). Then this operator is also symmetric with respect to the family of weight matrices

$$W_\zeta = e^{-t^2} \begin{pmatrix} 1+\zeta|a|^2t^2 & \zeta at \\ \zeta\bar{a}t & \zeta \end{pmatrix}, \quad \zeta > 0.$$

Hence, the orthogonal matrix polynomials with respect to each of these weight matrices are common eigenfunctions of the same operator. The explicit expression for the monic orthogonal polynomials $(P_{n,\zeta})_n$ with respect to W_ζ is given by

$$P_{n,\zeta} = \begin{pmatrix} 2^{-n} & 0 \\ 0 & \frac{2^{-n+1}}{2+\zeta|a|^2n} \end{pmatrix} \begin{pmatrix} h_n(t) & -anh_{n-1}(t) \\ -\zeta\bar{a}nh_{n-1}(t) & h_n(t)+\zeta|a|^2nth_{n-1}(t) \end{pmatrix}$$

(h_n stands for the nth Hermite polynomial). The convex hull of weight matrices $\Upsilon(D)$ associated to the operator D in (12.2.37) has dimension 2.

Now fix $a \in \mathbb{R} \setminus \{0\}$ and consider the second-order differential operator

$$\partial^2 \begin{pmatrix} 1 - at & a^2t^2 - 1 \\ -1 & 1 + at \end{pmatrix} + \partial \begin{pmatrix} -2a - 2t & 2a + 2\left(2 + a^2\right)t \\ 0 & -2t \end{pmatrix} + \begin{pmatrix} -1 & 2\frac{2+a^2}{a^2} \\ \frac{4}{a^2} & 1 \end{pmatrix}. \tag{12.2.44}$$

Notice that this operator is the sum of the operators defined in (12.2.37) and (12.2.42), and hence it is symmetric with respect to the weight matrix $W_{1,h}$ (12.2.36) ($a \in \mathbb{R}$). The convex hull of weight matrices $\Upsilon(D)$ associated to this operator has dimension 2, but the weight matrices in $\Upsilon(D)$ have a rather different structure to the ones in the previous example. Indeed, this operator is also symmetric with respect to the family of weight matrices (see Durán and de la Iglesia, 2008a)

$$W_\zeta = e^{-t^2} \begin{pmatrix} 1 + a^2t^2 & at \\ at & 1 \end{pmatrix} + \zeta\delta_0 \begin{pmatrix} 1 & 1 \\ 1 & 1 \end{pmatrix}.$$

Notice that the situation is absolutely different to that of the scalar case. When we add a mass point to any of the classical weights of Hermite, Laguerre, and Jacobi, the existence of a symmetric second-order differential operator automatically disappears. Only when the mass point is added at the endpoints of the support one gets the symmetry of a fourth- (or even larger) order differential operator which is not symmetric with respect to the original weight (see Durán and de la Iglesia, 2008a for more examples).

For weights containing mass points and higher-order differential operators see Grünbaum (2011).

12.2.5 A Method for Solving the Differential Equations for the Weight Matrix When F_2 Is Not Scalar

We now show how the method described in Theorem 12.2.4 can be modified to find symmetric second-order differential operators in the case when F_2 is properly a matrix function (see Durán, 2009b).

Theorem 12.2.8 *Let Ω be an open set of the real line. Let ρ, F_2, F_1, and F_0 be a (real) scalar function and three matrix functions defined in Ω, respectively. We assume that $\rho(t) \neq 0$, $t \in \Omega$, and that all the functions are twice differentiable in Ω. Define the matrix function C as*

$$C(t) = \frac{(\rho(t)F_2(t))'}{\rho(t)}, \qquad t \in \Omega. \tag{12.2.45}$$

We assume that the matrix equation

$$F_1(t) = F_2(t)F(t) + F(t)F_2(t) + C(t) \tag{12.2.46}$$

has a solution $F(t)$ for each $t \in \Omega$ and that the matrix function F is differentiable in Ω. Write T for a solution of the first-order differential equation

$$T'(t) = F(t)T(t),$$

with $T(t_0)$ nonsingular for certain $t_0 \in \Omega$. Define the matrix function W as

$$W(t) = \rho(t)T(t)T^*(t),$$

and assume finally that $F_2(t)W(t) = W(t)F_2^*(t)$, $t \in \Omega$. Then

(1) *the weight matrix W satisfies in Ω the first-order differential equation*

$$2(F_2W)' = F_1W + WF_1^*;$$

(2) *the weight matrix W satisfies in Ω the second-order differential equation*

$$(F_2W)'' - (F_1W)' + F_0W = WF_0^*$$

if and only if the matrix function

$$\chi = T^{-1}(-FF_2F - F'F_2 - FC - FF_2R + F_0)T \tag{12.2.47}$$

is Hermitian at each point of Ω.

Using this method one can produce examples with a different structure from those displayed in the previous section: namely, in all those given above only one of the classical scalar weights of Hermite, Laguerre, or Jacobi appears in the weight matrix which, by the way, always has one symmetric second-order differential operator with F_2 scalar (and others with F_2 nonscalar). In the example we display below, Hermite and Laguerre weights appear together in the weight matrix, which does not have any symmetric differential operators of order 2 with F_2 scalar (it actually has, up to multiplicative constant and modulo the identity matrix, only one symmetric differential operator of order 2). We give here the example for size 2×2 (see Durán, 2009b for the extension of this example to arbitrary size $N \times N$). The weight matrix W is defined by $W = TT^*$ where the function T is

$$T(t) = \begin{pmatrix} e^{-t^2/2} & ve^{-t/2}t_+^{1+\alpha} \\ 0 & e^{-t/2}t_+^{\alpha} \end{pmatrix}, \quad v \in \mathbb{C},\ \alpha > 0,$$

where $t_+^\alpha = \left\{\begin{smallmatrix} t^\alpha, & t>0, \\ 0, & t\le 0, \end{smallmatrix}\right.$ so that

$$W(t) = \begin{pmatrix} e^{-t^2} + |v|^2 t_+^{2+2\alpha}e^{-t} & vt_+^{1+2\alpha}e^{-t} \\ \bar{v}t_+^{1+2\alpha}e^{-t} & t_+^{2\alpha}e^{-t} \end{pmatrix}.$$

This weight matrix has a symmetric second-order differential operator of the form (12.2.2) with coefficients given by

$$F_2(t) = \begin{pmatrix} 1 & vt(2t-1) \\ 0 & 2t \end{pmatrix}, \quad F_1(t) = \begin{pmatrix} -2t & 2vt(3+2\alpha) \\ 0 & -2t+4\alpha+2 \end{pmatrix},$$

$$F_0(t) = \begin{pmatrix} -2 & v(4\alpha+2) \\ 0 & 0 \end{pmatrix}.$$

12.2.6 Rodrigues Formulas

The final section of this contribution is devoted to Rodrigues formulas for the explicit computation of the families of orthogonal matrix polynomials satisfying second-order differential equations. By a Rodrigues formula for the polynomials $(P_n)_n$ orthogonal with respect to the matrix weight W, we mean an expression of the form

$$P_n = (R_n)^{(n)} W^{-1}, \quad n \geq 1,$$

where the functions R_n are simple enough to allow for an explicit computation of the polynomials P_n.

In the scalar case, these functions R_n always have the same expression: $R_n = f_2^n w$, where f_2 is the coefficient of the second derivative in the corresponding symmetric second-order differential operator with respect to the classical weight w. But this canonical form does not work for orthogonal matrix polynomials (Durán and Grünbaum, 2005b).

In the matrix case, such a closed form and canonical expression for the functions R_n seems not to exist. However, these functions R_n have to satisfy certain second-order differential equations related to the one satisfied by the polynomials $(P_n)_n$ (see Durán, 2010).

Lemma 12.2.9 *Let F_2, F_1, and F_0 be matrix polynomials of degrees not larger than 2, 1, and 0, respectively. Let W, R_n be $N \times N$ matrix functions twice and n times differentiable, respectively, in an open set Ω of the real line. Assume that $W(t)$ is nonsingular for $t \in \Omega$ and that it satisfies the identity* (12.2.7), *and the differential equations* (12.2.8) *and* (12.2.9). *Define the functions P_n, $n \geq 1$ by*

$$\mathrm{P}_n = R_n^{(n)} W^{-1}. \tag{12.2.48}$$

If for a matrix Λ_n, the function R_n satisfies

$$(R_n F_2^*)'' - \left(R_n \left[F_1^* + n\,(F_2^*)'\right]\right)' + R_n \left[F_0^* + n\,(F_1^*)' + \binom{n}{2}(F_2^*)''\right] = \Lambda_n R_n, \tag{12.2.49}$$

then the function P_n satisfies

$$\mathrm{P}_n''(t) F_2(t) + \mathrm{P}_n'(t) F_1(t) + \mathrm{P}_n(t) F_0 = \Lambda_n \mathrm{P}_n(t). \tag{12.2.50}$$

Moreover, if $R_n = \rho Z_n$ where Z_n is a matrix polynomial and ρ is equal to e^{-t^2}, $t^\alpha e^{-t}$, or $(1-t)^\alpha (1+t)^\beta$ (in this last case with α or β noninteger), the converse is also true.

For an extension of this result to orthogonal matrix polynomials satisfying higher-order differential equations see Durán (2011b).

When one applies this lemma to find Rodrigues formulas for orthogonal matrix polynomials, we have to do more than just solve the differential equation (12.2.49): we have to look for solutions R_n so that $R_n^{(n)} W^{-1}$ is a polynomial of degree n with nonsingular leading coefficient. Eventually, the examples known up to now show that there are solutions R_n of (12.2.49) simple enough to provide an efficient way to compute the sequence of orthogonal polynomials $(P_n)_n$ with respect to W by using the Rodrigues formula $P_n = R_n^{(n)} W^{-1}$.

Here we give a couple of illustrative examples.

Orthogonal polynomials with respect to the weight matrix (see (12.2.36))

$$W = e^{-t^2}\begin{pmatrix} 1+|a|^2t^2 & at \\ \bar{a}t & 1 \end{pmatrix}$$

can be defined by the Rodrigues formula (see Durán and Grünbaum, 2005c)

$$P_n(t) = (-1)^n\left[e^{-t^2}\left(\begin{pmatrix} 1+|a|^2t^2 & at \\ \bar{a}t & 1 \end{pmatrix}\right) - X_n\right]^{(n)} W^{-1}, \quad n \geq 1,$$

where

$$X_n = \begin{pmatrix} -|a|^2 n/2 & 0 \\ 0 & 0 \end{pmatrix}.$$

A sequence of orthogonal polynomials for the weight matrix

$$W = t^{\alpha}e^{-t}\begin{pmatrix} t^{1/2}+(t-1)^2 & t-1 \\ t-1 & 1 \end{pmatrix}$$

(this is the particular case of example (12.2.41) for $a = 1/4$, $b = 1/2$, and $c = 0$) can be defined by means of the following Rodrigues formula (see Durán and Grünbaum, 2007):

$$P_n(t) = \left[t^{\alpha+n}e^{-t}\left(\begin{pmatrix} t^{1/2}+(t-1)^2 & t-1 \\ t-1 & 1 \end{pmatrix} - X_n(t)\right)\right]^{(n)} W^{-1}, \tag{12.2.51}$$

where

$$X_n(t) = \begin{pmatrix} \alpha t - n - \alpha & n+\alpha \\ -\frac{n}{\xi_n}\phi_n(t)t^{-\alpha-n}e^t - \frac{n}{\xi_n}\sqrt{t} & 0 \end{pmatrix},$$

$\phi_n(t) = \int_0^t x^{\alpha+n+1/2}e^{-x}\,dx - \xi_n \int_0^t x^{\alpha+n}e^{-x}\,dx$, and $\xi_n = \Gamma(\alpha+n+3/2)/\Gamma(\alpha+n+1)$ (so that $\phi_n(+\infty) = 0$).

A closed and canonical expression for the functions R_n exists for certain groups of families. We illustrate this situation with three different families of weight matrices. To introduce them, we consider the diagonal matrix $\mathscr{J}$ and the nilpotent matrix $\mathscr{N}$ defined in (12.2.14) and (12.2.13), respectively (to simplify we consider positive parameters in the matrix $\mathscr{N}$). We assume here that the parameters $v_i > 0$, $i = 1, \ldots, N-1$ of the nilpotent matrix $\mathscr{N}$ satisfy the constraints

$$(N-i-1)v_i^2 v_{N-1}^2 - \omega_i v_{N-1}^2 + \omega_{N-1}v_i^2 = 0, \quad 1 \leq i \leq N-2,$$

where the sequence ω_i depends on the corresponding family. The constraints allow us to define all the parameters from v_{N-1}.

The three families are defined by the formula

$$W = \rho T_\rho T_\rho^*,$$

where ρ is one of the classical scalar weights and T_ρ is certain matrix function.

The first example is the one-parametric family obtained by taking

$$\rho = e^{-t^2}, \quad T = e^{\mathcal{N}t}, \quad \omega_i = 2i(N-i), \quad 1 \leq i \leq N-1.$$

This is a particular case of the weight matrix involved in (12.2.22) and (12.2.23).

The second example is the two-parametric family defined by taking

$$\rho = t^{\alpha}e^{-t}, \quad T = e^{\mathcal{N}t}t^{\mathcal{J}/2}, \quad \omega_i = i(N-i), \quad 1 \leq i \leq N-1.$$

This is a particular case of the example displayed in Theorem 12.2.6 when $v_0 = 1/2$.

The third example is the three-parametric family defined by taking

$$\rho = t^{\alpha}(1-t)^{\beta}, \quad T = (2(1-t))^{C}(2t)^{\mathcal{J}/2},$$

with $C = (N-1)I - \mathcal{J} - \mathcal{N}$ and the constraints given by the sequence $\omega_i = 2i(N-i)(\beta + i - \frac{2(N-1)(\beta+N-1)}{2(N-1)+v_{N-1}^2})$ with $v_{N-1}^2 > \frac{2(N-1)(N-2)}{\beta+1}$. This weight matrix is a normalization of the weight matrix introduced in Pacharoni and Tirao (2007a) in connection with spherical functions associated to the complex projective space $\mathbb{P}_n(\mathbb{C})$. Observe that $T_\rho T_\rho^*$ is always a polynomial of degree $2N-2$.

The constraints on the parameters have important consequences. One is that each family goes along with two linearly independent second-order symmetric differential operators. Another is that they also satisfy a very simple Rodrigues formula of the same form. Indeed, define

$$f_2 = \begin{cases} 1 & \text{for the first weight matrix,} \\ t & \text{for the second weight matrix,} \\ t(1-t) & \text{for the third weight matrix.} \end{cases}$$

We point out that in each case the weight matrix has a symmetric second-order differential operator with leading differential coefficient equal to the corresponding $f_2 I$. Then a sequence of orthogonal polynomials with respect to W can be defined by using the Rodrigues formula

$$P_n(t) = \left(f_2^n \rho T_\rho L_n T_\rho^*\right)^{(n)} W^{-1},$$

where L_n is the diagonal matrix, independent of t, with entries

$$L_{n,k,k} = \prod_{l=k}^{N-1}\left(1 + \frac{n|v_l|^2}{\omega_l}\right),$$

and $(\omega_l)_l$ is the sequence that defines the constraints of the parameters in the nilpotent matrix $\mathcal{N}$.

If we remove the constraints on the parameters in the definitions of these three families, the resulting orthogonal polynomials appear not to satisfy a Rodrigues formula like the one above.

The classical families $(p_n)_n$ of orthogonal polynomials (Hermite, Laguerre, and Jacobi) satisfy a large number of formal properties, relationships, and structural formulas (see, for instance, Erdélyi et al., 1953b; Szegő, [1939] 1975; or Ismail, 2005b). This seems also to be the case for the families of orthogonal matrix polynomials satisfying second-order differential equations.

In some of the examples given in this contribution and in some others, Rodrigues formulas have allowed us to find many of these formal properties, relationships, and structural properties, such as the explicit computations of the three-term recurrence coefficients, expansions of the orthogonal matrix polynomials in terms of the classical families of the scalar case, generating functions, etc. (see Durán, 2009b, 2010; Durán, 2011a; Durán and de la Iglesia, 2008b; Durán and Grünbaum, 2005c, 2007; Durán and López-Rodríguez, 2007).

13

Some Families of Matrix-Valued Jacobi Orthogonal Polynomials

F. A. Grünbaum, I. Pacharoni, and J. A. Tirao

13.1 Introduction

Among the classical (scalar-valued) families of orthogonal polynomials with rich and deep connections to several branches of mathematics, the Jacobi polynomials occupy a distinguished role.

In this contribution we describe a way of obtaining some families of matrix-valued orthogonal polynomials of arbitrary dimension and depending on two parameters α, β, which extends the scalar theory in many respects. We will achieve this goal by focusing on a group representation approach. In the scalar case the Jacobi polynomials appeared in several concrete mathematical physics problems in the hands of people like Laplace and Legendre. The group-theoretical interpretation, in the hands of E. Cartan and H. Weyl, is of more recent vintage.

It is clear that in the matrix case the historical path is reversed and it remains an interesting challenge to find good concrete applications of these families of matrix-valued polynomials which satisfy three-term recursions as well as differential equations.

In the very last section we refer to one recent use of the recursion relation satisfied by the matrix Jacobi polynomials. An approach that would attach a physical meaning to these matrix differential equations could make these new special functions into a powerful tool in different areas.

13.2 Spherical Functions

The well-known Legendre polynomials are a special case of **spherical harmonics**. These can be defined as follows. A function $f(x, y, z)$ is harmonic if it satisfies Laplace's equation

$$\frac{\partial^2 f}{\partial x^2} + \frac{\partial^2 f}{\partial y^2} + \frac{\partial^2 f}{\partial z^2} = 0.$$

It is homogeneous of degree n if

$$f(\lambda x, \lambda y, \lambda z) = \lambda^n f(x, y, z).$$

The homogeneous harmonic polynomials can be considered as functions on the unit sphere S^2 in $\mathbb{R}^3$. They are called spherical harmonics.

Let (r, θ, ϕ) be ordinary spherical coordinates in $\mathbb{R}^3$:

$$x = r\sin\theta\cos\phi, \quad y = r\sin\theta\sin\phi, \quad z = r\cos\theta.$$

In terms of these coordinates the Riemannian structure of $\mathbb{R}^3$ is given by the symmetric differential form

$$ds^2 = dr^2 + r^2 d\theta^2 + r^2(\sin\theta)^2 d\phi^2,$$

and the Laplace operator is

$$\Delta = \frac{\partial^2}{\partial r^2} + \frac{1}{r^2}\frac{\partial^2}{\partial\theta^2} + \frac{1}{r^2(\sin\theta)^2}\frac{\partial^2}{\partial\phi^2} + \frac{2}{r}\frac{\partial}{\partial r} + \frac{\cos\theta}{r^2\sin\theta}\frac{\partial}{\partial\theta}.$$

If f is a homogeneous harmonic polynomial of degree n which does not depend on the variable ϕ, then

$$\frac{d^2 f}{d\theta^2} + \frac{\cos\theta}{\sin\theta}\frac{df}{d\theta} + n(n+1)f = 0.$$

By making the change of variables $y = (1-\cos\theta)/2$ we get

$$y(1-y)\frac{d^2 f}{dy^2} + (1-2y)\frac{df}{dy} + n(n+1)f = 0.$$

The bounded solution at $y = 0$, up to a constant, is ${}_2F_1(-n, n+1, 1; y)$. Since the Legendre polynomial of degree n is given by

$$P_n(x) = {}_2F_1\left(\begin{matrix}-n, n+1\\ 1\end{matrix}; (1-x)/2\right),$$

we get that $f(\theta) = P_n(\cos\theta)f(0)$.

Let $o = (0, 0, 1)$ be the north pole of S^2, and let $d(o, p)$ be the geodesic distance from a point $p \in S^2$ to o. Let $\phi(p) = P_n(\cos(d(o, p)))$. Then we have proved that ϕ is the unique spherical harmonic of degree n, constant along parallels and such that $\phi(o) = 1$. Moreover, the set of all complex linear combinations of translates $\phi_g(p) = \phi(g \cdot p)$, $g \in \mathrm{SO}(3)$ is the linear space of all spherical harmonics of degree n.

Legendre and Laplace found that the Legendre polynomials satisfy the following addition formula:

$$\begin{aligned} &P_n(\cos\alpha\cos\beta + \sin\alpha\sin\beta\cos\phi) \\ &\quad = P_n(\cos\alpha)P_n(\cos\beta) + 2\sum_{k=1}^{n}\frac{(n-k)!}{(n+k)!}P_n^k(\cos\alpha)P_n^k(\cos\beta)\cos k\phi, \end{aligned} \tag{13.2.1}$$

where the P_n^k are the associated Legendre polynomials. By integrating (13.2.1) we get

$$P_n(\cos\alpha)P_n(\cos\beta) = \frac{1}{2\pi}\int_0^{2\pi} P_n(\cos\alpha\cos\beta + \sin\alpha\sin\beta\cos\phi)\, d\phi. \tag{13.2.2}$$

Moreover, the Legendre polynomials can be characterized as solutions to (13.2.2). This integral equation can now be expressed in terms of the function ϕ on SO(3) defined by $\phi(g) = \phi(g \cdot o) = P_n(\cos(d(o, g \cdot o)))$. One can see that (13.2.2) is equivalent to

$$\phi(g)\phi(h) = \int_K \phi(gkh)\, dk, \tag{13.2.3}$$

where $K = \mathrm{SO}(2)$ denotes the compact subgroup of SO(3) of all elements which fix the north pole o, and dk denotes the Haar measure of K.

In fact, let A denote the subgroup of all elements of SO(3) which fix the point $(0, 1, 0)$. Then $\mathrm{SO}(3) = KAK$, that is, the expression of any rotation in $\mathbb{R}^3$ in terms of Euler angles. Thus to prove (13.2.3) it is enough to consider rotations g and h around the y-axis through the angles α and β, respectively. Then if k denotes rotation of angle ϕ around the z-axis we have

$$gkh \cdot o = (-\cos\alpha\cos\phi\sin\beta + \sin\alpha\cos\beta, -\sin\phi\sin\beta, \sin\alpha\cos\phi\sin\beta + \cos\alpha\cos\beta)^t.$$

Thus $\cos(d(o, g \cdot o)) = \cos\alpha\cos\beta + \sin\alpha\sin\beta\cos\phi$ and

$$\phi(gkh) = P_n(\cos\alpha\cos\beta + \sin\alpha\sin\beta\cos\phi).$$

Then equation (13.2.3) becomes (13.2.2).

The functional equation (13.2.3) has been generalized to many different settings. One is the following. Let G be a locally compact unimodular group and let K be a compact subgroup. A nontrivial complex-valued continuous function ϕ on G is a **zonal spherical function** if (13.2.3) holds for all $g, h \in G$. Note that then $\phi(k_1 g k_2) = \phi(g)$ for all $k_1, k_2 \in K$ and all $g \in G$, and that $\phi(e) = 1$, where e is the identity element of G. This is the viewpoint of E. Cartan and H. Weyl.

The example above arises when $G = \mathrm{SO}(3)$, $K = \mathrm{SO}(2)$, and $S^2 = G/K$. The other compact connected rank-one symmetric spaces have zonal spherical functions which are orthogonal polynomials in an appropriate variable. These polynomials are special cases of Jacobi polynomials and can be given explicitly as hypergeometric functions.

The set of all complex lines through the origin in $\mathbb{C}^3$ known as the complex projective plane $P_2(\mathbb{C}) = \mathrm{SU}(3)/\mathrm{U}(2)$, is another example of a rank-one symmetric space. In this case the zonal spherical functions are $P_n^{(0,1)}(\cos\phi)$, an instance of the Jacobi polynomials with $\alpha = 0$ and $\beta = 1$. By considering the d-dimensional complex projective space $P_d(\mathbb{C}) = \mathrm{SU}(d+1)/\mathrm{U}(d)$ we will obtain matrix-valued extensions of these polynomials.

A very fruitful generalization of the functional equation (13.2.3) is the following (see Tirao, 1977 and Gangolli and Varadarajan, 1988). Let G be a locally compact unimodular group and let K be a compact subgroup of G. Let $\widehat{K}$ denote the set of all equivalence classes of complex finite-dimensional irreducible representations of K. For each $\delta \in \widehat{K}$, let ξ_δ and $d(\delta)$ denote, respectively, the character and the dimension of any representation in the class δ, and set $\chi_\delta = d(\delta)\xi_\delta$. We shall denote by V a finite-dimensional complex vector space and by $\mathrm{End}(V)$ the space of all linear transformations of V into V.

A **spherical function** Φ on G of type $\delta \in \widehat{K}$ is a continuous function $\Phi\colon G \to \mathrm{End}(V)$ such that $\Phi(e) = I$ (where I = identity transformation) and

$$\Phi(x)\Phi(y) = \int_K \chi_\delta\left(k^{-1}\right)\Phi(xky)\,dk \quad \text{for all } x, y \in G. \tag{13.2.4}$$

When δ is the class of the trivial representation of K and $V = \mathbb{C}$, the corresponding spherical functions are precisely the zonal spherical functions. From the definition it follows that $\pi(k) = \Phi(k)$ is a representation of K, equivalent to the direct sum of n representations in the class δ, and that $\Phi(k_1 g k_2) = \pi(k_1)\Phi(g)\pi(k_2)$ for all $k_1, k_2 \in K$ and all $g \in G$. The number n is the height of Φ. The height and the type are uniquely determined by the spherical function.

By considering the pair $(G, K) = (\mathrm{SU}(d+1), \mathrm{U}(d))$ we will see that the solutions of (13.2.4) give rise to matrix-valued Jacobi orthogonal polynomials expressible in terms of the matrix-valued hypergeometric function ${}_2F_1\left({A;B \atop C}; z\right)$.

13.3 Matrix-Valued Spherical Functions Associated to $P_2(\mathbb{C})$

In Grünbaum, Pacharoni, and Tirao (2002) the problem of determining all irreducible spherical functions associated to the complex projective plane $P_2(\mathbb{C})$ was considered. This space can be realized as the homogeneous space G/K, $G = \mathrm{SU}(3)$, and $K = \mathrm{S}((\mathrm{U}(2)\times\mathrm{U}(1)) \simeq \mathrm{U}(2)$. In this case all irreducible spherical functions are of height 1. Let (V_π, π) be any irreducible representation of K in the class δ. Then an irreducible spherical function can be characterized as a function $\Phi\colon G \longrightarrow \mathrm{End}(V_\pi)$ such that

(i) Φ is analytic,
(ii) $\Phi(k_1 g k_2) = \pi(k_1)\Phi(g)\pi(k_2)$ for all $k_1, k_2 \in K$, $g \in G$, and $\Phi(e) = I$,
(iii) $[\Delta\Phi](g) = \lambda(\Delta)\Phi(g)$ for all $g \in G$ and $\Delta \in D(G)^G$.

Here $D(G)^G$ denotes the algebra of all left- and right-invariant differential operators on G. In our case it is known that the algebra $D(G)^G$ is a polynomial algebra in two algebraically independent generators Δ_2 and Δ_3, explicitly given in Grünbaum, Pacharoni, and Tirao (2002, Proposition 3.1).

The set $\widehat{K}$ can be identified with $\mathbb{Z} \times \mathbb{Z}_{\geq 0}$ in the following way: if $k \in K$ then

$$\pi(k) = \pi_{n,\ell}(k) = (\det k)^n k^\ell,$$

where k^ℓ denotes the ℓ-symmetric power of the matrix k.

For any $g \in \mathrm{SU}(3)$ we denote by $A(g)$ the upper-left 2×2 block of g, and we consider the open dense subset $\mathcal{A} = \{g \in G\colon \det(A(g)) \neq 0\}$. Then $\mathcal{A}$ is left and right invariant under K. For any $\pi = \pi_{n,\ell}$ we introduce the following function defined on $\mathcal{A}$:

$$\Phi_\pi(g) = \pi(A(g)),$$

where π above denotes the unique holomorphic representation of $\mathrm{GL}(2, \mathbb{C})$ which extends the given representation of $\mathrm{U}(2)$.

To determine all irreducible spherical functions $\Phi\colon G \longrightarrow \mathrm{End}(V_\pi)$ of type $\pi = \pi_{n,\ell}$, we use the function Φ_π in the following way: in the open set $\mathcal{A} \subset G$ we define the function H by

$$H(g) = \Phi(g)\Phi_\pi(g)^{-1}, \tag{13.3.1}$$

where Φ is a spherical function of type δ. Then H satisfies

(i) $H(e) = I$,
(ii) $H(gk) = H(g)$ for all $g \in \mathcal{A}$, $k \in K$,
(iii) $H(kg) = \pi(k)H(g)\pi(k^{-1})$ for all $g \in \mathcal{A}$, $k \in K$.

The canonical projection $p\colon G \longrightarrow P_2(\mathbb{C})$ defined by $p(g) = g \cdot o$, where $o = (0,0,1)$, maps the open dense subset $\mathcal{A}$ onto the affine space $\mathbb{C}^2$ of those points in $P_2(\mathbb{C})$ whose last homogeneous coordinate is not zero. Then property (ii) says that H may be considered as a function on $\mathbb{C}^2$ and, moreover, from (iii) it follows that H is determined by its restriction $H = H(r)$ to the cross section $\{(r,0) \in \mathbb{C}^2 : r \geq 0\}$ of the K-orbits in $\mathbb{C}^2$, which are the spheres of radius $r \geq 0$ centered at the origin. That is, H is determined by the function $r \mapsto H(r) = H(r,0)$ on the interval $[0,+\infty)$. Let M be the closed subgroup of K of all diagonal matrices of the form $\mathrm{diag}\,(e^{i\theta}, e^{-2i\theta}, e^{i\theta})$, $\theta \in \mathbb{R}$. Then M fixes all points $(r,0) \in \mathbb{C}^2$. Therefore (iii) also implies that $H(r) = \pi(m)H(r)\pi(m^{-1})$ for all $m \in M$. Since any V_π as an M-module is multiplicity-free, it follows that there exists a basis of V_π such that $H(r)$ is simultaneously represented by a diagonal matrix for all $r \geq 0$. Thus, if $\pi = \pi_{n,\ell}$, we can identify $H(r) \in \mathrm{End}\,(V_\pi)$ with a vector

$$H(r) = (h_0(r), \ldots, h_\ell(r))^t \in \mathbb{C}^{\ell+1}.$$

The fact that Φ is an eigenfunction of Δ_2 and Δ_3 makes $H = H(r)$ into an eigenfunction of certain differential operators $\tilde{D}$ and $\tilde{E}$ on $(0,\infty)$. Making the change of variables $u = r^2/(1+r^2) \in (0,1)$ these operators become

$$\bar{D}H = u(1-u)H'' + (2 - uA_1)H' + \frac{1}{u}(B_0 - B_1 + uB_1)H, \tag{13.3.2}$$

$$\bar{E}H = u(1-u)MH'' + (C_1 - C_0 - uC_1)H' + \frac{1}{u}(D_0 + D_1 - uD_1)H. \tag{13.3.3}$$

If we denote by $E_{i,j}$ the $(\ell+1)\times(\ell+1)$ matrix with entry (i,j) equal to 1 and 0 elsewhere, then the coefficient matrices are

$$A_1 = \sum_{i=0}^{\ell}(n+\ell-i+3)E_{i,i},$$

$$B_0 = \sum_{i=0}^{\ell}(i+1)(\ell-i)(E_{i,i+1} - E_{i,i}), \quad B_1 = \sum_{i=0}^{\ell} i(\ell-i+1)(E_{i,i} - E_{i,i-1}),$$

$$M = \sum_{i=0}^{\ell}(n-\ell+3i)E_{i,i},$$

$$C_0 = \sum_{i=0}^{\ell}(n-\ell+3i)(n+\ell-i+1)E_{i,i} - 3\sum_{i=0}^{\ell}(i+1)(\ell-i)E_{i,i+1},$$

$$C_1 = \sum_{i=0}^{\ell}(n-\ell+3i)(n+\ell-i+3)E_{i,i} - 3\sum_{i=0}^{\ell} i(\ell-i+1)E_{i,i-1},$$

$$D_0 = \sum_{i=0}^{\ell}(n+2\ell-3i)(i+1)(\ell-i)\left(E_{i,i+1}-E_{i,i}\right)$$

$$\quad - 3\sum_{i=0}^{\ell}(n+\ell-i+1)i(\ell-i+1)\left(E_{i,i}-E_{i,i-1}\right),$$

$$D_1 = (2n+\ell+3)B_1.$$

The following result is taken from Román and Tirao (2006, Theorem 3.8); see also Grünbaum, Pacharoni, and Tirao (2002).

Theorem 13.3.1 *The irreducible spherical functions* Φ *of* SU(3) *of type* (n,ℓ) *correspond precisely to the simultaneous* $\mathbb{C}^{\ell+1}$*-valued polynomial eigenfunctions* H *of the differential operators* $\bar{D}$ *and* $\bar{E}$*, introduced in* (13.3.2) *and* (13.3.3)*, such that* $h_i(u) = (1-u)^{i-n-\ell} g_i(u)$ *for all* $n+\ell+1 \le i \le \ell$ *with* g_i *polynomial and* $H(0) = (1,\dots,1)^t$.

In Grünbaum, Pacharoni, and Tirao (2002) the systems of equations alluded to above are given in terms of the variables $r > 0$ and $t = 1/(1+r^2)$. Explicitly, we have that the function $H(r) = (h_0(r),\dots,h_\ell(r))$ satisfies $(\bar{D}H)(r) = \lambda H(r)$ if and only if

$$\left(1+r^2\right)^2 h_i'' + \frac{\left(1+r^2\right)}{r}\left(3+r^2-2r^2(n+\ell-i)\right)h_i' - 4i(\ell-i+1)\left(h_{i-1}-h_i\right)$$
$$+ \frac{4\left(1+r^2\right)}{r^2}\left((i+1)(\ell-i)\left(h_{i+1}-h_i\right) + i(\ell-i+1)\left(h_{i-1}-h_i\right)\right) = \lambda h_i$$

for all $i = 0,\dots,\ell$.

Likewise one has $(\bar{E}H)(r) = \mu H(r)$ if and only if

$$(n-\ell+3i)\left(1+r^2\right)^2 h_i'' + 6(i+1)(\ell-i)\frac{\left(1+r^2\right)^2}{r} h_{i+1}'$$
$$- 6i(\ell-i+1)\frac{\left(1+r^2\right)}{r} h_{i-1}' + (n-\ell+3i)\frac{\left(1+r^2\right)}{r}\left(3+r^2-2r^2(n+\ell-i)\right)h_i'$$
$$+ 4(n+2\ell-3i)\frac{\left(1+r^2\right)}{r^2}\left((i+1)(\ell-i)\left(h_{i+1}-h_i\right) + i(\ell-i+1)\left(h_{i-1}-h_i\right)\right)$$
$$+ 4i(\ell-i+1)(2n+\ell+3)\left(h_{i-1}-h_i\right) = \mu h_i$$

for all $i = 0,\dots,\ell$. In terms of the variable t these systems become

$$t(1-t)h_i'' + \left((n+\ell-i+1) - t(n+\ell-i+3)\right)h_i'$$
$$+ (\ell-i)(i+1)\frac{h_{i+1}-h_i}{1-t} - i(\ell-i+1)t\frac{h_i-h_{i-1}}{1-t} = \frac{\lambda}{4}h_i$$

and

$$\begin{aligned}(n-\ell+3i)t(1-t)h_i'' &- 3(i+1)(\ell-i)h_{i+1}' \\ &+ (n-\ell+3i)\,(n+\ell-i+1-t(n+\ell-i+3))\,h_i' + 3i(\ell-i+1)th_{i-1}' \\ &+ (n+2\ell-3i)(\ell-i)(i+1)\frac{h_{i+1}-h_i}{1-t} - (n+2\ell-3i)i(\ell-i+1)\frac{h_i-h_{i-1}}{1-t} \\ &- (2n+\ell+3)i(\ell-i+1)\,(h_i-h_{i-1}) = \frac{\mu}{4}h_i.\end{aligned}$$

Notice that none of the systems above is yet in the celebrated hypergeometric form. In the next section we will see how (13.3.2) can be conjugated into this form.

13.4 The Spherical Functions as Matrix-Valued Hypergeometric Functions

A key result to characterize the spherical functions of SU(3) of type (n, ℓ) is the existence of a matrix-valued polynomial function ψ of degree ℓ such that if we let $F(u) = \psi(u)^{-1}H(u)$, then the eigenfunctions H of (13.3.2) analytic at $u = 0$ are in one-to-one correspondence with the eigenfunctions F analytic at $u = 0$ of a hypergeometric operator. Such a function is given by $\psi(u) = XT(u)$ where X is the Pascal matrix given by $X_{i,j} = \binom{i}{j}$ and $T(u)$ is the diagonal matrix such that $T(u)_{i,i} = u^i$. If we define $D = \psi^{-1}\bar{D}\psi$ and $E = \psi^{-1}\bar{E}\psi$ then

$$D = u(1-u)\frac{d^2}{du^2} + (C - uU)\frac{d}{du} - V \tag{13.4.1}$$

and

$$E = (1-u)\,(Q_0 + uQ_1)\,\frac{d^2}{du^2} + (P_0 + uP_1)\,\frac{d}{du} - (n+3+2\ell)V, \tag{13.4.2}$$

where the coefficient matrices are

$$\begin{aligned}C &= \sum_{i=0}^{\ell} 2(i+1)E_{i,i} + \sum_{i=0}^{\ell} iE_{i,i-1},\\ U &= \sum_{i=0}^{\ell}(n+\ell+i+3)E_{i,i},\\ V &= \sum_{i=0}^{\ell} i(n+i+1)E_{i,i} - \sum_{i=0}^{\ell}(\ell-i)(i+1)E_{i,i+1},\\ Q_0 &= \sum_{i=0}^{\ell} 3iE_{i,i-1},\\ Q_1 &= \sum_{i=0}^{\ell}(n-\ell+3i)E_{i,i},\\ P_0 &= \sum_{i=0}^{\ell}\left(2(i+1)(n+2\ell) - 3(\ell-i) - 3i^2\right)E_{i,i} - \sum_{i=0}^{\ell} i(3i+3+\ell+2n)E_{i,i-1},\end{aligned}$$

$$P_1 = \sum_{i=0}^{\ell} -(n-\ell+3i)(n+\ell+i+3)E_{i,i} + \sum_{i=0}^{\ell} 3(i+1)(\ell-i)E_{i,i+1}.$$

The matrix-valued hypergeometric function was introduced in Tirao (2003). Let W be a finite-dimensional complex vector space, and let $A,\ B,\ C \in \mathrm{End}(W)$. The hypergeometric equation is

$$z(1-z)F'' + (C - z(I+A+B))F' - ABF = 0. \tag{13.4.3}$$

If the eigenvalues of C are not in $-\mathbf{N}_0$ we define the function

$${}_2F_1\left({A;B \atop C};z\right) = \sum_{m=0}^{\infty} \frac{z^m}{m!}(C;A;B)_m,$$

where the symbol $(C;A;B)_m$ is defined inductively by $(C;A;B)_0 = I$ and

$$(C;A;B)_{m+1} = (C+m)^{-1}(A+m)(B+m)(C;A;B)_m, \quad m \geq 0,$$

for all $m \geq 0$. The function ${}_2F_1({A;B \atop C};z)$ is analytic on $|z| < 1$ with values in $\mathrm{End}(W)$. Moreover, if $F_0 \in W$ then $F(z) = {}_2F_1({A;B \atop C};z)F_0$ is a solution of the hypergeometric equation (13.4.3) such that $F(0) = F_0$. Conversely, any solution F, analytic at $z = 0$ is of this form.

In the scalar case the differential operator (13.4.1) is always of the form given in (13.4.3); after solving a quadratic equation we can find A, B such that $U = 1 + A + B$ and $V = AB$. This is not necessarily the case when $\dim(W) > 1$. In other words, a differential equation of the form

$$z(1-z)F'' + (C - zU)F' - VF = 0, \tag{13.4.4}$$

with $U, V, C \in \mathrm{End}(W)$, cannot always be reduced to one of the form of (13.4.3), because a quadratic equation in a noncommutative setting as $\mathrm{End}(W)$ may have no solutions. Thus it is also important to give a way to solve (13.4.4).

If the eigenvalues of C are not in $-\mathbb{N}_0$ we introduce the sequence $[C, U, V]_m \in \mathrm{End}(W)$ by defining inductively $[C;U;V]_0 = I$ and

$$[C;U;V]_{m+1} = (C+m)^{-1}\left(m^2 + m(U-1) + V\right)[C;U;V]_m$$

for all $m \geq 0$. Then the function

$${}_2H_1\left({U;V \atop C};z\right) = \sum_{m=0}^{\infty} \frac{z^m}{m!}[C;U;V]_m$$

is analytic on $|z| < 1$ and it is the unique solution of (13.4.4) analytic at $z = 0$, with values in $\mathrm{End}(W)$, whose value at $z = 0$ is I.

According to Theorem 13.3.1 the irreducible spherical functions of SU(3) of type (n, ℓ) are in one-to-one correspondence with certain simultaneous $\mathbb{C}^{\ell+1}$-polynomial eigenfunctions $H(u)$ of the differential operators $\bar{D}$ and $\bar{E}$; see (13.3.2) and (13.3.3). Then the functions $F(u) = \psi(u)^{-1}H(u)$ introduced above are simultaneous eigenfunctions of the differential operators D and E given in (13.4.1) and (13.4.2).

A delicate fact established in Román and Tirao (2006, Theorem 3.9) is that the functions $F(u)$ associated to irreducible spherical functions of SU(3) are indeed polynomials. Therefore

$$F(u) = {}_2H_1\left(\begin{smallmatrix} U;V+\lambda \\ C \end{smallmatrix}; u\right) F(0),$$

for some eigenvalues $\lambda \in \mathbb{C}$. Thus for each F associated to an irreducible spherical function of type (n, ℓ) there exists an integer $w \geq 0$ such that $[C; U; V]_{w+1}$ is singular and $[C; U; V]_w$ is not singular. This implies that

$$\lambda = -w(w + n + \ell + k + 2) - k(n + k + 1),$$

for some integer $0 \leq k \leq \ell$. See Román and Tirao (2006, Corollary 3.14).

The goal now is to describe all simultaneous $\mathbb{C}^{\ell+1}$-valued eigenfunctions of the differential operators D and E analytic on the open unit disk Ω. We let

$$V_\lambda = \{F = F(u) \colon DF = \lambda F,\ F \text{ analytic on } \Omega\}.$$

Since the initial value $F(0)$ determines $F \in V_\lambda$, we have that the linear map $\nu\colon V_\lambda \to \mathbb{C}^{\ell+1}$ defined by $\nu(F) = F(0)$ is a surjective isomorphism. Because Δ_2 and Δ_3 commute, both being in the center of the algebra of all left-invariant differential operators on G, then D and E also commute. Moreover, since E has polynomial coefficients, E restricts to a linear operator of V_λ. Thus we have the commutative diagram

$$\begin{array}{ccc} V_\lambda & \xrightarrow{\ E\ } & V_\lambda \\ \nu\downarrow & & \downarrow\nu \\ \mathbb{C}^{\ell+1} & \xrightarrow{\ M(\lambda)\ } & \mathbb{C}^{\ell+1} \end{array} \qquad (13.4.5)$$

where $M(\lambda)$ is the $(\ell + 1) \times (\ell + 1)$ matrix given by

$$M(\lambda) = Q_0(C + 1)^{-1}(U + V + \lambda)C^{-1}(V + \lambda) + P_0C^{-1}(V + \lambda) + R. \qquad (13.4.6)$$

This diagram will be put to practical use in the examples of the next two sections involving the generalized hypergeometric functions ${}_{p+1}F_p$.

Although the matrix $M(\lambda)$ has a complicated form, from Grünbaum, Pacharoni, and Tirao (2002, Theorem 10.3) it follows that for any $(n, \ell) \in \mathbb{Z} \times \mathbb{Z}_{\geq 0}$ the characteristic polynomial of the matrix $M(\lambda)$ is given by

$$\det(\mu - M(\lambda)) = \prod_{k=0}^{\ell} (\mu - \mu_k(\lambda)),$$

where $\mu_k(\lambda) = \lambda(n - \ell + 3k) - 3k(\ell - k + 1)(n + k + 1)$. Moreover, all eigenvalues $\mu_k(\lambda)$ of $M(\lambda)$ have geometric multiplicity 1. In other words, all eigenspaces are one-dimensional. Moreover, if $v = (v_0, \ldots, v_\ell)^t$ is a nonzero $\mu_k(\lambda)$-eigenvector of $M(\lambda)$, then $v_0 \neq 0$. Then a simultaneous $\mathbb{C}^{\ell+1}$-valued eigenfunction of the differential operators D and E analytic on Ω is a scalar multiple of

$$F_k(u) = {}_2H_1\left(\begin{smallmatrix} U;V+\lambda \\ C \end{smallmatrix}; u\right) F_{0,k},$$

where $F_{0,k}$ is the unique $\mu_k(\lambda)$-eigenvector of $M(\lambda)$ normalized by $F_{0,k} = (v_0, \dots, v_\ell)^t$ with $v_0 = 1$, for some $0 \le k \le \ell$. Notice that $DF_k = \lambda F_k$ and $EF_k = \mu_k(\lambda) F_k$.

Now Theorem 13.3.1 is considerably improved and takes the following form; see Grünbaum, Pacharoni, and Tirao (2002) and Román and Tirao (2006):

Theorem 13.4.1 *There is a bijective correspondence between the equivalence classes of all irreducible spherical functions of* SU(3) *of type* (n, ℓ) *and the set of pairs* $(\lambda_k, \mu_k) \in \mathbb{C} \times \mathbb{C}$ *where*

$$\lambda_k = -w(w + n + \ell + k + 2) - k(n + k + 1),$$
$$\mu_k = \lambda_k(n - \ell + 3k) - 3k(\ell - k + 1)(n + k + 1),$$

with w *a nonnegative integer,* $0 \le k \le \ell$*, and* $0 \le w + n + k$*. A representative of such a class can be obtained explicitly from the following vector-valued function:*

$$F_{w,k}(u) = {}_2H_1\left({}^{U;V+\lambda_k}_{\ \ C} ; u \right) F_{0,k}, \tag{13.4.7}$$

where $F_{0,k}$ *is the unique* μ_k*-eigenvector of* $M(\lambda_k)$ *normalized by* $F_{0,k} = (1, v_1, \dots, v_\ell)^t$.

We notice that the function $F = F_{w,k}$ considered above is a polynomial function of degree w such that

$$DF = \lambda_k F, \quad EF = \mu_k F.$$

Since the differential operators Δ_2 and Δ_3 are symmetric with respect to the L^2-inner product among matrix-valued functions on G, given by

$$\langle \Phi_1, \Phi_2 \rangle = \int_G \operatorname{tr}(\Phi_1(g)\Phi_2(g)^*)\, dg,$$

it follows that the differential operators D and E are symmetric with respect to the following inner product among vector-valued functions on the interval $[0, 1]$:

$$\langle F_1, F_2 \rangle = \int_0^1 F_2(u)^* W(u) F_1(u)\, du,$$

where the weight matrix $W(u)$ is defined by

$$W(u) = \sum_{i,j=0}^{\ell} \left(\sum_{r=0}^{\ell} \binom{r}{i}\binom{r}{j} (1-u)^{n+\ell-r} u^{i+j+1} \right) E_{i,j}. \tag{13.4.8}$$

Moreover, these F are orthogonal with respect to the weight function $W(u)$, that is,

$$\int_0^1 F_{w,k}(u)^* W(u) F_{w',k'}(u)\, du = 0 \quad \text{if } (w,k) \neq (w',k'), \tag{13.4.9}$$

since the pairs (λ,μ) of eigenvalues of the symmetric operators D and E characterize these F; see Theorem 13.4.1.

We point out that the weight function $W(u)$ has finite moments of all orders if and only if $n \geq 0$.

13.5 Matrix Orthogonal Polynomials Arising from Spherical Functions

From now on we shall assume that $n \geq 0$. We define the matrix polynomial P_w as the $(\ell+1)\times(\ell+1)$ matrix whose k-row is the polynomial $F_{w,k}$. In other words,

$$P_w = (F_{w,0},\ldots,F_{w,\ell})^t. \qquad (13.5.1)$$

We observe that an explicit expression for the columns of $(P_w)^t$ in terms of the matrix hypergeometric function introduced in Tirao (2003) follows from (13.4.7), namely

$$F_{w,k}(u) = {}_2H_1\left({U;V+\lambda_k \atop C};u\right)F_{0,k}, \qquad (13.5.2)$$

where its value $F_{0,k}$ at $u=0$ is the μ_k-eigenvector of $M(\lambda_k)$ properly normalized.

We notice that from (13.4.9) it follows that $(P_w)_w$ is a sequence of orthogonal polynomials with respect to the weight matrix $W(u)$:

$$\langle P_w, P_{w'}\rangle = \int_0^1 P_w(u)W(u)P_{w'}(u)^*\,du = 0$$

for all $w \neq w'$. The matrix-valued inner product defined above is an instance of a more general notion due to M. G. Krein (Krein, 1949a, b). A careful look at the definition shows that P_w is a matrix polynomial of degree w whose leading coefficient is a triangular nonsingular matrix. In other words, this sequence of matrix-valued polynomials fits squarely within Krein's theory.

The origin of this construction can be traced back to Grünbaum, Pacharoni, and Tirao (2002, Section 12). There one finds a sequence of matrix-valued orthogonal polynomials $\Phi(w,t)$ of size $(\ell+1)\times(\ell+1)$ obtained by "packaging," for each w, $\ell+1$ vector-valued eigenfunctions H alluded to in Theorem 13.3.1. The polynomials $\Phi(w,t)$ do not satisfy the condition $\deg \Phi(w,t) = w$ as required in Krein's theory. In Grünbaum, Pacharoni, and Tirao (2004, Section 5) the simple replacement of $\Phi(w,t)$ by $\Phi(0,t)^{-1}\Phi(w,t)$ takes care of this point.

Any sequence of matrix orthogonal polynomials $(P_w)_w$ satisfies a three-term recursion relation

$$uP_w(u) = A_w P_{w-1}(u) + B_w P_w(u) + C_w P_{w+1}(u), \quad w \geq 0 \qquad (13.5.3)$$

where we put $P_{-1}(u) = 0$.

A new theme enters below leading to the notion of classical orthogonal polynomials, first considered by Routh (1884) and Bochner (1929) in the scalar case, and then by Durán (1997) in the matrix-valued context; see Chapter 12.

A differential operator with matrix coefficients acting on matrix-valued functions could be made to act either on the left or on the right. One finds a discussion of these two actions in Durán (1997); see Chapter 12. The conclusion there is that if one wants to have matrix weights $W(u)$ that are not direct sums of scalar ones and that have matrix polynomials as their eigenfunctions, one should settle for right-hand-side differential operators. We agree now to say that D given by

$$D = \sum_{i=0}^{s} \partial^i F_i(u), \qquad \partial = \frac{d}{du}$$

acts on $P(u)$ by means of

$$PD = \sum_{i=0}^{s} \partial^i(P)(u) F_i(u).$$

The three-term recursion mentioned above gives rise to a difference operator L. This operator, which acts on the variable w of our family of polynomials $P_w(u)$ is acting on the left. As in any bispectral situation (Duistermaat and Grünbaum, 1986) there is another operator, acting in the variable u and its action should commute with that of L. In the matrix case the only way to get L and a differential operator D to commute for sure when acting on $P_w(u)$ is to make them act on different sides.

One could make D act on P on the right as defined above, and still write down the symbol DP for the result. The advantage of using the notation PD is that it respects associativity: if D_1 and D_2 are two differential operators we have $P(D_1 D_2) = (PD_1) D_2$.

Returning to our sequence of matrix polynomials (13.5.1) we consider the transposes of the operators D and E appearing in (13.4.1) and (13.4.2), namely let

$$D^t = \partial^2 u(1-u) + \partial\left(C^t - uU^t\right) - V^t$$

and

$$E^t = \partial^2(1-u)\left(Q_0^t + uQ_1^t\right) + \partial\left(P_0^t + uP_1^t\right) - (n+3+2\ell)V^t.$$

Now we observe that

$$\begin{aligned}
P_w D^t &= u(1-u)P_w'' + P_w'\left(C^t - uU^t\right) - P_w V^t \\
&= u(1-u)\left(F_{w,0}'', \ldots, F_{w,\ell}''\right)^t + \left(F_{w,0}', \ldots, F_{w,\ell}'\right)^t (C-uU)^t - (F_{w,0}, \ldots, F_{w,\ell})^t\, V^t \\
&= \left(u(1-u)\left(F_{w,0}'', \ldots, F_{w,\ell}''\right) + (C-uU)\left(F_{w,0}', \ldots, F_{w,\ell}'\right) - V\left(F_{w,0}, \ldots, F_{w,\ell}\right)\right)^t \\
&= (DF_{w,0}, \ldots, DF_{w,\ell})^t = (\lambda_{w,0}F_{w,0}, \ldots, \lambda_{w,\ell}F_{w,\ell})^t = \Lambda_w\left(D^t\right)P_w,
\end{aligned}$$

where $\Lambda_w(D^t) = \operatorname{diag}(\lambda_{w,0}, \ldots, \lambda_{w,\ell})$. Similarly, $P_w E^t = \Lambda_w(E^t)\, P_w$ where $\Lambda_w(E^t) = \operatorname{diag}(\mu_{w,0}, \ldots, \mu_{w,\ell})$.

In Durán (1997) a sequence of matrix orthogonal polynomials is called classical if there exists a symmetric differential operator of order 2 that has these polynomials as eigenfunctions with a matrix eigenvalue. One can dispense of the requirement of symmetry; see Grünbaum and Tirao (2007, Section 4).

Since in our case we have

$$P_w D^t = \Lambda_w \left(D^t\right) P_w, \quad P_w E^t = \Lambda_w \left(E^t\right) P_w,$$

we have exhibited a family of classical orthogonal polynomials featuring two algebraically independent differential operators of order 2.

In general, given a matrix weight $W(u)$ and a sequence of matrix orthogonal polynomials P_w one can consider the algebra of all matrix differential operators D such that

$$P_w D = \Lambda_w(D) P_w$$

where each Λ_w is a matrix. The set of these D will be denoted by $\mathcal{D}(W)$.

Starting with Grünbaum, Pacharoni, and Tirao (2002, 2004), Grünbaum (2003), and Durán and Grünbaum (2004) one has a growing collection of weight matrices $W(u)$ for which the algebra $\mathcal{D}(W)$ is not trivial, that is, does not consist only of scalar multiples of the identity operator. The study of this question starts with Durán (1997). A first attempt to go beyond the issue of the existence of one nontrivial element in $\mathcal{D}(W)$ and to study (experimentally, with assistance from symbolic computation) the full algebra is undertaken in Castro and Grünbaum (2006). An analytical proof of some of the results conjectured in Castro and Grünbaum (2006) appears in Tirao (2011).

To illustrate these results we will take $\ell = 1$; this means that the size of our matrices will be 2×2. The matrix $M(\lambda)$ (see (13.4.6)) is

$$M(\lambda) = \begin{pmatrix} \lambda(n + \frac{1}{2}) & \frac{9}{2} \\ \lambda(\frac{\lambda}{2} - n - 2) & \lambda(n + \frac{1}{2}) - 3(n + 2) \end{pmatrix}.$$

Using Theorem 13.4.1 we take $\lambda = -w(w + n + \ell + k + 2) - k(n + k + 1)$ for $k = 0, 1$ and obtain $\lambda_0 = -w(w + n + 3)$ and $\lambda_1 = -w(w + n + 4) - (n + 2)$, respectively. The desirable eigenvalues of $M(\lambda_0)$ and $M(\lambda_1)$ are

$$\mu_0 = \lambda_0(n - 1), \quad \mu_1 = (n + 2)(\lambda_1 - 3),$$

and the respective normalized eigenvectors are

$$F_{0,0} = \begin{pmatrix} 1 \\ -\frac{\lambda_0}{3} \end{pmatrix}, \quad F_{0,1} = \begin{pmatrix} 1 \\ \frac{\lambda_1 - 2(n+2)}{3} \end{pmatrix}.$$

From (13.4.7) we get

$$F_{w,0}(u) = \begin{pmatrix} {}_3F_2\left({-w, w+n+3, 2 \atop 3, 1} ; u \right) \\ \frac{w(w+n+3)}{3} \, {}_2F_1\left({-w+1, w+n+4 \atop 4} ; u \right) \end{pmatrix}$$

and

$$F_{w,1}(u) = \begin{pmatrix} {}_2F_1\left({-w, w+n+4 \atop 3} ; u \right) \\ -\frac{s_w}{3} \, {}_3F_2\left({-w, w+n+4, s_w+1 \atop 4, s_w} ; u \right) \end{pmatrix},$$

where $s_w = w(w+n+4)+3(n+2)$. Therefore

$$P_w(t) = \begin{pmatrix} {}_3F_2\left(\begin{smallmatrix} -w,w+n+3,2 \\ 3,1 \end{smallmatrix}; u\right) & \frac{w(w+n+3)}{3}\, {}_2F_1\left(\begin{smallmatrix} -w+1,w+n+4 \\ 4 \end{smallmatrix}; u\right) \\ {}_2F_1\left(\begin{smallmatrix} -w,w+n+4 \\ 3 \end{smallmatrix}; u\right) & -\frac{s_w}{3}\, {}_3F_2\left(\begin{smallmatrix} -w,w+n+4,s_w+1 \\ 4,s_w \end{smallmatrix}; u\right) \end{pmatrix}.$$

As we said above, $(P_w)_w$ is a sequence of matrix-valued orthogonal polynomials with respect to the weight matrix (13.4.8)

$$W(u) = u(1-u)^n \begin{pmatrix} 2-u & u \\ u & u^2 \end{pmatrix}$$

supported on the interval (0, 1). The differential operator D^t becomes

$$D^t = \partial^2 u(1-u) + \partial \begin{pmatrix} 2-(n+4)u & 1 \\ 0 & 4-(n+5)u \end{pmatrix} + \begin{pmatrix} 0 & 0 \\ 1 & -n-2 \end{pmatrix},$$

and

$$P_w D^t = \begin{pmatrix} -w(w+n+3) & 0 \\ 0 & -w(w+n+4)-n-2 \end{pmatrix} P_w.$$

The differential operator E^t becomes

$$\begin{aligned} E^t &= \partial^2 (1-u) \begin{pmatrix} (n-1)u & 3 \\ 0 & (n+2)u \end{pmatrix} \\ &+ \partial \begin{pmatrix} 2n+1-(n-1)(n+1)u & 2n+7 \\ 3u & 4n+5-(n+2)(n+5)u \end{pmatrix} \\ &+ (n+5) \begin{pmatrix} 0 & 0 \\ 1 & -n-2 \end{pmatrix}, \end{aligned}$$

and

$$P_w E^t = \begin{pmatrix} -w(w+n+3)(n-1) & 0 \\ 0 & -(n+2)(w(w+n+4)+n+5) \end{pmatrix} P_w.$$

Summing up, for an arbitrary choice of the nonnegative integer parameter n we have a weight matrix $W(u)$, supported on the interval (0, 1) of arbitrary size $\ell+1$ given by (13.4.8), such that the algebra $\mathcal{D}(W)$ is nontrivial. In the next section we will enrich this picture substantially by including two continuous free parameters α, β in the definition of $W(u)$. The continuous parameter α will be an extension of the discrete parameter n while β will be a completely new free continuous parameter. In the work so far β is hidden from view since in hindsight we will see that it corresponds to $\beta = 1$. For size 2 this step was taken in Grünbaum (2003) and later extended to arbitrary sizes in Pacharoni and Tirao (2007a). In the case of $\ell = 0$ this will retrieve the time-honored scalar-valued Jacobi polynomials $P_w^{(\alpha,\beta)}$.

13.6 The Matrix Jacobi Polynomials Arising from $P_d(\mathbb{C})$

All the work up to this point has a direct and fruitful generalization by replacing the complex projective plane by the d-dimensional complex projective space, $d \geq 2$, that is, the set of all complex lines through the origin in $\mathbb{C}^{d+1}$. This space is denoted by $P_d(\mathbb{C})$ and can be realized as the homogeneous space G/K, where $G = \mathrm{SU}(d+1)$ and $K = \mathrm{S}(\mathrm{U(d)} \times \mathrm{U}(1)) \simeq \mathrm{U}(d)$.

In this case, the finite-dimensional irreducible representations of K are parametrized by the d-tuples of integers

$$\mathbf{k} = (k_1, k_2, \dots, k_d) \in \widehat{K} \quad \text{such that} \quad k_1 \geq k_2 \geq \cdots \geq k_d. \tag{13.6.1}$$

We will restrict our attention to the irreducible spherical functions of type

$$\mathbf{k} = \Big(\underbrace{m+\ell, \dots, m+\ell}_{k}, \underbrace{m, \dots, m}_{d-k}\Big), \tag{13.6.2}$$

with $1 \leq k \leq d-1$ and $m \geq 0$; see Pacharoni and Tirao (2013).

By going over all the steps undertaken in the previous sections for $d = 2$ one arrives at a pair (D, E) of matrix second-order differential operators symmetric with respect to a matrix weight $W(u)$. They are given below and appeared first in Pacharoni and Tirao (2007a) and Pacharoni and Román (2008):

$$\begin{aligned}
D &= u(1-u)\frac{d^2}{du^2} + (C - uU)\frac{d}{du} - V,\\
E &= (1-u)(Q_0 + uQ_1)\frac{d^2}{du^2} + (P_0 + uP_1)\frac{d}{du} - (m + 2\ell + 3k)V,\\
W(u) &= (1-u)^m u^{d-1} \sum_{i,j=0}^{\ell} \left(\sum_{r=0}^{\ell} \binom{r}{i}\binom{r}{j}\binom{\ell+k-r-1}{\ell-r}\binom{d-k+r-1}{r}(1-u)^{\ell-r}u^{i+j} \right) E_{i,j},
\end{aligned}$$

where

$$\begin{aligned}
C &= \sum_{i=0}^{\ell}(d+2i)E_{ii} + \sum_{i=1}^{\ell} iE_{i,i-1},\\
U &= \sum_{i=0}^{\ell}(m+d+\ell+i+1)E_{i,i},\\
V &= \sum_{i=0}^{\ell} i(m+d+i-k)E_{i,i} - \sum_{i=0}^{\ell-1}(\ell-i)(i+d-k)E_{i,i+1},\\
Q_0 &= \sum_{i=1}^{\ell} 3iE_{i,i-1},\\
Q_1 &= \sum_{i=0}^{\ell}(m-\ell+3i)E_{i,i},
\end{aligned}$$

$$P_0 = \sum_{i=0}^{\ell} ((m+2\ell)(d+2i) - 3k(\ell - i) - 3i(d-k+i-1))\, E_{i,i}$$
$$- \sum_{i=1}^{\ell} i(3i+3d-3k+\ell+2m)E_{i,i-1},$$
$$P_1 = \sum_{i=0}^{\ell} -(m-\ell+3i)(m+d+\ell+i+1)E_{i,i}$$
$$+ \sum_{i=0}^{\ell-1} 3(d-k+i)(\ell-i)E_{i,i+1}.$$

A sequence of matrix orthogonal polynomials $(P_w)_w$ going along with $(W, \mathcal{D}(W))$ has an explicit expression for the columns $P_{w,j}$ of $(P_w)^t$ in terms of the matrix hypergeometric function introduced in Tirao (2003), namely

$$P_{w,j}(u) = {}_2H_1\left({U;V+\lambda_j \atop C} ; u\right) P_{w,j}(0), \qquad (13.6.3)$$

where $P_{w,j}(0)$ is the μ_j-eigenvector of $M(\lambda_j)$ properly normalized. Here

$$\lambda_j = -w(w+m+d+\ell+j) - j(m+d-k+j),$$
$$\mu_j = -w(w+m+d+\ell+j)(m-\ell+3j) - j(j+m+d-k)(m+2\ell+3k),$$

with $0 \le w$, $0 \le j \le \ell$, and $m > -1$. See Pacharoni and Román (2008).

For each fixed k this group-theoretical construction gives us a pair (W, D) for each choice of the **group parameters** m, d. Eventually we will replace the pair $(m, d-1)$ by the pair (α, β) of continuous real parameters with $\alpha, \beta > -1$. Before taking this step we consider the recursion relation satisfied by the sequence $(P_w)_w$.

For appropriate matrices A_w, B_w, C_w to be defined below one has

$$(1-u)\, P_w(u) = A_w P_{w-1}(u) + B_w P_w(u) + C_w P_{w+1}(u). \qquad (13.6.4)$$

The above three-term recursion relation which holds for all $w \ge 0$ can be written in the following way:

$$(1-u)\begin{pmatrix} P_0 \\ P_1 \\ P_2 \\ P_3 \\ \cdot \end{pmatrix} = \begin{pmatrix} B_0 & C_0 & 0 & & & \\ A_1 & B_1 & C_1 & 0 & & \\ 0 & A_2 & B_2 & C_2 & 0 & \\ & 0 & A_3 & B_3 & C_3 & 0 \\ & & \cdot & \cdot & \cdot & \cdot & \cdot \end{pmatrix} \begin{pmatrix} P_0 \\ P_1 \\ P_2 \\ P_3 \\ \cdot \end{pmatrix}. \qquad (13.6.5)$$

To introduce the square matrices A_w, B_w, C_w of size $\ell + 1$ we need the following notation: given a d-tuple $\mathbf{k} = (k_1, \ldots, k_d)$ as in (13.6.1) and a $(d+1)$-tuple of integers

$$\mathbf{m} = (m_1, \ldots, m_{d+1}) \quad \text{such that } m_1 \ge k_1 \ge m_2 \ge \cdots \ge k_d \ge m_{d+1},$$

let

$$
\begin{aligned}
a_i(\mathbf{m},\mathbf{k}) &= \left|\frac{\prod_{j=1}^{d}\left(k_j - m_i - j + i - 1\right)}{\prod_{j\neq i}\left(m_j - m_i - j + i\right)}\right|^{1/2}, \\
b_i(\mathbf{m},\mathbf{k}) &= \left|\frac{\prod_{j=1}^{d}\left(k_j - m_i - j + i\right)}{\prod_{j\neq i}\left(m_j - m_i - j + i\right)}\right|^{1/2}.
\end{aligned}
\tag{13.6.6}
$$

Then one can prove that

$$
\sum_{i=1}^{d+1} a_i^2(\mathbf{m},\mathbf{k}) = \sum_{i=1}^{d+1} b_i^2(\mathbf{m},\mathbf{k}) = 1. \tag{13.6.7}
$$

Since we are restricting our attention to spherical functions of type $\mathbf{k}$ as in (13.6.2) we will only be considering $(d+1)$-tuples $\mathbf{m}$ of the form

$$
\mathbf{m}(w,r) = (w+m+\ell, \underbrace{m+\ell,\dots,m+\ell}_{k-1}, m+r, \underbrace{m,\dots,m}_{d-k-1}, -w-r), \tag{13.6.8}
$$

with $1 \le k \le d-1$, $0 \le w$, and $0 \le r \le \ell$. Then the (r,s)-entries of the matrices A_w, B_w, and C_w are

$$
(A_w)_{r,s} = \begin{cases} a_{d+1}^2(\mathbf{m}(w,r))b_1^2\left(\mathbf{m}(w,r)+\mathbf{e}_{d+1}\right) & \text{if } s = r, \\ a_{k+1}^2(\mathbf{m}(w,r))b_1^2\left(\mathbf{m}(w,r)+\mathbf{e}_{k+1}\right) & \text{if } s = r+1, \\ 0 & \text{otherwise,} \end{cases}
$$

$$
(C_w)_{r,s} = \begin{cases} a_1^2(\mathbf{m}(w,r))b_{d+1}^2\left(\mathbf{m}(w,r)+\mathbf{e}_1\right) & \text{if } s = r, \\ a_1^2(\mathbf{m}(w,r))b_{k+1}^2\left(\mathbf{m}(w,r)+\mathbf{e}_1\right) & \text{if } s = r-1, \\ 0 & \text{otherwise,} \end{cases}
$$

$$
(B_w)_{\mathbf{r},\mathbf{s}} = \begin{cases} \sum_{1\le j\le d+1} a_j^2\left(\mathbf{m}(w,r)\right) b_j^2\left(\mathbf{m}(w,r)+\mathbf{e}_j\right) & \text{if } s = r, \\ a_{k+1}^2(\mathbf{m}(w,r))b_{d+1}^2\left(\mathbf{m}(w,r)+\mathbf{e}_{k+1}\right) & \text{if } s = r+1, \\ a_{d+1}^2(\mathbf{m}(w,r))b_{k+1}^2\left(\mathbf{m}(w,r)+\mathbf{e}_{d+1}\right) & \text{if } s = r-1, \\ 0 & \text{otherwise,} \end{cases}
$$

where $a_i^2(\mathbf{m}(w,\mathbf{r})) = a_i^2(\mathbf{m}(w,r),\mathbf{k})$, $b_i^2\left(\mathbf{m}(w,r)+\mathbf{e}_j\right) = b_i^2\left(\mathbf{m}(w,r)+\mathbf{e}_j,\mathbf{k}\right)$ for $1 \le j \le d+1$; see (13.6.6). For the benefit of the reader we include the following simplified expression of these coefficients:

$$
\begin{aligned}
a_1^2(\mathbf{m}(w,r)) &= \frac{(w+k)(w+\ell+d)}{(w+\ell-r+k)(2w+m+d+\ell+r)}, \\
a_{k+1}^2(\mathbf{m}(w,r)) &= \frac{(\ell-r)(r+d-k)}{(w+\ell-r+k)(w+m+d+2r-k)}, \\
a_{d+1}^2(\mathbf{m}(w,r)) &= \frac{(w+m+d+\ell+r-k)(w+m+r)}{(w+m+d+2r-k)(2w+m+d+\ell+r)};
\end{aligned}
$$

all the other $a_j^2(\mathbf{m}(w,r))$ are zero. The remaining coefficients are

$$b_1^2(\mathbf{m}(w,r)+\mathbf{e}_1) = \frac{(w+1)(w+\ell+k+1)}{(w+\ell-r+k+1)(2w+m+d+\ell+r+1)},$$
$$b_1^2(\mathbf{m}(w,r)+\mathbf{e}_{k+1}) = \frac{w(w+\ell+k)}{(w+\ell-r+k-1)(2w+m+d+\ell+r)},$$
$$b_1^2(\mathbf{m}(w,r)+\mathbf{e}_{d+1}) = \frac{w(w+\ell+k)}{(w+\ell-r+k)(2w+m+d+\ell+r-1)},$$

$$b_{k+1}^2(\mathbf{m}(w,r)+\mathbf{e}_1) = \frac{r(\ell-r+k)}{(w+\ell-r+k+1)(w+m+d+2r-k)},$$
$$b_{k+1}^2(\mathbf{m}(w,r)+\mathbf{e}_{k+1}) = \frac{(r+1)(\ell-r+k-1)}{(w+\ell-r+k-1)(w+m+d+2r-k+1)},$$
$$b_{k+1}^2(\mathbf{m}(w,r)+\mathbf{e}_{d+1}) = \frac{r(\ell-r+k)}{(w+\ell-r+k)(w+m+d+2r-k-1)},$$

$$b_{d+1}^2(\mathbf{m}(w,r)+\mathbf{e}_1) = \frac{(w+m+d+\ell+r)(w+m+d+r-k)}{(w+m+d+2r-k)(2w+m+d+\ell+r+1)},$$
$$b_{d+1}^2(\mathbf{m}(w,r)+\mathbf{e}_{k+1}) = \frac{(w+m+d+\ell+r)(w+m+d+r-k)}{(w+m+d+2r-k+1)(2w+m+d+\ell+r)},$$
$$b_{d+1}^2(\mathbf{m}(w,r)+\mathbf{e}_{d+1}) = \frac{(w+m+d+\ell+r-1)(w+m+d+r-k-1)}{(w+m+d+2r-k-1)(2w+m+d+\ell+r-1)}.$$

Notice that A_w is an upper bidiagonal matrix, C_w is a lower bidiagonal matrix, and B_w is a tridiagonal matrix.

From (13.6.7) one gets

$$\sum_{s=0}^{\ell}\Big((A_w)_{r,s}+(B_w)_{r,s}+(C_w)_{r,s}\Big)=1,$$

proving that the square semi-infinite matrix M appearing in (13.6.5) is stochastic.

The results above can be obtained by specializing the results in Grünbaum, Pacharoni, and Tirao (2011) to the case when $\mathbf{k}$ becomes the d-tuple given in (13.6.2). See also Pacharoni and Tirao (2007b; 2013).

In the expressions for D, E, W, P_w, and M the discrete parameters (m,d) enter in a simple analytical fashion. Appealing to some version of analytic continuation, it is clear that this entire edifice remains valid if one allows $(m,d-1)$ to range over a continuous set of real values (α,β). The requirement that W retains the property of having finite moments of all orders translates into the conditions $\alpha,\beta>-1$. We will denote the corresponding weight and the orthogonal polynomials by $W^{\beta,\alpha}$ and $P_w^{\beta,\alpha}$. These are the objects alluded to in the title of this contribution.

We illustrate the results above taking $\ell = 1$ (the special case $k = 1$ was considered in Pacharoni, 2009). Then the weight matrix becomes

$$W^{\beta,\alpha}(u) = u^{\beta}(1-u)^{\alpha}\begin{pmatrix} k+\beta-ku & \beta u \\ \beta u & \beta u^2 \end{pmatrix}$$

supported on the interval $(0,1)$. The matrix orthogonal polynomials are given by

$$P_w^{\beta,\alpha}(u) = \begin{pmatrix} {}_3F_2\left(\begin{smallmatrix} -w,w+\alpha+\beta+2,k+1 \\ \beta+2,k \end{smallmatrix};u\right) & \frac{w(w+\alpha+\beta+2)}{k(\beta+2)}\,{}_2F_1\left(\begin{smallmatrix} -w+1,w+\alpha+\beta+3 \\ \beta+3 \end{smallmatrix};u\right) \\ {}_2F_1\left(\begin{smallmatrix} -w,w+\alpha+\beta+3 \\ \beta+2 \end{smallmatrix};u\right) & -\frac{s_w}{\beta+2}\,{}_3F_2\left(\begin{smallmatrix} -w,w+\alpha+\beta+3,s_w+1 \\ \beta+3,s_w \end{smallmatrix};u\right) \end{pmatrix}, \tag{13.6.9}$$

where $s_w = (w(w+\alpha+\beta+3)+(\beta+2)(\alpha+\beta-k+2))\,/(\beta-k+1)$.

The differential operator D^t becomes

$$D^t = \partial^2 u(1-u) + \partial\begin{pmatrix} \beta+1-(\alpha+\beta+3)u & 1 \\ 0 & \beta+3-(\alpha+\beta+4)u \end{pmatrix} + \begin{pmatrix} 0 & 0 \\ \beta-k+1 & -\alpha-\beta+k-2 \end{pmatrix}$$

and

$$P_w^{\beta,\alpha}D^t = \begin{pmatrix} -w(w+\alpha+\beta+2) & 0 \\ 0 & -w(w+\alpha+\beta+3)-\alpha-\beta+k-2 \end{pmatrix} P_w^{\beta,\alpha}. \tag{13.6.10}$$

The differential operator E^t becomes

$$\begin{aligned} E^t &= \partial^2(1-u)\begin{pmatrix} (\alpha-1)u & 3 \\ 0 & (\alpha+2)u \end{pmatrix} \\ &+ \partial\begin{pmatrix} (\alpha+2)(\beta+1)-3k-(\alpha-1)(\alpha+\beta+3)u & 2\alpha+3\beta-3k+7 \\ 3(\beta-k+1)u & \gamma-(\alpha+2)(\alpha+\beta+4)u \end{pmatrix} \\ &+ (\alpha+3k+2)\begin{pmatrix} 0 & 0 \\ \beta-k+1 & -\alpha-\beta+k-2 \end{pmatrix}, \end{aligned}$$

with $\gamma = (\alpha-1)\beta+3(\alpha+k+1)$. Moreover,

$$P_w E^t = \begin{pmatrix} -w(w+\alpha+\beta+2)(\alpha-1) & 0 \\ 0 & \delta \end{pmatrix} P_w,$$

where $\delta = -w(w+\alpha+\beta+3)(\alpha+2)-(\alpha+\beta-k+2)(\alpha+3k+2)$.

In order to express the orthogonal polynomials $P_w^{\beta,\alpha}$ in terms of a single matrix-valued hypergeometric function we start by taking the transpose of the differential equation (13.6.10) satisfied by our function $P_w^{\beta,\alpha}$. Thus $F(u) = P_w^{\beta,\alpha}(u)^t$ satisfies

$$u(1-u)F'' + (C-uU)F' - (VF+F\Lambda) = 0, \tag{13.6.11}$$

with

$$C = \sum_{i=0}^{1} (\beta + 1 + 2i) E_{i,i} + E_{1,0},$$

$$U = \sum_{i=0}^{1} (\alpha + \beta + 3 + i) E_{i,i},$$

$$V = (\alpha + \beta + 2 - k) E_{1,1} - (\beta + 1 - k) E_{0,1},$$

$$\Lambda = \sum_{i=0}^{1} \lambda_i E_{i,i},$$

where $\lambda_1 = -w(w + \alpha + \beta + 2)$ and $\lambda_2 = -w(w + \alpha + \beta + 3) - \alpha - \beta + k - 2$.

The term $F\Lambda$ in (13.6.11) forces us to consider this equation as a differential equation on functions taking values on $\mathbb{C}^4$ and to consider left and right multiplication by matrices in $M(2, \mathbb{C})$ as linear maps in $\mathbb{C}^4$; a matrix T is identified with the column vector $(T_{11}, T_{12}, T_{21}, T_{22})^t$. Thus instead of equation (13.6.11) we shall consider

$$u(1 - u)\Phi'' + (L_C - uL_U)\,\Phi' - (L_V + R_\Lambda)\,\Phi = 0,$$

where

$$L_C = \begin{pmatrix} (\beta + 1)I & 0 \\ I & (\beta + 3)I \end{pmatrix}, \qquad L_U = \begin{pmatrix} (\alpha + \beta + 3)I & 0 \\ 0 & (\alpha + \beta + 4)I \end{pmatrix},$$

$$L_V + R_\Lambda = \begin{pmatrix} \Lambda & -(\beta - k + 1)I \\ 0 & \Lambda + (\alpha + \beta - k + 2)I \end{pmatrix},$$

and the letter I in the matrices above denotes the 2×2 identity matrix.

To find square matrices of size 4 with A, B such that

$$A + B + 1 = L_U, \qquad AB = L_V + R_\Lambda,$$

we need to solve the quadratic equation $B^2 + (1 - L_U)B + L_V + R_\Lambda = 0$. It is easy to verify that

$$B = \begin{pmatrix} -w & 0 & \frac{(\beta-k+1)(w+y)}{y-\alpha-\beta+k-2} & 0 \\ 0 & x & 0 & -\frac{(\beta-k+1)(w+x)}{-w-\alpha-\beta+k-2} \\ 0 & 0 & y & 0 \\ 0 & 0 & 0 & -w \end{pmatrix},$$

with $x^2 - (\alpha + \beta + 2)x + \lambda_2 = 0$ and $y^2 - (\alpha + \beta + 3)y + \lambda_1 + \alpha + \beta - k + 2 = 0$, satisfy such an equation. Then we take $A = L_U - B - 1$. Thus

$$A = \begin{pmatrix} w + \alpha + \beta + 2 & 0 & -\frac{(\beta-k+1)(w+y)}{y-\alpha-\beta+k-2} & 0 \\ 0 & \alpha + \beta + 2 - x & 0 & \frac{(\beta-k+1)(w+x)}{-w-\alpha-\beta+k-2} \\ 0 & 0 & \alpha + \beta + 3 - y & 0 \\ 0 & 0 & 0 & w + \alpha + \beta + 3 \end{pmatrix}.$$

Then $P_w^{\beta,\alpha}(u)^t$ as a column vector of dimension 4 is given by

$$P_w^{\beta,\alpha}(u)^t = {}_2F_1\left(\begin{smallmatrix} A;B \\ C \end{smallmatrix}; u\right) P_w^{\beta,\alpha}(0)^t.$$

13.7 Miscellanea

In this last section we give a brief guide to some recent application of the material discussed in the previous section to the construction of two stochastic processes. There is a long historical tradition of finding that certain families of special functions give the tools to solve certain time-honored physical models. The two main examples involve a model studied by D. Bernoulli around 1770 and independently by S. Laplace around 1810, and a model studied by P. and T. Ehrenfest in 1907. In the first case, the dual Hahn polynomials, and in the second case the Krawtchouk polynomials have been recognized rather recently as the eigenfunctions of the corresponding one-step transition probability kernel.

As we observed in the introduction, the study of matrix-valued orthogonal polynomials did not arise from the consideration of a good collection of physical or geometrical problems. In proposing the stochastic models mentioned below, one is thus reversing the path taken in the scalar-valued case.

In Grünbaum, Pacharoni, and Tirao (2011) we consider several random walks whose configuration spaces are subsets of $\widehat{\mathrm{U}}(d+1)(\mathbf{k})$, the so-called $\mathbf{k}$-spherical dual of $\mathrm{U}(d+1)$, and whose one-step transition matrices come from the stochastic matrix M that appears in (13.6.5). The dual of $\mathrm{U}(d+1)$ is the set $\widehat{\mathrm{U}}(d+1)$ of all equivalence classes of finite-dimensional irreducible representations of $\mathrm{U}(d+1)$. These equivalence classes are parametrized by the $(d+1)$-tuples of integers $\mathbf{m} = (m_1, \ldots, m_{d+1})$ subject to the conditions $m_1 \geq \cdots \geq m_{d+1}$. If $\mathbf{k} = (k_1, \ldots, k_d) \in \widehat{\mathrm{U}}(d)$, the $\mathbf{k}$-spherical dual of $\mathrm{U}(d+1)$ is the subset $\widehat{\mathrm{U}}(d+1)(\mathbf{k})$ of $\widehat{\mathrm{U}}(d+1)$ of the representations of $\mathrm{U}(d+1)$ whose restriction to $\mathrm{U}(d)$ contains the representation $\mathbf{k}$. Then it is well known (see Vilenkin, 1968) that $\widehat{\mathrm{U}}(d+1)(\mathbf{k})$ corresponds to the set of all $\mathbf{m}$ as above that satisfy the extra constraints

$$m_i \geq k_i \geq m_{i+1} \quad \text{for all } i = 1, \ldots, d. \tag{13.7.1}$$

Starting from the stochastic matrix M, we describe a random mechanism that gives rise to a Markov chain whose state space is the subset of $\widehat{\mathrm{U}}(d+1)(\mathbf{k})$ of all $\mathbf{m} \in \widehat{\mathrm{U}}(d+1)(\mathbf{k})$ such that $s_{\mathbf{m}} = s_{\mathbf{k}}$ and $k_d \geq 0$ ($s_{\mathbf{m}} = m_1 + \cdots + m_{d+1}$, $s_{\mathbf{k}} = k_1 + \cdots + k_d$), and whose one-step transition matrix coincides with the one we started from. One step of the Markov evolution will consist of two substeps taken in succession. In the first substep one of the values of m_i increases by 1, subject to the constraints (13.7.1). In the second substep one of the new values of our m_i decreases by 1, and again this is subject to the same constraints.

In Grünbaum, Pacharoni, and Tirao (2011) we construct a factorization of the stochastic matrix M that defines the three-term recursion relation for the sequence of matrix-valued orthogonal polynomials given in the previous section. The factorization $M = M_1 M_2$ into two stochastic matrices leads to the two substeps mentioned above.

The definition of the stochastic matrix M alluded to above, as well as its factorization, makes sense for any $\mathbf{m} \in \widehat{\mathrm{U}}(d+1)(\mathbf{k})$. To each configuration $m_1 \geq m_2 \geq \cdots \geq m_d \geq 0$ of d integer numbers we associate its Young diagram, a combinatorial object which has m_1 boxes in the first row, m_2 boxes in the second row, and so on down to the last row which has m_d boxes.

Young diagrams and their relatives the Young tableaux are very useful in representation theory. They provide a convenient way to describe the group representations of the symmetric and general linear groups and to study their properties. In particular, Young diagrams are in one-to-one correspondence with the irreducible representations of the symmetric group over the complex numbers and the irreducible polynomial representations of the general linear groups. They were introduced by Alfred Young in 1900. They were then applied to the study of the symmetric group by Georg Frobenius in 1903. Their theory and applications were further developed by many mathematicians and there are numerous and interesting applications, beyond representation theory, in combinatorics and algebraic geometry.

To each $\mathbf{m} \in \widehat{\mathrm{U}}(d+1)(\mathbf{k})$ such that $m_{d+1} \geq 0$ we have a configuration

$$m_1 \geq k_1 \geq m_2 \geq \cdots \geq m_d \geq k_d \geq m_{d+1} \geq 0.$$

Thus it is natural to represent such a state of our Markov chain by its Young diagram, which has m_1 boxes in the first row, k_1 boxes in the second row, and so on down to the last row which has m_{d+1} boxes.

In Grünbaum, Pacharoni, and Tirao (2011) we describe a random mechanism based on Young diagrams that gives rise to a random walk in the set of all Young diagrams of $2d+1$ rows and whose $2j$th row has k_j boxes $1 \leq j \leq d$, whereby in one unit of time one of the m_i is increased by 1 with probability $a_i^2(\mathbf{m}, \mathbf{k})$; see (13.6.6).

An alternative random mechanism involving a set of urns, a way of picking balls of different colors from these urns and moving them among the different urns, is also given in Grünbaum, Pacharoni, and Tirao (2011). This model is much more involved than those of Laplace–Bernoulli and Ehrenfest, but the punch line is that the Jacobi matrix-valued polynomials constructed in the previous sections play here the same role that the Krawtchouk and dual Hahn polynomials played in the classical cases. We anticipate that other people will find more applications for this new kind of special functions.

For an analysis of several of the topics above, as well as a few related ones, see also Grünbaum (2010), Grünbaum, Pacharoni, and Tirao (2001, 2003, 2005), Pacharoni and Tirao (2013), Román and Tirao (2012), and Pacharoni, Tirao, and Zurrián (2014).

References

Abdi, W. H. 1965. A basic analogue of the Bessel polynomial. *Math. Nachr.*, **30**, 209–219.

Ablowitz, M. J. and Ladik, J. F. 1976a. A nonlinear difference scheme and inverse scattering. *Studies in Appl. Math.*, **55**(3), 213–229.

Ablowitz, M. J. and Ladik, J. F. 1976b. Nonlinear differential-difference equations and Fourier analysis. *J. Math. Phys.*, **17**(6), 1011–1018.

Abramov, S. A. 1989. Problems in computer algebra that are connected with a search for polynomial solutions of linear differential and difference equations. *Moscow Univ. Comput. Math. Cybernet.*, **3**, 63–68. Transl. from Vestn. Moskov. Univ. Ser. XV Vychisl. Mat. Kibernet. 3, 56–60.

Abramowitz, M. and Stegun, I. A. (eds). 1965. *Handbook of Mathematical Functions, with Formulas, Graphs, and Mathematical Tables*. National Bureau of Standards Applied Mathematics Series, vol. 55. Superintendent of Documents, US Government Printing Office, Washington, DC. Third printing, with corrections.

Ahmed, S. and Muldoon, M. E. 1983. Reciprocal power sums of differences of zeros of special functions. *SIAM J. Math. Anal.*, **14**(2), 372–382.

Ahmed, S., Bruschi, M., Calogero, F., Olshanetsky, M. A., and Perelomov, A. M. 1979. Properties of the zeros of the classical polynomials and of the Bessel functions. *Nuovo Cimento B (11)*, **49**(2), 173–199.

Ahmed, S., Laforgia, A., and Muldoon, M. E. 1982. On the spacing of the zeros of some classical orthogonal polynomials. *J. London Math. Soc. (2)*, **25**(2), 246–252.

Akhiezer, N. I. 1965. *The Classical Moment Problem and Some Related Questions in Analysis*. New York: Hafner. Translated by N. Kemmer.

Akhiezer, N. I. and Glazman, I. M. 1950. *Theory of Linear Operators in Hilbert Space*. Reprint, New York: Dover Publications, Two volumes bound as one, 1993.

Al-Salam, W. A. and Carlitz, L. 1965. Some orthogonal q-polynomials. *Math. Nachr.*, **30**, 47–61.

Al-Salam, W. A. and Chihara, T. S. 1976. Convolutions of orthonormal polynomials. *SIAM J. Math. Anal.*, **7**(1), 16–28.

Al-Salam, W. A. and Chihara, T. S. 1987. q-Pollaczek polynomials and a conjecture of Andrews and Askey. *SIAM J. Math. Anal.*, **18**(1), 228–242.

Al-Salam, W. A. and Ismail, M. E. H. 1983. Orthogonal polynomials associated with the Rogers–Ramanujan continued fraction. *Pacific J. Math.*, **104**(2), 269–283.

Al-Salam, W. A. and Verma, A. 1983. q-analogs of some biorthogonal functions. *Canad. Math. Bull.*, **26**(2), 225–227.

Al-Salam, W. A., Allaway, W. R., and Askey, R. A. 1984. Sieved ultraspherical polynomials. *Trans. Amer. Math. Soc.*, **284**(1), 39–55.

Alfaro, M. P. and Vigil, L. 1988. Solution of a problem of P. Turán on zeros of orthogonal polynomials on the unit circle. *J. Approx. Theory*, **53**(2), 195–197.

Alfaro, M. P., Bello Hernández, M., Montaner, J. M., and Varona, J. L. 2005. Some asymptotic properties for orthogonal polynomials with respect to varying measures. *J. Approx. Theory*, **135**(1), 22–34.

Ammar, G. S. and Gragg, W. B. 1994. Schur flows for orthogonal Hessenberg matrices. Pages 27–34 of Bloch, Anthony (ed), *Hamiltonian and Gradient Flows, Algorithms and Control*. Fields Inst. Commun., vol. 3. Providence, RI: American Mathematical Society.

Andrews, G. E. 1976. *The Theory of Partitions*. Encyclopedia of Mathematics and its Applications, vol. 2. Reading, Mass.-London-Amsterdam: Addison-Wesley.

Andrews, G. E. 1981. Ramunujan's "lost" notebook. III. The Rogers–Ramanujan continued fraction. *Adv. Math.*, **41**(2), 186–208.

Andrews, G. E. 1986. *q-Series: Their Development and Application in Analysis, Number Theory, Combinatorics, Physics, and Computer Algebra*. CBMS Regional Conference Series in Mathematics, vol. 66. Washington, DC: Published for the Conference Board of the Mathematical Sciences.

Andrews, G. E. 1990. A page from Ramanujan's lost notebook. *Indian J. Math.*, **32**(3), 207–216.

Andrews, G. E. 2005. Ramanujan's "lost" notebook. VIII. The entire Rogers–Ramanujan function. *Adv. Math.*, **191**(2), 393–407.

Andrews, G. E. and Askey, R. A. 1985. Classical orthogonal polynomials. Pages 36–62 of Brezinski, C., Draux, A., Magnus, A. P., Maroni, P., and Ronveaux, A. (eds), *Orthogonal Polynomials and Applications (Bar-le-Duc, 1984)*. Lecture Notes in Math., vol. 1171. Berlin: Springer.

Andrews, G. E., Askey, R. A., and Roy, R. 1999. *Special Functions*. Encyclopedia of Mathematics and Its Applications, vol. 71. Cambridge: Cambridge University Press.

Andrews, G. E., Berndt, B. C., Sohn, J., Yee, A. J., and Zaharescu, A. 2003. On Ramanujan's continued fraction for $(q^2;q^3)_\infty/(q;q^3)_\infty$. *Trans. Amer. Math. Soc.*, **355**(6), 2397–2411 (electronic).

Andrews, G. E., Berndt, B. C., Sohn, J., Yee, A. J., and Zaharescu, A. 2005. Continued fractions with three limit points. *Adv. Math.*, **192**, 231–258.

Annaby, M. H. and Mansour, Z. S. 2005. Basic Sturm–Liouville problems. *J. Phys. A*, **38**(17), 3775–3797.

Appell, P. and Kampé de Fériet, J. 1926. *Fonctions Hypergéométriques et Hypersphérique; Polynomes d'Hermite*. Paris: Gauthier-Villars.

Area, I., Dimitrov, D. K., Godoy, E., and Ronveaux, A. 2004. Zeros of Gegenbauer and Hermite polynomials and connection coefficients. *Math. Comp.*, **73**(248), 1937–1951 (electronic).

Askey, R. A. 1971. Orthogonal expansions with positive coefficients. II. *SIAM J. Math. Anal.*, **2**, 340–346.

Askey, R. A. 1975a. *A Note on the History of Series*. Tech. rept. 1532. Mathematics Research Center, University of Wisconsin.

Askey, R. A. 1975b. *Orthogonal Polynomials and Special Functions*. Philadelphia, PA: Society for Industrial and Applied Mathematics.

Askey, R. A. 1980. Ramanujan's extensions of the gamma and beta functions. *Amer. Math. Monthly*, **87**(5), 346–359.

Askey, R. A. 1983. An elementary evaluation of a beta type integral. *Indian J. Pure Appl. Math.*, **14**(7), 892–895.

Askey, R. A. 1985. Continuous Hahn polynomials. *J. Phys. A*, **18**, L1017–L1019.

Askey, R. A. 1989a. Beta integrals and the associated orthogonal polynomials. Pages 84–121 of Alladi, K. (ed), *Number Theory, Madras 1987*. Lecture Notes in Math., vol. 1395. Berlin: Springer.

Askey, R. A. 1989b. Continuous q-Hermite polynomials when $q > 1$. Pages 151–158 of Stanton, D. (ed), *q-Series and Partitions*. IMA Volumes in Mathematics and Its Applications. New York, NY: Springer.

Askey, R. A. 1990. Graphs as an aid to understanding special functions. Pages 3–33 of Wong, R. (ed), *Asymptotic and Computational Analysis*. New York, NY: Marcel Dekker.

Askey, R. A. and Gasper, G. 1977. Convolution structures for Laguerre polynomials. *J. Analyse Math.*, **31**, 48–68.

Askey, R. A. and Ismail, M. E. H. 1976. Permutation problems and special functions. *Canadian J. Math.*, **28**, 853–874.

Askey, R. A. and Ismail, M. E. H. 1983. A generalization of ultraspherical polynomials. Pages 55–78 of Erdős, P. (ed), *Studies in Pure Mathematics*. Basel: Birkhäuser.

Askey, R. A. and Ismail, M. E. H. 1984. Recurrence relations, continued fractions and orthogonal polynomials. *Memoirs Amer. Math. Soc.*, **49**(300), iv + 108 pp.

Askey, R. A. and Roy, R. 1986. More q-beta integrals. *Rocky Mountain J. Math.*, **16**(2), 365–372.

Askey, R. A. and Wilson, J. A. 1979. A set of orthogonal polynomials that generalize the Racah coefficients or $6 - j$ symbols. *SIAM J. Math. Anal.*, **10**(5), 1008–1016.

Askey, R. A. and Wilson, J. A. 1982. A set of hypergeometric orthogonal polynomials. *SIAM J. Math. Anal.*, **13**(4), 651–655.

Askey, R. A. and Wilson, J. A. 1985. Some basic hypergeometric orthogonal polynomials that generalize Jacobi polynomials. *Memoirs Amer. Math. Soc.*, **54**(319), iv + 55 pp.

Askey, R. A. and Wimp, J. 1984. Associated Laguerre and Hermite polynomials. *Proc. Roy. Soc. Edinburgh*, **96A**, 15–37.

Askey, R. A., Ismail, M. E. H., and Koornwinder, T. 1978. Weighted permutation problems and Laguerre polynomials. *J. Comb. Theory Ser. A*, **25**(3), 277–287.

Askey, R. A., Rahman, M., and Suslov, S. K. 1996. On a general q-Fourier transformation with nonsymmetric kernels. *J. Comp. Appl. Math.*, **68**(1-2), 25–55.

Atakishiyev, N. M. and Klimyk, A. U. 2004. On q-orthogonal polynomials, dual to little and big q-Jacobi polynomials. *J. Math. Anal. Appl.*, **294**(1), 246–257.

Atakishiyev, N. M. and Suslov, S. K. 1985. The Hahn and Meixner polynomials of imaginary argument and some of their applications. *J. Phys. A*, **18**, 1583–1596.

Atakishiyev, N. M. and Suslov, S. K. 1992. Difference hypergeometric functions. Pages 1–35 of Gonchar, A. A. and Saff, E. B. (eds), *Progress in Approximation Theory (Tampa, FL, 1990)*. Springer Ser. Comput. Math., vol. 19. New York: Springer.

Atakishiyev, N. M., Frank, A., and Wolf, K. B. 1994. A simple difference realization of the Heisenberg q-algebra. *J. Math. Phys.*, **35**(7), 3253–3260.

Atkinson, F. V. 1964. *Discrete and Continuous Boundary Problems*. Mathematics in Science and Engineering, vol. 8. New York: Academic Press.

Atkinson, F. V. and Everitt, W. N. 1981. Orthogonal polynomials which satisfy second order differential equations. Pages 173–181 of *E. B. Christoffel (Aachen/Monschau, 1979)*. Basel: Birkhäuser.

Azor, R., Gillis, J., and Victor, J. D. 1982. Combinatorial applications of Hermite polynomials. *SIAM J. Math. Anal.*, **13**(5), 879–890.

Badkov, V. M. 1985. *Uniform Asymptotic Representations of Orthogonal Polynomials*. Sverdlovsk: Ural. Nauchn. Tsentr Akad. Nauk SSSR. Pages 41–53.

Badkov, V. M. 1987. Systems of orthogonal polynomials expressed in explicit form in terms of Jacobi polynomials. *Math. Notes*, **42**, 858–863.

Baginski, F. E. 1991. Comparison theorems for the ν-zeroes of Legendre functions $P_\nu^m(z_0)$ when $-1 < z_0 < 1$. *Proc. Amer. Math. Soc.*, **111**(2), 395–402.

Baik, J., Deift, P., and Johansson, K. 1999. On the distribution of the length of the longest increasing subsequence of random permutations. *J. Amer. Math. Soc.*, **12**, 1119–1178.

Bailey, W. N. 1935. *Generalized Hypergeometric Series*. Cambridge: Cambridge University Press.

Bannai, E. and Ito, T. 1984. *Algebraic Combinatorics I: Association Schemes*. Menlo Park: Benjamin/Cummings.

Baratella, P. and Gatteschi, L. 1988. The bounds for the error terms of an asymptotic approximation of Jacobi polynomials. Pages 203–221 of Alfaro, M., Dehesa, J. S., Marcellán, F. J., Rubio de Francia, J. L., and Vinuesa, J. (eds), *Orthogonal Polynomials and Their Applications (Segovia, Spain, 1986)*. Lecture Notes in Math., vol. 1329. New York: Springer.

Barrios, D. and López, G. 1999. Ratio asymptotics of polynomials orthogonal on arcs of the unit circle. *Constr. Approx.*, **15**, 1–31.

Basu, S. and Bose, N. K. 1983. Matrix Stieltjes series and network models. *SIAM J. Math. Anal.*, **14**(2), 209–222.

Bateman, H. 1905. A generalization of the Legendre polynomials. *Proc. London Math. Soc.*, **3**(2), 111–123.

Bateman, H. 1932. *Partial Differential Equations*. Cambridge: Cambridge University Press.

Bauldry, W. 1990. Estimates of asymmetric Freud polynomials on the real line. *J. Approximation Theory*, **63**, 225–237.

Beckenbach, E. F., Seidel, W., and Szász, O. 1951. Recurrent determinants of Legendre and ultraspherical polynomials. *Duke Math. J.*, **18**, 1–10.

Bello, M. and López, G. 1998. Ratio and relative asymptotics of polynomials orthogonal on an arc of the unit circle. *J. Approx. Theory*, **92**, 216–244.

Berezanskii, Ju. M. 1968. *Expansions in Eigenfunctions of Selfadjoint Operators*. Translations of Mathematical Monographs, vol. 17. Providence, RI: American Mathematical Society. Translated from the Russian by R. Bolstein, J. M. Danskin, J. Rovnyak, and L. Shulman.

Berg, C. 1994. Markov's theorem revisited. *J. Approx. Theory*, **78**, 260–275.

Berg, C. 1995. Indeterminate moment problems and the theory of entire functions. *J. Comput. Appl. Math.*, **65**(1-3), 27–55. Proceedings of the International Conference on Orthogonality, Moment Problems and Continued Fractions (Delft, 1994).

Berg, C. 1998. From discrete to absolutely continuous solutions of indeterminate moment problems. *Arab J. Math. Sci.*, **4**(2), 1–18.

Berg, C. 2004. Private communication.

Berg, C. and Christensen, J. P. R. 1981. Density questions in the classical theory of moments. (French summary). *Ann. Inst. Fourier (Grenoble)*, **31**(3), 99–114.

Berg, C. and Durán, A. J. 1995. The index of determinacy for measures and the l^2-norm of orthonormal polynomials. *Trans. Amer. Math. Soc.*, **347**, 2795–2811.

Berg, C. and Durán, A. J. 1996. When does a discrete differential perturbation of a sequence of orthonormal polynomials belong to ℓ^2? *J. Funct. Anal.*, **136**, 127–153.

Berg, C. and Ismail, M. E. H. 1996. q-Hermite polynomials and classical orthogonal polynomials. *Canad. J. Math.*, **48**, 43–63.

Berg, C. and Pedersen, H. L. 1994. On the order and type of the entire functions associated with an indeterminate Hamburger moment problem. *Ark. Mat.*, **32**, 1–11.

Berg, C. and Pedersen, H. L. 2007. Logarithmic order and type of indeterminate moment problems. Pages 51–79 of *Difference Equations, Special Functions and Orthogonal Polynomials*. Hackensack, NJ: World Scientific Publisher. With an appendix by Walter Hayman.

Berg, C. and Szwarc, R. 2011. The smallest eigenvalue of Hankel matrices. *Constr. Approx.*, **34**(1), 107–133.

Berg, C. and Szwarc, R. 2014. On the order of indeterminate moment problems. *Adv. Math.*, **250**, 105–143.

Berg, C. and Szwarc, R. 2017. *Symmetric moment problems and a conjecture of Valent.* Mat. Sb. **208** (2017), no. 3, 28–53; translated in Sb. Math. **208** (2017), no. 3–4, 335–359.

Berg, C. and Thill, M. 1991. Rotation invariant moment problems. *Acta Math.*, **167**, 207–227.

Berg, C. and Valent, G. 1994. The Nevanlinna parameterization for some indeterminate Stieltjes moment problems associated with birth and death processes. *Methods and Applications of Analysis*, **1**, 169–209.

Berg, C. and Valent, G. 1995. Nevanlinna extremal measures for some orthogonal polynomials related to birth and death processes. *J. Comput. Appl. Math.*, **57**(1-2), 29–43. Proceedings of the Fourth International Symposium on Orthogonal Polynomials and Their Applications (Evian-Les-Bains, 1992).

Berg, C., Christensen, J. P. R., and Ressel, P. 1984. *Harmonic Analysis on Semigroups*. Graduate Texts in Mathematics, vol. 100. New York: Springer. Theory of positive definite and related functions.

Berg, C., Chen, Y., and Ismail, M. E. H. 2002. Small eigenvalues of large Hankel matrices: the indeterminate case. *Math. Scand.*, **91**(1), 67–81.

Bergweiler, W. and Hayman, W. K. 2003. Zeros of solutions of a functional equation. *Comput. Methods Funct. Theory*, **3**(1-2), 55–78.

Berndt, B. C. 2016. Integrals associated with Ramanujan and elliptic functions. *Ramanujan J.*, **41**(1), 369–389.

Berndt, B. C. and Sohn, J. 2002. Asymptotic formulas for two continued fractions in Ramanujan's lost notebook. *J. London, Math. Soc.*, **65**, 271–284.

Biedenharn, L. and Louck, J. 1981. *The Racah–Wigner Algebra in Quantum Theory*. Reading: Addison-Wesley.

Boas, R. P., Jr. 1939. The Stieltjes moment problem for functions of bounded variation. *Bull. Amer. Math. Soc.*, **45**, 399–404.

Boas, R. P., Jr. 1954. *Entire Functions*. New York: Academic Press.

Bochkov, I. 2019. Polynomial birth-death processes and the second conjecture of Valent. *C.R. Acad. Sci. Paris, Ser. I*, **357** (2019), 247–251.

Bochner, S. 1929. Über Sturm–Liouvillesche polynomsysteme. *Math. Zeit.*, **29**, 730–736.

Bochner, S. 1954. Positive zonal functions on spheres. *Proc. Nat. Acad. Sci. USA*, **40**, 1141–1147.

Bonan, S., Lubinsky, D. S., and Nevai, P. 1987. Orthogonal polynomials and their derivatives. II. *SIAM J. Math. Anal.*, **18**(4), 1163–1176.

Bonan, S. S. and Clark, D. S. 1990. Estimates of the Hermite and the Freud polynomials. *J. Approximation Theory*, **63**, 210–224.

Bonan, S. S. and Nevai, P. 1984. Orthogonal polynomials and their derivatives. I. *J. Approx. Theory*, **40**, 134–147.

Bourget, J. 1866. Mémoire sur le mouvement vibratoire des membranes circulaires (June 5, 1865). *Ann. Sci. Éc. Norm. Supér.*, **III**(5), 5–95.

Braaksma, B. L. J. and Meulenbeld, B. 1971. Jacobi polynomials as spherical harmonics. *Indag. Math.*, **33**, 191–196.

Bressoud, D. 1981. On partitions, orthogonal polynomials and the expansion of certain infinite products. *Proc. London Math. Soc.*, **42**, 478–500.

Brezinski, C. 1980. *Padé-Type Approximation and General Orthogonal Polynomials*. International Series of Numerical Mathematics, vol. 50. Basel: Birkhäuser.

Brown, B. M. and Ismail, M. E. H. 1995. A right inverse of the Askey–Wilson operator. *Proc. Amer. Math. Soc.*, **123**, 2071–2079.

Brown, B. M., Evans, W. D., and Ismail, M. E. H. 1996. The Askey–Wilson polynomials and q-Sturm–Liouville problems. *Math. Proc. Cambridge Phil. Soc.*, **119**, 1–16.

Bryc, W., Matysiak, W., and Szabłowski, P. J. 2005. Probabilistic aspects of Al-Salam–Chihara polynomials. *Proc. Amer. Math. Soc.*, **133**, 1127–1134.

Buchwalter, H. and Cassier, G. 1984a. La paramétrisation de Nevanlinna dans le problème des moments de Hamburger. *Exposition. Math.*, **2**(2), 155–178.

Buchwalter, H. and Cassier, G. 1984b. Mesures canoniques dans le problème classique des moments. *Ann. Inst. Fourier (Grenoble)*, **34**(2), 45–52.

Bueno, M. I. and Marcellán, F. 2004. Darboux transformation and perturbation of linear functionals. *Linear Algebra and Its Applications*, **384**, 215–242.

Bunse-Gerstner, A. and Elsner, L. 1991. Schur parameter pencils for the solution of the unitary eigenvalue problem. *Linear Algebra Appl.* **154/156**, 741–778.

Burchnall, J. L. 1951. The Bessel polynomials. *Canad. J. Math.*, **3**, 62–68.

Burchnall, J. L. and Chaundy, T. W. 1931. Commutative ordinary differential operators. II. The identity $P^n = Q^m$. *Proc. Roy. Soc. London (A)*, **34**, 471–485.

Bustoz, J. and Ismail, M. E. H. 1982. The associated classical orthogonal polynomials and their q-analogues. *Canad. J. Math.*, **34**, 718–736.

Cachafeiro, A. and Marcellán, F. 1988. Orthogonal polynomials and jump modifications. *Lecture Notes Math.*, **1329**, 236–240.

Cachafeiro, A. and Marcellán, F. 1993. Modifications of Toeplitz matrices: jump functions. *Rocky Mountain J. Math.*, **23**, 521–531.

Calogero, F. 2001. *Classical Many-Body Problems Amenable to Exact Treatments*. Lecture Notes in Physics. New Series: Monographs, vol. 66. Berlin: Springer. (Solvable and/or integrable and/or linearizable...) in one-, two- and three-dimensional space.

Cantero, M. J., Moral, L., and Velázquez, L. 2003. Five-diagonal matrices and zeros of orthogonal polynomials on the unit circle. *Linear Algebra Appl.*, **362**, 29–56.

Cantero, M. J., Moral, L., and Velázquez, L. 2007. Matrix orthogonal polynomials whose derivatives are also orthogonal. *J. Approx. Theory*, **146**(2), 174–211.

Carlitz, L. 1958. On some polynomials of Tricomi. *Boll. Un. Mat. Ital. (3)*, **13**, 58–64.

Carlitz, L. 1959. Some formulas related to the Rogers–Ramanujan identities. *Ann. Mat. (IV)*, **47**, 243–251.

Cartier, P. and Foata, D. 1969. *Problèmes Combinatoires de Commutation et Réarrangements*. Lecture Notes in Mathematics, vol. 85. Berlin: Springer.

Castro, M. M. and Grünbaum, F. A. 2005. Orthogonal matrix polynomials satisfying first order differential equations: a collection of instructive examples. *J. Nonlinear Math. Phys.*, **12** (suppl. 2), 63–76.

Castro, M. M. and Grünbaum, F. A. 2006. The algebra of differential operators associated to a family of matrix-valued orthogonal polynomials: five instructive examples. *Int. Math. Res. Not.*, **2006**, Art. ID 47602, 33.

Charris, J. A. and Ismail, M. E. H. 1986. On sieved orthogonal polynomials. II: Random walk polynomials. *Canad. J. Math.*, **38**(2), 397–415.

Charris, J. A. and Ismail, M. E. H. 1987. On sieved orthogonal polynomials. V: Sieved Pollaczek polynomials. *SIAM J. Math. Anal.*, **18**(4), 1177–1218.

Charris, J. A., Ismail, M. E. H., and Monsalve, S. 1994. On sieved orthogonal polynomials. X: General blocks of recurrence relations. *Pacific J. Math.*, **163**(2), 237–267.

Chen, Y. and Ismail, M. E. H. 1997. Ladder operators and differential equations for orthogonal polynomials. *J. Phys. A*, **30**, 7817–7829.

Chen, Y. and Ismail, M. E. H. 1998. Some indeterminate moment problems and Freud-like weights. *Constr. Approx.*, **14**(3), 439–458.

Chen, Y. and Ismail, M. E. H. 2005. Jacobi polynomials from compatibility conditions. *Proc. Amer. Math. Soc.*, **133**(2), 465–472 (electronic).

Chihara, T. S. 1962. Chain sequences and orthogonal polynomials. *Trans. Amer. Math. Soc.*, **104**, 1–16.

Chihara, T. S. 1968. On indeterminate Hamburger moment problems. *Pacific J. Math.*, **27**, 475–484.

Chihara, T. S. 1970. A characterization and a class of distribution functions for the Stieltjes–Wigert polynomials. *Canad. Math. Bull.*, **13**, 529–532.

Chihara, T. S. 1978. *An Introduction to Orthogonal Polynomials*. New York: Gordon and Breach.

Chihara, T. S. 1982. Indeterminate symmetric moment problems. *J. Math. Anal. Appl.*, **85**(2), 331–346.

Chihara, T. S. 1989. Hamburger moment problems and orthogonal polynomials. *Trans. Amer. Math. Soc.*, **315**(1), 189–203.

Chihara, T. S. and Ismail, M. E. H. 1993. Extremal measures for a system of orthogonal polynomials. *Constructive Approximation*, **9**, 111–119.

Christiansen, J. S. 2003a. The moment problem associated with the q-Laguerre polynomials. *Constr. Approx.*, **19**(1), 1–22.

Christiansen, J. S. 2003b. The moment problem associated with the Stieltjes–Wigert polynomials. *J. Math. Anal. Appl.*, **277**, 218–245.

Christiansen, J. S. 2004. *Indeterminate Moment Problems within the Askey-Scheme*. Ph.D. thesis, University of Copenhagen.

Christiansen, J. S. 2005. Indeterminate moment problems related to birth and death processes with quartic rates. *J. Comp. Appl. Math.*, **178**, 91–98.

Christiansen, J. S. and Ismail, M. E. H. 2006. A moment problem and a family of integral evaluations. *Trans. Amer. Math. Soc.*, **358**(9), 4071–4097.

Christiansen, J. S. and Koelink, E. 2006. Self-adjoint difference operators and classical solutions to the Stieltjes–Wigert moment problem. *J. Approx. Theory*, **140**(1), 1–26.

Christiansen, J. S. and Koelink, E. 2008. Self-adjoint difference operators and symmetric Al-Salam–Chihara polynomials. *Constr. Approx.*, **28**(2), 199–218.

Ciccoli, N., Koelink, E., and Koornwinder, T. H. 1999. q-Laguerre polynomials and big q-Bessel functions and their orthogonality relations. *Methods Appl. Anal.*, **6**(1), 109–127. Dedicated to Richard A. Askey on the occasion of his 65th birthday, Part I.

Cohen, M. E. 1977. On Jacobi functions and multiplication theorems for integral Bessel functions. *J. Math. Anal. Appl.*, **57**, 469–475.

Connett, W. C. and Schwartz, A. L. 2000. Measure algebras associated with orthogonal polynomials. Pages 127–140 of Ismail, M. E. H. and Stanton, D. W. (eds), *q-Series from a Contemporary Perspective (South Hadley, MA, 1998)*. Contemp. Math., vol. 254. American Mathematical Society.

Cooper, S. 2002. The Askey–Wilson operator and the ${}_6\psi_5$ summation formula. *South East Asian J. Math. Math. Sci.*, **1**(1), 71–82.

Cycon, H. L., Froese, R. G., Kirsch, W., and Simon, B. 1987. *Schrödinger Operators with Application to Quantum Mechanics and Global Geometry*. Berlin: Springer.

Damanik, D., Pushnitski, A., and Simon, B. 2008. The analytic theory of matrix orthogonal polynomials. *Surv. Approx. Theory*, **4**, 1–85.

de Boor, C. and Saff, E. B. 1986. Finite sequences of orthogonal polynomials connected by a Jacobi matrix. *Linear Algebra Appl.*, **75**, 43–55.

de Bruin, M. G., Saff, E. B., and Varga, R. S. 1981a. On the zeros of generalized Bessel polynomials. I. *Nederl. Akad. Wetensch. Indag. Math.*, **43**(1), 1–13.

de Bruin, M. G., Saff, E. B., and Varga, R. S. 1981b. On the zeros of generalized Bessel polynomials. II. *Nederl. Akad. Wetensch. Indag. Math.*, **43**(1), 14–25.

Deaño, A., Gil, A., and Segura, J. 2004. New inequalities from classical Sturm theorems. *J. Approx. Theory*, **131**, 208–230.

DeFazio, M. V., Gupta, D. P., and Muldoon, M. E. 2007. Limit relations for the complex zeros of Laguerre and q-Laguerre polynomials. *J. Math. Anal. Appl.*, **334**(2), 977–982.

Deift, P. 1999. *Orthogonal Polynomials and Random Matrices: A Riemann–Hilbert Approach*. Courant Lecture Notes in Mathematics, vol. 3. New York, NY: New York University Courant Institute of Mathematical Sciences.

Denisov, S. 2004. On Rakhmanov's theorem for Jacobi matrices. *Proc. Amer. Math. Soc.*, **132**, 847–852.

Denisov, S. and Kupin, S. 2006. Asymptotics of the orthogonal polynomials for the Szegő class with a polynomial weight. *J. Approx. Theory*, **139**(1-2), 8–28.

Dette, H. and Studden, W. J. 2003. Quadrature formulas for matrix measures—a geometric approach. *Linear Algebra Appl.*, **364**, 33–64.

Diaconis, P. and Graham, R. L. 1985. The Radon transform on Z_2^k. *Pacific J. Math.*, **118**(2), 323–345.

Dickinson, D. J. 1954. On Lommel and Bessel polynomials. *Proc. Amer. Math. Soc.*, **5**, 946–956.

Dickinson, D. J., Pollack, H. O., and Wannier, G. H. 1956. On a class of polynomials orthogonal over a denumerable set. *Pacific J. Math*, **6**, 239–247.

Dickson, L. E. 1939. *New Course on the Theory of Equations*. New York: Wiley.

Dilcher, K. and Stolarsky, K. 2005. Resultants and discriminants of Chebyshev and related polynomials. *Trans. Amer. Math. Soc.*, **357**, 965–981.

Dimitrov, D. K. 2003. Convexity of the extreme zeros of Gegenbauer and Laguerre polynomials. *J. Comput. Appl. Math.*, **153**(1-2), 171–180. Proceedings of the Sixth International Symposium on Orthogonal Polynomials, Special Functions and Their Applications (Rome, 2001).

Dimitrov, D. K. and Nikolov, G. P. 2010. Sharp bounds for the extreme zeros of classical orthogonal polynomials. *J. Approx. Theory*, **162**(10), 1793–1804.

Dimitrov, D. K. and Rafaeli, F. R. 2007. Monotonicity of zeros of Jacobi polynomials. *J. Approx. Theory*, **149**(1), 15–29.

Dimitrov, D. K. and Rafaeli, F. R. 2009. Monotonicity of zeros of Laguerre polynomials. *J. Comput. Appl. Math.*, **233**(3), 699–702.

Dimitrov, D. K. and Rodrigues, R. O. 2002. On the behaviour of zeros of Jacobi polynomials. *J. Approx. Theory*, **116**(2), 224–239.

Dominici, D. 2008. Asymptotic analysis of the Krawtchouk polynomials by the WKB method. *Ramanujan J.*, **15**(3), 303–338.

Donoghue, W. F., Jr. 1974. *Monotone Matrix Functions and Analytic Continuation*. New York: Springer. Die Grundlehren der mathematischen Wissenschaften, Band 207.

Draux, A. 1983. *Polynômes Orthogonaux Formels – Applications*. Lecture Notes in Mathematics, vol. 974. Berlin: Springer.

Driver, K. and Duren, P. 2000. Zeros of the hypergeometric polynomials $F(-n, b; 2b; z)$. *Indag. Math. (N.S.)*, **11**(1), 43–51.

Driver, K. and Duren, P. 2001. Zeros of ultraspherical polynomials and the Hilbert–Klein formulas. *J. Comput. Appl. Math.*, **135**(2), 293–301.

Duistermaat, J. J. and Grünbaum, F. A. 1986. Differential equations in the spectral parameter. *Comm. Math. Phys.*, **103**(2), 177–240.

Dulucq, S. and Favreau, L. 1991. A combinatorial model for Bessel polynomials. Pages 243–249 of Brezinski, C., Gori, L., and Ronveau, A. (eds), *Orthogonal Polynomials and Their Applications*. Basel, Switzerland: J. C. Baltzer AG Scientific.

Durán, A. J. 1989. The Stieltjes moments problem for rapidly decreasing functions. *Proc. Amer. Math. Soc.*, **107**(3), 731–741.

Durán, A. J. 1993. Functions with given moments and weight functions for orthogonal polynomials. *Rocky Mountain J. Math.*, **23**(1), 87–104.

Durán, A. J. 1993. A generalization of Favard's theorem for polynomials satisfying a recurrence relation. *J. Approx. Theory*, **74**(1), 83–109.

Durán, A. J. 1995. On orthogonal polynomials with respect to a positive definite matrix of measures. *Canad. J. Math.*, **47**(1), 88–112.

Durán, A. J. 1996. Markov's theorem for orthogonal matrix polynomials. *Canad. J. Math.*, **48**(6), 1180–1195.

Durán, A. J. 1997. Matrix inner product having a matrix symmetric second order differential operator. *Rocky Mountain J. Math.*, **27**(2), 585–600.

Durán, A. J. 1999. Ratio asymptotics for orthogonal matrix polynomials. *J. Approx. Theory*, **100**(2), 304–344.

Durán, A. J. 2009a. Generating orthogonal matrix polynomials satisfying second order differential equations from a trio of triangular matrices. *J. Approx. Theory*, **161**(1), 88–113.

Durán, A. J. 2009b. A method to find weight matrices having symmetric second-order differential operators with matrix leading coefficient. *Constr. Approx.*, **29**(2), 181–205.

Durán, A. J. 2010. Rodrigues' formulas for orthogonal matrix polynomials satisfying second-order differential equations. *Int. Math. Res. Not.*, **2010**(5), 824–855.

Durán, A. J. 2011a. A miraculously commuting family of orthogonal matrix polynomials satisfying second order differential equations. *J. Approx. Theory*, **163**(12), 1815–1833.

Durán, A. J. 2011b. Rodrigues' formulas for orthogonal matrix polynomials satisfying higher order differential equations. *Experimental Mathematics*, **20**(1), 15–24.

Durán, A. J. and de la Iglesia, M. D. 2008a. Second-order differential operators having several families of orthogonal matrix polynomials as eigenfunctions. *Int. Math. Res. Not.*, **2008**, Art. ID rnn 084, 24.

Durán, A. J. and de la Iglesia, M. D. 2008b. Some examples of orthogonal matrix polynomials satisfying odd order differential equations. *J. Approx. Theory*, **150**(2), 153–174.

Durán, A. and Daneri-Vias, E. 2001. Ratio asymptotics for orthogonal matrix polynomials with unbounded recurrence coefficients. *J. Approx. Theory* **110**, 1–17.

Durán, A. J. and Defez, E. 2002. Orthogonal matrix polynomials and quadrature formulas. *Linear Algebra Appl.*, **345**, 71–84.

Durán, A. J. and Grünbaum, F. A. 2004. Orthogonal matrix polynomials satisfying second-order differential equations. *Int. Math. Res. Not.*, **2004**(10), 461–484.

Durán, A. J. and Grünbaum, F. A. 2005a. A characterization for a class of weight matrices with orthogonal matrix polynomials satisfying second-order differential equations. *Int. Math. Res. Not.*, **2005**(23), 1371–1390.

Durán, A. J. and Grünbaum, F. A. 2005b. Orthogonal matrix polynomials, scalar-type Rodrigues' formulas and Pearson equations. *J. Approx. Theory*, **134**(2), 267–280.

Durán, A. J. and Grünbaum, F. A. 2005c. Structural formulas for orthogonal matrix polynomials satisfying second-order differential equations. I. *Constr. Approx.*, **22**(2), 255–271.

Durán, A. J. and Grünbaum, F. A. 2007. Matrix orthogonal polynomials satisfying second-order differential equations: coping without help from group representation theory. *J. Approx. Theory*, **148**(1), 35–48.

Durán, A. J. and López-Rodríguez, P. 1996. Orthogonal matrix polynomials: zeros and Blumenthal's theorem. *J. Approx. Theory*, **84**(1), 96–118.

Durán, A. J. and López-Rodríguez, P. 2007. Structural formulas for orthogonal matrix polynomials satisfying second-order differential equations. II. *Constr. Approx.*, **26**(1), 29–47.

Durán, A. J. and Polo, B. 2002. Gaussian quadrature formulae for matrix weights. *Linear Algebra Appl.*, **355**, 119–146.

Durán, A. J., López-Rodríguez, P., and Saff, E. B. 1999. Zero asymptotic behaviour for orthogonal matrix polynomials. *J. Anal. Math.*, **78**, 37–60.

Durand, L. 1975. Nicholson-type integrals for products of Gegenbauer functions and related topics. Pages 353–374 of *Theory and Application of Special Functions (Proc. Advanced Sem., Math. Res. Center, Univ. Wisconsin, Madison, Wis., 1975)*. New York, NY: Academic Press. Math. Res. Center, Univ. Wisconsin, Publ. No. 35.

Durand, L. 1978. Product formulas and Nicholson-type integrals for Jacobi functions. I. Summary of results. *SIAM J. Math. Anal.*, **9**(1), 76–86.

Elbert, Á. and Laforgia, A. 1986. Some monotonicity properties of the zeros of ultraspherical polynomials. *Acta Math. Hungar.*, **48**, 155–159.

Elbert, Á. and Laforgia, A. 1987. Monotonicity results on the zeros of generalized Laguerre polynomials. *J. Approx. Theory*, **51**, 168–174.

Elbert, Á. and Laforgia, A. 1990. Upper bounds for the zeros of ultraspherical polynomials. *J. Approx. Theory*, **61**, 88–97.

Elbert, Á. and Muldoon, M. E. 1994. On the derivative with respect to a parameter of a zero of a Sturm-Liouville function. *SIAM J. Math. Anal.*, **25**(2), 354–364.

Elbert, Á. and Muldoon, M. E. 1999. Inequalities and monotonicity properties for zeros of Hermite functions. *Proc. Roy. Soc. Edinburgh Sect. A*, **129**(1), 57–75.

Elbert, Á. and Muldoon, M. E. 2008. Approximations for zeros of Hermite functions. Pages 117–126 of Dominici, D. and Maier, R. S. (eds), *Special Functions and Orthogonal Polynomials*. Contemp. Math., vol. 471. Providence, RI: American Mathematical Society.

Elbert, Á. and Siafarikas, P. D. 1999. Monotonicity properties of the zeros of ultraspherical polynomials. *J. Approx. Theory*, **97**, 31–39.

Erdélyi, A. 1938. The Hankel transform of a product of Whittaker functions. *J. London Math. Soc.*, **13**, 146–154.

Erdélyi, A., Magnus, W., Oberhettinger, F., and Tricomi, F. G. 1953a. *Higher Transcendental Functions*. Vol. 1. New York: McGraw-Hill.

Erdélyi, A., Magnus, W., Oberhettinger, F., and Tricomi, F. G. 1953b. *Higher Transcendental Functions*. Vol. 2. New York: McGraw-Hill.

Erdélyi, A., Magnus, W., Oberhettinger, F., and Tricomi, F. G. 1955. *Higher Transcendental Functions*. Vol. 3. New York: McGraw-Hill.

Even, S. and Gillis, J. 1976. Derangements and Laguerre polynomials. *Math. Proc. Camb. Phil. Soc.*, **79**, 135–143.

Exton, H. 1983. *q-Hypergeometric Functions and Applications*. Ellis Horwood Series: Mathematics and Its Applications. Chichester: Ellis Horwood. With a foreword by L. J. Slater.

Favard, J. 1935. Sur les polynômes de Tchebicheff. *C. R. Acad. Sci. Paris*, **131**, 2052–2053.

Faybusovich, L. and Gekhtman, M. 1999. On Schur flows. *J. Phys. A*, **32**(25), 4671–4680.

Fejér, L. 1922. Über die Lage der Nullstellen von Polynomen, die aus Minimumforderungen gewisser Art entspringen. *Math. Ann.*, **85**, 41–48.

Feldheim, E. 1941. Sur les polynômes généralisés de Legendre. *Bull. Acad. Sci. URSS. Sér. Math. [Izvestia Akad. Nauk SSSR]*, **5**, 241–248.

Fields, J. L. and Ismail, M. E. H. 1975. Polynomial expansions. *Math. Comp.*, **29**, 894–902.

Fields, J. L. and Wimp, J. 1961. Expansions of hypergeometric functions in hypergeometric functions. *Math. Comp.*, **15**, 390–395.

Filaseta, M. and Lam, T.-Y. 2002. On the irreducibility of the generalized Laguerre polynomials. *Acta Arith.*, **105**(2), 177–182.

Foata, D. 1981. Some Hermite polynomial identities and their combinatorics. *Adv. in Appl. Math.*, **2**, 250–259.

Foata, D. and Strehl, V. 1981. Une extension multilinéaire de la formule d'Erdélyi pour les produits de fonctions hypergéométriques confluentes. *C. R. Acad. Sci. Paris Sér. I Math.*, **293**(10), 517–520.

Forrester, P. J. and Rogers, J. B. 1986. Electrostatics and the zeros of the classical orthogonal polynomials. *SIAM J. Math. Anal.*, **17**, 461–468.

Forsyth, A. R. 1918. *Theory of Functions of a Complex Variable*. Third edn. Vols. 1, 2. Cambridge: Cambridge University Press.

Foster, W. H. and Krasikov, I. 2002. Inequalities for real-root polynomials and entire functions. *Adv. in Appl. Math.*, **29**(1), 102–114.

Frenzen, C. L. and Wong, R. 1985. A uniform asymptotic expansion of the Jacobi polynomials with error bounds. *Canad. J. Math.*, **37**(5), 979–1007.

Frenzen, C. L. and Wong, R. 1988. Uniform asymptotic expansions of Laguerre polynomials. *SIAM J. Math. Anal.*, **19**(5), 1232–1248.

Freud, G. 1971. *Orthogonal Polynomials*. New York: Pergamon Press.

Freud, G. 1976. On the coefficients in the recursion formulae of orthogonal polynomials. *Proc. Roy. Irish Acad. Sect. A (1)*, **76**, 1–6.

Gabardo, J.-P. 1992. A maximum entropy approach to the classical moment problem. *J. Funct. Anal.*, **106**(1), 80–94.

Gangolli, R. and Varadarajan, V. S. 1988. *Harmonic Analysis of Spherical Functions on Real Reductive Groups*. Ergebnisse der Mathematik und ihrer Grenzgebiete [Results in Mathematics and Related Areas], vol. 101. Berlin: Springer.

Gantmacher, F. R. 1959. *The Theory of Matrices*. Vol. 1. Translated from the Russian by K. A. Hirsch. Reprint, Providence, RI: AMS Chelsea Publishing, 1998.

Garrett, K., Ismail, M. E. H., and Stanton, D. 1999. Variants of the Rogers–Ramanujan identities. *Adv. in Appl. Math.*, **23**, 274–299.

Gasper, G. 1970. Linearization of the product of Jacobi polynomials. II. *Canad. J. Math.*, **22**, 582–593.

Gasper, G. 1971. Positivity and the convolution structure for Jacobi series. *Ann. Math.*, **93**, 112–118.

Gasper, G. 1972. Banach algebras for Jacobi series and positivity of a kernel. *Ann. Math.*, **95**, 261–280.

Gasper, G. 1983. A convolution structure and positivity of a generalized translation operator for the continuous q-Jacobi polynomials. Pages 44–59 of *Conference on Harmonic Analysis in Honor of Antoni Zygmund, Vols. I, II (Chicago, Ill., 1981)*. Wadsworth Math. Ser. Belmont, CA: Wadsworth.

Gasper, G. and Rahman, M. 1983. Positivity of the Poisson kernel for the continuous q-ultraspherical polynomials. *SIAM J. Math. Anal.*, **14**(2), 409–420.

Gasper, G. and Rahman, M. 2004. *Basic Hypergeometric Series*. Second edn. Cambridge: Cambridge University Press.

Gatteschi, L. 1949a. Approssimazione asintotica degli zeri dei polinomi ultrasferici. *Univ. Roma. Ist. Naz. Alta Mat. Rend. Mat. e Appl. (5)*, **8**, 399–411.

Gatteschi, L. 1949b. Una formula asintotica per l'approssimazione degli zeri dei polinomi di Legendre. *Boll. Un. Mat. Ital. (3)*, **4**, 240–250.

Gatteschi, L. 1950. Sull'approssimazione asintotica degli zeri dei polinomi sferici ed ultrasferici. *Boll. Un. Mat. Ital. (3)*, **5**, 305–313.

Gatteschi, L. 1952. Limitazione dell'errore nella formula di Hilb e una nuova formula per la valutazione asintotica degli zeri dei polinomi di Legendre. *Boll. Un. Mat. Ital. (3)*, **7**, 272–281.

Gatteschi, L. 1967/1968. Una nuova rappresentazione asintotica dei polinomi di Jacobi. *Univ. e Politec. Torino Rend. Sem. Mat.*, **27**, 165–184.

Gatteschi, L. 1972. Sugli zeri dei polimoni ultrasferici. Pages 111–122 of *Studi in onore di Fernando Giaccardi*. Torino: Baccola & Gili.

Gatteschi, L. 1985. On the zeros of Jacobi polynomials and Bessel functions. *Rend. Sem. Mat. Univ. Politec. Torino*, 149–177. International conference on special functions: Theory and computation (Turin, 1984).

Gatteschi, L. 1987. New inequalities for the zeros of Jacobi polynomials. *SIAM J. Math. Anal.*, **18**, 1549–1562.

Gatteschi, L. 1988a. Some new inequalities for the zeros of Laguerre polynomials. Pages 23–38 of *Numerical Methods and Approximation Theory, III (Niš, 1987)*. Niš: Univ. Niš.

Gatteschi, L. 1988b. Uniform approximations for the zeros of Laguerre polynomials. Pages 137–148 of *Numerical Mathematics, Singapore 1988*. Internat. Schriftenreihe Numer. Math., vol. 86. Basel: Birkhäuser.

Gatteschi, L. 2002. Asymptotics and bounds for the zeros of Laguerre polynomials: A survey. *J. Comput. Appl. Math.*, **144**(1-2), 7–27.

Gatteschi, L. and Pittaluga, G. 1985. An asymptotic expansion for the zeros of Jacobi polynomials. Pages 70–86 of *Mathematical Analysis*. Teubner-Texte Math., vol. 79. Leipzig: Teubner.

Gautschi, W. 1967. Computational aspects of three-term recurrence relations. *SIAM Rev.*, **9**, 24–82.

Gautschi, W. 2009. New conjectured inequalities for zeros of Jacobi polynomials. *Numer. Algorithms*, **50**(3), 293–296.

Gautschi, W. and Giordano, C. 2008. Luigi Gatteschi's work on asymptotics of special functions and their zeros. *Numer. Algorithms*, **49**(1-4), 11–31.

Gegenbauer, L. 1874. Uber einige bestimmte Integrale, *Sitz. math. natur. Klasse Akad. Wiss. Wien*, **70**, 433–443.

Gegenbauer, L. 1893. Das Additionstheorem der Functionen $C_n^\nu(x)$. *Sitz. math. natur. Klasse Akad. Wiss. Wien*, **103**, 942–950.

Geronimo, J. S. 1992. Polynomials orthogonal on the unit circle with random recurrence coefficients. *Lecture Notes in Math.*, **1550**, 43–61.

Geronimo, J. S. and Van Assche, W. 1986. Orthogonal polynomials on several intervals via a polynomial mapping. *Trans. Amer. Math. Soc.*, **308**, 559–581.

Geronimo, J. S., Gesztesy, F., and Holden, H. 2005. Algebro-geometric solution of the Baxter–Szegő difference equation. *Comm. Math. Phys.*, **258**, 149–177.

Geronimus, Ya. L. 1941. On the character of the solution of the moment problem in the case of the periodic in the limit associated fraction. *Bull. Acad. Sci. USSR Math.*, **5**, 203–210.

Geronimus, Ya. L. 1946. On the trigonometric moment problem. *Ann. of Math. (2)*, **47**, 742–761.

Geronimus, Ya. L. 1961. *Orthogonal Polynomials: Estimates, Asymptotic Formulas, and Series of Polynomials Orthogonal on the Unit Circle and on an Interval.* Authorized translation from the Russian. New York: Consultants Bureau.

Geronimus, Ya. L. 1962. *Polynomials Orthogonal on a Circle and Their Applications.* Amer. Math. Soc. Transl., vol. 3. Providence, RI: American Mathematical Society.

Geronimus, Ya. L. 1977. *Orthogonal Polynomials.* Amer. Math. Soc. Transl., vol. 108. Providence, RI: American Mathematical Society. Pages 37–130.

Gessel, I. M. and Stanton, D. 1983. Applications of q-Lagrange inversion to basic hypergeometric series. *Trans. Amer. Math. Soc.*, **277**(1), 173–201.

Gessel, I. M. and Stanton, D. 1986. Another family of q-Lagrange inversion formulas. *Rocky Mountain J. Math.*, **16**(2), 373–384.

Gil, A., Segura, J., and Temme, N. M. 2007. *Numerical Methods for Special Functions.* Philadelphia, PA: Society for Industrial and Applied Mathematics (SIAM).

Gilewicz, J., Leopold, E., and Valent, G. 2005. New Nevanlinna matrices for orthogonal polynomials related to cubic birth and death processes. *J. Comp. Appl. Math.*, **178**, 235–245.

Gilewicz, J., Leopold, E., Ruffing, A., and Valent, G. 2006. Some cubic birth and death processes and their related orthogonal polynomials. *Constr. Approx.*, **24**(1), 71–89.

Gillis, J., Reznick, B., and Zeilberger, D. 1983. On elementary methods in positivity theory. *SIAM J. Math. Anal.*, **14**, 396–398.

Gishe, J. and Ismail, M. E. H. 2008. Resultants of Chebyshev polynomials. *Z. Anal. Anwend.*, **27**(4), 499–508.

Godoy, E. and Marcellán, F. 1991. An analog of the Christoffel formula for polynomial modification of a measure on the unit circle. *Boll. Un. Mat. Ital.*, **4**(7), 1–12.

Gohberg, I., Lancaster, P., and Rodman, L. 1982. *Matrix Polynomials.* Computer Science and Applied Mathematics. New York: Academic Press. [Harcourt Brace Jovanovich Publishers].

Goldberg, J. 1965. Polynomials orthogonal over a denumerable set. *Pacific J. Math.*, **15**, 1171–1186.

Golinskii, B. 1958. The analogue of Cristoffel formula for orthogonal polynomials on the unit circle and some applications. *Izv. Vuz. Mat.*, **1**(2), 33–42.

Golinskii, B. 1966. On certain estimates for Cristoffel kernels and moduli of orthogonal polynomials. *Izv. Vuz. Mat.*, **1**(50), 30–42.

Golinskii, B. 1967. On the rate of convergence of orthogonal polynomials sequence to a limit function. *Ukrain. Mat. Zh.*, **19**, 11–28.

Golinskii, B. and Golinskii, L. 1998. On uniform boundedness and uniform asymptotics for orthogonal polynomials on the unit circle. *J. Math. Anal. Appl.*, **220**, 528–534.

Golinskii, L. 2000. Operator theoretic approach to orthogonal polynomials on an arc of the unit circle. *Mat. Fiz., Analiz, Geometriya*, **7**(1), 3–34.

Golinskii, L. and Khrushchev, S. 2002. Cesàro asymptotics for orthogonal polynomials on the unit circle and classes of measures. *J. Approx. Theory*, **115**, 187–237.

Golinskii, L. and Zlatoš, A. 2007. Coefficients of orthogonal polynomials on the unit circle and higher-order Szegő theorems. *Constr. Approx.*, **26**(3), 361–382.

Golinskii, L. B. 2006. Schur flows and orthogonal polynomials on the unit circle. *Mat. Sb.*, **197**(8), 41–62.

Golinskii, L. B. and Nevai, P. 2001. Szegő difference equations, transfer matrices and orthogonal polynomials on the unit circle. *Comm. Math. Phys.*, **223**, 223–259.

Golinskii, L. B., Nevai, P., and Van Assche, W. 1995. Perturbation of orthogonal polynomials on an arc of the unit circle. *J. Approx. Theory*, **83**, 392–422.

Gómez, R. and López-García, M. 2007. A family of heat functions as solutions of indeterminate moment problems. *Int. J. Math. Math. Sci.*, Art. ID 41526, 11.

Gorska, K. 2016. Private communication.

Grenander, U. and Szegő, G. 1958. *Toeplitz Forms and Their Applications*. Berkeley, CA: University of California Press. Reprint, Bronx, NY: Chelsea, 1984.

Groenevelt, W. 2015. Orthogonality relations for Al-Salam-Carlitz polynomials of type II. *J. Approx. Theory*, **195**, 89–108.

Grosjean, C. C. 1987. A property of the zeros of the Legendre polynomials. *J. Approx. Theory*, **50**, 84–88.

Grosswald, E. 1978. *The Bessel Polynomials*. Lecture Notes in Mathematics, vol. 698. Berlin: Springer.

Grünbaum, F. A. 2003. Matrix valued Jacobi polynomials. *Bull. Sci. Math.*, **127**(3), 207–214.

Grünbaum, F. A. 2010. An urn model associated with Jacobi polynomials. *Commun. Appl. Math. Comput. Sci.*, **5**, 55–63.

Grünbaum, F. A. 2011. The Darboux process and a noncommutative bispectral problem: Some explorations and challenges. Pages 161–177 of Kolk, J. A. C. and van den Ban, E. P. (eds), *Geometric Aspects of Analysis and Mechanics*. Progress in Mathematics, vol. 292. New York, NY: Birkhäuser/Springer. In Honor of the 65th Birthday of Hans Duistermaat.

Grünbaum, F. A. and Tirao, J. 2007. The algebra of differential operators associated to a weight matrix. *Integral Equations Operator Theory*, **58**(4), 449–475.

Grünbaum, F. A., Pacharoni, I., and Tirao, J. 2001. A matrix-valued solution to Bochner's problem. *J. Phys. A*, **34**(48), 10647–10656. Symmetries and integrability of difference equations (Tokyo, 2000).

Grünbaum, F. A., Pacharoni, I., and Tirao, J. 2002. Matrix valued spherical functions associated to the complex projective plane. *J. Funct. Anal.*, **188**(2), 350–441.

Grünbaum, F. A., Pacharoni, I., and Tirao, J. 2003. Matrix valued orthogonal polynomials of the Jacobi type. *Indag. Math. (N.S.)*, **14**(3-4), 353–366.

Grünbaum, F. A., Pacharoni, I., and Tirao, J. 2004. An invitation to matrix-valued spherical functions: Linearization of products in the case of complex projective space $P_2(\mathbb{C})$. Pages 147–160 of *Modern Signal Processing*. Math. Sci. Res. Inst. Publ., vol. 46. Cambridge: Cambridge University Press.

Grünbaum, F. A., Pacharoni, I., and Tirao, J. 2005. Matrix valued orthogonal polynomials of Jacobi type: The role of group representation theory. *Ann. Inst. Fourier (Grenoble)*, **55**(6), 2051–2068.

Grünbaum, F. A., Pacharoni, I., and Tirao, J. 2011. Two stochastic models of a random walk in the $U(n)$-spherical duals of $U(n+1)$. *Ann. Mat. Pura Appl.*, 1–27.

Gupta, D. P. and Muldoon, M. E. 2007. Inequalities for the smallest zeros of Laguerre polynomials and their q-analogues. *JIPAM. J. Inequal. Pure Appl. Math.*, **8**(1), Article 24, 7 pp. (electronic).

Habsieger, L. 2001a. Integer zeros of q-Krawtchouk polynomials in classical combinatorics. *Adv. in Appl. Math.*, **27**(2-3), 427–437. Special issue in honor of Dominique Foata's 65th birthday (Philadelphia, PA, 2000).

Habsieger, L. 2001b. Integral zeroes of Krawtchouk polynomials. Pages 151–165 of *Codes and Association Schemes (Piscataway, NJ, 1999)*. DIMACS Ser. Discrete Math. Theoret. Comput. Sci., vol. 56. Providence, RI: American Mathematical Society.

Habsieger, L. and Stanton, D. 1993. More zeros of Krawtchouk polynomials. *Graphs Combin.*, **9**(2), 163–172.

Hayman, W. K. 2005. On the zeros of a q-Bessel function. Pages 205–216 of *Complex Analysis and Dynamical Systems II*. Contemp. Math., vol. 382. Providence, RI: American Mathematical Society.

Hayman, W. K. and Ortiz, E. L. 1975/76. An upper bound for the largest zero of Hermite's function with applications to subharmonic functions. *Proc. Roy. Soc. Edinburgh Sect. A*, **75**.

Heine, E. 1961. *Handbuch der Kugelfunctionen. Theorie und Anwendungen. Band I, II.* Zweite umgearbeitete und vermehrte Auflage. Thesaurus Mathematicae, No. 1. Würzburg, 1961: Physica.

Helgason, S. 1978. *Differential Geometry, Lie Groups, and Symmetric Spaces*. Corrected reprint, Graduate Studies in Mathematics, vol. 34. Providence, RI: American Mathematical Society, 2001.

Hille, E. 1993. Über die Nullstellen der Hermiteschen Polynome. *Jahresber. Deutsch Math.-Verein.*, **44**, 162–165.

Hong, Y. 1986. On the nonexistence of nontrivial perfect e-codes and tight $2e$-designs in Hamming schemes $H(n, q)$ with $e \geq 3$ and $q \geq 3$. *Graphs Combin.*, **2**(2), 145–164.

Horn, R. A. and Johnson, C. R. 1992. *Matrix Analysis*. Cambridge: Cambridge University Press.

Ismail, M. E. H. 1981. The basic Bessel functions and polynomials. *SIAM J. Math. Anal.*, **12**, 454–468.

Ismail, M. E. H. 1982. The zeros of basic Bessel functions, the functions $J_{\nu+ax}(x)$, and associated orthogonal polynomials. *J. Math. Anal. Appl.*, **86**(1), 1–19.

Ismail, M. E. H. 1985a. On sieved orthogonal polynomials I: Symmetric Pollaczek polynomials. *SIAM J. Math. Anal.*, **16**, 1093–1113.

Ismail, M. E. H. 1985b. A queueing model and a set of orthogonal polynomials. *J. Math. Anal. Appl.*, **108**, 575–594.

Ismail, M. E. H. 1986a. Asymptotics of the Askey–Wilson polynomials and q-Jacobi polynomials. *SIAM J. Math. Anal.*, **17**, 1475–1482.

Ismail, M. E. H. 1986b. On sieved orthogonal polynomials II: Orthogonality on several intervals. *Trans. Amer. Math. Soc.*, **294**, 89–111.

Ismail, M. E. H. 1987. The variation of zeros of certain orthogonal polynomials. *Adv. in Appl. Math.*, **8**(1), 111–118.

Ismail, M. E. H. 1993. Ladder operators for q^{-1}-Hermite polynomials. *Math. Rep. Royal Soc. Canada*, **15**, 261–266.

Ismail, M. E. H. 1995. The Askey–Wilson operator and summation theorems. Pages 171–178 of Ismail, M., Nashed, M. Z., Zayed, A., and Ghaleb, A. (eds), *Mathematical Analysis, Wavelets and Signal Processing*. Contemporary Mathematics, vol. 190. Providence, RI: American Mathematical Society.

Ismail, M. E. H. 1998. Discriminants and functions of the second kind of orthogonal polynomials. *Results Math.*, **34**, 132–149.

Ismail, M. E. H. 2000a. An electrostatic model for zeros of general orthogonal polynomials. *Pacific J. Math.*, **193**, 355–369.

Ismail, M. E. H. 2000b. More on electronic models for zeros of orthogonal polynomials. *Numer. Funct. Anal. and Optimiz.*, **21**(1-2), 191–204.

Ismail, M. E. H. 2003. Difference equations and quantized discriminants for q-orthogonal polynomials. *Adv. in Appl. Math.*, **30**(3), 562–589.

Ismail, M. E. H. 2005a. Asymptotics of q-orthogonal polynomials and a q-Airy function. *Internat. Math. Res. Notices*, **2005**(18), 1063–1088.

Ismail, M. E. H. 2005b. *Classical and Quantum Orthogonal Polynomials in One Variable*. Cambridge: Cambridge University Press.

Ismail, M. E. H. 2005c. Determinants with orthogonal polynomial entries. *J. Comp. Appl. Anal.*, **178**, 255–266.

Ismail, M. E. H. and Jing, N. 2001. q-discriminants and vertex operators. *Adv. in Appl. Math.*, **27**, 482–492.

Ismail, M. E. H. and Kelker, D. 1976. The Bessel polynomial and the student t-distribution. *SIAM J. Math. Anal.*, **7**(1), 82–91.

Ismail, M. E. H. and Li, X. 1992a. Bounds for extreme zeros of orthogonal polynomials. *Proc. Amer. Math. Soc.*, **115**, 131–140.

Ismail, M. E. H. and Li, X. 1992b. On sieved orthogonal polynomials. IX. Orthogonality on the unit circle. *Pacific J. Math.*, **153**, 289–297.

Ismail, M. E. H. and Masson, D. R. 1991. Two families of orthogonal polynomials related to Jacobi polynomials. *Rocky Mountain J. Math.*, **21**(1), 359–375.

Ismail, M. E. H. and Masson, D. R. 1994. q-Hermite polynomials, biorthogonal rational functions, and q-beta integrals. *Trans. Amer. Math. Soc.*, **346**, 63–116.

Ismail, M. E. H. and Muldoon, M. 1991. A discrete approach to monotonicity of zeros of orthogonal polynomials. *Trans. Amer. Math. Soc.*, **323**, 65–78.

Ismail, M. E. H. and Muldoon, M. E. 1995. Bounds for the small real and purely imaginary zeros of Bessel and related functions. *Methods Appl. Anal.*, **2**(1), 1–21.

Ismail, M. E. H. and Mulla, F. S. 1987. On the generalized Chebyshev polynomials. *SIAM J. Math. Anal.*, **18**(1), 243–258.

Ismail, M. E. H. and Rahman, M. 1991. Associated Askey–Wilson polynomials. *Trans. Amer. Math. Soc.*, **328**, 201–237.

Ismail, M. E. H. and Rahman, M. 1998. The q-Laguerre polynomials and related moment problems. *J. Math. Anal. Appl.*, **218**(1), 155–174.

Ismail, M. E. H. and Ruedemann, R. 1992. Relation between polynomials orthogonal on the unit circle with respect to different weights. *J. Approximation Theory*, **71**, 39–60.

Ismail, M. E. H. and Simeonov, P. 1998. Strong asymptotics for Krawtchouk polynomials. *J. Comput. Appl. Math.*, **100**, 121–144.

Ismail, M. E. H. and Simeonov, P. 2015. Complex Hermite polynomials: Their combinatorics and integral operators. *Proc. Amer. Math. Soc.*, **143**, 1397–1410.

Ismail, M. E. H. and Stanton, D. 1988. On the Askey–Wilson and Rogers polynomials. *Canad. J. Math.*, **40**, 1025–1045.

Ismail, M. E. H. and Stanton, D. 1997. Classical orthogonal polynomials as moments. *Canad. J. Math.*, **49**, 520–542.

Ismail, M. E. H. and Stanton, D. 2002. q-Integral and moment representations for q-orthogonal polynomials. *Canad. J. Math.*, **54**, 709–735.

Ismail, M. E. H. and Stanton, D. 2003a. Applications of q-Taylor theorems. *J. Comp. Appl. Math.*, **153**(1-2), 259–272.

Ismail, M. E. H. and Stanton, D. 2003b. q-Taylor theorems, polynomial expansions and interpolation of entire functions. *J. Approx. Theory*, **123**, 125–146.

Ismail, M. E. H. and Stanton, D. 2003c. Tribasic integrals and identities of Rogers–Ramanujan type. *Trans. Amer. Math. Soc.*, **355**, 4061–4091.

Ismail, M. E. H. and Stanton, D. 2006. Ramanujan's continued fractions via orthogonal polynomials. *Adv. Math.*, **203**, 170–193.

Ismail, M. E. H. and Tamhankar, M. V. 1979. A combinatorial approach to some positivity problems. *SIAM J. Math. Anal.*, **10**, 478–485.

Ismail, M. E. H. and Valent, G. 1998. On a family of orthogonal polynomials related to elliptic functions. *Illinois J. Math.*, **42**(2), 294–312.

Ismail, M. E. H. and Wilson, J. 1982. Asymptotic and generating relations for the q-Jacobi and the ${}_4\phi_3$ polynomials. *J. Approx. Theory*, **36**, 43–54.

Ismail, M. E. H. and Wimp, J. 1998. On differential equations for orthogonal polynomials. *Methods Appl. Anal.*, **5**, 439–452.

Ismail, M. E. H. and Witte, N. 2001. Discriminants and functional equations for polynomials orthogonal on the unit circle. *J. Approx. Theory*, **110**, 200–228.

Ismail, M. E. H. and Zhang, C. 2007. Zeros of entire functions and a problem of Ramanujan. *Adv. Math.*, **209**, 363–380.

Ismail, M. E. H. and Zhang, R. 1988. On the Hellmann–Feynman theorem and the variation of zeros of special functions. *Adv. in Appl. Math.*, **9**, 439–446.

Ismail, M. E. H. and Zhang, R. 1994. Diagonalization of certain integral operators. *Adv. Math.*, **109**, 1–33.

Ismail, M. E. H. and Zhang, R. 2005. New proofs of some q-series results. Pages 285–299 of Ismail, M. E. H. and Koelink, E. H. (eds), *Theory and Applications of Special Functions: A Volume Dedicated to Mizan Rahman*. Developments in Mathematics, vol. 13. New York: Springer.

Ismail, M. E. H., Stanton, D., and Viennot, G. 1987. The combinatorics of the q-Hermite polynomials and the Askey–Wilson integral. *European J. Combinatorics*, **8**, 379–392.

Ismail, M. E. H., Letessier, J., and Valent, G. 1988. Linear birth and death models and associated Laguerre and Meixner polynomials. *J. Approx. Theory*, **55**, 337–348.

Ismail, M. E. H., Valent, G., and Yoon, G. J. 2001. Some orthogonal polynomials related to elliptic functions. *J. Approx. Theory*, **112**(2), 251–278.

Ismail, M. E. H., Nikolova, I., and Simeonov, P. 2004. Difference equations and discriminants for discrete orthogonal polynomials. *Ramanujan J.*, **8**, 475–502.

Ismail, M. E. H. and Zeng, J. 2010. Addition theorems via continued fractions. *Trans. Amer. Math. Soc.*, **362**(2), 957–983.

Jackson, F. H. 1903. On generalized functions of Legendre and Bessel. *Trans. Royal Soc. Edinburgh*, **41**, 1–28.

Jackson, F. H. 1903–1904. The application of basic numbers to Bessel's and Legendre's functions. *Proc. London Math. Soc. (2)*, **2**, 192–220.

Jackson, F. H. 1904–1905. The application of basic numbers to Bessel's and Legendre's functions, II. *Proc. London Math. Soc. (2)*, **3**, 1–20.

Jensen, J. L. W. V. 1913. Recherches sur la théorie des équations. *Acta Math.*, **36**(1), 181–195.

Jones, W. B. and Thron, W. 1980. *Continued Fractions: Analytic Theory and Applications*. Reading, MA: Addison-Wesley.

Jordan, C. 1965. *Calculus of Finite Differences*. New York: Chelsea.

Kadell, K. W. J. 2005. The little q-Jacobi functions of complex order. Pages 301–338 of Ismail, M. E. H. and Koelink, E. H. (eds), *Theory and Applications of Special Functions: A Volume Dedicated to Mizan Rahman*. Developments in Mathematics, vol. 13. New York: Springer.

Kalnins, E. G. and Miller, W. 1988. q-series and orthogonal polynomials associated with Barnes' first lemma. *SIAM J. Math. Anal.*, **19**, 1216–1231.

Kaplansky, I. 1944. Symbolic solution of certain problems in permutations. *Bull. Amer. Math. Soc.*, **50**, 906–914.

Karlin, S. and McGregor, J. 1957a. The classification of birth and death processes. *Trans. Amer. Math. Soc.*, **86**, 366–400.

Karlin, S. and McGregor, J. 1957b. The differential equations of birth and death processes and the Stieltjes moment problem. *Trans. Amer. Math. Soc.*, **85**, 489–546.

Karlin, S. and McGregor, J. 1958. Many server queuing processes with Poisson input and exponential service time. *Pacific J. Math.*, **8**, 87–118.

Karlin, S. and McGregor, J. 1959. Random walks. *Illinois J. Math.*, **3**, 66–81.

Karlin, S. and Szegő, G. 1960/1961. On certain determinants whose elements are orthogonal polynomials. *J. Anal. Math.*, **8**, 1–157.

Karp, D. 2001. Holomorphic spaces related to orthogonal polynomials and analytic continuation of functions. Pages 169–187 of Saitoh, S., Hayashi, N., and Yamamoto, M. (eds), *Analytic Extension Formulas and Their Applications (Fukuoka, 1999/Kyoto, 2000)*. Int. Soc. Anal. Appl. Comput., vol. 9. Dordrecht, The Netherlands: Kluwer Academic Publisher.

Khruschev, S. 2001. Schur's algorithm, orthogonal polynomials, and convergence of Wall's continued fractions in $L^2(\mathbb{T})$. *J. Approx. Theory*, **108**, 161–248.

Khruschev, S. 2002. Classification theorems for general orthogonal polynomials on the unit circle. *J. Approx. Theory*, **116**, 268–342.

Khruschev, S. 2003. Turán measures. *J. Approx. Theory*, **122**, 112–120.

Kibble, W. F. 1945. An extension of theorem of Mehler on Hermite polynomials. *Proc. Cambridge Philos. Soc.*, **41**, 12–15.

Killip, R. and Nenciu, I. 2005. CMV: The unitary analogue of Jacobi matrices. Preprint arXiv:math.SG/0508113.

Koekoek, R. and Swarttouw, R. 1998. *The Askey-Scheme of Hypergeometric Orthogonal Polynomials and Its q-Analogues*. Reports of the Faculty of Technical Mathematics and Informatics 98-17. Delft University of Technology, Delft.

Koekoek, R., Lesky, P. A., Swarttouw, R. F. 2010. *Hypergeometric Orthogonal Polynomials and their q-Analogues*. Springer Monographs in Mathematics, Springer-Verlag: Berlin Heidelberg.

Koelink, E. 1997. Addition formulas for q-special functions. Pages 109–129 of Ismail, M. E. H., Masson, D. R., and Rahman, M. (eds), *Special Functions, q-Series and Related Topics (Toronto, ON, 1995)*. Fields Inst. Commun., vol. 14. Providence, RI: American Mathematical Society.

Koelink, H. T. 1996. On Jacobi and continuous Hahn polynomials. *Proc. Amer. Math. Soc.*, **124**(3), 887–898.

Koelink, H. T. and Van der Jeugt, J. 1998. Convolutions for orthogonal polynomials from Lie and quantum algebra representations. *SIAM J. Math. Anal.*, **29**(3), 794–822.

Koornwinder, T. H. 1978. Positivity proofs for linearization and connection coefficients for orthogonal polynomials satisfying an addition formula. *J. London Math. Soc.*, **18**(2), 101–114.

Koornwinder, T. H. 1981. Clebsch–Gordan coefficients for SU(2) and Hahn polynomials. *Nieuw Arch. Wisk. (3)*, **29**(2), 140–155.

Koornwinder, T. H. 1982. Krawtchouk polynomials, a unification of two different group theoretic interpretations. *SIAM J. Math. Anal.*, **13**(6), 1011–1023.

Koornwinder, T. H. 1984a. Jacobi functions and analysis on noncompact semisimple Lie groups. Pages 1–85 of Askey, R. A., Koornwinder, T. H., and Schempp, W. (eds), *Special Functions: Group Theoretical Aspects and Applications*. Dordrecht: Reidel.

Koornwinder, T. H. 1984b. Orthogonal polynomials with weight function $(1-x)^\alpha(1+x)^\beta + M\delta(x+1) + N\delta(x-1)$. *Canad. Math Bull.*, **27**, 205–214.

Koornwinder, T. H. 1990. Orthogonal polynomials in connection with quantum groups. Pages 257–292 of Nevai, P. (ed), *Orthogonal Polynomials; Theory and Practice*. Nato ASI Series C: Mathematical and Physcal Science, vol. 294. Dordrecht: Kluwer Academic Publishers.

Koornwinder, T. H. 1993. Askey–Wilson polynomials as zonal spherical functions on the SU(2) quantum group. *SIAM J. Math. Anal.*, **24**(3), 795–813.

Koornwinder, T. H. 2004. *On q^{-1}-Al-Salam–Chihara Polynomials*. Informal note.

Koornwinder, T. H. 2005. A second addition formula for continuous q-ultraspherical polynomials. Pages 339–360 of Ismail, M. E. H. and Koelink, E. (eds), *Theory and Applications of Special Functions*. Developments in Mathematics, vol. 13. New York: Springer.

Koornwinder, T. H. 2006. Lowering and raising operators for some special orthogonal polynomials. Pages 227–238 of *Jack, Hall–Littlewood and Macdonald Polynomials*. Contemp. Math., vol. 417. Providence, RI: American Mathematical Society.

Krall, H. L. and Frink, O. 1949. A new class of orthogonal polynomials. *Trans. Amer. Math. Soc.*, **65**, 100–115.

Krasikov, I. 2003. Bounds for zeros of the Laguerre polynomials. *J. Approx. Theory*, **121**, 287–291.

Krasikov, I. 2007. Inequalities for orthonormal Laguerre polynomials. *J. Approx. Theory*, **144**(1), 1–26.

Krasikov, I. and Litsyn, S. 1996. On integral zeros of Krawtchouk polynomials. *J. Combin. Theory Ser. A*, **74**(1), 71–99.

Krein, M. 1949a. Infinite J-matrices and a matrix-moment problem. *Doklady Akad. Nauk SSSR (N.S.)*, **69**, 125–128.

Krein, M. G. 1949b. Fundamental aspects of the representation theory of Hermitian operators with deficiency index (m,m). *Ukrain. Math. Zh.*, **1**, 3–66. Amer. Math. Soc. Transl. (2) 97 (1970), 75–143.

Krein, M. G. and Nudel'man, A. A. 1977. *The Markov Moment Problem and Extremal Problems*. Translations of Mathematical Monographs, vol. 50. Providence, RI: American Mathematical Society.

Kuznetsov, A. 2016. *Constructing Measures with Identical Moments*. arXiv:1607.08003.

Kuznetsov, A. 2017. Solving the mystery integral. In Li, X. and Nashed, M. Z. (eds), *Frontiers in Orthogonal Polynomials and q-Series*. World Scientific.

Kwon, K. H. 2002. *Orthogonal Polynomials I*. Lecture Notes. KAIST, Seoul.

Labelle, J. and Yeh, Y. N. 1989. The combinatorics of Laguerre, Charlier, and Hermite polynomials. *Stud. Appl. Math.*, **80**(1), 25–36.

Laforgia, A. 1981. A monotonic property for the zeros of ultraspherical polynomials. *Proc. Amer. Math. Soc.*, **83**(4), 757–758.

Laforgia, A. and Muldoon, M. E. 1986. Some consequences of the Sturm comparison theorem. *Amer. Math. Monthly*, **93**, 89–94.

Landau, H. J. 1987. Maximum entropy and the moment problem. *Bull. Amer. Math. Soc. (N.S.)*, **16**(1), 47–77.

Lanzewizky, I. L. 1941. Über die orthogonalität der Fejer–Szegöschen polynome. *C. R. Dokl. Acad. Sci. URSS (N.S.)*, **31**, 199–200.

Laplace, P. S. 1782. Théorie des attractions des sphéroides et de la figure des planètes. *Mémoires de Mathématique et de Physique tirés des registres de l'Académie royale des Sciences, Paris*, 113–196.

Legendre, A. M. 1785. Recherches sur l'attraction des sphéorides homogènes. *Mémoires de Mathématique et de Physique presentés à l'Académie royale des Sciences par divers savans, Paris*, **10**, 411–434.

Legendre, A. M. 1789. Suite des recherches sur la figure des planètes. *Mémoires de Mathématique et de Physique tirés des registres de l'Académie royale des Sciences, Paris*, 372–454.

Leipnik, R. 1981. The lognormal distribution and strong nonuniqueness of the moment problem. *Teor. Veroyatnost. i Primenen.*, **26**(4), 863–865.

Lew, J. S. and Quarles, D. A., Jr. 1983. Nonnegative solutions of a nonlinear recurrence. *J. Approx. Theory*, **38**(4), 357–379.

Li, X. and Wong, R. 2000. A uniform asymptotic expansion for Krawtchouk polynomials. *J. Approx. Theory*, **106**(1), 155–184.

Lorch, L. 1977. Elementary comparison techniques for certain classes of Sturm–Liouville equations. Pages 125–133 of Berg, G., Essén, M., and Pleijel, A. (eds), *Differential Equations (Proc. Conf. Uppsala, 1977)*. Stockholm: Almqvist and Wiksell.

Lorentzen, L. and Waadeland, H. 1992. *Continued Fractions with Applications*. Amsterdam: North-Holland.

Louck, J. D. 1981. Extension of the Kibble–Slepian formula for Hermite polynomials using Boson operator methods. *Adv. in Appl. Math.*, **2**, 239–249.

Lubinsky, D. S. 1994. Zeros of orthogonal and biorthogonal polynomials: Some old, some new. Pages 3–15 of *Nonlinear Numerical Methods and Rational Approximation, II (Wilrijk, 1993)*. Math. Appl., vol. 296. Dordrecht: Kluwer Academic Publisher.

Makai, E. 1952. On monotonicity property of certain Sturm–Liouville functions. *Acta Math. Acad. Sci. Hungar.*, **3**, 15–25.

Marcellán, F. and Maroni, P. 1992. Sur l'adjonction d'une masse de Dirac á une forme régulière et semi-classique. *Ann. Mat. Pura Appl.*, **162**, 1–22.

Marden, M. 1966. *Geometry of Polynomials*. Second edn. Mathematical Surveys, No. 3. Providence, RI: American Mathematical Society.

Maroni, P. 1987. Prolégomènes à l'étude des polynômes orthogonaux semi-classiques. *Ann. Mat. Pura Appl. (4)*, **149**, 165–184.

Máté, A. and Nevai, P. G. 1982. Remarks on E. A. Rakhmanov's paper: "The asymptotic behavior of the ratio of orthogonal polynomials" [Mat. Sb. (N.S.) **103**(**145**) (1977), no. 2, 237–252; MR **56** #3556]. *J. Approx. Theory*, **36**(1), 64–72.

Máté, A., Nevai, P., and Totik, V. 1985. Asymptotics for the ratio of leading coefficients of orthonormal polynomials on the unit circle. *Constr. Approx.*, **1**, 63–69.

Máté, A., Nevai, P., and Totik, V. 1987a. Extensions of Szegő's theory of orthogonal polynomials, II. *Constr. Approx.*, **3**(1), 51–72.

Máté, A., Nevai, P., and Totik, V. 1987b. Strong and weak convergence of orthogonal polynomials. *Amer. J. Math.*, **109**, 239–281.

Máté, A., Nevai, P., and Totik, V. 1991. Szegő's extremum problem on the unit circle. *Ann. Math.*, **134**, 433–453.

Mazel, D. S., Geronimo, J. S., and Hayes, M. H. 1990. On the geometric sequences of reflection coefficients. *IEEE Trans. Acoust. Speech Signal Process.*, **38**, 1810–1812.

Mehta, M. L. 2004. *Random Matrices*. Third edn. San Diego: Elsevier.

Meixner, J. 1934. Orthogonale Polynomsysteme mit einer besonderen Gestalt der erzeugenden Funktion. *J. London Math. Soc.*, **9**, 6–13.

Meixner, J. 1942. Unformung gewisser Reihen. deren Gleider Produkte hypergeometischer Funktionen sind. *Deutsch. Math.*, **6**, 341–349.

Mhaskar, H. 1990. Bounds for certain Freud polynomials. *J. Approx. Theory*, **63**, 238–254.

Mhaskar, H. and Saff, E. 1990. On the distribution of zeros of polynomials orthogonal on the unit circle. *J. Approx. Theory*, **63**, 30–38.

Mikaelyan, L. 1978. The analogue of Cristoffel formula for orthogonal polynomials on the unit circle. *Dokl. Akad. Nauk Arm. SSR*, **67**(5), 257–263.
Milne-Thomson, L. M. 1933. *The Calculus of Finite Differences*. New York: Macmillan.
Moak, D. 1981. The q-analogue of the Laguerre polynomials. *J. Math. Anal. Appl.*, **81**(1), 20–47.
Mukaihira, A. and Nakamura, Y. 2000. Integrable discretization of the modified KdV equation and applications. *Inverse Problems*, **16**, 413–424.
Mukaihira, A. and Nakamura, Y. 2002. Schur flow for orthogonal polynomials on the unit circle and its integrable discretization. *J. Comput. Appl. Math.*, **139**, 75–94.
Muldoon, M. E. 1993. Properties of zeros of orthogonal polynomials and related functions. *J. Comput. Appl. Math.*, **48**(1-2), 167–186.
Muldoon, M. E. 2008. Continuous ranking of zeros of special functions. *J. Math. Anal. Appl.*, **343**, 436–445.
Naimark, M. A. 1947. Extremal spectral functions of a symmetric operator. *Izv. Akad. Nauk SSSR ser matem*, **11**, 327–344.
Nenciu, I. 2005. Lax pairs for the Ablowitz–Ladik system via orthogonal polynomials on the unit circle. *IMRN*, **11**, 647–686.
Nevai, P. 1983. Orthogonal polynomials associated with $\exp(-x^4)$. *Canadian Math. Soc. Conference Proceedings*, 263–285.
Nevai, P. 1991. Weakly convergent sequences of functions and orthogonal polynomials. *J. Approx. Theory*, **65**, 322–340.
Nevai, P. and Totik, V. 1989. Orthogonal polynomials and their zeros. *Acta Sci. Math. (Szeged)*, **53**, 99–104.
Nikolov, G. and Uluchev, R. 2004. Inequalities for real-root polynomials. Proof of a conjecture of Foster and Krasikov. Pages 201–216 of *Approximation Theory: A Volume Dedicated to Borislav Bojanov*. Prof. M. Drinov Academic Publ. House, Sofia.
Olver, F. W. J., Lozier, D. W., Boisvert, R. F., and Clark, C. W. (eds). 2010. *NIST Handbook of Mathematical Functions*. Cambridge: Cambridge University Press.
Ortiz, E. L. and Rivlin, T. J. 1983. Another look at the Chebyshev polynomials. *Amer. Math. Monthly*, **90**(1), 3–10.
Pacharoni, I. 2009. Matrix spherical functions and orthogonal polynomials: An instructive example. *Rev. Un. Mat. Argentina*, **50**(1), 1–15.
Pacharoni, I. and Román, P. 2008. A sequence of matrix valued orthogonal polynomials associated to spherical functions. *Constr. Approx.*, **28**(2), 127–147.
Pacharoni, I. and Tirao, J. 2007a. Matrix valued orthogonal polynomials arising from the complex projective space. *Constr. Approx.*, **25**(2), 177–192.
Pacharoni, I. and Tirao, J. 2007b. Three term recursion relation for spherical functions associated to the complex hyperbolic plane. *J. Lie Theory*, **17**(4), 791–828.
Pacharoni, I. and Tirao, J. 2013. One step spherical functions of the pair $(\mathrm{SU}(n+1), \mathrm{U}(n))$. In Huckleberry, A., Penkov, I., and Zuckerman, G. (eds), *Lie Groups: Structures, Actions and Representations*. Progress in Mathematics, vol. 306. Basel: Birkhäuser. Also available at http://arxiv.org/abs/1209.4500.
Pacharoni, I., Tirao, J., and Zurrián, I. 2014. Spherical functions associated to the 3-dimensional sphere. *Ann. Mat. Pura Appl.*, **193**, no. 6, 1727–1778.
Parlett, B. N. 1980. *The Symmetric Eigenvalue Problem*. Corrected reprint, Classics in Applied Mathematics, vol. 20. Philadelphia, PA: Society for Industrial and Applied Mathematics (SIAM), 1998.
Pedersen, H. L. 1995. Stieltjes moment problems and the Friedrichs extension of a positive definite operator. *J. Approx. Theory*, **83**(3), 289–307.

Pedersen, H. L. 1997. La paramétrisation de Nevanlinna et le problème des moments de Stieltjes indéterminé. *Exposition. Math.*, **15**(3), 273–278.

Pedersen, H. L. 1998. On Krein's theorem for indeterminacy of the classical moment problem. *J. Approx. Theory*, **95**(1), 90–100.

Pedersen, H. L. 2009. Logarithmic order and type of indeterminate moment problems. II. *J. Comput. Appl. Math.*, **233**(3), 808–814.

Peherstorfer, F. and Steinbauer, R. 1995. Characterization of general orthogonal polynomials with respect to a functional. *J. Comp. Appl. Math.*, **65**, 339–355.

Peherstorfer, F. and Steinbauer, R. 1999. Mass-points of orthogonality measures on the unit circle. *East J. Approx.*, **5**, 279–308.

Periwal, V. and Shevitz, D. 1990. Unitary-matrix models as exactly solvable string theories. *Phys. Rev. Lett.*, **64**, 1326–1329.

Pollaczek, F. 1949. Sur une généralisation des polynômes de Legendre. *C. R. Acad. Sci. Paris*, **228**, 1363–1365.

Pollaczek, F. 1956. *Sur une Généralisation des Polynômes de Jacobi*. Memorial des Sciences Mathematique, vol. 131. Paris: Gauthier-Villars.

Pólya, G. and Szegő, G. 1976. *Problems and Theorems in Analysis. II*. Reprint, Classics in Mathematics. Berlin: Springer. Theory of functions, zeros, polynomials, determinants, number theory, geometry. Translated from the German by C. E. Billigheimer, 1998.

Qiu, W.-Y. and Wong, R. 2004. Asymptotic expansion of the Krawtchouk polynomials and their zeros. *Comp. Meth. Func. Theory*, **4**(1), 189–226.

Rahman, M. 1981. The linearization of the product of continuous q-Jacobi polynomials. *Canad. J. Math.*, **33**(4), 961–987.

Rahman, M. 1984. A simple evaluation of Askey and Wilson's q-beta integral. *Proc. Amer. Math. Soc.*, **92**(3), 413–417.

Rahman, M. 1988. A generalization of Gasper's kernel for Hahn polynomials. *Canad. J. Math.*, **30**(133-146), 373–381.

Rahman, M. and Tariq, Q. 1997. Poisson kernels for associated q-ultrasperical polynomials. *Methods and Applications of Analysis*, **4**, 77–90.

Rahman, M. and Verma, A. 1986a. Product and addition formulas for the continuous q-ultraspherical polynomials. *SIAM J. Math. Anal.*, **17**(6), 1461–1474.

Rahman, M. and Verma, A. 1986b. A q-integral representation of Rogers' q-ultraspherical polynomials and some applications. *Constructive Approximation*, **2**, 1–10.

Rainville, E. D. 1960. *Special Functions*. New York: Macmillan.

Rakhmanov, E. A. 1977. On the asymptotics of the ratio of orthogonal polynomials. *Mat. Sb.*, **103**, 237–252. English translation in *Math. USSR Sb.* **32** (1977), 199–213.

Rakhmanov, E. A. 1980. Steklov's conjecture in the theory of orthogonal polynomials. *Math. USSR Sb.*, **36**, 549–575.

Rakhmanov, E. A. 1982a. Estimates of the growth of orthogonal polynomials whose weight is bounded away from zero. *Math. USSR Sb.*, **42**, 237–263.

Rakhmanov, E. A. 1982b. On the asymptotics of the ratio of orthogonal polynomials. II. *Mat. Sb.*, **118**, 104–117. English translation in *Math. USSR Sb.* **47** (1983), 105–117.

Riesz, M. 1923. Sur le problème des moments et le théorème de Parseval correspondant. *Acta Litt. Ac. Sci. Szeged*, **1**, 209–225.

Rogers, L. J. 1894. Second memoir on the expansion of certain infinite products. *Proc. London Math. Soc.*, **25**, 318–343.

Rogers, L. J. 1895. Third memoir on the expansion of certain infinite products. *Proc. London Math. Soc.*, **26**, 15–32.

Román, P. and Tirao, J. 2006. Spherical functions, the complex hyperbolic plane and the hypergeometric operator. *Internat. J. Math.*, **17**(10), 1151–1173.

Román, P. and Tirao, J. 2012. The spherical transform of any K-type in a locally compact group. *J. Lie Theory*, **22**, 361–395.

Roman, S. and Rota, G.-C. 1978. The umbral calculus. *Adv. Math.*, **27**, 95–188.

Romanov, R. 2017. Order problem for canonical systems and a conjecture of Valent. *Trans. Amer. Math. Soc.*, **369** (2017), no. 2, 1061–1078.

Routh, E. 1884. On some properties of certain solutions of a differential equation of the second order. *Proc. London Math. Soc.*, **16**, 245–261.

Rui, B. and Wong, R. 1994. Uniform asymptotic expansion of Charlier polynomials. *Methods Appl. Anal.*, **1**(3), 294–313.

Rui, B. and Wong, R. 1996. Asymptotic behavior of the Pollaczek polynomials and their zeros. *Stud. Appl. Math.*, **96**(3), 307–338.

Saff, E. B. and Totik, V. 1992. What parts of a measure's support attract zeros of the corresponding orthogonal polynomials. *Proc. Amer. Math. Soc.*, **114**, 185–190.

Saff, E. B. and Varga, R. S. 1977. On the zeros and poles of Padé approximants to e^z. II. Pages 195–213 of Saff, E. B. and Varga, R. S. (eds), *Padé and Rational Approximations: Theory and Applications*. New York: Academic Press.

Sarmanov, I. O. 1968. A generalized symmetric gamma-correlation. *Dokl. Akad. Nauk SSSR*, **179**, 1279–1281.

Sarmanov, O. V. and Bratoeva, Z. N. 1967. Probabilistic properties of bilinear expansions of Hermite polynomials. *Theor. Probability Appl.*, **12**, 470–481.

Schur, I. 1929. Einige Sätze über Primzahlen mit Anwendungen auf Irreduzibilitätsfragen, I. *Sitzungsber. Preuss. Akad. Wissensch. Phys.-Math. Kl.*, **23**, 125–136.

Schur, I. 1931. Affektlose Gleichungen in der Theorie der Laguerreschen und Hermiteschen Polynome. *J. Reine Angew. Math.*, **165**, 52–58.

Schwartz, H. M. 1940. A class of continued fractions. *Duke J. Math.*, **6**, 48–65.

Segura, J. 2003. On the zeros and turning points of special functions. *J. Comput. Appl. Math.*, **153**, 433–440.

Sharapudinov, I. I. 1988. Asymptotic properties of Krawtchouk polynomials. *Mat. Zametki*, **44**(5), 682–693, 703.

Shohat, J. A. 1936. The relation of the classical orthogonal polynomials to the polynomials of Appell. *Amer. J. Math.*, **58**, 453–464.

Shohat, J. A. 1938. Sur les polynômes orthogonèaux généraliséès. *C. R. Acad. Sci.*, **207**, 556–558.

Shohat, J. A. 1939. A differential equation for orthogonal polynomials. *Duke Math. J.*, **5**, 401–417.

Shohat, J. A. and Tamarkin, J. D. 1950. *The Problem of Moments*. revised edn. Providence, RI: American Mathematical Society.

Siegel, C. L. 1929. *Über einige Anwendungen Diophantischer Approximationery*. Abh. der Preuss. Akad. der Wissenschaften. Phys-math. Kl. Nr. 1.

Simon, B. 2004a. *Orthogonal Polynomials on the Unit Circle. Part 1: Classical Theory*. Vol. 54, part 1. Providence, RI: American Mathematical Society Colloquium Publications.

Simon, B. 2004b. *Orthogonal Polynomials on the Unit Circle. Part 2: Spectral Theory*. Vol. 54, part 2. Providence, RI: American Mathematical Society Colloquium Publications.

Simon, B. 2005a. Fine structure of the zeros of orthogonal polynomials. II. OPUC with competing exponential decay. *J. Approx. Theory*, **135**(1), 125–139.

Simon, B. 2005b. Sturm oscillation and comparison theorems. Pages 29–43 of *Sturm–Liouville Theory*. Basel: Birkhäuser.

Simon, B. 2006. Fine structure of the zeros of orthogonal polynomials. I. A tale of two pictures. *Electron. Trans. Numer. Anal.*, **25**, 328–368 (electronic).

Simon, B. 2007a. CMV matrices: Five years after. *J. Comput. Appl. Math.*, **208**(1), 120–154.

Simon, B. 2007b. Zeros of OPUC and the long time asymptotics of Schur and related flows. *Inverse Probl. Imaging*, **1**(1), 189–215.

Simon, B. and Totik, V. 2005. Limits of zeros of orthogonal polynomials on the circle. *Math. Nachr.*, **278**(12-13), 1615–1620.

Simon, B. and Zlatoš, A. 2005. Higher-order Szegő theorems with two singular points. *J. Approx. Theory*, **134**(1), 114–129.

Sinap, A. 1995. Gaussian quadrature for matrix valued functions on the real line. *J. Comput. Appl. Math.*, **65**(1-3), 369–385. Proceedings of the International Conference on Orthogonality, Moment Problems and Continued Fractions (Delft, 1994).

Sinap, A. and Van Assche, W. 1994. Polynomial interpolation and Gaussian quadrature for matrix-valued functions. *Linear Algebra Appl.*, **207**, 71–114.

Slater, L. J. 1966. *Generalized Hypergeometric Functions*. Cambridge: Cambridge University Press.

Slepian, D. 1972. On the symmetrized Kronecker power of a matrix and extensions of Mehler's formula for Hermite polynomials. *SIAM J. Math. Anal.*, **3**, 606–616.

Srivastava, H. M. and Singhal, J. P. 1973. New generating functions for Jacobi and related polynomials. *J. Math. Anal. Appl.*, **41**, 748–752.

Stanton, D. 1984. Orthogonal polynomials and Chevalley groups. Pages 87–128 of Askey, R. A., Koornwinder, T. H., and Schempp, W. (eds), *Special Functions: Group Theoretical Aspects and Applications*. NATO Sci. Ser. II Math. Phys. Chem. Dordrecht: D. Reidel.

Stieltjes, T.-J. 1885a. Sur les polynômes de Jacobi. *C.R. Acad. Sci. Paris*, **100**, 620–622. Reprinted in *Œuvres Complètes*, vol. 1, pp. 442–444.

Stieltjes, T.-J. 1885b. Sur quelques théorèmes d'algèbre. *C. R. Acad. Sci. Paris*, **100**, 439–440. Reprinted in *Œuvres Complètes*, vol. 1, pp. 440–441.

Stieltjes, T.-J. 1887. Sur les racines de l'équation $X^n = 0$. *Acta Math.*, **9**(1), 385–400.

Stieltjes, T.-J. 1894. Recherches sur les fractions continues. *Ann. Fac. Sci. Toulouse Sci. Math. Sci. Phys.*, **8, 9**, 1–122, 1–47.

Stieltjes, T.-J. 1993. *Œuvres Complètes/Collected papers*. Vols. I, II. Berlin: Springer. Reprint of the 1914–1918 edition.

Stone, M. H. 1932. *Linear Transformations in Hilbert Space*. Reprint, American Mathematical Society Colloquium Publications, vol. 15. Providence, RI: American Mathematical Society, 1990.

Szász, O. 1950. On the relative extrema of ultraspherical polynomials. *Boll. Un. Mat. Ital. (3)*, **5**, 125–127.

Szász, O. 1951. On the relative extrema of the Hermite orthogonal functions. *J. Indian Math. Soc. (N.S.)*, **15**.

Szegő, G. 1915. Ein Grenzwertsatz über die Toeplitzschen Determinanten einer reellen positiven Funktion. *Math. Ann.*, **76**(4), 490–503.

Szegő, G. 1920. Beiträge zur Theorie der Toeplitzschen Formen. *Math. Z.*, **6**(3-4), 167–202.

Szegő, G. 1921. Beiträge zur Theorie der Toeplitzschen Formen. II. *Math. Z.*, **9**(3-4), 167–190.

Szegő, G. 1926. Beiträge zur Theorie der Thetafunktionen. *Sitz. Preuss. Akad. Wiss. Phys. Math. Kl.*, **XIX**, 242–252. Reprinted in Collected Papers, (R. A. Askey, ed.), vol. I, Boston: Birkhauser, 1982.

Szegő, G. 1936. On some Hermitian forms associated with two given curves of the complex plane. *Trans. Amer. Math. Soc.*, **40**(3), 450–461.

Szegő, G. 1950a. On certain special sets of orthogonal polynomials. *Proc. Amer. Math. Soc.*, **1**, 731–737.

Szegő, G. 1950b. On the relative extrema of Legendre polynomials. *Boll. Un. Mat. Ital. (3)*, **5**.

Szegő, G. [1939] 1975. *Orthogonal Polynomials*. Fourth edn. Providence, RI: American Mathematical Society. American Mathematical Society, Colloquium Publications, vol. XXIII.

Szwarc, R. 1992. Connection coefficients of orthogonal polynomials. *Canad. Math. Bull.*, **35**(4), 548–556.

Szwarc, R. 2005. Orthogonal polynomials and Banach algebras. Pages 103–139 of *Inzell Lectures on Orthogonal Polynomials*. Adv. Theory Spec. Funct. Orthogonal Polynomials, vol. 2. Hauppauge, NY: Nova Sci. Publ.

Tirao, J. A. 1977. Spherical functions. *Rev. Un. Mat. Argentina*, **28**(2), 75–98.

Tirao, J. A. 2003. The matrix-valued hypergeometric equation. *Proc. Natl. Acad. Sci. USA*, **100**(14), 8138–8141 (electronic).

Tirao, J. A. 2011. The algebra of differential operators associated to a weight matrix: a first example. In Milies, C. P. (ed), *Groups, Algebras and Applications*. Contemporary Mathematics, vol. 537. Providence, RI: American Mathematical Society. XVIII Latin American Algebra Colloquium, August 3–8, 2009, in São Paulo, Brazil.

Toda, M. 1989. *Theory of Nonlinear Lattices*. Second edn. Springer Series in Solid-State Sciences, vol. 20. Berlin: Springer.

Todd, J. 1950. On the relative extrema of the Laguerre orthogonal functions. *Boll. Un. Mat. Ital. (3)*, **5**, 122–125.

Tricomi, F. G. 1947. Sugli zeri delle funzioni di cui si conosce una rappresentazione asintotica. *Ann. Mat. Pura Appl. (4)*, **26**, 283–300.

Tricomi, F. G. 1954. *Funzioni Ipergeometriche Confluenti*. Roma: Edizioni Cremonese.

Tricomi, F. G. 1957. *Integral Equations*. Reprint, New York: Dover Publications, 1985.

Turán, P. 1980. On some open problems of approximation theory. *J. Approx. Theory*, **29**(1), 23–85. P. Turán memorial volume. Translated from the Hungarian by P. Szusz.

Tyan, S. and Thomas, J. B. 1975. Characterization of a class of bivariate distribution functions. *J. Multivariate Anal.*, **5**, 227–235.

Tyan, S. G., Derin, H., and Thomas, J. B. 1976. Two necessary conditions on the representation of bivariate distributions by polynomials. *Ann. Statist.*, **4**(1), 216–222.

Underhill, C. 1972. *On the Zeros of Generalized Bessel Polynomials*. Internal note. University of Salford.

Uvarov, V. B. 1959. On the connection between polynomials, orthogonal with different weights. *Dokl. Acad. Nauk SSSR*, **126**, 33–36.

Uvarov, V. B. 1969. The connection between systems of polynomials that are orthogonal with respect to different distribution functions. *Ž. Vyčisl. Mat. i Mat. Fiz.*, **9**, 1253–1262.

Valent, G. 1994. Asymptotic analysis of some associated orthogonal polynomials connected with elliptic functions. *SIAM J. Math. Anal.*, **25**(2), 749–775.

Valent, G. 1995. Associated Stieltjes-Carlitz polynomials and a generalization of Heun's differential equation. *J. Comput. Appl. Math.*, **57**(1-2), 293–307. Proceedings of the Fourth International Symposium on Orthogonal Polynomials and Their Applications (Evian-Les-Bains, 1992).

Valent, G. 1996a. Co-recursivity and Karlin–McGregor duality for indeterminate moment problems. *Constr. Approx.*, **12**(4), 531–553.

Valent, G. 1996b. Exact solutions of some quadratic and quartic birth and death processes and related orthogonal polynomials. *J. Comput. Appl. Math.*, **67**(1), 103–127.

Valent, G. 1999. Indeterminate moment problems and a conjecture on the growth of the entire functions in the Nevanlinna parametrization. Pages 227–237 of *Applications and Computation of Orthogonal Polynomials (Oberwolfach, 1998)*. Internat. Ser. Numer. Math., vol. 131. Basel: Birkhäuser.

Valent, G. and Van Assche, W. 1995. The impact of Stieltjes' work on continued fractions and orthogonal polynomials: Additional material. *J. Comput. Appl. Math.*, **65**(1-3), 419–447. Proceedings of the International Conference on Orthogonality, Moment Problems and Continued Fractions (Delft, 1994).

Van Assche, W. 1993. The impact of Stieltjes' work on continued fractions and orthogonal polynomials. Pages 5–37 of Dijk, G. Van (ed), *Thomas Jan Stieltjes, Œuvres Complètes/Collected papers*. Vol. I. Berlin: Springer.

Van Assche, W. 2007. Rakhmanov's theorem for orthogonal matrix polynomials on the unit circle. *J. Approx. Theory*, **146**(2), 227–242.

Van Assche, W. and Magnus, A. P. 1989. Sieved orthogonal polynomials and discrete measures with jumps dense in an interval. *Proc. Amer. Math. Soc.*, **106**(1), 163–173.

Van der Jeugt, J. 1997. Coupling coefficients for Lie algebra representations and addition formulas for special functions. *J. Math. Phys.*, **38**(5), 2728–2740.

Van der Jeugt, J. and Jagannathan, R. 1998. Realizations of $\mathrm{su}(1,1)$ and $U_q(\mathrm{su}(1,1))$ and generating functions for orthogonal polynomials. *J. Math. Phys.*, **39**(9), 5062–5078.

Van Deun, J. 2007. Electrostatics and ghost poles in near best fixed pole rational interpolation. *Electron. Trans. Numer. Anal.*, **26**, 439–452.

van Eijndhoven, S. J. L. and Meyers, J. L. H. 1990. New orthogonality relations for the Hermite polynomials and related Hilbert spaces. *J. Math. Anal. Appl.*, **146**, 89–98.

Verblunsky, S. 1935. On positive harmonic functions: A contribution to the algebra of Fourier series. *Proc. London Math. Soc.*, **38**, 125–157.

Verblunsky, S. 1936. On positive harmonic functions (second paper). *Proc. London Math. Soc.*, **40**, 290–320.

Verma, A. 1972. Some transformations of series with arbitrary terms. *Ist. Lombardo Accad. Sci. Lett. Rend. A*, **106**, 342–353.

Viennot, G. 1983. *Une théorie combinatoire de polynômes orthogonaux generaux*. Université de Québec à Montréal. Lecture notes.

Vilenkin, N. Ja. 1968. *Special Functions and the Theory of Group Representations*. Translations of Mathematical Monographs, vol. 22. Providence, RI: American Mathematical Society. Translated from the Russian by V. N. Singh.

Vilenkin, N. Ja. and Klimyk, A. U. 1991–1993. *Representation of Lie Groups and Special Functions*, vols. 1–3. Mathematics and Its Applications (Soviet Series), vols. 72, 74, 75. Dordrecht, The Netherlands: Kluwer Academic Publishers Group.

Vinet, L. and Zhedanov, A. 2004. A characterization of classical and semiclassical orthogonal polynomials from their dual polynomials. *J. Comput. Appl. Math.*, **172**(1), 41–48.

Volkmer, H. 2008. Approximation of eigenvalues of some differential equations by zeros of orthogonal polynomials. *J. Comput. Appl. Math.*, **213**(2), 488–500.

Wall, H. S. 1948. *Analytic Theory of Continued Fractions*. New York, NY: D. Van Nostrand.

Wall, H. S. and Wetzel, M. 1944. Quadratic forms and convergence regions for continued fractions. *Duke Math. J.*, **11**, 89–102.

Wallisser, R. 2000. On Lambert's proof of the irrationality of π. Pages 521–530 of *Algebraic Number Theory and Diophantine Analysis (Graz, 1998)*. Berlin: de Gruyter.

Watkins, D. S. 1993. Some perspectives on the eigenvalue problem. *SIAM Rev.*, **35**(3), 430–471.

Watson, G. N. 1944. *A Treatise on the Theory of Bessel Functions*. Cambridge: Cambridge University Press.

Wendroff, B. 1961. On orthogonal polynomials. *Proc. Amer. Math. Soc.*, **12**, 554–555.

Whittaker, E. T. and Watson, G. N. 1927. *A Course of Modern Analysis*. An introduction to the general theory of infinite processes and of analytic functions; with an account of the principal transcendental functions. Fourth edn. Reprint, Cambridge Mathematical Library, Cambridge: Cambridge University Press, 1996.

Widder, D. V. 1941. *The Laplace Transform*. Princeton Mathematical Series, vol. 6. Princeton, NJ: Princeton University Press.

Widom, H. 1967. Polynomials associated with measures in the complex plane. *J. Math. Mech.*, **16**, 997–1013.

Wigert, S. 1923. Sur les polynomes orthogonaux et l'approximation des fonctions continues. *Ark. Mat. Astronom. Fys.*, **17**(18), 15 pages.

Wilson, J. A. 1980. Some hypergeometric orthogonal polynomials. *SIAM J. Math. Anal.*, **11**(4), 690–701.

Wilson, J. A. 1982. *Hypergeometric Series Recurrence Relations and Properties of Some Orthogonal Polynomials*. (Preprint).

Wilson, J. A. 1991. Asymptotics for the ${}_4F_3$ polynomials, *J. Approx. Theory*, **66**, no. 1, 58–71.

Wilson, M. W. 1970. Nonnegative expansions of polynomials. *Proc. Amer. Math. Soc.*, **24**, 100–102.

Wimp, J. 1985. Some explicit Padé approximants for the function Φ'/Φ and a related quadrature formula involving Bessel functions. *SIAM J. Math. Anal.*, **16**(4), 887–895.

Wimp, J. 1987. Explicit formulas for the associated Jacobi polynomials and some applications. *Canad. J. Math.*, **39**(4), 983–1000.

Wintner, A. 1929. *Spektraltheorie der unendlichen Matrizen*. Leipzig: S. Hirzel.

Wong, R. and Zhang, J.-M. 1994a. Asymptotic monotonicity of the relative extrema of Jacobi polynomials. *Canad. J. Math.*, **46**(6), 1318–1337.

Wong, R. and Zhang, J.-M. 1994b. On the relative extrema of the Jacobi polynomials $P_n^{(0,-1)}(x)$. *SIAM J. Math. Anal.*, **25**(2), 776–811.

Zhani, D. 1984. *Problème des Moments Matriciels sur la Droite: Construction d'une Famille de Solutions et Questions D'unicité*. Publications du Département de Mathématiques. Nouvelle Série. D [Publications of the Department of Mathematics. New Series. D], vol. 84. Lyon: Université Claude-Bernard Département de Mathématiques. Dissertation, Université de Lyon I, Lyon, 1983.

Index